NUTRITION AND FOOD SCIENCE
Present Knowledge and Utilization

VOLUME 2
Nutrition Education and Food Science and Technology

NUTRITION AND FOOD SCIENCE
Present Knowledge and Utilization

Volume 1   FOOD AND NUTRITION POLICIES AND PROGRAMS

Volume 2   NUTRITION EDUCATION AND FOOD SCIENCE
           AND TECHNOLOGY

Volume 3   NUTRITIONAL BIOCHEMISTRY AND PATHOLOGY

NUTRITION AND FOOD SCIENCE
Present Knowledge and Utilization

VOLUME 2
Nutrition Education and Food Science and Technology

Edited by

Walter Santos, Nabuco Lopes, J. J. Barbosa, and Dagoberto Chaves
Brazilian Nutrition Society
Rio de Janeiro, Brazil

and

José Carlos Valente
National Food and Nutrition Institute and
Associação Brasiliense de Nutrologia
Rio de Janeiro, Brazil

PLENUM PRESS · NEW YORK AND LONDON

Library of Congress Cataloging in Publication Data

International Congress of Nutrition, 11th Rio de Janeiro, 1978.
  Food and nutrition programs and policies.

  (Nutrition and food science; v. 1)
  Includes index.
  1. Nutrition policy — Congresses. 2. Food supply — Congresses. 3. Food relief — Congresses. I. Santos, Walter II. Title. III. Series.
  TX359.I57 1978   362.5   79-27952
  ISBN 0-306-40343-9 (v. 2)

Proceedings of the Eleventh International Congress on Nutrition, organized by the Brazilian Nutrition Society, held in Rio de Janeiro, Brazil, August 27—September 1, 1978, and published in three volumes of which this is Volume 2.

© 1980 Plenum Press, New York
A Division of Plenum Publishing Corporation
227 West 17th Street, New York, N.Y. 10011

All rights reserved

No part of this book may be reproduced, stored in a retrieval system, or transmitted, in any form or by any means, electronic, mechanical, photocopying, microfilming, recording, or otherwise, without written permission from the Publisher

Printed in the United States of America

FOREWORD

The Brazilian Society of Nutrition, through the present publication, brings to the attention of the world scientific community the works presented at the XI INTERNATIONAL CONGRESS OF NUTRITION which, promoted by this Society and under the sponsorship of the International Union of Nutritional Science, was held in the city of Rio de Janeiro from August 27th to September 1st, 1978.

The publication, edited by Plenum Publishing Corporation, is titled "Nutrition and Food Science: Presented Knowledge and Utilization" and appears in three volumes, under the following titles and sub-titles:

Vol. I - FOOD AND NUTRITION POLICIES AND PROGRAMS
- Planning and Implementation of National Programs
- The role of International and Non-governmental Agecies
- The role of the Private Sector
- Program Evaluation and Nutritional Surveillance
- Nutrition Intervention Programs for Rural and Urban Areas
- Mass Feeding Programs
- Consumer Protection Programs

Vol. II - NUTRITION EDUCATION AND FOOD SCIENCE AND TECHNOLOGY
- Animal and Vegetable Resources for Human Feeding
- Food Science and Technology
- Research in Food and Nutrition
- Nutrition Education

Vol. III - BIOCHEMICAL AND PATHOLOGICAL NUTRITION
- Nutritional Biochemistry
- Pathological and Chemical Nutrition
- Nutrition, Growth and Human Development

It is hoped that this publication may prove useful to all those who are interested in the different aspects of Nutrition Science.

Editorial Committee:

Walter J. Santos
J.J. Barbosa
Dagoberto Chaves
José Carlos Valente
Nabuco Lopes

PREFACE

The XI INTERNATIONAL CONGRESS OF NUTRITION — XI ICN —, promoted by the INTERNATIONAL UNION OF NUTRITIONAL SCIENCES — IUNS —, was carried out by the BRAZILIAN NUTRITION SOCIETY — BNS —, in the Convention Center of the Hotel Nacional, in the city of Rio de Janeiro, Brazil, from August 27th to September 1st, 1978.

Taking place for the first time in the southern hemisphere, the XI ICN received the collaboration and participation of several international agencies, including the World Health Organization (WHO), the Pan American Health Organization (PAHO), the United Nations Children's Fund (UNICEF), the Food and Agriculture Organization (FAO), the International Fund for Agricultural Development (IFAD), the United Nations Educational, Scientific and Cultural Organization (UNESCO), the World Food Program (WFP) and the World Food Council (WFC).

The meeting had a multi-disciplinary character, with the participation of professionals and students from the different sectors related to the field of food and nutrition, and was characterized by the interest it arose, which was demonstrated by the presence of 5,026 participants from 92 countries, and the presentation of more than 1,200 scienfific papers.

During the period prior to the Congress, the different committees, aside from the regular work such as receiving requests for registration and selection of the scientific papers to be presented, also developed an intensive activity of divulgation, both in Brazil and in the other countries, by means of written, verbal and televised press, distribution of circular letters, and the publication of a monthly, bilingual (Portuguese and English) Information Bulletin.

In addition, there were several preparatory meetings, not only in Brazil but also in the exterior, in which members of the Executive Committee participated.

In this Congress, besides the scientific activities other events were considered such as:

- Parallel Meetings promoted by scientific entities related to food and nutrition, trough a formal request to the Congress Executive Committee;

- Intensive Courses about different subjects, especially about Diet Therapy and physiopatology;

- The V International Congress of Paraenteral Nutrition, promoted by the International Society of Paraenteral Nutrition.

We feel the need to clarify that the works were not grouped in accordance to the various events of the Congress, which were organized by the classifications as per argument elaborated with the objective of providing the specialists with correlated matters in each one of the three volumes of the Proceedings.

Since many of the works were not presented in the English language, the Brazilian Society of Nutrition had them translated to the referred idiom, trying, as much as possible, to homogenize the scientific nomenclature of these works.

Lastly we would like to render account for two facts: one, in respect to the exclusion of works related to Paraenteral Nutrition, originated by the fact that the Society never received the respective original works, although these were insistently requested. The other, in reference to the choice made by Plenum Publishing Corporation for the editing and distribution at international level of the Proceedings, given the international tradition of this Publishing House in works of this nature.

Hopefully, the Proceedings will be usefull to all those who are interested in the different aspects of the science of Nutrition.

> Editorial Committee:
>
> Walter J. Santos
> J.J. Barbosa
> Dagoberto Chaves
> José Carlos Valente
> Nabuco Lopes

CONTENTS

I. ANIMAL AND VEGETABLE RESOURCES FOR HUMAN FEEDING

Nutrition Priorities in the International Agricultural Research Network - Introductory Remarks . . . 3
M. Milner

How Nutrition Priorities can be Integrated into Crop Improvement Programs 5
J.H. Hulse

Progress in Nutritional Improvement of Maize and Triticale 19
E.M. Villegas

Improvement of Nutritional Quality of Sorghum and Pearl Millet 39
R. Jambunathan

The Need for Food Utilization and Processing Studies to Supplement Nutritional Evaluation 55
R.A. Luse

Modern Practices in Animal Feeding - Relationship to Human Nutrition - Introductory Remarks 65
A. Rerat

Modern Practices in Pig Feeding and Husbandry: Relationship to Human Nutrition 73
A. Rerat

Perspectives of Milk/Beef Production Systems in the Tropics: Relationship to Human Nutrition 101
F. Pérez-Gil Romo

Modern Practices in Cattle Feeding for
  Meat Production - Relationship to
  Human Nutrition 109
  C.J. Kercher

Modern Practices in Cattle Feeding for Milk
  Production - Relationship to Human
  Nutrition 113
  C.O. Claesson

Unexpected Nutritional Benefit from a Dairy
  Project 125
  M. Jul

Utilization of Waste for Food Production Using
  Animals: Application to Human Nutrition 141
  W.J. Pigden

Report and Recommendations of the Workshop
  on Modern Practices of Animal Feeding:
  Relationship to Animal Nutrition 153
  V.A. Oyenuga

Feed Formulation and Feed Technology 157
  O.R. Braekkan

Fish Feeding Technology 161
  T. Nose

Protein and Energy Requirements 171
  C.B. Cowey

Vitamin, Fat and Mineral Requirements 177
  J.E. Halver

Vitamin Requirements of Finfish 181
  J.E. Halver

Status of Fish Food Formulation and Fish
  Nutrition Research - in Brazil 199
  N. Castagnolli

II. FOOD SCIENCE AND TECHNOLOGY

Potential Utilization of Unconventional Food
  Sources and New Food Products -
  Introductory Remarks 207
  J.C. Sanahuja

CONTENTS

The Nutritional Value of Winged Bean
     (*Psophocarpus tetragonolobus* L. DC),
     with Special Reference to Five Varieties
     (cultivars) Grown in Sri Lanka 211
  N.S. Hettiarachchy, H.M.W. Herath, and T.W.W. Wickramanayake

Prospects and Problems of Cultivation of
     Lupins in South America 221
  R. Gross

Vegetable Protein 227
  J. Mauron

Studies Done in Ecuador on the Potential
     Utilization of Non-conventional
     Sources of Foods 237
  L.P. de Benitez

B.P.C. - a Fish Protein Concentrate for
     Human Consumption 249
  A.H. Delfino

Utilization of Non-Conventional Foods in Brazil 269
  R.A. Salum

The Experience in Mexico on the Utilization of
     Non-Conventional Protein Sources 277
  H. Bourges and J.C. Morales de León

Agricultural Production of Soybeans: Relation
     Between Varieties and Nutritive Values 289
  A. Lam-Sánchez

Types of Soy Protein Products 293
  J. Wapinski

Conversion of Raw Soybeans into High-Quality
     Protein Products 299
  J.J. Rackis

Studies on the Nutritive Values of Mixed Protein
     (50% Soy and 50% Fish) and Soy Protein
     Isolate in Japanese Adults 325
  G. Inoue

The Technology of Iron Fortification 333
  D.S Titus

The Technology of Vitamin A 337
  J.C. Bauernfeind

Biological and Chemical Products Applied
    in Agriculture: Their Influence
    on Food Quality   375
    H. Schmidt-Hebbel

Environmental Contaminants   381
    I.C. Munro

Contaminants and Additives in Animal Feeding:
    Relationship to Human Nutrition   393
    R. Ferrando

Contamination of the Argentinian Family
    Basket by Chlorinated Defensives
    Residues – Biological Repercussion   403
    E. Astolfi

Protection of the Consumer from Aflatoxins
    in Food   417
    H. Schulze

Nutritional Losses in Storage and Processing
    of Legumes   421
    K. Saio

Role of Post-Harvest Conservation of Foods in
    Achieving Nutritional Goals   451
    R.V. Bhat

Significance of Nutritional Losses in the
    Processing of Grains   457
    D.A.V. Dendy

III. RESEARCH IN FOOD NUTRITION

Report and Recommendations of the Workshop on
    Priorities in Nutrition Research –
    Panel Discussion   467
    G.H. Beaton

Approaches to Setting Priorities   471
    G.H. Beaton

Priorities for Nutrition Research   477
    J.M. Gurney

Priority for Applied Nutrition Research in
    Developing Countries   483
    R. Orraca-Tetteh

CONTENTS

Priorities in Nutrition Research: Remarks
    from the Standpoint of the Italian
    Experience 489
  A. Mariani

Priorities for Nutrition Research in a
    Developing Country: Turkey 495
  O. Köksal

Nutrition Research Priorities for North Africa 503
  Z. Kallal

New Zealand National Diet Survey: Prelude to
    a National Nutrition Policy 505
  J.A. Birkbeck

Critical Considerations for National Nutrition
    Surveys in Africa 511
  A. Zerfas

The Need for Periodical Revision of the National
    Nutritional Status 515
  F.A. Goncalves Ferreira

Assessment and Monitoring of Nutritional Health 519
  A.E. Schaefer

New Approaches to the Assessment of Nutritional
    Status - Selection and Utilization of Nutri-
    tional Indicators - The Jamaican Experience 525
  V.S. Campbell

Report and Recommendations of the Workshop on
    "New Approaches to the Assessment of
    Nutritional Status. Selection and
    Utilization of Nutritional Indicators" 533
  A.E. Schaefer

The Food and Nutrition Situation and the
    Perspectives for Food Demand in Morocco 537
  M'Bareck Essatara

Contribution of Anthropology to Nutrition:
    Introductory Remarks 551
  N.W. Jerome

An Anthropological Perspective on Nutrition
    Program Evaluation 553
  G.H. Pelto and N.W. Jerome

Anthropological Perspectives in Nutrition
    and Health Programmes. The Khombole
    Project, Senegal, 1957-78 573
  I. de Garine

Evaluation in a Caribbean Context: Socio-
    Cultural Factors Affecting a Community
    Health Aide Program 593
  T.J. Marchione

Evaluation in a Middle Eastern Context: A
    Case Study 611
  C. Geissler

Evaluation in an African Context: Special Em-
    phasis on the Woman Producer and Repro-
    ducer - Some Theoretical Considerations 619
  W.B. Eide

The Experience of Greece 637
  A. Polychronopoulou-Trichopoulou

Differences in Energy Expenditure in Selected
    Work Under Control and *ad libitum* 641
  W. Wirths

A Study of the Energy Expenditure, Dietary
    Intake and Pattern of Daily Activities
    Among Various Occupational Groups
    IV: Weightlifters 657
  Ma. P.E. de Guzman

Water and Electrolyte Balance in Exercise 667
  D. Costill

Predicting Nutritive Value of Food Protein by
    in vitro Method 683
  D. Petitclerc, G. Goulet, and G.J. Brisson

Protein Quality of Food Measured with
    tetrahymena thermophila and Rat 689
  H. Baker

Enzymatic Hydrolysis of Food Proteins for
    Amino Acid Analysis 693
  R. Öste and B.M. Nair

Protein Quality Measurements on a Bacterial Single-
    Cell Protein (SCP) Grown on Methanol 705
  H.F. Erbersdobler and R. Müller-Landau

CONTENTS

Methionine-Sulphoxide Content of Fish Protein
    Concentrate (FPC) and Available Methionine
    Determined Microbiologically with
    streptococcus zymogenes   713
  L.R. Njaa and E. Lied

Defining Dietary Fiber in Food Technology   719
  D.R. Schaller

Fractionation and Chemical Characterization of
    Dietary Fibre Components   727
  O. Theander

Evaluation of Methods Suggested for Assay of
    Dietary Fibre   741
  N.-G. Asp

IV. NUTRITION EDUCATION

Food and Nutrition Education within National
    Education Systems   747
  F. Aylward

What can be Expected from Nutritional Education
    Directed towards the Population of the
    Third World   759
  M.M. de Chávez

Evaluation of Nutrition Education Programs   769
  R. Wolf

Reactor's Comments on Nutrition Education   775
  W.J. Darby

Changing Nutrition and Health Behavior Through
    Mass Media: The Philippine Experience   779
  D.B. Aguillon

Nutrition Education in the Early Childhood
    Grade Levels in Israel   795
  R. Lipsky

Case Study on Nutrition Education in the
    United Kingdom   805
  D.F. Hollingsworth

Nutrition Education in the Netherlands   811
  K. Clay

Community Nutrition Education 817
  M.M. de Medina

The Marketing Approach to Nutrition Education
    in Developing Countries 823
  R.K. Manoff, T.M. Cooke, and D. Aguillon

Nutrition Training of Health Workers in
    Developing Countries 835
  K. Bagchi

Nutritional Education for Medical and Other
    Health Science Professionals in East
    Europe 843
  E. Morava

Nutrition Education for Medical and Other
    Health Science Profession in Asia 847
  Aree Valyasevi

The Tanzania Experience 861
  T.N. Maletnlema

Nutrition and Primary Health Care Services
    II: Training 871
  E.F.P. Jelliffe

Report on Workshop on Nutrition Education for
    Medical and Other Health Sciences 877

The Nutritionist-Dietitian in Latin-American 883
  M. Beaudry - Darismé

Communications Competencies of the Nutritionist-
    Dietitian 887
  S.J. Icaza

The Role of the Nutritionist in Expanding
    Community Nutrition Education Programs 895
  H.D. Ullrich

Qualification of the Nutritionist-Dietitian
    in Order to Participate in Community
    Based Health Programmes 899
  S.M. Alves de Souza

Adaptation of the Training of Dietitians
    to Their Roles 905
  Y. Serville

CONTENTS

Dietetics around the World - 1977 Survey of
    the Work and Training of Dietitians 913
  J. Woodhill

Thoughts on the Ethics of the Dietetic
    Profession 917
  W. Aign

Author Index 921

Subject Index 923

I. Animal and Vegetable Resources for Human Feeding

NUTRITION PRIORITIES IN THE INTERNATIONAL AGRICULTURAL

RESEARCH NETWORK - INTRODUCTORY REMARKS

        Max Milner - Chairman

        Mass. Inst. of Technology

        Cambridge, Mass. 021 39 - U.S.A.

    The United Nations World Food Council closed its fourth session in Mexico City several weeks ago voicing what is termed "fears for the future", because (and I paraphrase closely) world food production has grown more slowly in this decade than in the 1960's; because per capita food production has declined in food priority countries in this decade; because the number of undernourished people in the world continues to increase each year; because external assistance for increasing food production declined sharply in 1976; because many countries have not been able to increase their priorities for food production and nutrition; because a new international wheat agreement has not yet been formulated, which would include a world cereal reserve; and because trade barriers, trade instability and mounting protecionism handicap food trade and food production of many developing countries.

    Perhaps we can find some encouragement from the fact that in this somber statement the World Food Council did not suggest that inadequate research in food production and nutrition were in part responsible for the global food and nutrition crisis. Or perhaps the Council may not have thought about the research problem in the context of their report.

    In any event, international research and development in crop production, although still limited in scope, is alive and well in the regional institutions which have come to be called collectively The International Agricultural Research System.

    Materialization of the idea that research into increasing food crop production at the international level could be stimulated by comprehensive and competent regional centers, appeared in 1960 with establishment of the International Rice Research Institute in the Philippines, with support at that time principally from the Ford and Rockfeller Foundations. This Center's mandata was to develop improved agricultural production systems. IRRI was followed into existence by the International Maize and Wheat Improvement Center in Mexico in 1966, to emphasize maize and wheat improvement. Many will recall that Dr. Norman Borlaug received the Nobel Peace Prize in 1970 for his leadership in CIMMYT's wheat improvement activities. Already in the 1960's the world began to hear about so-called "miracle rice" and new high-yielding varieties of wheat, as well as the widely quoted term "The Green Revolution".

Other international centers appeared thereafter in rapid succession. The Center for Tropical Agriculture (CIAT) whose mandate it was to improve agriculture in the humid lowland tropics, through improvement of cassava, field beans and livestock animals, began operations in Colombia in 1967. The International Institute of Tropical Agriculture (IITA) in Ibadan, Nigeria, started its program in 1968, emphasizing legumes and root and tuber crops. The International Potato Center (CIP) in Lima, Peru, was established in 1971, to develop new and improved potato varieties and to stimulate the expansion of potato production to new areas. In 1972, the International Crops Research Institute for the Semi-arid Tropics (ICRISAT) came into existence in Hyderabad, India, to develop superior varieties of sorghums, millets, food legumes and groundnuts, as well as improved farming and irrigation systems. Other centers now in existence include the International Laboratory for Research on Animal Diseases (ILRAD) in Nairobi, Kenya, the International Livestock Center for Africa (ILCA), the West African Rice Development Association (WARDA), and the International Center for Agricultural Research in the Dry Areas (ICARDA). Establishment of additional centers, such as one for aquaculture, is under consideration.

Centralized and coordinated funding for the system appeared in 1971 with the creation of the Consultative Group for International Agricultural Research (CGIAR) with headquarters at the World Bank in Washington. This group administers and allocates funds for all the network which are received from various governments, international agencies and foundations. CGIAR functions with scientific and technical advice provided by a Technical Advisory Committee (TAC) from FAO in Rome, whose members are proeminent agriculture and food scientists. In 1972, the first year of operations, $15 million was authorized through CGIAR, and it is anticipated that about $90 million will be committed in 1978, originating from about 30 donors.

It may be emphasized that increased food crop production has always been the number-one priority of these institutes, and notwithstanding the disturbing pronouncement of the World Food Council, these centers have been productive and effective in terms of their research mandates. One has little difficulty in agreeing that increased production is the first priority, but it may be said candidly that considerations in these programs for nutritional quality have not always been primary priorities. I am happy to report today, however, that this situation is changing, and as we shall see from the papers we are about to hear, nutrition considerations are now being given attention at a number of the centers. Incidentally, this session is the first at an International Nutrition Congress to deal with progress in this international system.

As noted in your program, we will begin with a statement from the International Development Research Center, which has been a major supporter of CGIAR, in which Mrs. Odette Pearson will suggest how nutrition priorities can be integrated into crop improvement programs. Dr. R. Jambunathan will follow with a paper from ICRISAT on how the nutritional qualities of sorghum and millet are being improved. Representing CIMMYT will be Dr. Evangelina Villegas who will report on progress in nutritional improvement of maize and triticale. Our co-chairman, Dr. Bressani of INCAP will review progress in bettering the nutritional quality of food legumes, an activity in which INCAP collaborates with CIAT. Our final speaker, Dr. Robert Luse representing CIAT, will discuss relationships between food processing and utilization practices and nutrition considerations.

May I emphasize that what you will hear today is only a sampling of research on the nutritional aspects of new food crop development which is in progress today throughout this international system. I hope there will be time for useful discussion of the various papers.

HOW NUTRITION PRIORITIES CAN BE INTEGRATED

INTO CROP IMPROVEMENT PROGRAMS

        J.H. Hulse*

        International Development Research Centre

        Ottawa, Canada

THE WORLD FOOD PROBLEM

    Nutritional survey statistics are a constant reminder that the world food situation is serious, even precarious. By recent estimates[1], 500 million people live on the edge of starvation and 1.2 to 1.3 billion would benefit from a more varied diet. The greatest majority of these people live in Asia, Southeast Asia and sub-Saharan Africa. Clinical surveys and hospital records indicate that malnutrition wherever it exists is severest among infants, preschool children and pregnant and lactating women; that it is most prevalent in depressed rural areas and the slums of large cities; and that the problem is lack of calories as much as lack of protein.

    The developing countries are generally characterized by high population growth rates showing little tendency to slacken and lagging food production which has become more pronounced in recent years. By recent estimates, the territories of the lesser developed countries account for two thirds of the world population but only one third of the world food production. During the past two decades, recorded world food production increased by an average of about 69 percent. Food production increased by 65 percent in the developed countries and by 75 percent in the less developed countries. Worldwide, the annual rate of increase was approximately 2.8 percent for food production and 2 percent per annum for population. On this

---

* The views expressed in this paper are those of the author and do not necessarily represent those of the International Development Research Centre.

basis, it would appear that each of the 3.8 billion people alive in 1973 had nearly 20 percent more food to eat than did the 2.7 billion in 1954. However, because of markedly different population growth rates, food production per capita rose at an annual rate of only 0.4 percent in the developing countries as compared with 1.5 percent in the developed regions. Although most of the major developing areas experienced some advances in per capita food production during the last half of the 1960's, in Africa, a major downtrend has been observed since 1961.

In Africa, the food supply problem is most critical in the low income food deficit countries of the sub-Saharan region in which the per capita GNP is less than US $200. In the five Sahelian countries (Tchad, Mali, Niger, Senegal and Upper Volta), staple crop production trends were negative for the whole of the 1960-75 period. During the most recent drought of 1970-75, cereal production in the Sahel was 8 percent below the production average for the 1960-70 period.

According to a recent report of the International Food Policy Research Institute (IFPRI)[2], to provide the undernourished in the developing countries with 100 percent of the caloric levels recommended by the 1971 Joint FAO/WHO ad hoc Expert Committee on Protein and Energy Requirements, would have required in 1975 additional imports of 45-70 million tons of cereal equivalent, over and above the 31 million tons of cereal actually imported. Asia would have required some 60 percent of these additional imports, sub-Saharan Africa 25 percent, with lesser amounts (5-10 million tons) going to each of North Africa/Middle East and Latin American regions. In order to balance indigenous production with projected demand by 1985, food production in the developing countries will have to increase by 4.2 percent anually or again by 5.5 percent per annum if internal production is to bridge the caloric gap[3].

THE PROTEIN CONTROVERSY

Estimates of the caloric gap based on global production and population statistics must be carefully interpreted, in that production figures are not always a true measure of food supplies available to meet human nutritional requirements. In the first place, available data assumes total consumption of whole grain. However most cereals and legumes are processed by dehulling, soaking, boiling, milling and cooking, all of which result in significant nutrient losses. Second, it is a known fact that many of the seeds of edible subsistence plants including cereals and legumes, contain substances which seriously interfere with the nutritional quality of protein and other nutrients. Sorghum, for instance, which covers a world acreage larger than maize and is the primary subsistence crop of millions of the poorest people of the semi-arid tropics, contains tannins which reduce the digestibility of the cereal protein. Third, national averages are also misleading in that they take little account of the problem of the maldistribution of food supplies among different regions of the world, among countries within regions, among families within a community and even among members within a family. Finally, further studies are needed to determine the nutritional requirements of young children, particularly those suffering or recovering from infectious diseases.

The subject of nutrient requirements has fascinated physicians and nutritionists for over 100 years and continues to be the subject of much debate. The first comprehensive set of recommended allowances for 12 nutrients, including energy and protein, was produced by the Technical Commission on Nutrition of the League of Nations in 1935. The difficulties encountered in defining recommended dietary levels which describe an "optimum diet" cannot be overlooked. In setting nutritional requirements it is necessary to decide on an average per capita requirement for a nutrient based on available information and then to set recommended daily allowances at levels intended to cover at least 97.5 percent of the population. Adequate intake standards for energy and protein vary among different regions of the world, reflecting average body size, climate, physical activity, differences in dietary protein quality and other factors.

Since the mid-1950's, successive expert committees on energy and protein requirements from the Food and Agricultural Organization and World Health Organization of the United Nations, have convened to reevaluate estimates of adequate energy and protein intakes for humans. In April 1971, the Joint FAO/WHO Expert Committee lowered the levels of dietary protein intake for adults by 20 percent and thereby redefined dietary protein malnutrition. Nutritionists working with the planning commissions of several developing countries suddenly found themselves preparing a vigorous attack on a problem that no longer existed as a priority issue. Many nutritionists, economists and planners concluded from comparison with per capita dietary intakes that emphasis on protein could now be dropped or at least greatly reduced in the formulation of agricultural, educational, health and economic policy and in overall nutritional planning. The lowering of the protein requirement led to the conclusion that except in areas where roots and tubers are major dietary components, protein levels are adequate if caloric needs are met. According to Dr. Nevin Scrimshaw of the Massachusetts Institute of Technology, however, there is increasing evidence that these revised recommended levels for protein may be too low[4].

Infants and preschool children and pregnant and lactating women are the primary target groups of nutrition intervention programs. In response to the statement that protein-calorie malnutrition can be prevented by increasing the quantity of the traditional diet consumed by children, Scrimshaw replies that it is useless to suggest that a child can get sufficient protein and calories from a cereal diet if he merely eats more of it, when he is unable to do so. The traditional diet of the developing countries is frequently bulky and young children do not have the capacity to ingest the large quantities that would be necessary to satisfy their full nutrient requirements. Furthermore, the dietary protein[5] needs of the very young in developing countries are increased, not only by the poorer absorption of dietary protein caused by intestinal parasites and chronic damage to the gastrointestinal tract from repeated infections, but also by the extra protein lost from the body during the acute infections so commonly experienced by these children. This loss is induced by the stress response which causes amino acids to be mobilized from the protein in the lean body tissues for use by the liver in making glucose. The need for dietary calories may also be increased by impaired intestinal absorption and by fever, but the stress response is specific for protein. Moreover, whereas a deficit in calories may be compensated in part by reduced activity, the body has no comparable mechanism for protein deficiency. Dr. Scrimshaw concludes that:

> "There is no doubt that good nutrition requires a balanced complement of protein and calories, and neither can be neglected in the diets of the underprivileged and vulnerable. To the extent that the pendulum swung too far in emphasizing protein in the 1960s, and too far in emphasizing calories in the 1970s, it must come to a more appropriate intermediate position for the 1980s and beyond"[6].

SIGNIFICANCE OF CEREAL GRAINS AND LEGUMES IN HUMAN NUTRITION

Among the food crops of the world, cereal grains contribute more than any other single group of food staples to both calories and protein in the human diet. In its major document on world agricultural development plans, published in 1969, the FAO stated that cereals, particularly wheat, rice, maize, sorghum, millet and barley provided more than 50 percent of calories and protein for the people of the Sahelian region, more than 60 percent of calories and protein for the people of Asia and more than 65 percent of calories and protein for the Near East. In Central America, maize provided 57 percent of the daily intake of calories and 45 percent of the daily protein for adult population[7].

Dietary surveys indicate that the diet of the majority of the population in developing countries is based on a cereal grain and food legume combination. In Southeast Asia, soybeans and mung beans supplement the rice staple; in Africa, cowpeas and pigeon peas supplement sorghum and millets, while in Latin America maize and beans are a familiar dietary combination.

NUTRITION PRIORITIES AND CROP IMPROVEMENT PROGRAMS

Nutritionally, cereals and legumes are complementary. Whereas most common varieties of cereal grains are deficient in certain essential amino acids (primarily lysine) and are relatively good sources of sulfur-containing amino acids, legume grains contain twice as much protein as cereal grains and are a rich source of lysine although relatively low in total sulfur-containing amino acids. The optimum nutritional combination is provided by a diet composed of roughly 65 percent cereal and 35 percent legume. However only in Latin America does the ratio of cereals to legume production approach the desirable 2:1 ratio. In South and Southeast Asia, because of the significant decline in per capita legume production over the past 25 years, the ratio of cereals to legumes produced is of the order of 9:1. The production of legumes throughout most of the developing world has been steadily declining over the past two decades in relation both to cereal production and to population increase. During the past two decades in Asia, population increased by about 51 percent, total food production by 65 percent but legume production by little more than 20 percent. Consequently unless some significant changes occur, we may over the next twenty years be witness to a seriously inadequate food production in developing countries, both in terms of quantity and nutritional balance.

PRIOTITIES OF CROP IMPROVEMENT PROGRAMS

Keeping in mind the world food situation, agricultural scientists have focused most of their attention on increasing the total production and productivity of cereal grains and food legumes. To achieve increased productivity, scientists are attempting to maximize the efficiency of the plant to utilize energy, carbon dioxide, water and soil nutrients; attention is also being given to increasing the availability and efficiency of limiting soil nutrients and to biological processes dealing with a more efficient control of plant diseases and pests.

However in any major international breeding program, attention must not only be given to the quantitative aspects of production. Overall, the more efficient use of available land is defined by three factors[8]:

(a) Yield (kg/ha) - New varieties must be bred so as to give higher yields on the lands they now occupy. They must also be adapted for other areas that can be economically opened up for cropping.

(b) Nutritive value - Nutritional considerations such as protein content, amino acid balance and digestibility also contribute to improve the efficiency of utilization of foods.

(c) Technical value - Technological value refers to the attributes related to consumer acceptability including milling and cooking characteristics.

THE INTERNATIONAL AGRICULTURAL RESEARCH SYSTEM

Although agricultural research of one form or another has existed for centuries, publicly supported research is little more than a century old. Until recently, almost all investment in agricultural research was made in North America, Japan and Northern Europe. While the number of agricultural scientists and annual expenditures on agricultural research are increasing in developing countries, in 1974 only 25 percent of the world's public expenditure on agricultural research occurred in Africa, Latin America and Asia, despite the fact that these areas have more than 75 percent of the world's farm population[9].

A number of developing countries have made substantial progress in establishing a strong national research capability. These developing countries have attempted to increase research on specific problems by establishing regional research centres and cooperative regional programs. However many of the institutions in developing countries lack the human or financial resources necessary to undertake all of the research required by their countries in any given area of agricultural activity. At the same time, several countries, particularly those in the same agro-climatic zone, demonstrate similar conditions, opportunities and needs.

In contributing to the advancement of agricultural research in developing countries, the International Development Research Centre has played a catalytic role in establishing research networks by bringing together scientists from developing regions who have undertaken to identify regional research priorities, and then, with IDRC financial support, to define and develop cooperative research programs based on these priorities. The scientists are able to map out a comprehensive program of research which is of interest to a great many countries but which is more diverse and demanding of greater resources than any single country can provide. In several instances, IDRC is providing a technical advisor or network coordinator who acts as a focal point for information exchange and technical advice and support.

One of the most positive developments in the last decade in increasing cooperative international agricultural research has been the establishment of the International Agricultural Research Centres. The first two centres, the International Rice Research Institute (IRRI) and the International Maize and Wheat Improvement Centre (CIMMYT) were established by the Ford and Rockefeller foundations in the 1960's. Recognition of the early successes of CIMMYT and IRRI in the production of new high-yielding varieties (HYV's) of wheat and rice led to the formation of a unique international organization known as the Consultative Group for International Agricultural Research (CGIAR). The CGIAR is a permanent body of international financial support for the international agricultural research centres. The official sponsors of the CGIAR are the World Bank, the Food Agriculture Organization of the United Nations, and the United Nations Development Program (UNDP). Membership in the CGIAR which is entirely voluntary, includes a number of governments, international and regional organizations and private foundations. In 1972, its first year of funding, 15 donor members supported the work of five IARCs with a total budget of $15 million. In 1977, 29 donor members supported eleven IARCs with a total budget in excess of $80 million.

A major advantage of the IARCs is their ability to stimulate and support national agricultural research programs in developing countries. The IARCs provide national programs with a wider range of technological expertise in the form of improved germ plasm and agronomic practices which the lesser developed countries can adapt to their own agro-climatic and economic conditions. Links with developing countries are also strengthened by active training programs run by the centres. Over 3,000 scientists and production specialists have been trained in these centres to data.

IDRC'S CONTRIBUTION TO AGRICULTURAL RESEARCH

The final portion of this paper will deal more specifically with IDRC's contribution to agricultural research.

IDRC's contribution to agricultural research has taken many forms. Principally, however, IDRC's mandate is to initiate, encourage and support research programs which produce tangible results in terms of practical solutions to specific agricultural problems. Most of IDRC's financial resources are directed to networks of applied research projects in developing countries or to research programs in regional centres and IARCs. IDRC's research networks emphasize the following criteria:

(1) cooperation between international and national research programs;
(2) an interdisciplinary approach to agricultural research; and,
(3) research design and testing in association with the intended beneficiary - the farmer.

SORGHUM AND MILLETS RESEARCH NETWORK

The research network which focuses on the crops of the semi-arid tropics is a functioning example of the interrelationships outlined above.

The semi-arid tropics include most of the countries surrounding the Sahara, much of East Africa, a significant area of Central India and parts of Southeast Asia and South America. The principal cereal crops of these regions are sorghum and millets. Total world acreage of sorghum and millets exceeds 70 million hectares. In the US where sorghum is grown as a feed grain, average yields are seven times those attained in India and Africa. Clearly the opportunity exists for increasing yields in Asia and Africa through applied research and improved agronomic practices.

The hub of semi-arid crops research is ICRISAT, the International Crops Research Institute for the Semi-Arid Tropics. ICRISAT has primary responsibility for improving sorghum, millets, groundnuts, chickpeas and pigeon peas. The project network supported by the IDRC is linked with the central ICRISAT program which provides breeding materials and technical support to regional research programs.

In Ethiopia, IDRC has been supporting an important sorghum improvement program at the Agriculture Faculty of the University of Addis Ababa. The Ethiopian researchers have identified two particularly high-yielding cultivars that provide yields 7-10 times greater than the national farm average of approximately one ton per hectare. In addition several Ethiopian varieties have been found which contain more than 3 percent lysine. Normal sorghum protein contains 1.8-2.0 percent lysine. These new lines are now being distributed and crossed with other sorghums in many parts of the world to bring about an overall improvement in the nutritional quality of the sorghum grain.

IDRC has also established several contract research projects with research institutions in developed countries, conducting fundamental research on topics of importance to the sorghum, millet or legume improvement programs in the semi-arid tropics. At Canadian universities, scientists are studying the biological factors that influence the ability of sorghum to tolerate drought stress. At the University of Sussex, in England, the Centre is supporting research on striga and orobanche, the parasitic weeds which attack the root system of sorghum and legume crops. Trials are now underway in many countries of Africa, the Near East and Asia, using substances synthesized at Sussex which cause striga and orobanche seeds to germinate before the crops are planted. Finally, two scientists working at the University of Sheffield, in England, have made significant advancement in the identification of the "tannins", naturally occurring antinutritive factors present in the seed coats of many sorghums which seriously impair the digestibility of sorghum protein.

FOOD LEGUMES

In comparison to the cereal grains, the food legumes or pulses represent the most neglected food crops of the semi-arid region.

Much less research has been done on improving their yield potential and nutritional quality despite the fact that they are an important source of protein and are good sources of thiamine, niacin, tocopherol, calcium, iron and phosphorus.

IDRC is presently encouraging and supporting research to develop nutritionally improved legumes capable of giving higher yields. Since its inception in 1970, the Centre has supported the chickpea and pigeon pea improvement program carried out by scientists at ICRISAT. The purpose of this program is to breed and select for higher and more stable yields, higher protein content, and resistance to diseases and pests.

Other food legume research projects underway include a network of cowpea improvement projects involving several West African countries, the International Institute of Tropical Agriculture (IITA) in Nigeria and the International Fertilizer Development Centre (IFDC).

Food legumes are also a major component in Centre funded research on multiple cropping systems.

Multiple cropping is generally defined as growing more than one crop in the same year on the same piece of land. Multiple cropping systems include intercropping in which several different crops are grown simultaneously, and sequential or rotational cropping in which two or more additional crops are planted before or after the normal cropping season.

Intercropping or mixed cropping like plant breeding is a fairly ancient practice. In the lowland tropics of Africa one of the commonest cropping mixtures involves grain legumes such as cowpea or groundnut planted under cereals such as sorghum, millet or maize. This planting pattern is the result of hundreds of years of trial and error research by the small farmer.

Several advantages are attributed to mixed cropping as compared to pure stands of the component crops. Through mixed cropping, the farmer is provided withfood for his family and feed for his livestock. This system of farming offers insurance against poor harvests; it enables the farmer to utilize family labour more efficiently and provides family members with a more balanced diet. Above all however, there is evidence to suggest that combined yields per unit area from the intercrop are higher than the yields of either crop as a monoculture. Maize with mung bean intercrop trials, for example, have shown that the maize yield actually increased by 18 percent over maize grown alone as a result of the restriction of weed growth by the mung bean plant.

Recognition of these advantages has motivated renewed interest in the study of this traditional cropping method. IDRC is presently supporting an experimental program at IRRI which is examining the productivity of various crop combinations of corn, sorghum, legumes and vegetable crops with and around the main rice crop.

Further agricultural research is needed to gain a better understanding of the processes responsible for the attributes of mixed cropping so that proper screening of new high-yielding genotypes can be effected. Agriculturalists, nutritionists and economists should be working together in assessing the social, economic and nutritional impact of better cropping systems. Nutritionally, intercropping is destined to have a major impact on the nutritional status of the malnourished of the LDCs. For example in regions where cassava is a staple, the diet is generally low in essential vitamins, minerals and protein. Increasing the legume in the diet would overcome the vitamin and mineral deficiency and supply an increased amount of protein. To increase the available legumes, their production must be increased. This could be achieved through an intercropping system in which production of the major staple is maintained.

In view of the very important developments in agricultural research programs, there is little doubt that through the continued support of the international agricultural research centres and the expansion and strenghtening of national agricultural programs, significant improvements can be made both in the quantity and quality of the food crops of the lesser developed countries. However improving productivity and agronomic properties is only a first step in the integration of nutritional priorities into agricultural research programs.

In addition to yield and nutritional considerations, equal attention must be accorded to the selection of genotypes which meet consumer criteria in terms of grain size, colour, texture, milling and cooking characteristics. Clearly consumer acceptance of new varieties which cannot be milled using available tools or which require longer cooking may be delayed. Also there is little purpose in increasing grain production in order to increase the caloric intake of rodents. insects and microorganisms. For these reasons, post-harvest systems research which comprehends the total food system from the time of and including harvesting until the grain is delivered as food to the table must become part and parcel of research programs dedicated to the improvement of crop characteristics and cropping methods. As such, interdisciplinary approaches to agricultural research involving plant breeders, agronomists, food technologists, nutritionists and economists will bring us that much closer, that much sooner to alleviating the world food problem.

One might anticipate the reasonable question "How long will it take for these genetic and agronomic improvements to find their way into farmers' fields and the results into consumers' stomachs?" Though it is a difficult question to answer, there is no doubt that the lead time can be gradually and effectively reduced by involving the farmer in the early stages of the research process. In the research that it supports, IDRC encourages scientists to carry out at least part of their research in farmer's fields, where they can gain a better understanding of the farmer's primary constraints and his attitude to risk and change. Also only in on farm trials can research results be adjusted to the farmer's actual level or resource availability and management capabilities. Inasmuch as scientists strive to understand local, social and economic circumstances, if they comprehend why farmers do what they do, then their biological and technical research will be more likely to fit the farmers' needs.

The world's resent history of support for international agricultural development represents an important first step. There is yet a long distance to travel, however, before we can view with satisfaction the state of nutritional well-being among our poorest neighbours. If the distance between nutritional need and food supply is to be bridged during our lifetime, we must immediately move more swiftly, more imaginatively and more unselfishly in our support and encouragement for international food and agricultural research and development.

REFERENCES

[1] International Food Policy Research Institute, Recent and Prospective Developments in Food Consumption: Some Policy Issues, July, 1977.
[2] International Food Policy Research Institute, op. cit.
[3] Gavan, J.D. and Hathaway, D.E., Recent and Prospective Developments in Food Consumption: Some Policy Issues. PAG Bulletin, Vol. VII, No. 1-2, March-June 1977.
[4] Scrimshaw, N.S., Shattuck Lecture - Strengths and Weaknesses of the Committee Approach. An analysis of Past and Present Recommended Daily Allowances for Protein in Health and Disease, New Engl. J. Med. 294(3-4), Jan. 15 and 22, 1976.
[5] Protein Advisory Group Statement (No. 20) on the "Protein Problem", PAG Bulletin, Vol. III, No. 1, 1973.
[6] Scrimshaw, N.S., Through a Glass Darkly Discerning the Practical Implications of Human Dietary Protein-Energy Interrelationships. Nutrition Review, 25 (12), December 1977.
[7] FAO, Provisional Indicative World Plan for Agricultural Development, C69/4, Foor and Agriculture Organization, Rome, Italy, 1969.
[8] Bressani, R. and Elias, L.G., Tentative Nutritional Objectives in the Major Food Crops for Plant Breeders, in: Nutritional Standards and Methods of Evaluation for Food Legume Breeders, Hulse, J.H., Rachie, K.O., and Billingsley, L.W., International Development Research Centre, 1977.
[9] Boyce, J.K. and R.E.Evenson, National and International Agricult-

ural Research and Extension Programs. Agricultural Development Council Inc., New York, 1975.

PROGRESS IN NUTRITIONAL IMPROVEMENT

OF MAIZE AND TRITICALE

E.M. Villegas

International Maize and Wheat Improvement Center

Londres 40, Mexico 6, D.F.

INTRODUCTION

Considerable interest exists all around the world to upgrade the quality of protein in maize and other major cereal crops to improve their nutritional value. In some crops such as maize, barley and sorghum the quality of protein can be enhanced genetically by manipulation of known mutant genes while in other crops the search for such mutant genes is still underway. Man has developed a new cereal crop named Triticale. This cereal is produced by cross-breeding of wheat (Triticum) and rye (Secale). Under certain ecological conditions its yield outperforms that of wheat or rye. Although the cross of wheat and rye was demonstrated over 100 years ago, Triticale remained much of a laboratory curiosity until the techniques of embryo culture and doubling of the number of chromosomes through colchicine treatment were developed in 1940's. These techniques intensified and made the production of hexaploid triticales (durum wheat x rye) possible. The protein nitrogen content of triticale falls between the protein content of its two immediate parents. This appears to be also true of the proportion of essential aminoacids present, the lysine content in triticale protein being generally higher than in wheat but lower than in rye.

MAIZE

In maize only half of the actual protein content present in the endosperm is of importance from the nutritional standpoint. This is so because roughly 50% of the protein of maize endosperm is constituted by zein fraction which practically lacks lysine in its amino acid profile. The mutant genes that affect the quality of protein in maize can reduce the systhesis of zein in protein thereby resulting in increased proportion of other protein fractions that have good levels of lysine and tryptophan. This alteration in the proportion of different protein fractions in maize endosperm is thus responsible for giving a boost in protein quality in maize.

Breeding for improved protein quality of maize endosperm through the use of different mutant genes has been underway for the past thirteen years. Though several genes are known to increase the levels of lysine and tryptophan in protein to almost double, only opaque-2 gene has been used extensively to convert normal maize genotypes to opaque-2.

It may be of interest to point out some of the developments and the progress that has taken place ever since the biochemical effects of opaque-2 gene were first discovered[1]. Historically, the years 1963-1964 generated considerable interest among maize breaders all around the world to develop maize materials with superior protein quality of maize endosperm. Straight opaque-2 versions of normal open-pollinated varieties and the parental inbred lines involved in hybrid combination were obtained during the first 6-7 years of intensive research efforts. Some of these materials moved into commercial production in the early seventies in different countries but by and large these materials failed to give comparable performance with their normal counterpart maize materials. As soon as some of the opaque-2 materials moved into commercial production in some countries, those problems associated with opaque-2 maize became more evident. Some of the problems that need to be high-lighted include: i) reduced kernel weight; ii) unacceptable kernel appearence; iii) greater vulnerability to ear rot organisms; iv) more infestation by weevils during storage; v) slower drying of grain following physiological maturity[2]. Though minor problems still exist, a major breakthrough in remedying some of these problems has already been made. The period 1972-77 may be considered of tremendous significance in breeding of quality protein maize for the following reasons:

1) Attempts to solve problems associated with quality protein maize were initiated.
2) Basic information on the effects of opaque-2 gene was gathered in greater depth.
3) Genetical and biochemical information of opaque-2 modifiers in modifying undesirable effects of opaque-2 gene has been accumulated.
4) Interaction of opaque-2 gene with other endospermic mutants has been studied to solve problems confronting opaque-2 maize.
5) The relative importance of different problems affecting opaque-2 maize were further assessed.
6) Refinement in analytical techniques and new methods to detect the presence of protein quality were devised.

7) Protein quality laboratories have been established in many national programs to support breeding programs to develop quality protein maize materials and personal training has been provided.
8) Biological tests on acceptable type opaque-2 materials have been continued and produced encouraging results that protein quality was being maintained and was of superior biological value.
9) Seed increase and commercial production of opaque-2 materials started in some countries during this period.

Breeding of High Quality Maize

In breeding opaque-2 materials with acceptable characteristics, one must consider in the first place as to what kinds of materials can be accepted in different countries. Depending on the need, one must, therefore, put emphasis on the following points:

1) in areas where soft opaque-2 materials can be accepted without any problem (i.e. Andean Region), the major emphasis in the program should be placed on increased yield and greater resistance to ear rot organisms.
2) In areas where hard flints and dents are preferred, the emphasis should be to develop hard endosperm opaque-2 materials comparable in performance to their normal counterparts.

Development of Broad-Based Hard Endosperm Opaque-2 Source Populations and Other Opaque-2 Materials at Advanced Stages of Development

Intensive research efforts have been underway for the last four years to develop hard endosperm opaque-2 materials with modifiers accumulated from a wide range of maize materials being grown in different areas of the world, including CIMMYT's experiment stations in Mexico. The major emphasis in all such materials has been to increase kernel vitreosity while maintaining the same protein quality as that of soft opaque-2 materials. In the initial cycles all emphasis was put on improving protein quality. At present when kernel appearance has reached a point of acceptance, it is proposed to exert more pressure on the stability of opaque-2 modifiers responsible for changing soft endosperm to hard endosperm. Now the conversion of all tropical and temperate gene pools to hard endosperm opaque-2 is underway by the UNDP-CIMMYT Global Research Project. A special project is also being conducted to convert highland gene pools to opaque-2 and a sugary-2/opaque-2 conversion program is in very early stage.

The Protein Content and Quality of Opaque-2 Converted Materials

The ears selected in each generation of opaque-2 converted materials further undergo selection for best vitreous segregates that are available in each ear. In general, 10 seeds from each family are analyzed for protein and tryptophan content by the MicroKejeldah and Hernández and Bates procedures, respectively[3]. The families that do not meet the minimum acceptable quality levels are eliminated before pollination. This means that pollinations are restricted among selected families, for example, the mean values of several families analyzed in each population from different cycles for both protein content and quality are given in Table 1. It can be seen from this table that mean values for protein and tryptophan in protein of hard endosperm opaque-2 versions are in general fairly good in advanced materials. Lysine is determined in selected materials after tryptophan determination.

In some materials where the breeder is interested in the quality of the whole kernel, DBC analysis is performed[4]. The quality of protein in such materials is indicated by quality index (Q.I.) value which is calculated by dividing DBC value by percent protein in the whole grain. Values above 3.5 represent good quality of protein. This Q.I. correlates well with percent lysine in protein. Table 2 shows that most of the materials analyzed by this method had good quality of protein.

TRITICALE

CIMMYT started working on triticale during 1964-65 in collaboration with the University of Manitoba, Canada. During the 1970's this work was intensified with the financial assistance of CIDA, covering all aspects of triticale improvement. Presently, this new crop is being considered as an alternative source of human and animal food in many countries of the world.

Yield and Adaptation

Original triticale material received by CIMMYT from Canada had the problems of floret infertility, tall height and lateness (photosensitivity) which caused low yields. In 1968 a triticale line was observed to have high fertility which could be inherited in the segregating populations. This line was named Armadillo and has been used extensively in the breeding program. Good floret fertility combined with the dwarfness of Norin 10 gene (from wheat) has improved the yields of newer advanced lines considerably. However, under marginal conditions, where wheat culture is very poor, triticale has shown still a larger increase in terms of yield, introducing triti-

TABLE 1

MEAN VALUES FOR PROTEIN AND TRYPTOPHAN IN ENDOSPERM
OF QUALITY PROTEIN HARD ENDOSPERM OPAQUE-2 MATERIALS

PEDIGREE	ORIGIN	Nº OF FAMILIES	MEAN VALUES	
			Protein (%)	Tryptophan in protein (%)
Mezcla tropical blanca H.E.o_2	PR-76B 825	42	7.9	0.78
Ant. x Ver. 181 H.E.o_2	806	178	8.6	0.82
Mix.1-Col. Gpo.1 x Eto H.E.o_2	827	21	7.7	0.82
Mezcla Amarilla H.E.o_2	828	29	7.7	0.80
Amarillo Cristalino H.E.o_2	807	188	8.5	0.77
Tuxpeño Caribe H.E.o_2	808	144	8.0	0.86
Eto Blanco H.E.o_2	837	-	8.1	0.77
Ant. x Rep. Dominicana	PR-76B Lote 99	319	7.8	0.80
La Posta H.E.o_2	809	85	7.8	0.91

TABLE 2

MEAN VALUES FOR PROTEIN AND QUALITY INDEX
IN SOME HARD ENDOSPERM OPAQUE-2 MATERIALS

PEDIGREE	ORIGIN	N° of FAMILIES	MEAN VALUES	
			Protein (%)	Quality index
Mezcla tropical blanca H.E.o_2	PR-76B 803	134	9.8	4.0
Mix.1-Col. Gpo. 1xEto H.E.o_2	802	58	9.3	3.9
Mezcla Amarilla H.E.o_2	805	162	9.8	3.6
Amarillo Dentado H.E.o_2	801	96	7.8	4.1
White o_2 B.U. Pool	PR-76B Lote 91	443	7.0	3.7
Yellow o_2 B.U. Pool	Lote 92	430	9.3	4.7

cale has shown still a larger increase in terms of yield, introducing triticales to these non-wheat adapted areas first.

The transfer of photo-insensivity from Mexican wheats to triticale has improved the adaptation of triticale in regions within $30°N$ and $30°S$. In areas of higher latitude, where photo-sensitivity and/or vernalization is required, newer germplasm derived from winter x spring triticales, seems to be more adaptable. The results obtained from International Triticale Yield Nursery (ITYN) identify the following areas where triticale is highly successful:

a) Acid soils with aluminum, iron or manganese toxicity, i.e. Brazil, México, Ethiopia, etc.
b) Highland and montanous regions with low temperature, i.e. India, Pakistan, Nepal, Kenya, Andean Region of South America, etc.
c) Endemic areas for the diseases of wheat, i.e. Kenya (stem rust), Andean Region (stripe rust), North Africa (leaf-blotch), etc.

Progress on the development of high yielding and widely-adapted triticale strains at CIMMYT is summarized in Table 3.

Triticale is grown commercially in Hungary, U.S.S.R., China, Spain, South Africa, Argentina, Mexico, Canada and U.S.A. The crop is expected to continue to expand its production initially in the areas mentioned above.

Grain Type

A major problem in the improvement of triticale is that of overcoming the tendency of the seed to shrivelling at maturity. The degree of seed-shrivelling is reflected in the low test-weight and thereby in total grain yield and flour extraction during milling.

Although triticale lines tend to produce fair to good test weights under optimum crop conditions, under adverse growth conditions it drops sharply. During the 1970's a constant effort has been made to improve the grain-type in triticales. As a result, newer strains in 1977 have much better kernel density and test weight. Table 4 shows seven advanced lines with a yield over 7.0 tons/ha and test weight over 75.0 kg/hl. In addition, there is a total of 87 lines with yield over 6500 kg/ha and test-weight ranging between 70.0 to 77.0 kg/hl.

Industrial Quality

With the improvement in grain plumpness and test-weight, an ap-

TABLE 3

PROGRESS ON THE DEVELOPMENT OF HIGH YIELDING
TRITICALE STRAINS AT CIMMYT

ADVANCED TRIALS IN MEXICO				I T Y N		
YEAR	IDENTITY	Yield kg/ha	Test Wt. kg/hl	Year	Ave. Yield kg/ha	N° of Locations
1968-69	Bronco X224	2356	64.4	1969-70	2579	39
1969-70	Arm T909	3100	65.8	1970-71	3272	17
1970-71	Badger PM 122	4492	68.5	1971-72	3274	34
1971-72	Arm X208-14Y	5490	65.4	1972-73	3716	25
1972-73	Cinammon	5550	66.8	1973-74	4437	47
1973-74	MayaII-Arm X2802	6300	70.0	1974-75	4746	45
1974-75	Yoreme	7000	71.0	1975-76	4483	60
1975-76	Beagle	7500	68.0	1975-76	4483	60
1976-77	Mapache	8000	72.0			

TABLE 4

TRITICALE LINES ABOVE TEST WT. 72kg/hl
AND YIELD ABOVE 7,000 kg/ha   Y-76/77

VARIETY N?		CROSS AND PEDIGREE	TEST WT. kg/hl	YIELD kg/ha
19	4	CML-PATO SEL 495	76.9	7050
3	13	IA X1648-1N-4M-6Y-3M-1Y-0M	75.5	7067
*		MIA X2148-5N-2M-3Y-2M-0Y	76.3	7223
7	21	IRA-COQUENA X14595-2Y-2Y-2M-0Y	75.9	7067
10	28	IRA^2-M2A^2 X11308-B-2M-3Y-2Y-4M-0Y	77.5	7050
2	23	KLA-IA X8814-D-3Y-1M-1Y-0Y	75.3	7188
4	16	M2A-IRA X 11923-30M-1Y-0Y	76.0	7096

* Average of two or more tests.

preciable increase in the industrial quality has been achieved.

Flour extraction percent — an economic factor for milling industry — has improved from less than 60% to over 65% in general, and up to 71% in specific cases. This top level is comparable to that of bread-wheats.

Similarly, water absorption and loaf-volume are considered very important factors for baking industry. Until 1975, most triticale strains baked low loaf volume breads (less than 500 c.c. vs. more than 800 c.c. for good bread-wheats) with undesirable grain structure. At present a large number of triticale lines have a loaf-volume over 600 c.c. and up to 760 c.c. In addition, mixtures of triticale and bread wheat are suitable for bread-manufacturing. An excellent bread quality can be obtained with 60% triticale and 40% bread-wheat mixture.

Other food products such as flat unlaven bread, like tortilla and chapati, or cakes, cookies, etc., can be satisfactorily made from triticale flour using suitable technology. A group of selected advanced lines, based on industrial quality characteristics is presented in Table 5.

TABLE 5

INDUSTRIAL QUALITY CHARACTERISTICS OF SELECTED TRITICALE LINES

CROSS AND PEDIGREE	YIELD kg/ha	TEST WT. kg/hl	FALLING NUMBER	FLOUR EXT. %	PROTEIN %	SEDIMEN-TATION	BAKING	
							Water Abs.	Loaf Vol.
Maya - Arm "S" 106	7156	76.6	126	67.9	9.3	23	56.5	690
IA-IRA X13202-100Y-101B-100Y-OY	6042	69.6	96	65.4	9.5	20	59.8	680
BGL-Bulk F_2 X11066A$2^2$1M-100Y-101B-100Y-OY	6934	68.9	203	62.0	11.0	34	60.3	735
Cin-PI251923 x Pato X8061-2M-1Y-1M-3Y-2B-ON	6155	73.3	64	62.7	10.4	28	61.2	760
Kla x Octo-Hexa X7203-1M-4Y-100M-101Y	6074	74.9	68	62.2	9.4	24	57.9	705
MPE "S" X2802F-12M-1N-2M-OY	6758	74.4	96	67.4	9.3	19	57.6	715
Cin-Cno x Bgl X 16337	6681	72.9	99	68.1	9.8	26	60.2	680

NUTRITIONAL IMPROVEMENT OF MAIZE AND TRITICALE

Nutritional Quality

The emphasis put in developing highly fertile triticales with higher test-weight and plump kernel resulted in a decrease in the percentage of protein, from an average of 17.5% in 1968 to 15% in 1970. The level of protein content tended to drop with the increased yield levels. Although yields have continued to improve, the protein content has remained relatively stable at approximately 13.0% since 1973. This has resulted in a substantial increase in protein yield/hectare (Fig. 1).

Lysine is the first limiting amino acid in the cereal protein. Triticale protein is also deficient in lysine content; however, there is a large variability present with regard to this character which makes possible the selection and recombination of improved types. The average lysine content in 1968 crop was 2.83% ranging from 2.56 to 3.28. The 1976-77 crop averaged 3.4% lysine, ranging from 2.8 to 3.7%. Some data are presented in Table 6.

Potential Prospects of Triticale

The distribution of triticale nurseries internationally has exposed them to various environments, diseases, climatic and soil conditions, which differ from one place to another.

The soil conditions under which triticales are grown are an important factor: triticale seems to be highly adapted to low phosphorous acid soils much more than to saline soils. Areas having these conditions would be probably very important for the cultivation of triticales in the near future.

In such areas and others, where triticale has performed better than existing cereal crops, many tests on its utilization have been conducted with satisfactory results. This, combined with the superiority in protein and lysine content over some other cereal grains, opens an avenue for areas where cereals form a major part of the human diet. The nutritional quality and ability of tricicale to be used as forage crop is also being explored by many countries.

Biological Evaluation

Some of the most promising advanced materials (maize and triticale) that where previously selected through chemical analysis have been biologically evaluated at the National Institute of Animal Science at Copenhagen, Denmark. Amino acid composition of the materials tested are presented in Table 7 and 8.

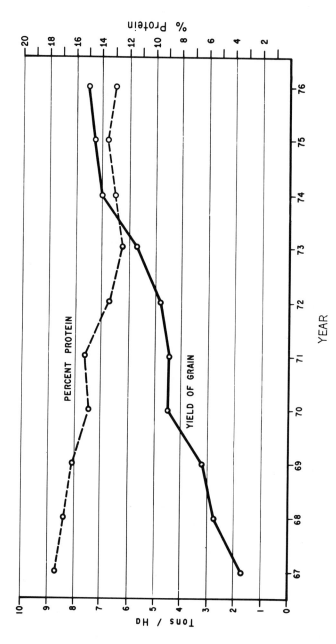

Fig. 1 - Grain yield vs percent protein in triticales at Cimmyt 1967-1976

TABLE 6

SELECTED TRITICALE STRAINS ON THE BASIS OF PROTEIN PERCENT AND PROTEIN QUALITY

CROSS AND PEDIGREE	PROTEIN %	QUALITY INDEX	D.B.C.	LYSINE % OF PROTEIN	Y77-78
IG-Octo x M2A X1109 3A-6M-1Y-2M-1Y-2M-0Y-0M	14.3	4.8	68	3.4	CB-705
M 2 A X2802-37N-1M-4N-2M-2Y-2M-1Y-0M-0Y-0M	14.6	4.8	70	3.4	CB-706
M 2 A^2 X8504C-2Y-2M-100M-103B-101Y-2M-1Y-4M-0Y-0M	12.9	5.0	64	3.7	CB-707
M 2 A^2 X85C4C-2Y-2M-100M-104B-102Y-0M-0Y-0M	14.3	4.9	70	3.6	CB-708
M 2 A- Cm1 X8543D-3Y-2M-0Y-102B-104Y-0M-0Y-0M	14.0	5.0	70	3.6	CB-709
M 2 A- IRA X12566-8Y-1Y-2M-2Y-0M-0Y-0M	15.2	4.9	74	3.4	CB-711

TABLE 7

AMINO ACID COMPOSITION OF MAIZE SAMPLES USED IN NUTRITION STUDIES

	SAMPLE						
	A (S)	B (HE)	C (HE)	D (HE)	E (HE)	F (HE)	G (N)
	AMINO ACIDS (g/16g NITROGEN)						
Valine	5.22	5.08	5.15	5.26	5.39	5.33	4.74
Isoleucine	3.29	3.08	3.25	3.23	3.29	3.25	3.59
Leucine	8.63	8.45	9.58	8.84	9.02	8.74	13.17
Tyrosine	3.55	3.40	3.56	3.54	3.62	3.47	4.31
Phenylalanine	3.99	3.80	4.01	4.02	4.11	3.99	4.78
Lysine	4.18	3.79	3.54	3.98	4.09	4.04	2.63
Methionine	1.88	1.65	1.78	1.79	1.82	1.76	2.16
Cystine	2.56	2.54	2.73	2.65	2.75	2.67	2.03
Tryptophan	0.96	0.91	0.95	1.35	1.16	1.03	0.64

TABLE 8

AMINO ACID COMPOSITION OF TRITICALE AND WHEAT SAMPLES

AMINO ACIDS (g/16 g Nitrogen)

	TRITICALES					WHEAT
	Mapache	Rahum	Beagle	Bacum	PC-297	Hermosi-llo-77
Valine	4.33	4.29	4.51	4.38	4.14	3.96
Isoleucine	3.60	3.37	3.40	3.48	3.52	3.36
Leucine	6.60	6.45	6.49	6.48	6.41	6.42
Tyrosine	3.14	3.21	3.10	3.13	3.02	3.43
Phenylalanine	4.56	4.15	4.23	4.13	4.41	4.37
Lysine	3.03	3.08	3.30	3.14	2.72	2.32
Methionine	1.73	1.67	1.82	1.79	1.69	1.61
Cystine	2.09	2.10	2.16	2.18	2.05	2.07
Tryptophan	1.04	1.10	1.00	1.00	0.92	1.02

Maize Evaluation

Groups of five Wistar male rats weighing approximately 75 g were used in these experiments in which a preliminary period of four days and a balance period of five days were employed. Each animal received 150 mg N in 10 g dry matter daily throughout the preliminary and balance periods. Feeding took place once a day. The N was adjusted by using N-free mixture[6].

The response criteria are True Digestibility (TD), Biological Value (BV) and Net Protein Utilization (NPU).

TD is the percentage of nitrogen intake which is absorbed by the organism. In cereal grain protein is known to be between 80 and 90%. The maize samples show higher values (95.3 to 96.6%). Less than 5% of the dietary nitrogen was not absorbed by the rats (Table 9).

Since the BV is the part of absorbed nitrogen which is retained in the organism, it indicated protein quality. The respective values are fairly high between 72 and 78%. The BV values agree very well with the lysine values of the respective samples, being higher in the samples with higher lysine content (more than 4.0%).

It shoulg be mentioned that in the group of the three top samples (A, E, F) one is a soft endosperm type (A), the other are hard endosperm types. The other hard endosperm samples also performed well. It is satisfying to see that the nutritional quality of the hard endosperm types is as good as the quality of the soft endosperm type. In addition, hard endosperm have more desirable kernel characteristics and consequently good acceptability, and are by far superior in nutritional quality to normal maize.

Five triticale and one bread wheat samples selected on the basis of good agronomic traits and adaptability and not for quality (through chemical analysis) were submitted for feeding studies.

Data of essential amino acids in the five triticale and wheat samples are presented in Table 8. In general the content of the essential amino acids is slightly higher in triticale and more considerably in the lysine content. However, as shown on data obtained in the nutrition studies, lysine is still the limiting amino acid in all samples.

In contrast to rye protein, which is known to be hardly digested only around 75% to 80%, wheat protein is digested highly. The values for TD of the wheat variety Hermosillo-77 as well as of the triticales are high and closed together between 91% and 93% which is normally the range of wheat protein (Table 10).

TABLE 9

TRUE DIGESTIBILITY (TD), BIOLOGICAL VALUE (BV) AND NET PROTEIN UTILIZATION (NPU) FOR SEVEN MAIZE TYPES

SAMPLE	IDENTIFICATION	RESPONSE CRITERIA					
		T D		B V		NPU	
		%	(s)	%	(s)	%	(s)
A	Tuxpeño O_2 (Soft E.) PR-76A (IPTT-37)	96.0	1.8	77.6	2.1	74.5	1.3
B	CIMMYT HEO_2 PR-76B, Bh^2-101	95.6	0.4	73.5	.8	70.2	.9
C	$PD(MS)_6$ HEO_2 IPTT-38	95.7	0.5	71.8	1.4	68.7	1.3
D	Yellow HEO_2 IPTT-39	95.3	1.0	74.0	1.4	70.5	1.9
E	Amarillo Dentado HEO_2 PR-76-13, 801	95.8	1.4	74.5	0.8	71.4	1.4
F	Ant. x Ver 181 HEO_2 806	96.6	1.1	76.2	1.5	73.6	1.2
G	White Maize Cr T1-77A	98.1	1.2	62.7	1.0	61.5	0.8

TABLE 10

TRUE DIGESTIBILITY (TD), BIOLOGICAL VALUE (BV) AND NET PROTEIN UTILIZATION FOR FIVE TRITICALE AND ONE WHEAT SAMPLES

SAMPLE	RESPONSE CRITERIA					
	T D		B V		N P U	
	(%)	(s)	(%)	(s)	(%)	(s)
Triticale:						
Mapache	92.7	1.9	66.1	0.8	61.3	1.0
Rahum	93.2	1.2	65.3	1.3	60.9	1.4
Beagle	91.0	0.9	69.9	1.2	63.7	1.0
Bacum	93.0	1.1	68.7	1.4	63.9	1.2
PC-297	91.5	1.9	59.3	1.9	54.2	1.3
Wheat:						
Hermosillo-77	92.0	1.0	57.6	1.4	52.9	1.8

The BV values indicating protein quality, show differences which agree strongly with the corresponding lysine levels; the wheat check as well as the triticale Panda, which are relatively low in lysine, show also lower BV values compared with Beagle which has 3.3% lysine in protein.

Concluding Remarks

Progress in quality protein maize and triticale during the last five years can be summarized as follows:

On quality protein maize:

- Considerable progress has already been made in developing quality protein materials that approach more nearly in kernel appearance and performance to normal materials, including yield performance.
- Simultaneously new laboratory techniques have been devised to assess quality characteristics. Laboratory activities are supporting breeding work in many countries in the world.

- Biological value of hard endosperm materials, as shown in previous section, is comparable to that of soft-endosperm opaque-2, and quite superior to normal maize.

On triticale:

- Total protein content of triticale is slightly higher than that of wheat grown under the same conditions, as well as its protein quality.
- Digestibility of triticale protein is comparable to that of wheat protein and superior in its biological value.
- Triticale is now a new crop alternative for food production in many regions in the world.

REFERENCES

[1] Mertz, E.T., Bates, L.S., and Nelson, O.E., 1964, Mutant gene that changes protein composition and increases lysine content of maize endosperm. Science, 145: 279-80.
[2] Vasal, S.K., 1977. CIMMYT Maize Annual Report (in press).
[3] Villegas, E. and Mertz, E.T., 1971, In Chemical screening methods for maize protein quality at CIMMYT. Res. Bull. No. 20, CIMMYT, London 40, Mexico 6, D.F. - Mexico.
[4] Mossberg, R., 1969, Evaluation of protein quality and quantity by dye-binding capacity: a tool in plant breeding. In: "New approaches to breeding for improved plant protein". IAEA, Vienna.
[5] Kohli, M.M. and Zillinsky, F., 1977, CIMMYT report on wheat improvement (in press).
[6] Eggum, B.O., 1973, A study of certain factors influencing protein utilization in rats and pigs. 406 beretning fra forsøgslaboratoriet. Udgivet of statens Husdyrbougsudvalg, København.

IMPROVEMENT OF NUTRITIONAL QUALITY OF SORGHUM AND PEARL MILLET

R. Jambunathan

International Crops Research Inst. for the Semi-Arid Tropics (ICRISAT)
Hyderabad - 500 016 - INDIA

INTRODUCTION

The discovery that the Opaque-2 gene in maize improves protein quality has stimulated great interest among breeders, nutritionists and biochemists and considerable progress has been made towards genetic improvement of plant protein quality in other cereals. Sorghum and pearl millet are two of the most important crops grown in the semi-arid tropics (SAT). If sorghum and millets are to retain their place, and to increase, as major cereals for human food in the SAT where they are more productive and reliable than other cereals, their grain quality is of paramount importance. The International Crops Research Institute for the Semi-Arid Tropics (ICRISAT), created in July 1972 with headquarters in Hyderabad, India, has four main objectives. One of its four main objectives is

> "To serve a world center to improve the genetic potential for for grain yield and nutritional quality of sorghum, pearl millet, pigeonpea, chickpea and groundnut".

A stable high yield of the crop is one of the main objectives of ICRISAT. At the same time we will exploit whatever latitude exists for other nutritional characters such as protein, lysine, starch types, oil, minerals, etc.

Grain quality can perhaps be considered to be made of two main parts: (i) evident quality which is based on appearance, flabour and cooking quality characteristics; (ii) cryptic quality based on nutritional value. In this paper, progress that has been made with regard to improvement of nutritional quality of sorghum and pearl millet will be discussed.

SORGHUM

Sorghum is largely a self pollinated crop and is grown on more than 40 million hectares in tropical and temperate zones. About 83 percent of the areas are semi-arid tropical in nature[1]. Yields in the more developed countries are five times those in the tropics, indicating that much improvement is possible in SAT.

High Lysine Gene

As in other cereals, lysine is the first limiting amino acid in sorghum. After screening more than 9000 accessions in the world germplasm collection, Singh & Axtell in 1973[2] reported that two sorghum lines of Ethiopian origin, IS 11167 and IS 11758, had exceptionally high lysine at relatively high levels of protein. Both lines were also high in oil percent (Table 1). The Protein Efficiency Ratio (PER) values obtained with IS 11167 and IS 11758 were 1.78 and 2.06, respectively, as compared with the PER of 0.86 obtained for normal sorghum. Inheritance studies suggested that the increased amount of lysine of each line was controlled by a single recessive gene and can be easily transferred by standard plant breeding procedures.

Protein Distribution in High Lysine Sorghum

The improved nutritional quality in Opaque-2 maize is due to the decreased prolamine (zein) and increased concentration of albumins, globulins and glutelins, resulting in a larger amount of lysine in the whole kernel.

Jambunathan et al.[3] fractionated the protein of the two high lysine (hl) sorghums and compared the distribution of protein with that of F_2 kernels obtained from a cross between Redlan (a normal variety) and high lysine lines. The distribution of proteins in these F_2 kernels and high lysine sorghums as shown in Table 2, indicate that high lysine sorghums have lower percentage of alcohol soluble fractions and higher percentage of saline soluble (albumins and globulins) fractions and the distribution pattern is similar to that of Opaque-2 maize. It is clear that the high lysine F_2 sorghum kernels have protein distribution patterns similar to that of high lysine sorghums.

As one of the ICRISAT's objective is to improve the nutritional quality of sorghum and as the two high lysine lines were available at the time when ICRISAT started to function, improvement of the nutritional quality of sorghum was included in the sorghum breeding program in 1973. As the kernels of the high lysine sorghums are floury in nature, are partially dented and have low seed weight re-

TABLE 1

CHEMICAL COMPOSITION AND SEED CHARACTERISTICS OF WHOLE GRAIN SAMPLE OF HIGH LYSINE AND NORMAL SORGHUM LINES

CHARACTER	HIGH LYSINE LINES		NORMAL SORGHUM
	IS 11167	IS 11758	
Protein composition			
Protein %	15.70	17.20	12.70
Lysine, g/100g protein	3.33	3.13	2.05
Lysine, % of sample	0.52	0.54	0.26
Chemical composition			
Oil %	5.81	6.61	3.32
Seed characteristics			
Percent germ	14.60	16.30	10.10
Seed weight, g/100 seeds	2.78	2.45	2.75
Carbohydrate composition			
Sucrose, % of sample	3.08	2.61	1.03
Starch, % of sample	58.90	57.80	60.80

Source: Singh and Axtell, Crop Science, 1973.

TABLE 2

NITROGEN DISTRIBUTION IN THE WHOLE KERNELS OF NORMAL AND HIGH LYSINE SORGHUMS[a]

FRACTION	REDLAN	REDLAN X IS 11758 F$_2$ KERNELS	IS 11758	IS 11167
I Saline	10.0	15.3[b] (22.4)	26.0	25.3
II Isopropanol	15.7	26.4 (13.7)	10.3	15.2
III Isopropanol + 2-Mercapto ethanol	31.3	26.5 (20.2)	19.6	19.3
IV Borate Buffer + 2-Mercapto ethanol	4.5	4.3 (4.3)	6.5	4.5
V Borate Buffer + 2-Mercapto ethanol + Sodium Dodecyl Sulfate	29.3	22.5 (33.5)	27.2	29.5
Total Nitrogen Extracted (%)	90.8	95.0 (94.1)	89.6	93.3
Protein %	13.53	13.0 (15.6)	18.5	16.3
Lysine (g/100g P)	1.56	1.85 (3.1)	3.27	3.10

[a] Percent of total nitrogen

[b] Average values of kernels obtained from five different heads (55073-55077). First value is the normal sample value followed by the high lysine sample value in brackets.

Source: Jambunathan et al., Cereal Chemistry, 1975.

sulting in low yield potential, attempts were made to transfer the shrunken high lysine (hl) grain of the Ethiopian cultivars to photoperiod insensitive genotypes with plump well-filled grains. Another chemically induced high lysine mutant, P721, discovered at Purdue University[4] was also used in the crossing program at ICRISAT. The amino acid composition and fractionation data on P721 has been recently reported by Guiragossian et al.[5]

One of the functions of the Biochemistry and Common Laboratory Services Unit at ICRISAT is to assist breeders in selecting the desired cultivars and progenies with improved lysine concentration. As we selected for vitreous plump grain, the screening method using a light box for Opaque character could not be employed. Therefore, in our laboratory, we have evaluated several methods for the rapid and accurate estimation of protein and lysine. Progress that has been made in identifying suitable methodology to screen thousands of samples is described below:

Methodology

i) <u>Protein Estimation</u> - We needed simple, rapid, inexpensive and reasonably accurate methods. Several methods are available for the estimation of proteins; some of the commonly used procedures include the micro or macro Kjeldahl method which is still used as a standard for crude protein estimation, biuret method, Lowry method, estimation of ammonia using the Technicon auto-analyser, and the near infrared reflectance method.

When the laboratory started to function in 1974, we tried the biuret procedure[6] for the estimation of protein in sorghum and obtained a correlation coefficient (r) of 0.91** with microKheldahl method. However, when we used this method for routine screening, the differences between biured protein value and microKjeldahl value were especially large in the low protein range as shown in Table 3. Later we tried the Technicon auto analyser (TAA) method and a very high correlation coefficient of 0.99** was obtained between the microKejeldahl and TAA protein values. During routine screening of samples using the TAA method, error percentage between TAA values and microKjeldahl values determined on random samples showed less than 2 percent variation over a wide range of protein content. Therefore, we are using the TAA method for the estimation of protein values in sorghum.

ii) <u>Lysine Estimation</u> - For estimation of lysine content we used the dye binding capacity (DBC) method[7]. This method is based on the principle that the basic amino acids (lysine, histidine and arginine) react in an acid medium with a monosulfonic acid azo dye (acid orange 12) to form an insoluble complex and results in a decreased intensity of the solution. The proportion of the dye bound

TABLE 3

DEVIATION OF BIURET (B) PROTEIN FROM MICROKJELDAHL (MKJ) PROTEIN VALUES IN SORGHUM SAMPLES

PROTEIN %	CLASS %	NUMBER OF SAMPLES	DEVIATION OF B FROM MKJ (%)
5-5.9	5	12	- 22.7
6-6.9	6	44	- 17.9
7-7.9	7	69	- 18.7
8-8.9	8	97	- 16.6
9-9.9	9	31	- 16.9
10-10.9	10	16	- 11.9
11-11.9	11	10	- 11.2
		279	- 16.6

is directly related to the total basic amino acids in the sample and the unbound dye can be conveniently measured colorimetrically as percent of transmission and is expressed as Udy instrument reading (UIR). The UIR depends on the total quantity of protein in the sample and also on the basic amino acid concentrations of the protein. After protein determination by TAA method, the weight of each of the sample was adjusted to contain 80 mg of protein and the UIR value was obtained on the sample. The UIR value represents the total amount of basic amino acids in the sample. In order to speed up the analysis and also to reduce the possible influence of sample size on UIR values the procedure was slightly modified as follows:

UIR values were taken on a one-gram weight of sample (instead of adjusting the sample weight to contain 80 mg of protein) and the readings (UIR) were divided by the percent protein (P) in grain sample to obtain a ratio (UIR/P). This ratio was compared with the lysine value determined by ion exchange chromatography using an amino acid analyzer. Using 58 sorghum samples, which had a range of 1.34 to 2.98 percent lysine and UIR/P range of 2.32 to 4.57, we obtained a correlation coefficient of 0.93** between actual lysine concentration and UIR/P ratio. A regression equation was obtained using this correlation and lysine values were predicted on routing samples by this method. We do check these estimated values by analyzing selected samples with the amino acid analyzer. We can now analyze using the above procedures about 140 samples for protein and lysine per day.

Based on laboratory values, ICRISAT breeders selected the best grain samples among progenies of random mating populations and crosses envolving the Ethiopian 'hl' gone. When the improvement of protein content and quality in the selected lines was followed year after year it was observed that some of the selections showed a large fluctuation in protein levels (Table 4). Also, we could not increase the frequency of occurance of such plants very much and this led us to question whether the high lysine (hl) gene was stable in a normal (plump seed) endosperm background. It was clear that we needed more data before this program could be continued.

Therefore, a systematic study is now being conducted by a research scholar on the variation of protein and lysine due to location, management and environment conditions and on the stability of the high lysine gene under these varying conditions. Preliminary results indicate that the crosses involving P721 are promising as they give rise to a much higher frequency of high lysine segregants. This study is still in progress.

Polyphenolic compounds, also known as tannins present in the grain of some sorghum cultivars substantially reduce the bio availability of protein and other nutrients which indirectly has a major negative effect on the nutritional quality of grain sorghum. At the same time many researchers have presented data supporting "bird resistant" qualities associated with the brown (high tannin) sorghum in areas where bird damage is severe[8]. Weathering, deterioration of seed quality due to weather conditions. including preharvest seed germination, is reportedly less serious in high tannin sorghums. More information need to be obtained to understand the role of tannin in bird and weather resistance of sorghum grain.

A major source of improving the nutritional quality of sorghum is the germplasm collection. The sorghum collection at ICRISAT exceeds 14,000 accessions. We have recently completed proximate and mineral analysis on 100 germplasm collections representing the following types — with lustre, with persistent sub-coat, completely corneous, almost corneous, intermediate, almost floury, completely floury, waxy endosperm and with white, yellow, straw, light brown, brown, reddish brown, light red, red, gray and purple seed coat colors (Tables 5 and 6). These were grown on red soil at ICRISAT farm in the 1976 Rabi season. The wide range obtained in minerals and trace elements indicate that it is possible for sorghum lines to contain various amounts of these elements. It is recognized that the mineral composition of grain is influenced by the environment, soil and management conditions. However, this observation draws our attention to the importance of analyzing the advanced elite lines in the breeding program for all the possible chemical constituents so that any cultivar having a very low amount of any of the important constituents could be identified at an early stage.

TABLE 4

VARIATION IN PROTEIN CONTENT OF SORGHUM LINES FROM SEASON TO SEASON

ENTRY	1974 RABI			1975 KHARIF			1975 RABI		
	Protein %	UDY[1]	N[2]	Protein %	UDY	N	Protein %	UDY	N
79337-2	7.2	30.0	1	7.8	34	1	12.0	24.5	20
79339-3	7.1	30.5	1	7.9	28	1	13.5	24.5	6
79337-4	7.2	31.5	1	7.2	28	1	12.7	23.0	8
79751-4	7.6	30.0	1	7.8	30	1	15.1	25.5	6
79337-2	7.2	30.0	1	7.5	31.5	1	12.0	23.0	4
79339-3	7.1	30.5	1	7.5	31.5	1	12.0	23.0	4

[1] UIR per 80 mg protein
[2] No. of entries used to calculate the mean

Kharif – Mansoon season from June to October
Rabi   – Winter season from October to March

TABLE 5

PROXIMATE ANALYSIS OF 100 SELECTED SORGHUM GERMPLASM SAMPLES[a]

	STARCH	PROTEIN	ETHER EXTRACT	CRUDE FIBRE	ASH
Range	55.6–75.2	10.6–18.5	2.1–7.1	1.0–3.4	1.6–3.3
Mean	70.8	14.1	3.3	1.9	2.1

	SUGAR	TANNIN	LYSINE[b] g/100g P	100 SEED wt (g)	GRAIN HARDNESS (kg)
Range	0.8–4.2	0.1–6.4	1.37–3.39	1.3–5.7	1.8–10.2
Mean	1.3	0.6	1.7	2.8	6.5

[a] 1977 data
[b] Estimated by DBC procedures

TABLE 6. Mineral and Trace Element Composition (mg/100g) of 99 Selected Sorghum Germ Plasm Samples

Element	Range	Mean
Phosphorus	388-756	526
Magnesium	167-325	212
Potassium	363-901	537
Iron	4.70-14.05	8.48
Copper	0.39-1.58	0.86
Zinc	2.49-6.78	3.91
Manganese	0.68-3.30	1.75

1977 data

Recently there have been several reports on protein calorie malnutrition and criticisms have been levelled against the protein quality work in food grain samples[9,10]. Studies conducted by the National Institute of Nutrition in Hyderabad, India, have shown that the primary deficiencies in the diet of people in India are mainly calories, vitamins and minerals[11]. More research is required to determine the extent to which vitamins and minerals are heritable before one could screen and select for these constituents.

One of our current interest in sorghum is to determine the factors that affect/relate to consumer acceptance of products prepared from sorghum. Our target population is sorghum and pearl millet consumers living in SAT regions. It is not uncommon to find farmers in India growing a local cultivar in a small area for their own family use and another high yielding cultivar or hybrid in a larger area for selling in the market. Therefore, evident quality characteristics deserve an important consideration in a breeding program and efforts need to be made to screen for any characteristics that might be associated with the preparation of food products. There are certain characteristics that are preferred by people and are associated with good food products. Some of the desired characteristics in sorghum are shown in Table 7. This data was obtained from the responses received from the participants attending the International Sorghum Workshop held at ICRISAT in 1977.

TABLE 7

PREFERRED CHARACTERISTICS IN SORGHUM:
RESPONSES RECEIVED FROM PARTICIPANTS[a]

Color	White/yellow
Size	Large/bold
Hardness	Hard/corneous
Nutritional qualities	High protein, high lysine, low tannin, high feeding value
Other characteristics	More dough elasticity, good Injera, Couscous, To, Chapathi, Sweet, to blend with wheat.

[a]International Sorghum Workshop, Hyderabad, 1977.

Our economists are making an attempt to relate the market price with some of the known characteristics that are said to be associated with better food products prepared in the traditional way. We are also analyzing some of the market samples for various chemical constituents in order to find out whether any relationship exists between the market price and evident or cryptic quality.

We are also concentrating on chapatti (an unleavened bread) making characteristics from sorghum and preliminary information has been obtained on evident quality characters by our sorghum breeding section. Contrasting samples from the preliminary screening will be taken up for chemical analysis. Very little work has been done in this area and we will be directing our attention in finding more about the characteristics affecting cooking quality in our Institute.

PEARL MILLET

Pearl millet is a highly cross pollinated crop which is extensively grown in semi arid tropical regions of the Indian subcontinent and Africa. Protein content of pearl millet vary from 8 to 23 percent, lysine 0.9 to 3.8 percent, oil 2.8 to 8.0 percent and carbohydrates 59.7 to 74.5 percent. Due to high oil content, pearl millet flour develop rancidity especially when stored under humid conditions. Many of the observations made for sorghum also apply for pearl millet. The methods tried for sorghum quality evaluation were also tried with pearl millet samples. The methodology is the same as explained under sorghum; results are reported here.

Protein Estimation

The biuret method gave a high correlation with microKjeldahl values, but again we observed that large variations occured between microKjeldahl and biuret values when this procedure was employed for routine screening. Therefore, the Technicon auto analyzer method (which gave a correlation of 0.97** with MKJ method) was used for protein estimation in pearl millet.

Lysine Estimation

We have obtained a correlation of 0.92** between the UIR values obtained on samples adjusted to contain 80 mg protein and actual lysine concentration determined by ion exchange chromatography. As described in the sorghum section, we are now relating the ratio of the UIR readings obtained on constant weight of the sample and the protein values (UIR/P) with actual lysine concentration.

High lysine pearl millet has not been reported so far. Also

TABLE 8

MAIN PREPARATIONS MADE FROM PEARL MILLET
GRAIN FOR HUMAN CONSUMPTION[a]

I	Chapathi	> 50%
	Porridge	
	Cous Cous	
	To, Tuwo	
II	Nadida - Thin Gruel	
	Kali - Cracked grain taken as Porridge & Dumpling	
	Bouille	
	Kune	
	Kichri, Rabri	
	Cuppasa - Thick Pancake	
	Loas	
	Loax	
	Burkutu - Liquor Drink	
	Fura	
	Gadogado - Nigerian Pancake	
	Waina - Fried Type Cake	

[a] International Pearl Millet Workshop 1977.

our breeders were interested in obtaining basic data on the relationship between protein, yield, seed size etc. However, protein content in several composites showed a wide range from 8 to 15 percent. Environment seems to play a major role in influencing protein content and the relationship between yield and protein content was not very strong. Therefore, it seems possible to increase the protein content of pearl millet cultivars without much affecting the yield. More attention will be paid to this possibility during the coming years.

As discussed for sorghum, we are also looking at the consumer acceptance characteristics in pearl millet. Table 8 shows a list of various preparations made from pearl millet. These were tabulated from the responses we had received from participants attending the International Pearl Millet Workshop in September of 1977. Therefore, our efforts in determining evident quality characteristics as well as improving the protein content and quality of pearl millet will continue.

CONCLUSION

For the rapid estimation of protein and lysine in large number of sorghum and pearl millet samples Technicon auto analyser method and dye binding capacity method were found to be most suitable.

We need more data to understand whether high lysine gene in sroghum is stable in normal plump seed endosperm background.

In pearl millet, relationship between yield and protein was not very strong. Therefore it seems possible to increase the protein content without much affecting the yield.

Evident quality characteristics in sorghum and pearl millet are very important. Attempts are being made to understand more about the factors that govern the cooking quality and consumer acceptance characteristics in food products.

REFERENCES

[1] Kanwar, J.S., Ryan, J.G., Recent trends in world sorghum and millet production and some possible future developments. Paper presented at the Symposium on Production, Processing and Utilization of Maize, Sorghum and Millets. Central Food Technological Research Institute, Mysore, India, December, 1976.
[2] Singh, R., Axtell, J.D., High lysine mutant gene (hl) that improves protein quality and biological value of grain sorghum. Crop Science, 13: 535-539, (1973).
[3] Jambunathan, R., Mertz; E.T., Axtell, J.D., Fractionation of soluble proteins of high lysine and normal sorghum grain. Cereal

Chemistry, 52: 119-121, (1975).
[4] Mohan, D.P., Chemically induced high lysine mutants in sorghum bicolor (L) Moench, Ph. D. Thesis, Purdue University, W. Lafayette, Indiana (1975).
[5] Guiragossian, V., Chibber, B.A.K., von Scoyoc, S., Jambunathan, R., Mertz, E.T., Axtell, J.D., Characteristics of proteins from normal, high lysine, and high tannin sorghums. J. of Agric. and Food Chem., 26: 219-223, (1978).
[6] Johnson, R.M., Craney, L.E., Rapid biuret method for protein content in grains. Cereal Chemistry, 48: 276-282, (1971).
[7] Udy, D.C., Improved dye method for estimating protein, J. of the Amer. Oil Chem. Soc. 48: 29A-33A, (1971).
[8] McMillen, W.W., Wiseman, B.R., Burns, R.E., Harris, H.B., Green, G.C., Bird resistance in diverse germplasm of sorghum. Agron. J., 64: 821-822, (1972).
[9] McLaren, D.S., The great protein fiasco, The Lancet, 2: 93-96, (1974).
[10] Payne, P.R., Nutritional criteria for breeding and selection of crops: with special reference to protein quality. Plant Foods for Man, 2: 95-112, (1976).
[11] National Institute of Nutrition: National nutrition monitoring bureau report for the year 1975. Indian Council of Medical Research, Hyderabad, (1976).

THE NEED FOR FOOD UTILIZATION AND PROCESSING STUDIES

TO SUPPLEMENT NUTRITIONAL EVALUATION

R.A. Luse

Centro Internacional de Agricultura Tropical (CIAT)

Apto. Aéreo 67-13 - Cali, COLOMBIA

INTRODUCTION

The multidisciplinary program to improve cowpea for West Africa carried out at the International Institute of Tropical Agriculture (IITA) included both nutritional evaluation (protein, surfur amino acids) and consumer acceptance factors (cooking time; taste; texture and appearance of prepared cowpea dishes). Such screening identified cowpea varieties combining superior yield potential, good nutritional value and high consumer acceptance. A similar program at CIAT to improve bean (Phaseolus vulgaris) will include screening for content of protein, methionine, lysine, tryptophan, and antinutritional factors, plus cooking time, seed preference characteristics, and suitability for processed foods.

It is felt that the international agricultural research centers should extend their present nutritional screening activities to include consumer utilization and processing studies. In developing countries consumer preference in such basic foodstuffs as beans are often sharply defined. Without good acceptability/preference characteristics, a new crop variety will find no market and hence low acceptance by the farmer producer. The international centers and national plant breeding programs should encourage food utilization and processing research in national nutrition and food science institutes and work cooperatively with them.

1. STATED INTEREST OF THE INTERNATIONAL CENTERS
   ON PRODUCT QUALITY

The two international agricultural research centers at which the author has worked have made very positive statements regarding their intent to improve the quality of the commodities with which they are concerned. The term quality has been interpreted to include both nutritional value and consumer acceptance factors. The Centro Internacional de Agricultura Tropical (CIAT) has stated its objectives as follows:

> "To generate and deliver, in collaboration with national institutions, improved technology which will contribute to increased production, productivity and quality of specific basic food commodities in the tropics — principally in countries of Latin America and the Caribbean — thereby enabling producers and consumers, especially those with limited resources, to increase their purchasing power and improve their nutrition".

The International Institute of Tropical Agriculture (IITA) has given as one of its major objectives the following:

> "To increase yields and improve the quality of food crops in the humid and sub-humid tropics through every available means, especially the development of high yielding and insect and disease-resistant plants".

These statements have been taken as a clear mandate to examine several aspects of quality in the food crop varieties being developed at these centers, especially cowpea (Vigna unguiculata), lima bean (Phaseolus lunatus) and cassava (Manihot esculenta) in IITA, and field bean (Phaseolus vulgaris) and cassava at CIAT.

2. PRESENT ACTIVITIES OF IITA AND CIAT ON PRODUCT QUALITY

International Institute of Tropical Agriculture (IITA)

A program to evaluate nutritional value and consumer acceptance of new grain legume cultivars was established and was closely integrated with the plant breeding activities in 1973. Evaluation extended from initial mass screening for seed protein content and protein quality, through determination of complete amino acid composition of selected lines, to measurement of consumer acceptance factores. As promising lines were identified in any of the other legume screening programs (e.g. pathology, entomology), these lines were evaluated for nutritional value and consumer acceptance. Of the over 8000 lines in the IITA world collection of cowpea germplasm, over 5000 were screening for protein content and protein quality. This first stage screening relied on determination of total nitrogen content to estimate seed protein and of total sulfur content to estimate the sulfur containing amino acids that limit protein quality. The result of this work was to identify four classes of cowpeas on the basis of their content and quality of protein, for use in subsequent plant breeding work. Two classes were of particular interest: the first consisted of lines having high sulfur and high nitrogen content and were found with a frequency of about 0.1%. The second contained lines, occurring with a frequency of about 0.06%, that had higher than average protein quality as indicated by their high sulfur to nitrogen ratios (S/N). This latter class is the likely source of genes for high protein quality and was used in subsequent plant breeding work. To confirm this initial screening, the amino acid composition of these cowpea lines was determinde by ion exchange chromatography.

In order to estimate the acceptance of new IITA cowpea cultivars by consumers in West Africa, Several factors that determine acceptance were measured. These include cooking time and water uptake for whole beans, and the taste, texture, and appearance of dishes prepared from ground beans. Cooking time and water uptake (or the ability to "fill the cooking pot") were measured by plotting increase of wet seed weight as a function of the time that beans were submerged in boiling water. The beans initially absorbed water lineraly with time but at the point where they were judged "cooked" by test panel members, water uptake stopped. Cooking time was thus taken as the time to reach this plateau, while swelling was directly related to the maximum water absorpiton. More than 100 cowpea lines were screened for these two factors. Cooking times ranged from 35 to 90 minutes and water uptake from 98-170%. Only those lines with short cooking time and high swelling capacity were acceptable to consumers.

Cowpeas are frequently consumed in West Africa as fried akara balls and steamed moin-moin, both of which are prepared from the ground beans. In order to estimate this aspect of consumer acceptance, these dishes were made from local recipes, using flour of IITA cowpea cultivars. This work was done in cooperation with the test kitchen of the University of Ibadan. Taste panels graded the product on the basis of taste, texture and appearance. A high-quality cultivar was always included in the test as standard, so that results were stated as preference or non-preference of teh IITA cultivar compared with the standard.

A limited number of informal dietary surveys were carried out by IITA staff as a part of our nutritional studies. A number of more formal surveys done by the Food Science and Applied Nutrition Unit of the University of Ibadan were made available to us as unpublished reports. On the basis of these reports and others, it was determined that grain legumes are an important source of protein in the diets of people living on villages in Sourthern Nigeria, but that green leafy vegetables and other components of the soups and stews eaten daily also contribute more protein than is usually realized.

In work done by the IITA plant physiologist, a large number of cassava plants were tested for level of cyanide content since this is an important factor of nutritional value for the so-called "bitter" varieties having high cyanogenic glucoside content. Nearly 100,00 plants were screened during the period of this work, using a very simple color test with sodium picrate paper. Frequency of plants having low cyanide content was judged to be about 3%, though in certain composites developed in the breeding program, the frequency of low cyanide cultivars reached over 7%. In other work, done by the IITA starch chemist, the rheological properties of cassava starch were studied, together with their relationship with consumer preference factors such as texture and gari manufacture.

Centro Internacional de Agricultura Tropical (CIAT)

Work on cassava quality at CIAT centers primarily on post harvest storage and processing. Cassava is a highly perishable crop with physiological deterioration of the roots often beginning within two days after harvest and with microbial deterioration starting in less than one week. The fresh roots are bulky to handle since they have about 65% water content. Simple methods are needed to solve these problems of perishability and the bulky nature of the root. Recent work at CIAT has shown that high quality stored roots may be achieved by either pre-harvest leaf pruning or by packing the roots subsequent to harvest in polyethylene bags containing a fungicide. The problem of residual fungicide toxicity may not be serious, since cassava is peeled before eating, but it is essential to search for less toxic materials and to determine any residues that are remaining in the roots. A number of research centers in the world are studying simple methods utilizing solar energy for the drying of cassava. CIAT scientists will be testing these methods and if necessary adapting them to fit the needs of the Latin American cassava producer.

At the present time the author is setting up a program at CIAT to screen for protein content all advanced Phaseolus bean lines coming from the breeding program. Cooking time in these lines will also be studied, although with emphasis primarily on ways to reduce cooking time through simple chemical treatment during soaking or through proper storage conditions. Here it may be noted that the treatment of seeds with edibel oil, a method found effective against weevil attack on cowpea and field bean, is being examined as to its possible reduction of hard seed coat development during poor storage conditions. As other methods are proved of value in determining nutritional value and for estimating consumer acceptance, they will be applied at CIAT in a full screening program for bean quality.

3. PROPOSED ACTIVITIES FOR QUALITY EVALUATION AND UTILIZATION

Nutritional Evaluation

Estimating the nutritional value of a foodstuff requires knowledge of three important factors: its role in actual dietary consumption, its protein digestibility, and how it is processed for food. While it is necessary to have a general idea of the amino acid composition of a foodstuff, there is a complementary balancing of amino acids between cereals and legumes (lysine from legume ⟶ methionine from cereal), so that simple amino acid scores are of limited value in estimating the true value of mixed diets. This point has been well made by Bressani in the case of maize and dried beans in Central America. An even more impressive case can be made for cowpea-sesame seed mixtures (see Table 1), where the complementarity is excellent and serves as the basis for the "benniseed" weaning food in Sierra Leone.* A less exact matching is that of pigeon pea and millet (Table 2), though here the chemical score may be nearly boubled over those of the components through an appropriate mixing. The factor of protein digestibility becomes important when the food legumes are involved, since they typically have digestibilities of only 69-83%. Lastly, processing of the foodstuff may have important effects on its nutritional value and digestibility, and can convert a poor foodstuff into a good food.

The above statements imply that when strategies for plant improvement include nutritional characters, the plant breeders need to have dietary consumption data for the target areas where the improved varieties will be used, plus additional information beyond the usual amino acid composition, preferably NPU or equivalent nutritional values. Obtaining such figures is outside the activities of the international centers and there is good case to be made for working cooperatively with national and regional institutes of nutrition and food science. To generate such information CIAT is at present proposing such cooperation with INCAP in Guatemala and with the Instituto Nacional de Nutrición in Caracas. Attached is a summary of the program that has been proposed for USAID funding. As can be seen, the program will be directed to gaining consumption information in Central America and to the development of methods suitable for estimating nutritional values in Phaseolus beans.

---

* Since rice is a major component of the benniseed product, sesame seed is only about one-fifth of the total weight of the blend.

TABLE 1

AMINO ACID PATTERNS IN COWPEA AND
SESAME SEED ANA THEIR MIXTURE

AMINO ACID	AMINO ACID CONTENT, mg/g N			FAO/WHO PATTERN, 1973
	Cowpea	Sesame	1:1 mixture	
Isoleucine	290	250	270	250
Leucine	530	440	485	440
Lysine	490	195	340	340
Methionine + Cystine	180	310	245	220
Phenylalanine + Tyrosine	640	550	600	380
Threonine	290	250	270	250
Tryptophan	100	110	105	60
Valine	370	310	340	310
Chemical Score	82	57	100	
Limiting EAA	Met + Cys	Lys	—	

TABLE 2

AMINO ACID PATTERN IN PIGEON PEA
AND MILLET SEED AND THEIR MIXTURE

AMINO ACID	AMINO ACID CONTENT, mg/g N			FAO/WHO PATTERN, 1973
	Pigeon pea	Millet	33-67% mixture	
Isoleucine	194	230	218	250
Leucine	394	783	655	440
Lysine	481	168	271	340
Methionine + Cystine	32, 61 ⊢ 93	120, 97 ⊢ 217	176	220
Phenylalamine + Tyrosine	517, 126 ⊢ 643	305, 239 ⊢ 544	577	380
Threonine	182	225	211	250
Tryptophan	-	-	-	60
Valine	225	303	277	310
Chemical score	42	49	80	-
Limiting EAA	Met + Cys	Lysine	Met + Cys ≡ Lys	

Food Preparation and Processing Studies

Included in the CIAT proposal is a small amount of work on the study of food preparation techniques (typically at the household and village level) and of commercial food processing techniques (typically at a national or regional level). This was done because it was felt there is a possibility to extend traditional food preparation methods so as to improve both the nutritional quality and the consumer acceptance of the food actually reaching the consumer. Here such techniques as chemical treatment during soaking to produce a "quick cooking" bean may be considered. Processing at the commercial level — and possibly even at the community level — also needs to be studied, with the idea of introducing modern techniques such as the low-cost food extruders. These types of studies might have a profound difference on the actual utilization of beans in Latin America.

Such studies are also generally outside of the proper scope of the international centers, yet there needs to be a dialogue between the plant breeders on the one hand, who know the genetic diversity within their crop, and the food technologists on the other hand, who know what their equipment can do. Hopefully, the food scientist can tell the plant breeder what characters are required for a given processing technology and the plant breeder can develop it. In many cases, changes in the processing steps may be necessary to utilize available genetic diversity, but it should be possible to end up with a compromise between what the processer demands and what the plant breeder can produce, to the ultimate benefit of the consumer. There is also a subtle benefit to the international centers here, since in the developing countries which the centers have as target areas, consumer preferences in such basic foodstuffs as beans are often sharply defined. Without good acceptability/preference characteristics, a new crop variety will find no market and hence low acceptance by the farmer producer. The centers would be well advised to know these preferences — and how they may be met by processing — and to carry out their plant improvement programs accordingly.

In summary, it is felt that the international centers should have staff with the following responsibilities:

1) To screen for nutritional quality and consumer acceptance;
2) To gather information on food utilization and nutritional value through cooperation with national institutes of nutrition and food science;
3) To conduct limited amounts of research into various phases of utilization and processing where the wide range of international center germplasm is of value.

NEED FOR FOOD UTILIZATION STUDIES

SUMMARY OF PROPOSAL

Title: Research program on factors affecting acceptability and nutritive value of food legumes of importance in Latin America.

Submitted to: Agency for International Development (AID)

Investigators: Dr. Robert A. Luse, CIAT Plant Biochemist (Coordinator); with Dr. Ricardo Bressani, INCAP; and Dr. Werner Jaffe, INN

Cooperators: CIAT; Institute of Nutrition for Central America and Panama (INCAP), Guatemala City; Instituto Nacional de Nutrición (INN), Caracas.

OBJECTIVES OF PROGRAM

Objective I

To gather information on the steps in the chain: production — marketing — utilization, for food legumes (primarily Phaseolus bean) in representative Latin countries (initially Honduras), in order to identify constraints to legume consumption by low-income populations, both rural and urban. When the constraints have been identified, action will be taken, as possible and appropriate, by the institutions cooperating in this program, plus national and regional agencies, to demonstrate methods by which to overcome these constraints.

There is a need to go beyond the usual production survey, so as to include those factors that determine the quantities of legumes actually reaching the household. In this way one can estimate the elasticity of demand by the consumer and the elasticity of production by the farmer, to answer such questions as: "If the farmer increases legume production by 50% using new agricultural technology, can he expect to see the increased production (or use it in his own household) and, if so, at what price?" "If the housewife finds 50% more beans on the market (probably at a lower price than previously), will she buy more and thus improve the nutrition of her family?" "What is the potential for increased consumption through new foods based on beans?"

Objective II

To develop methods suitable for the large-scale screening of food legume germplasm (initially Phaseolus bean) for those factors most important in legume consumption, vis: nutritional value, consumer acceptance, and suitability for food technology. Those methods determined to be most suitable will be utilized by CIAT to evaluate its extensive legume germplasm and newly developed varieties.

At the present time, the following methods seem suitable to estimate the various factors underlaying legume consumption by the consumer:

Factor	Methods
Nutritional value	Protein content (via total nitrogen content)
	Protein quality (via total sulfur and via methionine and cystine content)
	Lysine content
	Trypsin inhibitor level
	Hemagglutinin level
	Protein digestibility (via PER assay)
Consumer acceptance	Seed characteristics (color, size, hardness)
	Water absorption in cooking
	Thickness of broth
	Food preparation (cooking, frying, etc.)
Suitability for food technology	Predicated by the technology, e.g. extrudability if in Brady crop cooker.

However, in many cases these methods are empirical and related only to North American tastes and preferences, not to Latin American food habits. Further, the key aspect of protein digestibility (or availability) is poorly defined and is presently related only to animal feeding tests — hardly a mass screening technique.

MODERN PRACTICES IN ANIMAL FEEDING - RELATIONSHIP TO

HUMAN NUTRITION - INTRODUCTORY REMARKS

A. Rerat

I.N.R.A. - C.N.R.Z.

78350 Jouy-en-Josas - FRANCE

Protein consumption is very unevenly distributed over the world (Table 1). In developing countries it represents 3/5 of that of the industrialized countries of Europe and North America. This difference is not due to the intake of plant proteins, which is almost the same, but to that of animal protein which is five fold higher in the developed countries. Thus, animal proteins (Table 2) represent more than half of the protein supply in the developed countries versus 20% in the developing ones. The differences are even more marked if dividing these groups of countries into subgroups in which the development is more or less high (Table 3). The major supply of animal proteins proceeds from meat (except in the Near-East and Middle-East) and milk comes in the second position. These data show the importance of animal protein production and the interest of using the most efficient techniques and conditions. It has to be emphasized that the production costs of these proteins are very high and vary in very large proportions according to species and type of production. This is illustrated in Table 4 (Holmes, 1970).

In animal species the feed efficiency never exceeds 23%, either in terms of metabolizable energy (pig) or protein efficiency (milk production). In large and small sized slaughter ruminants this value is generally lower.

Accordingly, in terms of human food production per surface unit, plant cultures are superior to any form of animal production (Table 5, Holmes, 1970).

The high meat production costs are particularly marked when formerly grazing species such as ruminants are fed with diets based on cereals. According to Blaxter (1968), production of 1 kg muscle

TABLE 1

PROTEIN SUPPLIES (1963-1965) PER CAPITA AND PER DAY
Autret, 1970

	CALORIES	ANIMAL PROTEINS (g)	VEGETABLE PROTEINS (g)	TOTAL PROTEINS (g)
Developing regions	2.140	10.7	46.9	57.6
Developed regions	3.070	48.3	40.8	89.1
World	2.380	21.0	45.1	66.1

TABLE 2

PERCENTAGE CONTRIBUTION OF VARIOUS COMMODITIES TO PERCENTAGE SUPPLIES (PROTEIN SUPPLIES 1963-1965), Autret, 1970

	CEREALS	STARCHY ROOTS AND TUBERS	PULSE NUTS AND SEEDS	VEGET. AND FRUITS	VEGET. PROTEINS	MEAT	EGGS	FISH	MILK	ANIM. PROT.
Developing regions	57.2	3.8	16.8	3.3	81.4	8.3	0.9	4.0	5.4	18.6
Developed regions	31.9	4.7	3.9	5.3	45.8	25.4	4.3	3.9	20.4	54.2
World	47.9	4.1	12.1	3.9	68.2	14.7	2.1	3.9	10.9	31.8

TABLE 3

PROTEIN SUPPLIES 91963-1965) (g PER CAPITA/DAY)

	TOTAL PROT.	TOTAL ANIM. PROT.	ANIMAL PRODUCTS*				MEATS*				
			Meat	Eggs	Fish	Milk	Beef Veal	Mutton	Pork	Poultry	Others
Far East	54.8	8.6	3.6	0.4	2.5	2.1	0.79	0.25	1.62	0.54	0.40
Latin America	67.6	24.1	12.4	1.3	1.8	8.6	7.94	0.50	1.61	0.74	1.61
Near and Middle East	71.6	14.0	5.7	0.5	1.0	6.8	2.05	2.39	0	0.23	1.03
Africa	58.5	10.9	5.4	0.3	2.4	2.8	2.70	0.86	0.16	0.38	1.30
Europe	87.6	42.3	18.8	3.3	3.7	16.5	7.33	1.13	6.39	1.69	2.26
North America	93.1	65.3	33.8	5.4	2.7	23.2	16.9	0.68	7.44	6.76	2.03
World (g)	66.1	21.0	9.7	1.4	2.6	7.2	4.07	0.48	2.81	1.26	1.07

* Calculated from Autret (1970).

TABLE 4

FEED EFFICIENCY IN WHOLE FARM SITUATIONS (Holmes, 1970)

	ME[1] (%)	GE[2] (%)	PROTEIN[3] (%)	PROTEIN (g/Mcal ME)[4]	PROTEIN (g/Mcal GE)[5]
Dairy herd	21	12	23	10	5.4
Dairy + beef herd	20	11	20	9	4.7
Beef herd	7	4.5	6	2.6	1.5
Sheep flock	3	1.7	3	1.3	0.8
Pig herd	23	17	12	6	4.0
Broiler flock	13	10	20	11	7.7
Egg flock	15	11	18	11	8.0

Each enterprise allows for replacements.
[1] (Edible energy x 100) ÷ (total metabolizable energy consumed)
[2] (Edible energy x 100) ÷ (total gross energy consumed)
[3] (Edible protein x 100) ÷ (total food protein consumed)
[4] (Edible protein (g) ÷ (total metabolizable energy consumed (Mcal))
[5] (Edible protein (g) ÷ (total gross energy consumed (Mcal))

TABLE 5

ANNUAL YIELDS FROM ANIMALS AND FROM CROPS
(Holmes, 1970)

	ENERGY (Mcal/ha)	PROTEIN (kg/ha)
Dairy cows	2,500	115
Dairy + beef cattle	2,400	102
Beef cattle	750	27
Sheep	500	23
Pigs	1,900	50
Broilers	1,100	92
Eggs	1,150	88
Wheat	14,000	350
Peas	3,000	280
Cabbage	8,000	1,100
Potatoes	24,000	420

tissue in pigs requires 12 kg feed whereas 35 and 45 kg, respectively are necessary for beef cattle and sheep (Table 6). Thus, only 1 kg muscle protein is produced with 60, 170 and 220 kg of a cereal based diet in pig, cattle and sheep, respectively. On account of this very poor conversion yields of plants into edible animal protein and on the basis of the human mean consumptions reported in Tables 1 and 2, the intake of cereal equivalents ranges from more than 1 ton per year and per head in the developed countries to 250-300 kg in the developing ones. These data show the bad collective utilization of the world's food resources, depending to a small extent on a possible excess of animal protein consumption in the developed countries and to a larger extent on the recent development of ruminant feeding methods in these countries. To prevent a competition between species such as ruminants and man in the satisfaction of their nutritive requirements, it could be imagined that it would be necessary to feed again ruminants with diets mainly composed of roughages which are not consumed by man. Another important factor is that in the developed countries people ask for quality products and the satisfaction of such demands is expensive. Thus, according to Pearson (1974)

TABLE 6

FOOD COST (PER kg) USING AN ALL-CEREALS DIET FOR PRODUCING 1 kg OF MUSCLE TISSUE IN YOUNG FATTENED ANIMALS, BLAXTER (1968)

	DIRECT COST	INDIRECT COST		TOTAL FEED COST	
	Food required to fatten the young animal in addition to dam's milk	Maintenance of the breeding herd or flock	Replacement of the breeding herd or flock	Without consideration of meat from disposed animals	When meat from disposed animals considered
Sheep	15.0[a]	31.2[b]	13.0[c]	59.2	44.9
Pig	9.37	2.52	0.38	12.3	12.1
Cow	14.5	23.8	7.1	45.4	34.6

[a] Based on daily gains exceeding 300 g/day
[b] Based on average of 2.95 Mcal ME/day, with a minimum of 1.85 Mcal during reproductive rest and a maximum of 6.0 in the first month of lactation
[c] Based on gains costing 7.0 kg/kilogram (100/gain/day).

the change from "good quality" to "very high quality" products represents a supplementary expenditure of 500 kg feed for cattle slaughtered at 500 kg (Table 7).

Some prefectly edible feeds for man are used in animal feeding and this is well illustrated by the data concerning fish protein. The intensification of the fishery has led to an increase in the supply of fish protein from 4.3 g per head and per day in 1950 to 8.9 g in 1968-1970. A very large proportion (about 2/3) of this larger amount of fish is not directly used in human nutrition, but in the manufacturing of fish meal for animal feeding (Stilling, 1973).

The problem of the competition between animals and man in the use of plant resources becomes more and more important both because of changes in the feeding of some species and because of the systematic and expensive search for products of the best quality. This question is one of the points that should be discussed during this symposium.

TABLE 7

FEED SAVINGS OF MARKETING "GOOD" VS "CHOICE" BEEF (FROM A.M. PEARSON, PERSONAL COMMUNICATION, 1974) IN LEVEILLE (1975)

MARKET GRADE	YEARLING STEERS		STEER CALVES	
	Good	Choice	Good	Choice
Initial weight, lb	690	690	430	430
Market weight, lb	1,000	1,100	900	1,000
Gain necessary to reach grade, lb	310	410	470	570
Average daily gain, lb	2.6	2.4	2.3	2.2
Time on feed to reach gain, days	120	170	205	260
Feed required/lb gain, lbs	8.3	9.8	7.3	8.3
Total feed to reach gain	2,760	3,952	3,588	4,680
Feed savings by marketing at lower grade, lb	1,192		1,092	
%	30		23	

The problem can be partly solved not only by using roughages again in ruminants or industrial by-products and wastes in all species, but also by improving the animals through genetics and nutrition. In the latter case and especially in monogastric animals, the goal can be reached not only by establishing the best balance between the various nutrients in the diets, but also by using feed additives. However, the introduction of such substances into animal feeds and the eventual presence of contaminants give rise to various problems which will also be discussed during this symposium.

So, the various themes of the symposium are the following:

- pig meat and beef production;
- milk production in dairy cows;
- the competition between these two types of production in cattle;
- use of waste products in animal feeding; and
- role of feed additives.

BIBLIOGRAPHY

Autret, M., 1970, in: "proteins as Human Food", Proc. 16th Easter School in Agric. Science, Univ. Nottingham, Lawrie, R.A. (ed.), 3-19.
Blaxter, K.L., 1968, Proc. Second World Conf. on Animal Prod., Univ. Maryland, Bruce Publish. Comp., St. Paul, Minnesota, 31-40.
Holmes, W., 1970, Proc. Nutr. Soc., 29, 237-244.
Leveille, G.A., 1975, J. Anim. Sci., 41, 723-731.
Stilling, B.R., 1973, in "Proteins in Human Nutrition", Proc. NATO Advanced Study Institute in the Chemistry, Biology and Physics of Protein evaluation, J.W.G. Porter and B.A. Rolls (eds.), 11-33.

MODERN PRACTICES IN PIG FEEDING AND HUSBANDRY:

RELATIONSHIP TO HUMAN NUTRITION

A. Rerat

I.N.R.A. - C.N.R.Z.

78350 Jouy-en-Josas - FRANCE

I. PROBLEMS RELATED TO PIG MEAT
   AND HUMAN NUTRITION

In 1965, the total amount of animal protein available per head and per day in man was 21 g, corresponding to 32 percent of the world population's protein feeding. Most of these animal proteins came from meat products (46 percent) and dairy products (34 percent) and the rest from eggs (7 percent) and fish (12 percent). The meat products were mainly composed of beef (42 percent), pig meat (29 percent) and poultry (13 percent) and the rest of sheep and goat meat (5 percent) as well as meats from other species (11 percent). These data clearly show the great interest of producing pigs for meeting the protein requirements in man.

However, the production of pigs is unevenly distributed over the world both in terms of populations and productivity (Table 1).

Thus, the South American pig population represents 17 percent of that of the world but corresponds only to 5 percent of the world's pig meat production. Conversely, the European and North American pig population represents 31 percent of the world's pig herd and 52 percent of the overall supply of pig meat. This leads to large disparities in the human consumption of pig meat which is almost inexistent in Africa, the near East and Far East and very high in Western Europe and North America.

A comparison of the data of 1968 with those of 1952 (Braude, 1970) (Table 2) clearly shows that the pig production has greatly increased during this period (101 percent, on an average), corresponding to a general rise in the world production of animal proteins

TABLE 1

HUMAN POPULATION, PIG POPULATION AND PIG MEAT SUPPLY (1968)
(Kroeske, 1972)

REGION	HUMAN POPULATION (millions)		PIG POPULATION (Millions)		PIG MEAT PRODUCTION (Thousands of tons)		PIG MEAT/ HEAD/DAY (Grams)
Europe	454	(13)*	125	(21)	11,213	(33)	68
U.S.S.R	238	(7)	51	(8)	3,075	(9)	36
North America	222	(6)	61	(10)	6,460	(19)	79
South America	268	(7)	101	(17)	1,536	(5)	16
Near East	161	(4)	0.1	(0)	15	(0)	0
Far East	1,106	(31)	44	(7)	1,873	(5)	5
China	815	(23)	213	(35)	9,430	(28)	33
Africa	290	(8)	7	(1)	193	(0.5)	3
South Sea Island	19	(1)	3	(0.5)	200	(0.5)	30
WORLD TOTAL	3,573	(100)	605	(100)	33,795	(100)	25

* All figures between brackets are the percentage of the world total.

TABLE 2

CHANGES IN THE PIG POPULATION (MILLIONS) IN THE WORLD
(1952 - 68)

REGION	AVERAGE OF 5 YEARS ENDING 1952	1968	% INCREASE
Europe	69.3	123.5	78
(U.K.)	(3.4)*	(7.8)	(140)
U.S.S.R.	19.7	50.8	158
North and Central America	75.5	83.4	10.5
South America	35.6	80.5	126
Asia	20.6	44.8	115.5
China	73.7	213.0	190
Africa	4.4	6.1	39
Oceania	1.9	2.9	52.5
TOTAL	300.8	605.1	101

* Low numbers following the reduction during and after the Second World War.

(55 percent for dairy products, 100 percent for eggs and meat products). It has to be emphasized that the production has only slightly increased in North America, corresponding to a stabilization of the individual consumption, and largely increased in China (almost 200 percent).

There has been a continuous development of the situation since 1968 and this is clearly shown by the statistical data supplied by the European countries of the E.E.C. (Table 3).

In these countries, the consumption of beef has only increased by 18 percent between 1965 and 1977 whereas that of pig meat has grown by 63 percent during the same period.

This increase in the overall consumption of pig meat in the world depending on the increase in the world population and on food habit changes (larger pig meat consumption in some countries), gives rise to both quantitative and qualitative problems for the farmers.

TABLE 3

CHANGES IN THE ANNUAL PER HEAD CONSUMPTION (kg)
OF CARCASS MEAT IN EEC BETWEEN 1965 AND 1977

YEAR	BEEF AND VEAL	PIG	MUTTON AND GOAT	HORSE	POULTRY	MISCEL-LANEOUS	OFFALS	TOTAL
1965	22	22	1.1	1.1	6.1	1.4	4.8	58.5
1971	25.2	25.9	1.2	1.1	11.1	2.5	5.5	72.5
1977	25.9	35.9	1.6	1.3	12.6	2.9	6.0	86.1

The quantitative problem can be solved by intensifying the production by means of a modification of management structures and methods, and by improving the productivity. The qualitative problem can be considered under two aspects:

One of these aspects is that people, for nutritional and medical reasons, are eating less and less fat (Table 4).

Now, the fat percentage of pig carcasses is much higher than in other species (Table 5). Thus, the fat content of pig carcass generally ranges about 40 percent at the usual commercialization stage versus 25 percent in beef. Let us, however, emphasize that the excess of fat is in particular located in a part of the carcass that can be separated from the rest so that at the consumption stage, pig meat does not differ much from the other kinds of meats (Table 6). However that may be, this gives rise to the problem concerning the control of fatness through rearing and breeding methods aiming at reducing the lipid production.

Furthermore, the various edible tissues, lean and fat, exhibit different physical and chemical characteristics which define their quality. It has been well established that backfat has a variable chemical composition on which its tast, flavour and preservability depend. The quality of lean depends on some characteristics: colour, flavour, tenderness, sapidity, pH, water holding capacity, which, if they are modified unfavourably, may be at the origin of the more and more frequent occurrence of exudative myopathies leading to the so-called PSE meat (pale, soft, exudative meat). According to Van Logtestijn (1970), 10.20 percent of the products supplied to the European slaughter houses can be ranked into that category of meat.

The other aspect concerning the quality of the product concerns

TABLE 4

PERSONAL PREFERENCES AMONG 504 PEOPLE FOR MEAT FAT

	% OF SAMPLE WHO	
	EAT fat	LEAVE on the plate
Men	55	45
Women	40	60
Boys	39	61
Girls	24	76
Adults	48	52
Children	31	69
WHOLE SAMPLE	46	54

Rhodes (1976).

TABLE 5

PIG CARCASS CHARACTERISTICS COMPARED WITH THOSE OF OTHER ANIMAL SPECIES (SCHON, 1973)

CARCASS %	MUSCLE		FAT		BONE	
	Min.	Max.	Min.	Max.	Min.	Max.
Cattle	48	- 82	0.5	- 35	11	- 35
Calf	58	- 74	0.5	- 18	14	- 30
Sheep	46	- 76	4	- 37	12	- 26
Pig	27	- 69	12	- 55	9	- 17

TABLE 6

LIPID LEVEL IN THE LONGISSIMUS
DORSI M. IN MATURE ANIMALS
(Laurie, 1966)

	% FRESH WEIGHT
Cattle	3.4
Sheep	7.9
Rabbit	2.0
Pig	2.9

the possible existence of more or less reduced amounts of substances liable to cause pathological disorders (toxicity, allergy, etc.) in the consumer. These substances are either residues of products included in the feeds of the animals to increase their value or animal feeding contaminants or their metabolites which, in some cases, are accumulating in animal tissues. We are not going to talk about this aspect which will be treated by Pr. Ferrando.

Thus, this paper is divided into several parts. In the first chapter, we shall examine the physiological factors, inherent to the animal. which are inducing variations in the body composition.

Some of the breeding and feeding techniques leading to modifications of the proportions of various edible tissues will be analysed in Chapter 2.

In the last chapter, we shall describe the factors involved in the quality of each of these tissues.

2. GROWTH AND BODY COMPOSITION:
   FACTORS OF VARIATION

During growth, the body composition of the animal changes in the proportions of the various tissues, genetic origin (breed and strain), sex, and environment may affect these characteristics.

<u>Changes in the body composition according to age and weight</u>

It has long been known that when the animal gains weight, it gets fatter; this can be seen from the results of a great number of experiments dealing with the dissection and chemical analysis of carcasses from animals of increasing weights. The results obtained by Richmond and Berg (1971) from pigs derived from a variety of crossings between Yorkshire, Duroc and Hampshire breeds, are shown in Figure 1. These results are expressed in terms of dissectable tissues; they are in agreement with those obtained earlier by other authors.

Bone grows relatively more slowly than muscle does. Muscle grows fairly rapidly until the weight reaches 90 kg, then the decrease in the growth rate of the muscle is compensated by an increase in the deposition of fatty tissue.

These changes are manifested chemically by a very rapid increase in the level of lipids in the carcass, and a gradual fall in the level of proteins as the weight of the animal increases (Table 7).

This fattening as the weight increases does not mean that no lean tissues are synthesized by the animal, but that the proportion of muscle in the weight gain tends to decrease and that of adipose tissue to increase.

Genetic variations of body composition

It is beyond doubt that there are important differences between breeds and within the same breed as regards growth performance and body composition. These differences may be judiciously exploited by means of selection. This is illustrated by Hetzer and Harvey (1967), who, by selection for back fat thickness only, obtained after 10 generations fat lines and lean lines, the characteristics of which have been defined more accurately by Davey et al. (1969) (Table 8).

Thus, careful selection may give profitable results, though the improvement may be slow.

Although selection can improve the body composition, the crossing of breeds does not give the same result. Sellier (1970) reviewed world literature on this topic and published a recapitulatory table (Table 9).

The results show that heterosis has a moderate effect on growth rate and feed conversion ratio and no effect on body characteristics, tissue composition or meat quality. Consequently, the body composition of the crossbred pig is exactly half-way between those of the two parent breeds. Nevertheless, this technique also gives favourable results as the growth performance of the crossbred pig is similar to that of the better parent breed. The main advantage obtained is probably an increase in the vitality of the animals and in their

TABLE 7

CHANGES IN THE BODY COMPOSITION WITH AGE
(Oslage, 1962)

WEIGHT (kg)	EMPTY WEIGHT (kg)	DRY MATTER (g)	% OF CARCASS			% OF DRY MATTER	
			Dry matter	Protein	Fat	Protein	Fat
25.1	22.7	8,046	35.4	15.7	16.4	44.8	45.9
40.1	36.9	14,086	38.2	16.1	19.2	42.1	50.2
60.3	56.0	24,450	43.6	15.0	26.7	34.6	61.1
89.8	83.9	41,671	49.6	14.4	33.1	29.1	66.6
110.4	103.2	52,689	52.1	14.3	34.5	28.0	67.5

TABLE 8

A COMPARISON OF FAT AND LEAN PIGS OBTAINED
BY SELECTION FROM THE SAME POPULATION

(Davey et al., 1969)

BREED	YORKSHIRE		DUROC	
TYPE	Fat	Lean	Fat	Lean
Average daily gain (g/day)	604	640	596	712
Feed conversion ratio	3.79	3.77	3.67	3.46
Lean cuts (kg)	23.7	30.2	21.7	28.0
Fat cuts (kg)	34.4	28.2	28.2	37.0

TABLE 9

HETEROSIS FOR TRAITS OF ECONOMIC INTEREST IN THE PIG: MEAN VALUES OBSERVED DURING CROSSING EXPERIMENTS BETWEEN PURE BREEDS (WITHOUT PREVIOUS INBREEDING) (Sellier, 1970)

Variables \ Dam / Offspring	TYPE OF MATING		
	Purebred animals	Simple cross	Triple cross
Dam:	Purebred animals	Purebred animals	Crossbred animals
Offspring:	Purebred animals	Crossbred animals	Crossbred animals
Performances of the Dam			
Number of piglets born alive	100	102	108
Proportion of weaned piglets	100	106	108
Number of weaned piglets	100	108	116
Litter weight at birth	100	104	110
Litter weight at weaning	100	115	125
Performances of the Offspring			
Individual weight at weaning	100	106	107
Daily mean weight after weaning	100	106	106
Individual weight at 154 days	100	110	110
Feed efficiency	100	103	103
Proportion of lean cuts in the carcass	100	100	100

Influence of Sex on Body Composition

It is well known that sex has a substantial effect on body composition. Thus, many experiments have shown that for a given weight, castrated males are fatter than females, whose fatness exceeds that of entire males. This is shown in Table 10 (Desmoulin, 1973).

It has to be pointed out that the higher anabolizing abilities of the entire males is particularly marked in the case of a feed restriction. In any case, the increase in the percentage of lean tissues in the entire male as compared to the castrated one, ranges around 25-30 percent.

Unfortunately, carcasses produced by entire males may exhibit more or less important defects. The disadvantage is that boar meat in some instances has a disagreable odour (Rhodes, 1969; Desmoulin, 1972), which seems to be caused, to a great extent, by androstenone (Patterson, 1968). This "sexual" odour (boar taint) mainly located in the fatty tissues of the male does not occur before 60 kg liveweight in the various breeds studied. From 90 kg liveweight, the proportion of animals exhibiting a marked boar taint increases with age (Patterson et al., 1971). This percentage is however relatively low (10-15 percent for animals slaughtered between 80 and 100 kg); it has not been established in all breeds and in all conditions. Accordingly, the risks of providing the consumer with a poorly edible meat are not very great, but have to be taken into account. This problem can be partly solved by detecting meats with boar taint at the slaughterhouse and then transferring them to the meat processing industry.

When using a rather low incorporation level, it does not seem that products prepared with such meats can be distinguished from those prepared with intact meats (Maarse et al., 1972). A previous selection between meats with and without boar taint requires use of rapid olfactory tests such as that of the soldering-iron (Jarmoluck et al., 1970). Using this method it is possible to detect pigs devoid of any odour defect (70-75 percent). Doubtful carcasses (25 percent) should be subjected to further tests in the laboratory, but those tests are slower and more laborious so that they cannot easily be used at the slaughterhouse.

Conclusions

The growth rate, nitrogen and energy retention, and body composition are affected by several physiological factors. The most important of these is age; some very accurate information has been

TABLE 10

INFLUENCE OF SEX AND CASTRATION ON GROWTH AND FEED EFFICIENCY AND VARIATIONS
ACCORDING TO THE RESTRICTION SCHEDULE (T = control; R = 25 percent restriction)
(Desmoulin, 1973)

Feeding level	SEX - CASTRATION							
	MALES		CASTRATED MALES		FEMALES		CASTRATED FEMALES	
	T	R	T	R	T	R	T	R
Mean daily feed intake* (kg)	2,22	1,77	2,30	1,80	2,12	1,68	2,13	1,67
Total feed intake (kg)	254a	252a	295b	308b	280b	297b	297b	327c
Growth rate (g/d)	682a	555b	627a	470c	610ab	440c	568b	408c
Fattening length (d)	115	143	129	171	132	176	137	196
Feed conversion ratio kg feed/kg gain	3,26 a	3,18* a	3,69 b	3,84 b	3,49 ab	3,85 b	3,74 b	4,19 c

* Feed 85 percent dry matter – The same letters indicate not significantly different values.

collected over the past few years which facilitates the calculation of the deposition of nitrogen adn of energy, and the rates of conversion. Genetic origin and sex are two other important factors. The elaboration of standards for nitrogen and energy must accomodate differences in the development of the carcass which depend on breed, strain, and sex. The uncastrated male has proved to be highly interesting, because of its high rate of conversion of the feed, but its meat may exhibit some defects whose conditions of occurrence and detection have not yet been perfectly controlled.

3. INFLUENCE OF DIETARY FACTORS ON GROWTH AND BODY COMPOSITION

It is well known since long that there is a close relationship between growth and body composition characteristics and the intake level and composition of energy and protein.

The production of lean pig is clearly a consequence of controlled nutrition. Sources of energy must be supplied in amounts which are just sufficient enough to ensure an optimum synthesis of protein, and consumption must never be excessive. Adequate amounts of properly balanced sources of nitrogen must also be supplied to allow an optimum synthesis of tissues.

Control of Energy Intake in Relation to Growth and Body Composition Characteristics

Many studies have been undertaken to establish the control of the energy intake necessary to produce lean carcasses as economically as possible. There are several ways of achieving this; they may be arranged into three groups:

- variations of the energy composition of the diet;
- variations of the feeding level;
- physical form of the feed and mode of feeding.

Only the first two points will be discussed.

The Influence of Energy Composition of the Diet on the Physiological Characteristics of the Growing Animal - When a pig is given a mixed feed ad libitum, the dry matter intake depends upon the composition of the diet: "energy-density", protein level, amino acid balance, and the levels of minerals and vitamins. The level of energy has been the subject of much detailed study.

There is general agreement that when the energy level of a diet is increased (by addition of lipids) the consumption of dry matter is reduced, the feed efficiency is however improved as well as very

TABLE 11

INFLUENCE OF THE LEVEL OF DIETARY LIPIDS ON GROWTH PERFORMANCES (20-90 kg) AND BODY COMPOSITION OF THE PIG (Henry and Rerat, 1964)

					STANDARD DEVIATION OF THE MEAN
Protein p. 100	16	16	8	8	
Lipids p. 100	5	15	5	15	(x)
					g
Growth (20-90 kg)					
Daily mean gain (g)	633	653	486	498	17,5 N^{xx}
Daily feed intake:					
- kilograms	1,81	1,69	1,68	1,56	0,04 $N^{xx}E^{xx}$
- therms	7,75	7,99	6,82	7,18	0,16 $N^{xx}E^{xx}$
Feed conversion ratio					
- kg feed/kg gain	2,87	2,59	3,50	3,15	
- therms/kg gain	12,28	12,25	14,21	14,49	0,21 N^{xx}
Body composition					
Yield ($\frac{\text{net weight}}{\text{live weight}}$ x 100)	72,9	73,5	75,3	75,1	0,4 N^{xx}
Lean cuts p. 100	51,4	50,0	48,7	47,5	0,5 $N^{xx}E^{xx}$
Fat cuts p. 100	18,8	20,6	22,5	23,7	0,6 $N^{xx}E^{xx}$
Mean backfat thickness $(\frac{\text{loin + back}}{2})_{mm}$	28,9	34,7	36,8	39,0	

often the growth rate and the carcasses are fatter. All these facts have been clearly demonstrated in an experiment by Henry and Rerat (1964) (See Table 11).

According to the results of this factorial experiment — two levels of protein (16 percent and 8 percent), two levels of fat (5 percent and 15 percent) and utilization of females and castrated males — an average reduction of 7 percent in the consumption of dry matter can be noticed when the level of energy increases. The fall in the dry matter intake was not, however, proportional to the rise in the level of energy, thus there was a rise in the energy consumption of about 5 percent. The reduced consumption depressed the consumption of protein, so aggravating the imbalance of protein and energy; the pigs grew fatter in consequence.

Contrary to what happens in the case of fats, the addition of cellulose to the diet reduces the "energy-density" of the latter, which results in an increase in the pigs' consumption of dry matter when they are fed ad libitum. However, this increase is not high enough to compensate for the reduction in the daily energy intake, resulting in reduced fatness (but poorer dressing percentage). This effect has been well illustrated by the experiments completed by Henry (1966, 1969) in which pigs were given diets with two different levels of protein and of cellulose (see Table 12).

The addition of cellulose appears to be beneficial, in terms of apparent energy efficiency; however, in relation to carcass net weight, energy efficiency remains unchanged.

Moreover, the digestibility of the organic matter is considerably reduced when the level of cellulose is increased, but there is no significant change in nitrogen retention efficiency (Henry, 1969).

Influence of the Feeding Level Upon the Physiological Characteristics of the Growing Animal - For several years now, interest has been growing in the control of feed intake by means of restricted feeding methods. The responses of various physiological criteria have been studied, especially the interactions between different aspects: age, sex, genetic origin, and the composition of the ration — particularly the quantity and quality of the sources of energy and protein.

A review has recently been made by Vanschoubroek et al. comprising the results of various authors. Vanschoubroek has studied the effects of feed restriction on the different performances of the pig. The severity of the dietary restriction was calculated taking the ad libitum consumption for reference (between 30 and 90 kg liveweight a daily consumption of 2.7 kg of a ration containing 3,000 digestible calories per kg, giving a daily gain of 750 g and a degree

TABLE 12

INFLUENCE OF VARIATIONS IN THE LEVELS OF CELLULOSE AND OF PROTEIN IN A RATION ON GROWTH, FEED CONSUMPTION AND BODY COMPOSITION OF GROWING PIGS (Henry, 1966)

CRUDE PROTEIN (%)	12		16		$S\bar{x}$	STATISTICAL SIGNIFICANCE (1)
Cellulose (%)	5	15	5	15		
Digestible energy (Kcal/kg of dry matter)	3 930	3 540	3 950	3 950		
Average gain (body weights between 20 kg and 90 kg) (g/day)	705	689	731	716	15.4	n.s.
Daily consumption:						
- of feed (kg/day)	2.20	2.27	2.21	2.30	0.032	Cellulose*
- of digestible energy (Mcal/day)	7.89	7.39	8.00	7.59	0.16	Cellulose**
Feed conversion ratio						
kg/kg gained	3.14	3.31	3.04	3.22	0.049	Cellulose**
Digestible Mcal/kg of gain	11.27	10.77	11.00	10.79	0.14	Cellulose**
Digestible Mcal/kg of net weight	15.00	14.84	14.63	14.85		
Average backfat thickness = $\frac{\text{Loin + back (mm)}}{2}$	33.9	30.0	32.7	28.6	1.01	Cellulose**

n.s. - non significant
* - 0.05
** - 0.01

(1) Threshold of significance
$S\bar{x}$ - standard deviation

of fatness corresponding to a back-fat thickness of 37 mm). The results of these calculations are shown in Table 13.

As the severity of restriction increases, there is:

- a depression in the growth rate which becomes progressively more marked as the restriction becomes more severe;
- an improvement in the feed conversion ratio which reaches the lowest level when the restriction is one-quarter of the ad libitum level; at higher levels it rises again;
- a linear reduction in back-fat thickness.

These variations arise from two conflicting phenomena. First, the prolongation of the fattening period causes an increase in the proportion of needs for maintenance relatively to the total needs for growth, but the production cost for every unit of gain declines because the pig produces more meat. Much more energy is required to produce fatty tissue than to produce muscle, and consequently the feed conservion ratio can be improved by reducing fatness in the pig. It follows from these three criteria that the best reduction is 20 to 25 percent of the amount consumed when feeding is ad libitum. This restriction corresponds to an average consumption of 2-2.2 kg of feed between live-weights of 50 and 90 kg and will produce a decline of 20 to 25 percent in the growth rate (600 to 650 g/day); an improvement of about 6 percent in feed conversion ratio (a saving of 0.2 kg of feed for every kilogram of weight-gain) and a reduction of 8 percent in back-fat thickness (only 3 mm).

However, the influence of feed restriction varies according to different factors among which sex and genetic origin are the main ones. It has been established that a moderate energy restriction applied to the female (Desmoulin, 1969) leads to a substantial improvement of the body composition characteristics. Conversely, the castrated animal has to be subjected to a very severe restriction during the finishing period for obtaining a reduction of fatness. This result is obtained at the expense of a large decrease in the feed efficiency. Likewise, the energy restriction intensity has to be adapted to the genetic abilities of the animals; in pigs of the fat type in which the fatty tissues develop early the energy restriction must be applied when the animals are young.

Influence of Protein Intake on Growth and Body Composition

Protein nutrition and energy nutrition have a common purpose: obtaining a lean carcass economically. The aim is to improve the efficiency and the quality of production by determining the conditions under which the dietary supply of protein can be adjusted so as to facilitate the best possible performances by the pig.

TABLE 13

MEAN EFFECTS OF THE FEED RESTRICTION ON GROWTH RATE, FEED CONVERSION RATIO AND BACKFAT THICKNESS (VAN-SCHOUBROEK, de WILDE AND LAMPO, 1967), ACCORDING TO BIBLIOGRAPHICAL DATA

DAILY RESTRICTION PERCENT AD LIBITUM LEVEL	PERCENT DAILY MEAN GAIN DECREASE	PERCENT FEED CONVERSION RATIO IMPROVEMENT	PERCENT BACKFAT THICKNESS DECREASE
5	4,1	0,7	3,6
10	7,0	3,3	4.9
15	10,5	5,2	6,1
20	14,7	6,3	7,4
25	19,6	6,6	8,7
30	25,1	6,2	9,9
35	31,3	5,0	11,2
40	38,1	3,0	12,4
45	45,6	0,3	13,7

Body composition, growth, feed intake and nitrogen retention depend on the qualitative and quantitative supply of nitrogenous matters. The animals according to sex and breed are not behaving in the same way towards these factors.

- Protein level: the percentage of lean tissue (X %) grow simultaneously with the dietary protein level (Z %) and as declines the percentage of fatty tissues (Y %) according to a linear regression obtained with the equations formulated by Cooke et al. (1972):

$$X = 0.388 Z + 39.16 \quad r = 0.976$$
$$Y = -0.468 Z + 33.96 \quad r = -0.992$$

These relations are clearly shown in Table 14.

- The quality of the proteins ingested also plays a part in the degree of fatness through the best amino acid pattern for covering the requirements for muscle synthesis and through its effect on the appetite.

TABLE 14

BODY COMPOSITION OF PIGS FED INCREASING
AMOUNTS OF PROTEIN BETWEEN 23 AND 59 kg LIVEWEIGHT
(Cooke et al., 1972)

	DIETARY CRUDE PROTEIN (%)						STD. DEVIATION
	15	17.5	20	22.5	25	27.5	
Lean tissue (%)	44.7 c	46.6 b	46.8 b	47.7 b	49.0 a	50.0 a	0.43
Fat tissue (%)	26.6 d	25.1 c	23.8 b	23.3 b	21.6 a	20.5 a	0.43
Bone (%)	9.8 bc	9.6 bc	10.0 ac	9.9 ac	10.1 ac	10.5 a	0.14
Skin (%)	3.9 a	3.9 a	4.2 ac	4.4 bc	4.5 bc	4.3 ac	0.12

- Like in the case of energy intake there is an interaction between protein supply (nature and amount) and protein synthesis potential of the animal depending on particular on its sex and genetic origin.

4. FACTORS OF VARIATION OF THE QUALITY OF EDIBLE TISSUES

Fatty Tissue Quality

The quality of fat is characterized by:

- its colour which must be white. Thus, feeds rich in carotenoids (yellow maize) must be avoided as they may bring about a formation of yellow fat.

- its firmness, determined by the relative proportion of unsaturated fatty acids (oleic, linoleic, linolenic acids) and of saturated fatty acids (palmitic, stearic acids...).

Soft and oily backfat contain a larger proportion of unsaturated fatty acids than firm backfat; the iodine index is then higher and the melting point lower. Although the consumption of unsaturated fatty acids instead of saturated ones seems to be preferable for health (American Heart Association, 1973), the meat packers now prefer a firm fat not only because of its aspect, but also because of its storage abitity. It is well known that fats containing a large proportion of unsaturated fatty acids become rancid, but this could be avoided by proper handling and refrigeration. Among the factors liable to change the quality of fat let us mention the nature of the dietary lipids and the feeding level.

<u>Influence of the Nature of the Dietary Lipids</u> - Much of the fat which is deposited does not come from exogenous fats, which never represent more than a very small fraction of the energy consumed by the pig. This can be conclusively demonstrated if pigs are maintained on a fat-free diet (Table 15), they synthesize new fatty acids from carbon-chains originating from the degradation of carbohydrates and nitrogenous substances; the lipids deposited essentially consist of oleic acid (about 50%), palmitic acid (about 25%) and stearic acid (about 16%).

It seems, however, that the pig is uncapable of synthesizing those fatty acids which contain more than one double bond (such as linoleic and linolenic acid); these apparently only come from external sources. If a pig is fed a ration which contains fat having the same composition as its own fat-lard for example, — the composition of the fat deposited has changed becoming more highly unsaturated because of an increase in the proportion of oleic acid — which is due to the preferential utilization of saturated fatty acids in metabolic processes (see Table 15).

The digestibility and metabolism of the fatty acids strongly influence the composition of the fat deposited; this is particularly evident when fats which contain fatty acids usually absent from these deposits are introduced into the diet (see Table 15). Thus, lauric acid is almost entirely catabolized; it is found only in small quantities in the deposits; in contrast, the pig stores about one half of the linoleic acid it consumes. Considering the more common fatty acids, it can be noticed that the level of palmitic acid in the deposits is almost constant (25%) — whatever the proportion of it in the feed; but the level of stearic acid fluctuates very much. The concentration of oleic acid in the fat deposited is almost the same as that in the mixed feed consumed. It may be concluded that the participation of oleic acid in the formation of deposits is proportional to the amount of fixed fats.

Conclusions may be drawn from all these facts which are very important with respect to feeding and body composition of the pig. In order to reduce the levels of unsaturated fatty acids in the backfat, responsible for "soft" fat, some of the saturated acids present in the feed must replace oleic acid because the latter accumulates

TABLE 15

VARIATIONS IN THE COMPOSITION OF LIPID DEPOSIT BETWEEN 45 AND 100 kg LIVE WEIGHT ACCORDING TO THE NATURE OF THE DIET (Flanzy and al., 1970)

DIET (1)	Lipid free basal diet			BD + lard (2)			BD+mixture (2) rich in C 18 : 0 (cocoa butter : 80%) (cotton oil : 20%)			BD+mixture (2) rich in C 12 : 0 (cocoa butter : 50%) (cotton oil : 15%) (coco-nut oil : 35%)		
	Diet	Mixture absorbed **	Depot	Diet	Mixture absorbed **	Depot	Diet	Mixture absorbed **	Depot	Diet	Mixture absorbed **	Depot
Fatty acids*												
C 12 lauric	—	—	—	—	—	—	—	—	—	15,3	19,2	1,3
C 14 myristic	—	—	—	—	—	—	—	—	—	6,2	5,8	4,9
C 16 palmitic	—	—	28,1	26,8	27,5	25,5	24,9	16,3	24,5	19,6	12,9	26,8
C 18 stearic	—	—	18,1	15,0	9,5	14,4	27,0	16,5	22,1	20,0	11,4	22,2
C 18:1 oleic	—	—	52,9	50,3	53,9	55,8	32,9	45,2	42,9	28,2	36,2	37,5
C 18:2 linoleic	—	—	0,9	7,9	9,1	4,3	15,2	22,0	10,5	10,7	14,5	7,3

(1) Semi-synthetic diet containing skim-milk spray powder and maize starch
(2) 13 p.100 of the diet
* Fatty acids in p.100 of total lipids ingested, absorbed or of the deposits
** On account of the digestibility of the various fatty acids.

in the body-fat at higher levels than those present in the feed. Furthermore, in order to reduce the deposition of fat, fats which are mainly composed of short-chain fatty acids, and possibly linoleic acid, must be added to the diet whilst the growth of the pig is retarded by restricting its feed intake. In this way, the degree of saturation of the fats which are deposited is simultaneously increased. But once again, it has to be underlined that it would be better, for the consumer's health, to increase the unsaturated fatty acid proportion in the fatty tissues. This may easily be achieved by increasing the proportions of these unsaturated acids in the fats of the pig's diet.

From a practical point of view, giving pigs vegetable oils which are rich in polyunsaturated fatty acids (such as soya or maize or peanut) produces soft back-fat which has a high Iodine index as compared with the backfat obtained with a diet containing animal fats (such as tallow or lard) (Brooks, 1971, 1972; Leat et al., 1964; Mason and Sewell, 1967; Barrick et al., 1953; Koch et al., 1968).

Among the animal fats, tallow as compared to lard produces a rise in the degree of unsaturation by increasing the oleic acid contents of the deposits (Brooks, 1971, 1972; Thrasher et al., 1959; Kuryvial and Bowland, 1962).

There are also difference between the vegetable oils according to their origin. Thus, the replacement of soyaben oil rich in unsaturated fatty acids by the same proportion of coconut oil, rich in saturated fatty acids leads to an elevation of the proportion of saturated fatty acids in the body fat. The earlier the replacement, the higher this proportion as shown by a trial made in the United States by Blumer et al. (1957) (Table 16).

The backfat is all the firmer as the animals have received coconut oil for a longer time.

Effect of Feed Restriction - A severe feed restriction leads to soft backfat whereas a liberal intake level results in formation of firm backfat. This is illustrated in Table 17 (Greer et al., 1965).

This can be easily explained as the animal only synthesizes saturated fatty acids whose proportion increases with the intensity of the lipogenesis. So, when the feeding level is reduced the deposition of exogenous fat becomes relatively more marked. This influence is all the more evident as the diet contains more polyunsaturated fatty acids (maize containing diet). Let us mention that an increase in the dietary crude fibre level, as it causes a restriction of the supply of energy, has the same effect on the composition of the fat deposits as a lowering of the feeding level.

Furthermore, let us mention that females generally have a softer

TABLE 16

BACKFAT QUALITY AS AFFECTED BY THE NATURE OF THE DIETARY LIPIDS
(Blumer et al., 1957)

DIET	CONTROL SOYBEAN OIL	COCONUT OIL FROM A LIVEWEIGHT OF			
		80kg	70kg	55kg	20kg
Iodine index	93.6	80.6	72.7	67.3	54.2
Saturated fatty acids, p.100	33.1	38.7	42.8	45.2	51.6
Backfat firmness	soft	moderately soft		moderately firm	firm

TABLE 17

INFLUENCE OF THE NATURE OF THE CEREAL AND OF THE FEEDING LEVEL ON FAT DEPOSIT COMPOSITION
(Greer et al., 1965)

Diet	MAIZE - SOYABEAN			BARLEY - SOYABEAN		
Fatty acid composition	Diet	Fat Deposit		Diet	Fat deposit[1]	
Feeding level		100	85		100	85
Lipids % dry matter	1.7			0.6		
Fatty acids, % total lipids:						
14 : 0	13	1.0	0.9	0.5	1.3	1.0
16 : 0	13.3	24.8	25.8	22.4	28.1	25.8
16 : 1	0.9	1.7	4.0	0.8	4.9	1.2
18 : 0	1.1	13.4	10.6	2.1	10.9	12.6
18 : 1	30.4	47.7	46.3	21.1	48.9	50.2
18 : 2	52.9	11.4	12.3	50.8	6.7	9.3

[1] External backfat layer

and more unsaturated backfat than castrated males (Koch et al., 1968). This can also be related with the lower feed consumption in the females.

Meat Quality

Meat palatability depends on various criteria: tenderness, juiciness, colour, aroma and flavour. The surface of pig meat must be firm, dry and of dark pink colour. The estimation of pig meat quality takes into account the colour, firmness and water holding capacity, the extremes being, on the one hand, pale, soft and exudative meat (PSE) and on the other, dark firm and dry meat (DFD). It has to be pointed out that PSE meat has exactly the same chemical composition as normal meat. The occurrence of defects depends on a more rapid post-mortem disappearance of muscle glycogen than normal, bringing about a fall in muscle pH related with the production of lactic acid. Now, the pH of the muscle is closely related with its water holding capacity (Herring et al., 1971). This syndrome partly depends on heredity and seems to be associated with the stress resistance in the living animals.

Although it has not been well established yet, it seems that selection might solve the problem of exudative meats. The heritability of most meat characteristics ranges between 0.20 and 0.40 (Christian, 1972). It may thus be concluded that pig meat quality can be improved by supplying suitable environmental conditions to the animals, but a genetic approach of this problem may be possible (Steinhauf, 1969). It has also to be mentioned that selection made to reduce backfat thickness (fatness criteria) does not seem to have contributed to increase the occurrence of PSE meats as Vos and Sybesma (1971) argue that there is an inverse relationship between backfat thickness and meat quality. On the other hand, there is no relationship between carcass "muscularity" and meat quality. Accordingly, the efforts made in genetics to increase the amount of lean tissue by decreasing the backfat thickness do not seem to have an unfavourable influence on meat quality. But, increasing the amount of lean by selecting large muscled animals leads to a decrease in the meat quality.

Beside the genetic route there are other means to reduce the incidence of PSE meats. It has been well established that manipulations of the animals during the period prior to slaughtering may have disastrous consequences. A high room temperature and the accumulation of CO_2 in the atmosphere accelerate the post mortem muscle glycolysis leading to the appearance of PSE meats in animals susceptible to stress (Wismer-Pedersen, 1959). Such unfavourable conditions are created by an overcrowding of the vehicles used to transport the animals to the slaughterhouse. Situations like that can naturally be prevented. At the moment of slaughtering, the CO_2 anaesthesia

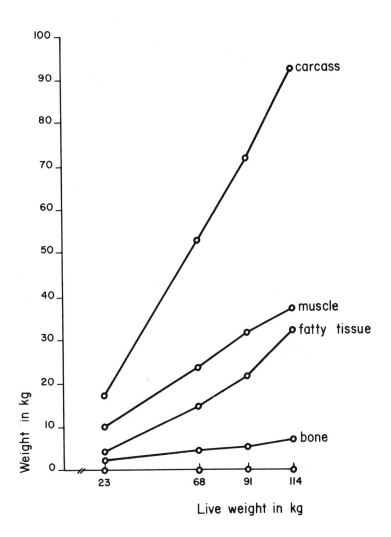

Fig 1 - Weights of carcass, muscle, fatty tissue, and bone according to live-weight.

creating an anoxia brings about an acceleration of the post mortem glycolysis and an increase in the appearance of PSE meats as compared with the results obtained with electrical anaesthesia (Sybesma and Groen, 1970). However, despite of this, both methods can be considered as valid for industrial slaughtering (Van der Wal, 1971).

A rapid cooling of the carcass after slaughter reduces the occurrence of PSE meats often resulting from the conjunction of a low pH and a high temperature. Besides the lowering of the muscle temperature, this cooling reduces the post mortem glycolysis and thus delays the appearance of a low pH.

CONCLUSIONS

Increase in pig meat consumption and changes in the taste of human consumers, characterized by a preference for leaner meats, have led to an intensification of the production and to a modification of the herd management and breeding methods.

The efforts made to reduce the production of fat are based on genetic and nutritional means. From a genetic point of view, selection most certainly represents an efficient means to decrease the volume of fatty tissue, but when this selection is realized with the aim of increasing the muscularity of the animal it may lead to defective meats and in particular to PSE meats. Breders of today have to pay a special attention to this problem as the incidence of this defect may be reduced by selection. From a nutritional point of view, the fatness factors are now very well controlled. By reducing the feed intake level, by diluting the diet with cellulose compounds, by using adequate levels of well balanced proteins it is now possible to obtain lean carcasses corresponding better to the taste of the consumer. It is also possible to control the quality of the fatty tissue through the feeding. It may therefore be concluded that despite the industrialization and intensification of pig production the means are now available for providing the consumer with a product corresponding to his taste and requirements.

BIBLIOGRAPHY

American Heart Association, 1973, "Diet and Coronary Heart Disease".
Barrick, E.R., Blumer, T.N., Brown, W.L., Smith, F.H., Tove, S.B., Lucas, H.L. and Stewart, H.A., 1953, J. Anim. Sci., 12, 899 (abstr.).
Blumer, T.N., Barrick, E.R., Brown, W.L., Smith, F.H., and Smart, W.W.G. Jr., 1957, J. Anim. Sci., 16, 68-73.
Braude, R., 1970, Proc. Nut. Soc., 29, 262-270.
Brooks, C.C., 1971, J. Anim. Sci., 33, 1224-1231.
Brooks, C.C., 1972, J. Anim. Sci., 34, 216-224.

Christian, L.L, 1972, in: "Proceedings on the Pork Quality Symposium", R.G. Lassens, F. Giesler and O. Kolb (eds.), p. 91, Univ. of Wisconsin Ext., Madison.
Cooke, R., Lodge, G.A., and Lewis, D., 1972, Anim. Prod., 14, 35-46.
Davey, R.J., Morgan, D.P., and Kincaid, C.M., 1969. J. Anim. Sci., 28, 197-203.
Desmoulin, B., 1972. Revue Franç. des Corps Gras, n° 7, 437-455.
Desmoulin, B., 1973. J. Rech. Porc. en France, p. 189-199, I.N.R.A.-I.T.P. (ed.)
Eurostat, 1977. Bilan d'approvisionnement n° 12, p. 114.
Flanzy, J., François, A.C. and Rerat, A., 1970. Ann. Biol. Anim. Bioch. Biophys., 10, 603-620.
Greer, S.A.N., Hays, V.W., Speer, V.C., McCall, J.T., and Hammond, E.G., 1965. J. Anim. Sci., 24, 1008-1013.
Henry, Y. and Rerat, A., 1964. Ann. Biol. Anim. Bioch. Biophys., 4, 263-271.
Henry, Y., 1966. 9th Int. Congress of Anim. Prod., pp. 9-10, Edinburgh, Oliver and Boyd Ltd. (ed.).
Herring, H.K., Haggard, J.H., and Hansen, L.J., 1971, J. Anim. Sci. 33, 578.
Hetzer, H.O. and Harvey, W.R., 1967. J. Anim. Sci., 26, 1244-1251.
Jarmoluk, L., Martin, A.H., and Fredeen, H.T., 1970, Detection of taint in pork. Canad. J. Anim. Sci., 50, 750-752.
Koch, D.E, Pearson, A.M., Magge, W.T., Hoefer, J.A., and Schweigert, B.S., 1968. J. Anim. Sci., 27, 360-365.
Kroeske, D., 1972. Rev. Mond. de Zootechnie, 2, 25-30.
Kuryvial, M.S. and Bowland, J.P., 1962. Canad. J. Anim. Sci., 42, 33-40.
Lawrie, R.A., 1966, "Meat Science", Pergamon Press Ltd., 368 pp.
Leat, W.M.F., Cuthbertson, A., Howard, H.N. and Gresham, G.A., 1964. J. Agric. Sci., 63, 311-317.
Maarse, H., Moerman, P.C., and Walstra, P., 1972, IVO Rapport C 180 and Rapport n° 3 - Researchgroep Vlees en Vleeswaren TNO.
Mason, J.V. and Sewell, R.F., 1967. J. Anim. Sci., 26, 1342-1347.
Oslage, H.J., 1962. Z. Tierphysiol. Tierernähr. Futtermittelk, 17, 350-357; 18, 14-17.
Patterson, R.L.S., 1968. J. Sci. Food Agric., 19, 31-38, 434.
Patterson, R.L.S. and Stinson, G.G., 1971, 17th Europ. Meet. of Meat Res. Workers, Bristol, England, B12, 148.
Rhodes, D.N., 1969, in: "Meat production from entire male animals", Rhodes, D.N. (ed.), 189, Churchill, London.
Rhodes, D.N., 1976, in: "Meat Animals, growth and productivity", E. Lister, D.N. Rhodes, V.R. Fowler and M.F. Fuller (eds.), Nato Advanced Study Institutes Series, Series A: Life Sciences, vol. 8, 9-24.
Richmond, R.J. and Beng, R.T., 1971. Canad. J. Anim. Sci., 51, 31-39.
Schön, I., 1973. World Rev. Anim. Prod., 9, 34-47.
Sellier, P., 1970. Ann. Genet. Sel. Anim., 2, 145-207.
Steinhauf, D., 1969, in: "Recent points of view on the condition and

meat quality of pigs for slaughter, Sybesma, W., Van der Wal, P.G., and Wlastra, P. (eds.), p. 283, I.V.O. Zeist the Netherlands.

Sybesma, W. and Groen, W., 1970. "Proc. Eur. Meet. Meat Res. Workers 16th"

Thrasher, D.M., Brown, R.E., Mullins, A.M., Hansard, S.L, and Brown, P.B., 1959. J. Anim. Sci., 18, 1494 (abst.).

Van der Wal, P.G., 1971, in: "The condition and meat quality of pigs", J.C.M. Hessel, de Heer, et al. (eds.), Proc. 2nd int. Symp. p. 145, Wageningen, Netherlands.

Van Logtestijn, J.G, Sybesma, W., and Van Gils, J.H.J., 1970. Arch. Lebensmittel. Hyg. 3, 55.

Vanschoubroek, F., de Wilde, R., and Lampo, Ph., 1967. Anim. Prod., 9, 67.

Vos, M.P.M. and Sybesma, W., 1971, in: "The condition and meat quality of pigs, J.C.M. Hessel de Heer et al., (eds.), p. 278, Wageningen, Netherlands.

Wismer-Pedersen, J., 1959. Food Res. 24, 711.

PERSPECTIVES OF MILK/BEEF PRODUCTION SYSTEMS IN THE TROPICS:

RELATIONSHIPS TO HUMAN NUTRITION

      Fernando Pérez-Gil Romo

      División de Nutrición, Instituto Nacional de la Nutrición,

      D.F. - MÉXICO

    Because of the requirement of equilibrium, the global biological system can be considered as being a continuous flow of energy and nutrients through a network of interlocking cycles. The function of agriculture is to divert this flow to the benefit of a single species. Natural forms of vegetation are replaced by cultivated varieties that have been selected for their efficiency in manufacturing foodstuff for man.

    Domesticated animals are introduced for a similar purpose. A third essential link in the food chain — the microorganisms — still consists mainly of wild species, but agricultural technology may eventually intervene in this part of the cycle as well. Hence, in order to feed the human population, we must ensure the nutrition of an assortment of plants, animals and microorganisms[10].

    In the past 20 years, Latin America has increased meat production by 64.5% and milk by 86%[8]. Milk and beef are the main animal products in Latin America. However, the annual increase in the ruminant population (2.7%) is below that for humans, while the demand for animal products increases at an estimated rate of 5% per year[13].

    In the developing countries of the world, man lives mainly on plant foods, but is frequently unable to meet his needs. According to the FAO (1975), 25% of the population of these countries received insufficient energy and/or protein in 1970[9].

    In the less developed areas of the world, cereals consumption corresponds to 182 kg per person per year. In the developed world this consumption reaches 910 kg per head per year. 90% of this, is consumed indirectly, because it is given to the animals, and is converted to animal products of high protein value: meat, milk, eggs; and in several products which are not eaten by men: wool, leather, fur, etc.[3,14].

The availability of meat from cattle, sheep and goats per person in Tropical America is 17.2 kg/year, compared with 49.1 kg. for the U.S.A.[5].

The area devoted to grasslands in Latin America in 1967 consisted of 446 million hectares, accounting for 79.9% of the agricultural land available[7]. However, the ratio of man to bovines, formerly 1:1, has decreased in recent years, because of the growth in human population, and will continue to diminish if adequate measure are not taken to stimulate livestock production[4].

In poor countries, livestock and man are in competition for the available grain. As a result, grain is fed only to animals that are efficient converters of calories to tissue, such as the chicken and the pig, and to animals needed for work[10]. Ruminants either graze or consume wastes.

Non-ruminant animals require a diet that in many respects does not differ much from the human diet. In the event of a food shortage it would theoretically be more efficient to bypass the animal and reserve the available food for human consumption.

In many countries, especially the smaller ones, there might be competition between humans and ruminants for use of land to produce food, specifically in those countries where tillable land is scarce and the best agricultural lands have been devoted to pastures[8].

Limited supplies of grains and vegetable proteins have led to competition between humans and animals for these products, so that the question of the efficiency of different kinds of animals in converting feed into high grade animal protein is vitally important. Where meat production efficiency is concerned, poultry are superior to pigs, which in turn are well ahead of ruminants.

From an agricultural point of view, the humid tropics are probably the richest regions in the world in terms of potential for both crop and animal production. Nevertheless, there has not so far been the development of intensive beef and milk production systems, as has occured in the temperate regions. More than any other factor, it is probably the scarcity of cereal grains, together with the fact that fast growing tropical grasses are usually of low nutritive value, which have been the reasons for the relative lack of development of intensive systems[11].

Because inefficient production methods are responsible for the low production in Latin America, a new technology has to be developed.

Animals can be classified in three groups based on their ability to assimilate cellulose, a classification that is reflected in the structure of their digestive apparatus. Carnivores, such as the dog, and ommivores, such as man and the pig, have simple stomachs and have difficulty digesting cellulose. Non-ruminant herbivores, such as the horse, the rabitt, and the guinea pig, can derive sustenance from cellolose, but they assimilate it less efficiently than the ruminants. Ruminant herbivores, such as cattle, sheep, goats, deer, buffalo and many others, can efficiently break down cellolose and extract a large proportion of their dietary calories from it[10].

Animals derive almost all their energy directly or indirectly from the decomposition of two polysacharides: cellulose and starch, which are closely related; both consist of long chains of the sugar glucose. They differ only in the geometry of the bonds between the glucose units. This small difference in structure, however, brings vast differences in the physical properties of the two molecules and in their suitability as a constituent of animal diets.

The superlative adaptation to the challenge of a diet rich in cellulose is found in these animals (ruminants). Bacteria fermentation of fibrous foods is concentrated in an extensive fermentation vat, the rumen, at the beginning of the digestive tract. There cellulose is broken down into smaller molecules, and the capabilities of the bacteria for synthesizing amino acids and certain vitamins are exploited.

The bacterial transformation is accomplished before the food enters the small intestine, so that the full length of the intestine is available for the absorption of nutrients extracted from the food and synthesized by the microorganisms.

Increasing ruminant production is a means of making this land more productive and helpful to the total economy of Latin America, without competing for man's food sources. Because forage is of no direct use to man, ruminant production is the best method for converting it into animal products for human food and other uses, and of adding quality and palatability to the diet[6].

If developing nations wish to enlarge their animal agriculture, it seems important that they should do so by utilizing existing reserves of pasture and forage crops, rather than by using cereals. The opportunities for applying improvements of this nature are present in almost every country, whereas developments in the utilization of concentrate feeds are restricted in their applications to countries with surplus cereals[9].

Animals convert plant products into human food with an efficiency that varies between 2 and 18%[10]. That inefficient conversion however, is not a loss but a gain, if the photosynthetic energy consumed by the animals could not have been recovered otherwise. The justification for animals in agriculture is their ability to transform products of little or no value into nutritious food[10].

There are now indications of a potential breakthrough in this situation, principally as a result of recent research, which indicates that sugar cane, coffee pulp. rice polishings, cottonseed meal, coconut meal, cassava, bananas, groundnut meal, etc., are the most productive crops in the tropics, and can be used as a basis of intensive animal production systems[11].

The justification for developing integrated milk/beef production systems with dual purpose cattle for tropical conditions, has been discussed by Preston, 1976[2].

The humid tropical region has the greatest productive potential, although a limiting factor is the severity of the constantly humid tropical climate, which completely excludes the possibility of their being used for livestock of specialized European breeds. However, on land which possesses an average fertility, or on the highly fertile river lowlands, very productive livestock raising is being developed, particularly that dedicated to the development and fattening of crossbred young bulls of zeby breeds. Productive animal loads of two animal units per hectare are the rule, and certain improved pastures allow for up to five animal units per hectate[1].

The alternative approach, which has been proposed for milk production in the tropics, envisages the use of adapted native cattle partially upgraded with European dairy breeds, such as Holstein, Brown Swiss or Simmental, to produce dual purpose animals with moderate milk production combined with the capacity to raise a calf of excellent attributes for beef production[12].

THE DUAL-PURPOSE SYSTEM IN THE TROPICS

This organization of livestock raising has evolved empirically in many tropical countries and appears to be an adaptation forced by the economic necessity of selling milk. However, the conditions of poor forage quality and scarcity during drought, force the producer to retain certain characteristics of a meat producing enterprise[1].

In Latin America, ruminant animals depend, to a large extent, on forage production.

Latin America has the advantage of plenty of sunlight, a long growing season, and a warm climate, all important in forage production[6].

Dairy cattle can grow, maintain themselves, reproduce and remain healthy, when fed entirely on forages. On a forage of average quality, however, cows have difficulty obtaining sufficient nutrients to produce more than 10 to 20 lbs of milk per day. High quality forages can support from 30 to 40 lbs, of milk per day. With liberal grain feeding, dairy cows can produce more than 100 lbs of milk per day[10].

Competition for protein suitable for human consumption could be reduced and perhaps eliminated in ruminants by replacing protein in the ruminant diet with other sources of nitrogen. A lactating cow on a diet of waste roughage and urea produces more protein than it consumes.

The dual-purpose system in the tropics is characterized by a rearing of the calf by the cow, and separation from the mother between 8 and 14 hours daily, equal raising of males and females, and short lactation periods, dictated principally by drought (seasonal production).

By using sugar-cane subproducts, we can get milk productions of 1,500 kg, which is equivalent to 5 kg/day, during a lactation period of 300 days, plus 2.5 kg/day as an average for the calf (restricted suckling), weight gains of 600g/day until weaning, with a weaning weight of 200 kg; from weaning to sacrifice 850 g/day as an average daily weight gain, with a final weight of 400 kg at 17 months of age[15].

Tropical livestock raising is adapted to these principal lines, especially that of zeby stock[1].

We know from experience that it is difficult to replace concentrates with forages without reducing production. In general, forages have a lower metabolizable energy concentration than cereals, and are physically more bulky. Satisfactory replacement of concentrates by forages requires a general improvement in the nutrient concentration of the latter feeds[3].

In Mexico, the most common method of improving yields is through cross-breeding with the Brown Swiss, which at the same time; improves the characteristics of the steer. The system is constantly criticized by livestock technicians, and it is often heart that, in reality, it is a no purpose system, since it produces poor-quality meat inefficiently, and very little milk. In the tropical zone, with its prolonged dry season, and total rainfall of less than 1200 mm, the system is defended by cattlemen as the only way to sustain the business, especially when prices decline for cattle on the hoof.

In the southeast, with precipitations of 2500 mm, the system becomes logical, due to the possibility of improving pastures with new grass species, and the introduction of legumes. The data obtained to date indicates that with such improved pastures, and with more specialized methods, it would be feasible to introduce the artificial breeding of calves and cattle with some Holstein blood. Longer lactation periods without the aid of concentrate seem feasible, and recent results await economic analysis of profits and losses on a long term basis. In the same way, on irrigated lands within the tropical climate zones, it seems feasible to develop dairy cattle more intensively, based precisely on direct grazing. The attempts of confinement, cut grasses/or silage have been an economic failure, as was expected, due to the scarcity of concentrates in the tropical zones and the costs of mechanization[1].

Achievements from a wide range of production practices are minimal, if one considers the statistics given for ruminant production in the tropics of Latin America as a whole. Improved practices have been imported for the most part, from developed countries without adaptation to the conditions prevailing here, and in many cases, trial error procedures have produced nothing but losses[8].

Before ruminant production can be greatly increased in Latin America, there needs to be a market and people must be able to purchase animal products. Equal in importance to solving animal production problems, are economic and social programs designed to make it possible for people to have better living standards, including ability to buy meat, milk, and other food they need[6].

The problems to be solved in order to increase milk/beef production in Latin America include: better quality cattle, improved breeding programs, proper feeding and nutrition, better disease and parasite control, improved management, and many other production problems that need to be solved or ameliorated[6].

As Dr. Wellhausen said once, "in order to increase our agricultural production, we need a more precise definition of highly profitable technological production techniques, for each of the many different ecological regions; more effective strategies for gaining widespread adoption of the techniques; a large experienced multi-disciplinary team of experts, who are silling to work in close association with the farmers, ranchers, teaching team, and encouraging them to new levels of animal productivity"[16].

REFERENCES

[1] Alba, M.J., "Current Status of Mexican Livestock Raising". Fondo de Garantia y Fomento para la Agricultura, Ganaderia y Avicultura (FIRA), 1976.

[2] Alvarez, F.J. and Preston, T.R., Leucaena leucocephala as a Protein Supplement for Dual Purpose Milk and Weaned Calf Production on Sugar Cane Based Rations. Trop. Anim. Prod. 1(2):

112-118 (1976).
[3] Brown, L. and Eckholm, E.P., "By Bread Alone", Praeger, New York, Washington, 1974.
[4] Cabellero, H., Producción de carne de res en América Latina, in: "Recursos Proteínicos en América Latina", INCAP, Guatemala City, Guatemala, pp. 145-159 (1971).
[5] Chicco, C.F. and Shultz, T.A., Sistemas Extensivos de Producción de Carne y Leche Usando Pastos Tropicales con y sin Suplementación, VII Reunión Interamericana sobre Control de la Fiebre Aftosa y otras Zoonosis. Puerto España, Trinidad, OPS, OMS, abril 1974.
[6] Cunha, T.J., Ruminant Production in Increasing Animal Foods in Latin America, in: "Nutrition and Agricultural Development", Scrimshaw and M. Behar, (eds.), Plenum Press, New York and London, pp. 355-362 (1976).
[7] FAO, United Nations, 1967, Prod. Yearbook, Vol. 21, Rome, Italy, 1968.
[8] Fonseca, H. The Role of Ruminants in Tropical America. In: "Nutrition and Agricultural Development". N.S. Scrimshaw and M. Behar (ed.), Plenum Press, New York and London, pp. 363-368 (1976).
[9] Greenhalgh, J.F.D., The Dilemma of Animal Feeds and Nutrition. Anim. Feed Sci. Technol., 1:1-7 (1976).
[10] Janick, J., Noller, C.H. and Rhykerd, C.J., The Cycles of Plant and Animal Nutrition. Scientific American, 235: 75-86 (1976).
[11] Leng, R.A. and Preston, T.R., Sugar Cane for Cattle Production: Present Constraints, Perspectives and Research Priorities. Trop. Anim. Prod., 1(1): 1-22 (1976).
[12] McLeod, N.A., Morales, S. and Preston, T.R., "Milk Production for Dual Purpose Cows Grazing Unsupplemented Pangola or Fed in Drylot on Sugarcane and Molasses/Urea Based Diets. Trop. Anim. Prod., 1: 128-137 (1976).
[13] Mitchell, C. and Shatan, J., La Agricultura en la América Latina: Perspectivas para su Desarrollo, in: "Desarrollo Agrícola de América Latina en la Próxima Década". Banco Internacional del Desarrollo, Washington, D.C., pp. 47-181 (1967).
[14] Pimentel, D., Dritschilo, W., Krummel, J. and Kutzman, J., Energy and Land Constraints in Food Protein Production, Science, 190: 754-761 (1975).
[15] Preston, T.R., Estrategia para la Producción Bovina en los Trópicos, Rev. Mundial Zootec., 21: 11-17 (1976).
[16] Wellhausen, E.J., The Agriculture of Mexico, Scientific American, 235(3): 128-153 (1976).

MODERN PRACTICES IN CATTLE FEEDING FOR MEAT PRODUCTION

- RELATIONSHIP TO HUMAN NUTRITION

       Conrad J. Kercher

       Animal Sci. Div. - University of Wyoming

       Laramie, Wyoming - USA

Nutritionists recommend a variety of foods in the diet of humans to insure providing the necessary nutrients and foods which are relished. Foods are commonly grouped according to the essential nutrients they provide as follows:

1) vegetables and fruits;
2) dairy products;
3) cereal products;
4) fats and sugars, and
5) meats, fish and eggs.

By eating some foods from each of these groups, it is possible to consume a nutritionally adequate diet. Meat products provide excellent quality protein, energy, minerals and vitamins, especially riboflavin and vitamin B_{12}. They are low in calcium and vitamin a (except for liver).

Humans have always enjoyed eating meat. Statistics show that per capita meat consumption decreases with increasing population density. As population increases more arable land is used for crop production; however, a considerable portion of the earth's surface (23% permanent pasture and rangeland — 30% forest and woodlands) is only suitable for forage production. In the United States 63% of the land is classified as grazing or rangeland.

Ruminant animals have contributed to the welfare of man for centuries by providing milk, meat, fuel, wool, hides and other by-products. They have the unique ability to convert feeds which can't be utilized by humans into high quality foods which humans enjoy. In addition to utilizing forages, ruminants utilize many crop residues and by-product feeds from the manufacture of human foods which would present a waste-disposal problem if they could not be fed to ruminants. Some people have referred to ruminants as scavengers.

Demographers have estimated that the world population will increase from 4 billion in 1975 to 7 billion in 2000 and 16 billion in 2035. As population increases we will need to produce more food. Since animal products are an important source of nutrients for humans we will also need to increase meat production. Animal agriculture has always been an important part of the economy. It is necessary for livestock producers to show a profit if they are expected to increase production.

The nutritional value of meat is relatively unaffected by the ration fed to the animal except for the amount of fat and hence energy in the meat. However, total meat production is affected by many factors including environmental, genetic and nutritional factors. The quality and quantity of food available to the ruminant has a pronounced effect on production.

In the past food production has been increased by:

1) increasing the number of acres of land under cultivation;
2) irrigating cultivated land to increase crop yields;
3) using fossil fuel to increase the size of agricultural units and production per capita, and
4) using modern technology to increase production.

It is obvious that many resources such as arable land, irrigation water, and energy are finite and can't be increased indefinitely to increase food production. However, man can manipulate animals and their food resources utilizing modern technology to increase food production.

Some of the modern practices which can be used to increase meat production are:

1) selection of animals which are productive in their environment;
2) use of crossbreeding to capitalize on heterosis;
3) proper feeding and management of breeding animals to increase reproductive performance;
4) use of management practices to minimize health problems and increase reproductive rate;
5) use of more forages and less concentrates in the production of red meat;
6) use of feed additives to stimulate production or minimize health problems;
7) development of high yielding, palatable and nutritious introduced forages;

8) forage and processing methods to increase nutrient value;
9) range or pasture improvement to increase forage quantity and quality,
10) better utilization of meat and meat by-products by preventing waste.

In selecting animals which are most productive in their environment, it may be desirable to utilize more than one species of animals depending upon the avilable feed resources. In some areas of the world, wild animals may be more efficient meat producers than domestic animals. Animal adaptation to heat or cold stress, disease tolerance or other environmental conditions are all important in determining maximum meat production. In order to select the most efficient animals for the environment, it will be necessary to keep adequate records so that meaningful decisions can be made.

In addition to selecting productive animals, it may be desirable to use crossbreeding to take advantage of heterosis as well as combining different desirable traits of two or more breeds. It generally is possible to improve meat production faster by crossbreeding and selection than by selection alone. Any crossbreeding program needs to be carefully planned.

It is well known that nutrition has an important role in initiating and maintaining reproductive functions in animals. To increase meat production it is important that the female conceives and rears an offspring as often as possible. Providing sufficient nutrients is important during puberty, breeding, late gestation and lactation. If nutrients are lacking, the beef cow may produce a calf once every two years instead of annually. It is important for the producer to recognize when proper feeding is most critical in the reproductive cycle so that he can feed properly to maximize production.

Herd health management can have a significant effect upon livestock production. Animals can be vaccinated for some diseases to minimize their effect. Other diseases can be minimized by good sanitation management and proper feeding to insure healthy animals.

In most areas of the world ruminants are fed primarily forages from weaning to slaughter. Even in the United States where large amounts of grain are fed to feedlot cattle, it has been estimated that forage still contributes about 75% of the nutrients that beef cattle consume. Animals fed only forage do require more time to reach market weight than animals fed forage and grain. It is important that we develop high yielding, palatable and nutritious forages to use in cultivated areas or where reseeding of native rangelands is feasible. We also need to utilize harvesting and processing methods which will economically preserve more nutrients for animal use.

Range improvement practices should be utilized to improve the carrying capacity of rangeland that is unsuitable for reseeding. Such practices as water spreading, removal of brush and toxic plants by mechanical or chemical means, proper stocking rates, rotational grazing, etc. will increase meat production per hectare. It is also important to know the nutritive value of the forages during the season of use so that supplements can be provided if they are needed.

There are many chemicals and drugs (feed additives) such as hormones and antibiotics which can be used to enhance production. It is important that they be used in the proper manner to insure maximum animal performance and safe meat products. These will be covered by another speak in this symposium.

Once the animal product is produced, we must market it in a way to minimize waste. Generally this involves a good distribution system including transportation and refrigeration.

To increase meat production we in the Animal Science profession must continue to search for new and innovative technological tools. As has been true in the past, the producer must be motivated by increasing profit to utilize the modern practices we now know will increase production as well as new practices we may devise in the future.

MODERN PRACTICE IN CATTLE FEEDING FOR MILK PRODUCTION

— RELATIONSHIP TO HUMAN NUTRITION

C.O. Claesson

Prof. of Animal Husbandry - Swedish Univ. of Agri. Sci.

750 07 - Uppsala 7, SWEDEN

1. APPLICABILITY TO HUMAN NUTRITION OF PROGRESSES IN DAIRY COW FEEDING

The basic differences in physiology of nutrition between man and ruminants create marked restrictions in the applicability to humans of new findings in the nutrition of the dairy cow (or even the small ruminant goat and ewe).

In addition the dairy cow is fed mainly for producing milk, and the intensity of lactation in relation to the metabolic weight is much higher than in women. For cows yielding 6000 kg/lactation (305 days) the energy intake for milk production is on an average about 1.65 times the maintenance (600 kg body w.) requirements with a maximum about 4 times the maintenance requirement. While the recommended energy intake for lactating women is not more than about 25-50 % of that for non-lactating ones. Thus also in the restricted field of nutrition for lactation the applicability to humans of results obtained in cow feeding is very limited.

However, in the field of calf-feeding some recent results seem to be of interest in the development of milk products for human consumption. In Sweden, like in many other countries, milk replacements are widely used for rearing of calves, and in these formulas some antibiotics are commonly added. Work in Sweden has shown that these antibiotics can favourably be replaced by gently dried yogurt culture, for maintaining a good intestinal health. These results have stimulated the research on the nutritive properties of various fermented milk products, and led to the development of a fermented milk product using the Lactobacillus acidophilus in an "Acidophilus" which, according to preliminary results of clinical studies, seems to have favourable effects on the intestinal functions of patients with some digestion disturbances.

2. EFFICIENCY OF MILK PRODUCTION IN THE HUMAN FOOD SUPPLY

For a satisfactory nutrition of the human population in the world in the year 2000 it is estimated that double the present amount of food will be needed. This means that within less than a quarter of a century, we have to increase the food production per year to double that amount which has taken the whole history of mankind to reach. Thus the necessity of improved efficiency of food production is obvious, and this relates particularly to animal production.

Because of its fundamental role in human nutrition, milk has to be carried over to the consumer, in high hygienic quality and a reasonable price. Therefore, despite the fact that milk, in relation to its nutritive value, is one of the cheapest foods, the milk price to the consumer will always be a very important political question. Consequently, efficient dairying is of mutual interest to both milk producers and milk consumers.

Input/Output Rates

Among the different types of animal production, milk production is the most efficient in converting feed energy and protein into edible food. In well managed animal husbandry the input-output detailed in Table 1 are realistic. The quality of the feed protein can be lower for the cattle production which is seen in Fig. 1.

In comparing efficiency of milk production with direct vegetable food production all the resources used before the food is consumed should be taken into account. The price in relation to the nutritive value when consumed could be used for the comparison, and then milk appears to be one of cheapest, i.e., most efficiently produced, food in most countries. This is the result of continuous work in improvement of milk production and dairy technology.

Important Improvements Obtained

A short review of the progresses made in practical milk production would be useful for judging various possibilities for improvement in efficiency of future milk production, and could also form the basis for the planning of present and future research and development. In principle these progresses have followed the same pattern in most countries. The improvement in Swedish milk produc-

TABLE 1

RATES OF CONVERSION OF FEED ENERGY
AND PROTEIN INTO EDIBLE FOOD

PRODUCTION	PERCENT RECOVERED IN THE PRODUCTS	
	ENERGY, METABOLIZABLE	PROTEIN, DIGESTIBLE
Milk	25 - 35	30 - 40
Beef	10 - 15	15 - 25
Pork	25	18
Eggs, poultry	20	30

tion efficiency may be chosen to illustrate the principles of the progress, which can be summarized as follows[1].

From 1930 up to the present day the number of cows has decreased from about 2 million to about 700.000. During the same period the average yield per cow has doubled. This has resulted in a decrease in the feed used per liter of milk by about 30%, which is a remarkble improvement in feed conversion efficiency. By means of mechanization and rationalization the amount of labour per cow and year has been reduced by 75% during this period of time.

Besides the increase in yield per cow, a more efficient utilization of milk for human nutrition has made it possible to meet the increased demand for milk and dairy products despite the marked reduction in the number of cows. This is illustrated by the fact that milk, particularly skim milk, for feeding has decreased from about 50% to only about 2% of the total milk delivered to the dairies.

The development of milk replacers for calf feeding has been very efficient in Scandinavia and today most of the milk replacers in Sweden are mainly based on waste dairy products without alternative food values. Despite the marked decrease in the number of cows the production of beef has increased by more than 50% since 1930 as a result of better utilization of the calf material, mainly by rearing to higher age and body weight before slaughter.

Prospects of Development

From the development of the milk production described above it may be concluded that for further improvements a reduction of the labour used is less important than an increased production per cow by better feeding and management. An improved stability of production with less disturbances seems also to be very important for increased productivity. This is influenced both by genetic factors of the cow as well as feeding, management and housing environment.

The greatest prospects for the increased efficiency of milk production are an improvement of the feed conversion into milk and an increased use of feed and land of low value for direct food production. Important factors for this improvement are:

a) a continued increase in the yield capacities of the cows;
b) a better control of feeding with regard to the capacity of the cow;
c) a more efficient use of land for fodder production, and
d) utilization of by-products from farms and food industry.

a) The Utilization of the Genetic Capacity of the Cow. Today the genetic capacity of the cows is far from being fully utilized in many countries. These unused possibilities for improvements are very promising and may be illustrated by the following example.

The average genetic capacity of the Swedish dairy cow is about 6,000 kg milk per year with very small variation between herds. But actual average production is only about 4,500 kg per cow and year. By increasing the average yield to 5,000 kg, i.e., the average for recorded cows, the present total milk production could be supplied by 150.000 fewer cows. This would save maintenance feed corresponding to 375 million kg hay per year or 68,000 hectares of grassland. The saving in labour and other cost are even higher than those in maintenance feed.

Current research in Sweden in herds on various productions levels (4,000-7,000 kg/year) has shown that about 65-70% of the differences between herd yields could be ascribed to differences in feed and feeding. Methods are being developed by means of which it will be possible to evaluate and improve the utilization of the cow's capacity in the individual herd.

b) Controlled Feeding. Utilization of the cow's yield capacity is mainly a problem of early lactation. Results of investigations, particularly in Sweden and Norway, have shown that the marginal productivity decreases with increasing feeding intensity much more rapidly than hitherto assumed. Therefore, after the peak of lactation an improved feed conversion efficiency can be achieved by a proper control of the feeding intensity of energy and protein, in relation to the yield of the cow and the price of feed and milk. The exploitation of these results in practical milk production is going on in all the Scandinavian countries.

Besides the control of the intensity of feeding, the constancy in the amount fed per day and meal has been shown in Swedish experiments to have a significant influence on milk yield, and possibly also on the fertility of the cow. Even the order in which the various feeds are fed has been shown to affect feed consumption and milk production.

c) Fodder Production. An improved utilization of the yield capacity of the cows increases the demand for high quality feed. This has to be met by an improvement of the domestic feed production, mainly with regard to a more efficient land utilization, and independence of imported protein feeds. This has been the objective of research cooperation between the Scandinavian countries, and in Sweden a five years research program on feed protein production has been carried out. The results of this work may be summarized as follows. The investigations have included improvements of grassland production, increase of the protein content of fodder grain, mainly barley, increased use of rapeseed meal and legumes.

The results have shown that by application of suitable techniques in plant husbandry, harvesting and conservation, there are very good possibilities of markedly increasing both the energy and protein yield per hectare of grassland. This means that the milk can be produced more extensively on grassland products, and independent of imported protein feeds. The total area of land used for fodder grain is possible by plant breeding and fertilizing, and the extra protein could be fully utilized by the cow.

Rather comprehensive investigations have been carried out on the use of rapeseed meal in the dairy feed ration, and the effect on the quality of milk. No adverse effects could be found by replacing half of the protein feed from oil cakes with rapeseed meal, which up to this substitution (10% of the concentrate mixture) can replace soybean meal. Altogether these results have shown that the energy and protein supply, even to high yield cows, can be covered by improved grassland production and the use of feed cereals and rapeseed meal, i.e., without any imported vegetable protein feeds.

d) By-Products as Feed. Investigations into the use of by-products or wastes from farms have been mainly concerned with re-cycling of dried manure from laying hens in cages, and the utilization of straw. New techniques have been developed in Sweden for drying poultry manure, and feeding experiments show that the dried product can replace the vegetable protein feed in the concentrate mixture to the dairy cows. By these new techniques about 50,000 tons of dried product may be available, which is equal to about half the amount of rapeseed meal now used for cattle production in Sweden[2].

During recent years the use of straw as a feed for dairy cows has attracted a considerable and widespread interest in all the Scandinavian countries. In Sweden large scale lactation experiments have been carried out using strawbased diets, supplemented by different formulations of concentrate. One important factor found was the optimization of the fat content in the ration. Current experiments with whole (pelleted) rapeseeds as a fat source in a straw-based ration indicates an unexpected positive effect and a very promising, economic ration.

In Norway the use of alkali-treated straw has been in practical use since the last World War. New techniques for on-farm use have been developed by means of which the water pollution is avoided.

Industrial processes for alkali or other treatments of straw have been developed in Denmark, and some of these products are commercially used. A completely new approach has been taken in Sweden. This includes a harvesting of the whole cereal plant and subsequent separation of grains and various fractions of straw in a process plant. The feed fraction, obtained from straw, seems to have such a value that alkali or other treatments can be left out. Further feeding experiments for evaluation of the alkali-treated straw particularly, seem to be necessary for a proper economic evaluation of the various treatments.

e) <u>Non-Protein Nitrogen Supplement</u>. The ability to use non-protein nitrogen (NPN-compounds) for protein synthesis, places ruminants in a unique position compared with other animals. A great deal of work has been carried out to evaluate various NPN-compounds in ruminant feeding. Among many possible nitrogen sources urea has been used most extensively. In order to improve the palatability and slow down the release of ammonia in the rumen, which is of importance for an efficient microbial protein synthesis and also for decreasing the toxicity, urea has been bond to various carbohydrates. These carbohydrates also serve as an energy source for the micro-organisms in the rumen. Starea is a urea-starch product developed at Kansas State University. This has been widely used in Sweden for several years. Another interesting urea product is lactosyl-urea, developed at the National Institute for Research in Dairying, England. This compound is processed from whey and it is now being tested in milk production experiments in Sweden[3].

The use of NPN-supplements is hardly justified when high quality grassland products are used as roughage but could possibly save some protein concentrate in corn or straw-based rations.

PRACTICES IN CATTLE FEEDING FOR MILK PRODUCTION

Milk Production in Developing Countries

The present situation of animal protein in the human diet is characterized by a considerable surplus in the industrialized part of the world, while the developing countries generally have a deficiency of animal protein. This is particularly obvious with regard to the milk protein. Three quarters of the world production of milk is produced in Europe, North America, Australia and New Zealand, where only one quarter of the world population is living.

The immediate suggestion for a solution of the problem is a distribution of the surplus milk solids from these areas to the deficient areas for recombination to milk and milk products. If produced in sufficient quantities this recombined milk will have an immediate effect on the most pronounced protein deficiency and induce a market demand for milk and milk products.

However, with regard to the social and economic development increased food production with less dependency on imports are of outmost importance for a peaceful, independent and harmonic development of the countries in the Third World. Therefore, the recombining dairy industry has to be accompanied by a development of the domestic milk production, which successively will replace the imports of milk solids.

This model of recombining dairy industry combined with the stimulation of domestic milk production has been successfully working in India and Algeria as reported to the last International Dairy Congress in Paris June 1978.

a) <u>Special aspects on the efficiency of milk production</u>. The development of the domestic milk production in developing countries has to follow the same principles as those discussed above, with the necessary adjustment to the local conditions. There is generally a poor land and feed utilization by overstocking and low yield. It can be shown that due to low yielding capacity and poor feeding the amount of feed energy used per litre of milk produced is very high. As an example we have estimated that the feed energy used to produce one litre of milk in India is on an average about 5 times that in Swedish milk production. Therefore, the number of cows must be reduced and the milk yielding capacity of the remaining ones must be increased by breeding.

Improved feeding should be based on fodder crops (favouring a good crop rotation) and an cash crops residues, where available.

b) <u>The keeping quality of milk</u>. In most developing countries the collection of the domestically produced milk is one of the most difficult limiting factors. High ambient temperature in combination with poor hygiene and lack of refrigerating facilities causes great losses of milk before it reaches the dairy or consumers. Based on scientific work on antimicrobial systems in milk[4] a method for improving the keeping quality of milk by activating the lactoperoxidase of milk has been developed in Sweden[5]. Preliminary results from warm climate (Kenya) are very promising.

Work is going on in order to use the activated antimicrobial systems of milk in feeds and food, particularly to newborn and growing mammals, including babies.

3. EFFECT OF FEEDING ON THE NUTRITIVE VALUE OF MILK AND DAIRY PRODUCTS

The effect of feeding on the composition and properties of milk from the nutritional point of view, may be divided into three groups, viz:

1. Effect on the chemical composition;
2. effect on the keeping quality;
3. effect on the wholesomeness.

No doubt the first group has attracted the most interest among research workers.

Effect on the Chemical Composition

Results of investigations during 1940-50 indicated that the variation in the composition of milk due to genetic factors and feeding would open possibilities for a regional specialization of milk production with regard to use of the milk, e.g., for liquid milk, cheese, butter, dried or condensed milk, etc.

However, because of the development within the dairy industry this is no longer feasible in most countries.

The milk delivered to the dairy must be regarded as a raw material for processing of various products. However, several properties of the milk are essential irrespective of the product to be manufactured. The most important variation is that in the content of the major milk constituents fat, protein and lactose, and their interrelations.

A great number of investigations have been carried out regarding the effect of feeding on the composition of milk.

It is well known that the feeding may have a drastic effect on the fat content of milk as well as on the composition and properties of the fat.

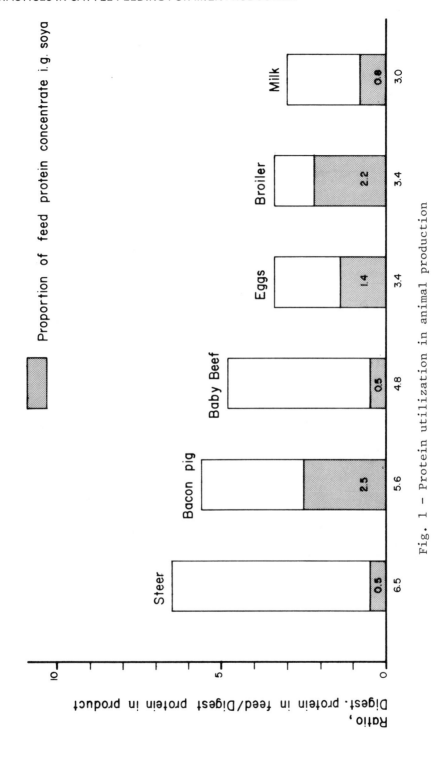

Fig. 1 – Protein utilization in animal production

Methods for coating polyunsaturated fats in the cow's diet have been developed. By this coating the fat is not saturated in the rumen, but is, to a high proportion, carried over to the milk, thus resulting in a butterfat with an increased proportion of these fatty acids. However, from the point of view of human nutrition the same results may be obtained by blending butter with a vegetable oil, e.g., soya-oil. This has been done for a long time in Sweden and it seems to be more economic, while composition and properties of the product are better controlled.

The composition and properties of the milk proteins are to some degree influenced by the feeding of the cow. This refers to both the interrelations between the various protein fractions as well as the properties of the casein-complex. Large variations have been found in the time for curd formation with rennin and pepsin. These properties have a marked effect on passage time through the stomach and may be of importance for the intestinal digestion.

Feeding also affects the heat stability of the milk, which is of importance at UHT-treatment and the subsequent stability of the milk during storage (after-coagulation).

Large variation occurs in the content of vitamins and minerals in the milk and these variations can be greatly influenced by the feeding of the cow.

To what extent should the milk producer influence the composition and properties by the feeding? In general it is much easier and more economic to obtain an optimum composition of a human diet by mixing various foods rather than trying to optimize the composition of each of them. This is also valid for milk and dairy products.

Therefore the goal for the composition of the milk produced should be to maintain or increase the content of total solids in the milk. Less water in the milk means less cost in cooling, transportation and processing. With increased total solids content the proportion of fat and protein increases and the proportion of lactose decreases. This may also have some favourable effect on the secretion of milk.

The Effect on the Keeping Quality of the Milk

The keeping quality of milk may be divided into bacteriological and chemical stability.

a) Bacteriological Stability. The effect of feed on the bacteriological quality of milk may be either by contamination in the barn from feed of poor quality, e.g., silage and hay or by promotion of the development of a specific flora in the milk due to other specific effects. Of special importance for the quality of the milk is the development of lipas-producing bacteria, particularly during cold storage. This lipas has been found to be quite heat resistant, even for UHT-treatment, and it causes rancidity of milk and other dairy products during storage. The above mentioned method of activating the lactoperoxidase of the milk prevents the growth of the lipas-producing flora.

b) Chemical Stability. The chemical stability of milk may be illustrated by resistance to develop oxidized flavour or spontaneous rancidity. Both these properties are influenced by the feeding, but more work has to be carried out before the relationships are fully understood.

The smell and flavours of milk are highly dependent on the feeding. It is well known that certain feeds have specific effects on the smell and flavour of milk. Modification in the feed technology as well as the feeding time of these feeds in relation to milking can largely reduce or eliminate these effects.

The Effect on the Wholesomeness of Milk

To this group may be assigned substances either from feed additives or metabolites from damaged or specific feeds, e.g., mycotoxins, metabolites from rapeseed meal. Although recent works indicate that the barrier in the cow udder against such substances is very high, such feed should be used with great care because of the risk of direct effects on the cow.

REFERENCES

[1] Claesson, C.O., 1977, Scand. Jour. of Dairy Technology, 4: 5/77, p. 267.
[2] Claesson, C.O. and Ahlström, B., 1976, "XIX Proc. Int. Dairy Congress", New Delhi, India, Vol. I-E, pp. 88-89.
[3] Nilsson, T. and Widell, S., 1977, Scand. Jour. of Dairy Technology, 4, p. 277.
[4] Reiter, B., 1978, Jour. of Dairy Research 45, pp. 131-147.
[5] Björck, L., 1978, Jour. of Dairy Research 45, pp. 109-118.

UNEXPECTED NUTRITIONAL BENEFIT

FROM A DAIRY PROJECT

      Mogens Jul

      Dept. of Food Preservation, Royal Vet. and Agric. Univ.

      Howitzvej 13, DK 2000 Copenhagen F, DENMARK

BACKGROUND

One of the world's most successful dairy developments has been that of the so-called AMUL dairy in Anand in the state of Gujarat in India. Its operation is based on daily milk collections from about 225,000 milk producers in the Kaira district. They are joined in cooperatives, which together own the dairy in Anand, a very large and modern milk processing plant. These activities have resulted in considerable income and employment generation in the area.

OPERATION FLOOD OBJECTIVES

Experience with the Anand dairy was so encouraging that the Indian government decided to embark on a project whereby the same successful system was to be introduced in the milkshed areas for Bombay, Calcutta, Delhi and Madras. The project started with the clear objective of furthering dairying in the areas concerned as indicated bellow:

Speed up dairy development, particularly

1) increase processing facilities;
2) create modern milk supplies;
3) resettlement of city cattle;
4) develop storage and transportation;
5) improve milk collection; and
6) improve dairy farming.

UN ASSISTANCE

For assistance in financing the project, the Indian government applied for UN assistance, especially assistance from the World Food Programme, WFP. WFP was able to donate considerable amounts of dried skimmed milk and butter oil to the project. These constituents could be recombined in existing dairies in India, the resulting milk sold to the public, and the proceeds from the sales could be used for financing the building up of a dairy system including improvement of milch cattle, building up collection systems and dairies in the countryside, and construction of transport facilities to the cities and distribution of the dairies in these.

WFP ASSISTANCE OBJECTIVES

Before any UN commitment, the UN reviewed the project. Because of the altruistic nature of UN, the objectives were changed as indicated above, a major emphasis now being placed on the supply of milk to vulnerable groups in Indian cities:

a) make milk available for vulnerable groups;
b) satisfy consumers' needs and get producers a larger share of price;
c) improve productivity of dairy farming;
d) remove cattle from cities; and
e) form a basis for national dairy industry.

After about six years of operation, the project was studied by an UN evaluation team. The data here presented were collected partly in connection with that evaluation.

MILK CONSUMPTION IN CITIES

Since the project had as one of its objectives the supply of milk to vulnerable groups in cities, it was appropriate in the course of the evaluation to study milk consumption among low income urban groups. Figure 1 suggests that in all the four cities concerned, milk consumption is very low among the lowest income segments. It seemed that the project had not met its specific objective in this regard; this matter, therefore, needed further study.

One may first look at income distribution in the cities. Table 1 for Calcutta is revealing. If one assumes a family size of six and two wage owners per family, the lowest income group will constitute about 25% of the population. This group earns less than 150 rupees (about US $21) per family per month or five rupees per day. If approximately 60% of the income is spent on food, this suggests a family food budget of 3 rupees per day.

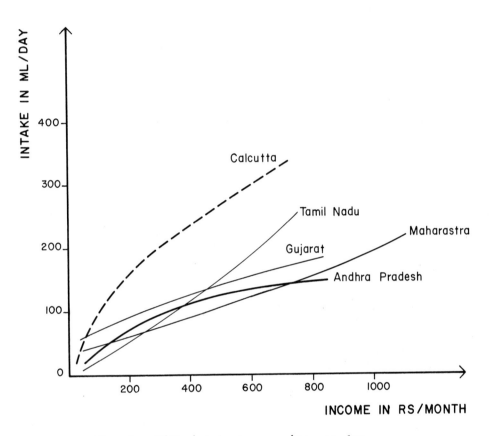

Fig. 1 - Milk intake per capita per day according to household income group.

TABLE 1

INCOME DISTRIBUTION IN CALCUTTA, 1970, MONTHLY, IN RUPEES

INCOME (RUPEES)	PERCENT
No income	67.08
Less than 75	8.56
75-150	7.42
150-300	7.03
300-500	4.02
500-750	1.52
750-1000	0.90
1000-1500	0.48
1500-2000	0.19
More than 2000	0.19
Undetermined	2.62

Source: Hindustan Thompson, 1972.

One may now refer to Table 2 giving food prices for the same period in Calcutta; the amount of energy in kJ which can be purchased per rupee as different food is indicated therein. If a six member family's food energy requirement is put at 50,000 kJ per day, it will be seen that only by buying wheat atta, the flour used for chapaties, will a family be able to obtain an energy intake as high as 47,000 kJ per day. Such a diet would be unreasonable, monotonous, and nutritionally not well-balanced and could not even be prepared without some oil. Yet, it would still not mean the satisfaction of energy requirements. If in this case the purchase of milk is proposed, the smallest unit, one half litre of bufallo milk, would cost approximately 1 rupee but would result in the daily energy intake per family being reduced to about 32,000 kJ, i.e., 64% of requirements. Clearly, in this desperate situation, populations could not divert any of their very modest resources to the purchase of milk. Nevertheless, one quarter of the population or more live under these or worse conditions. Their only recourse is to resort to a reasonably balanced diet, composed of the least expensive items available in the market, e.g. wheat atta, some cooking oil and some legumes.

TABLE 2

LOWEST OBSERVED FOOD PRICES IN CALCUTTA, 1970, and calculated energy in kJ

	Rs/kg	kJ/kg	kJ/Rs
Rice (open market)	1.97	15,050	7,640
Rice (control)	1.26	15,050	11,940
Wheat atta (open market)	1.03	15,050	14,610
Wheat atta (control)	0.96	15,050	15,675
Legumes	1.50	5,850	3,900
Mustard oil	5.12	37,000	7,230
Milk (buffalo)	2.02	2,720	1,350
Potatoes	0.65	2,930	4,500
Vegetables	0.70	1,000	1,430
Fish	2.35	4,200	1,790
Meat	2.83	8,700	3,090

REDUCING MILK PRICES

An obvious conclusion might seem to be that milk prices need be reduced. However, Table 2 illustrates that the reduction would have to be very substantial in order that milk could become competitive with other so-called protective foods, e.g., legumes or fish. However, the dairy system itself is designed to provide milk transport and distribution at the least possible cost. Therefore, any reduction of price would have to be achieved by reduced payments to milk producers in the country, and the price would have to be reduced very substantially, since that component of the price, which represents distribution costs, could not be reduced. Clearly, this would be to ask presumably poor rural producers to subsidize equally poor urban dwellers, but the price sacrifice would have to be so great that in all probability no milk would be collected. Besides, since milk consumption is much larger in the higher income groups, this would really mainly mean a sacrifice in income for the rural poor to subsidize foods for better off people in cities.

TABLE 3

PRICES FOR VARIOUS FOODS, IN RUPEES

ITEMS	2000 kJ	15 g protein
Standard milk	0.97	0.67
Double-toned milk	0.87	0.37
Extruded veg. food	0.61	0.50
Multipurpose food	0.30	0.11
Rice	0.28	0.46
Home-prepared balanced food	0.25	0.25
Bal Ahar	0.18	0.20
Wheat	0.15	0.21
Groundnut flour	0.14	0.06

FREE DISTRIBUTION OF MILK

One might then conclude that milk should be distributed free to the most vulnerable groups. Table 3 gives the prices for various foods which might be used in supplementary feeding schemes. The lowest amount of milk which could be considered for such distribution, i.e., a portion corresponding to 2,000 kJ, in the form of double toned milk, would cost 0.87 rupee per day. The population of the four cities here considered is approximately 20 million. The economic situation is probably worse in Calcutta than in the other three cities, yet an estimate of 10% of this population falling in the lowest income, very deprived groups, seems to be a conservative one. Distributing a portion of milk to 2 million people a day, would in milk and distribution cost amount to well over 1,000 million rupees annually, i.e., an expenditure considerably in excess of the total amount of resources expended over the first seven years of the duration of the project. Clearly, Indian public funds would not be sufficient for such an undertaking and foreign aid therefore would not be likely to be available.

Besides, one might seriously question whether in such a situation milk would be the preferred food for free distribution. Table 3 suggests that such foods as the Indian multi-purpose food or the Indian Bal Ahar might serve an equally useful purpose at a much lower

TABLE 4

FOOD INTAKES AMONG LOW-INCOME GROUPS IN CALCUTTA

	ENERGY			PROTEIN		
	Intake	Requirement	%	Intake	Requirement	%
	Kcal			g		
Pre-school child	500	1,200	42	15	18	54
School child	850	1,900	45	25	37	44
Adult	1,220	2,700	45	35	50	46
	VITAMIN A			CALCIUM		
	Intake	Requirement	%	Intake	Requirement	%
	I.U.			mg		
Pre-school child	300	900	33	200	450	44
School child	1,000	3,000	33	230	650	35
Adult	1,500	3,000	50	275	450	61

cost. In this, it must be considered that these stable products are much easier and cheaper to distribute than perishable and easily-spilled milk.

MALNOURISHMENT

One might of course argue that milk has a special nutritional value and that its content of calcium, vitamin A and protein would justify the extra expenditure for an individual in acquiring milk. However, Table 4 and Figure 2 suggest that the population groups here concerned are so desperately undernourished that their main requirement is for food energy. Besides, Table 4 and Figure 2 suggest that the diet is reasonably balanced since intakes low in energy are generaly equally but not more deficient in protein, vitamin A or calcium Thus, these people do not need a better balanced diet, they simply need more food.

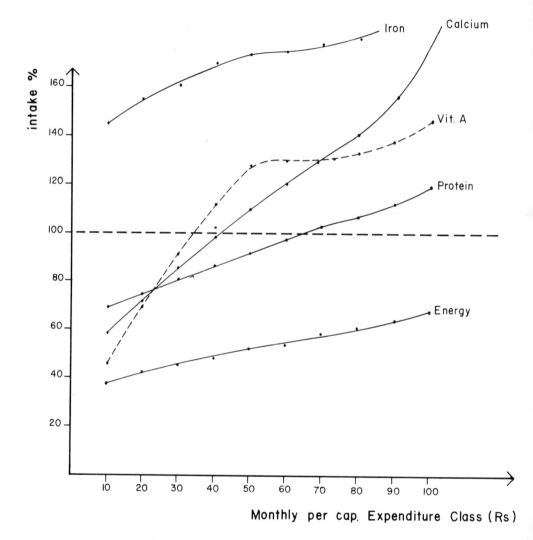

Fig. 2 - Daily per cap. % intake. Calcutta 1969.

TABLE 5

DISTRIBUTION OF DISPOSABLE PERSONAL INCOME IN INDIA, 1974

DISPOSABLE ANNUAL INCOME (rs)	RURAL SECTOR	URBAN SECTOR	ALL INDIA
	Percent of households		
< 1,000	23.5	8.5	18.5
1,000- 2,000	36.6	21.5	34.5
2,001- 4,000	27.4	33.2	30.2
4,001- 6,000	6.4	14.4	6.8
6,001-10,000	3.7	11.2	5.0
10,001-15,000	1.6	5.5	2.4
15,001-20,000	0.5	2.8	1.9
20,001-30,000	0.2	1.8	0.4
30,000 up	0.1	1.1	0.3

RURAL AREAS

At this point of the study some became quite disappointed with the performance of the project. However, it seemed necessary to consider also the effect of the project in rural areas. The populations in the milkshed areas concerned are about 12 million people. Table 5 suggests that the economic situation just as bad for rural populations in India as for urban groups. Even if some concession is made for some considerably intake of home grown foods where a person has access to any land at all or to food donated, pilfered, etc. Table 5 suggests that rural groups may be every bit as badly off nutritionally as urban populations.

However, one of the objectives of the project was to organize milk collection in rural areas. It is well known that India has a very large herd of cows and buffaloes. They are mainly used for their value as draft animals and for the value of the manure. However, where an organized milk collection system is put in, the milk also becomes of some value. A non-descript local cow may produce some 250 litres of milk per year. The procurement price may well be about 2 rupees or more. This means that just having one cow producing milk may add more than 50% to a person's income in the lowest income brackets.

TABLE 6

FARM INCOME, MADIAD VILLAGE (KAIRA DISTRICT)

		TOTAL	DIARY	%
		Rs/year		
Adopters				
Small	(<5 acres)	1,511	1,151	78
Medium	(5-10 acres)	3,353	1,917	57
Large	(>10 acres)	10,492	3,851	37
Average		4,097	1,985	48
Non-adopters				
Small	(<7.5 acres)	752	453	60
Medium	(7.5-15 acres)	2,081	1,212	58
Large	(>15 acres)	6,002	2,556	43
Average		1,192	675	57

Such economic improvements seem in fact to take place in areas with organized milk collection. Thus, Table 6 suggests that the income of those persons in rural areas, which adopted the new system of milk collection cooperatives, have doubled their dairy income. Peculiarly enough, they have also, as the display indicates, doubled their non-dairy income. The explanation for this is assumed to be that introducing organized dairying brings along with it a number of other economic and social improvements.

SCALE BIAS

Much agricultural development has been criticized because it is suspected of benefitting mainly large agricultural producers. For this reason, it became highly relevent to consider whether this project had such scale bias. Table 7 suggests that the project at least to a very large extend benefits small farmers and even landless farm labourers. Out of 328 persons in the village, 122 landless farm labourers or small farmers had milch animals. Table 7 illustrates similarly that by far the most landless farm labourers and small farmers keep milch animals.

TABLE 7

VILLAGE OF SABARKANTA DISTRICT IN GUJARAT: TOTAL NUMBER
OF HOUSEHOLDS AND HOUSEHOLDS WITH MILCH ANIMALS

	TOTAL NUMBER OF HOUSEHOLDS	NUMBER HAVING MILCH ANIMALS
Landless labourers	82	22 (27%)
Small farmers	131	100 (76%)
Medium and large farmers	115	100 (87%)

EFFECT ON NUTRITION

The above considerations, illustrated in Tables 6, 7 and 8, suggest that organized milk collection systems may double income in rural areas and that it benefits also the smallest producer and even the landless farm labourer. One may then question what the effect of this increased income is on nutrition.

Figure 3 seems to indicate that among very low rural income groups the income elasticity of the intake of food energy and some other nutrients is rather high, approximately 0.2 to 0.3. This means that doubling income will result in a 30% increase of food energy and most other nutrients. This, in effect, suggests that the project may bring about a considerable nutritional improvement among the rural producers and their families, affected by the system, and totalling 12 million individuals. The project may thus well be the development project with the greatest impact in form of nutritional improvement ever undertaken in India.

FOOD SUPPLY CONSIDERATIONS

In much of the project area one heard complaints that it was difficult to provide fodder to the dairy cattle and that green feed poducts should be produced on suitable, i.e., normally irrigated land. The rule of thumb to many was that 1/3 of an acre of irrigated land, grown with green fodder, e.g., alfalfa, per cow or buffalo, should be set aside for this purpose. The reason was obvious. An indigenous cow or a buffalo on good feed management or a cross-bred cow, which can thrive only on efficient feed, will produce much more milk than a dairy cattle left to its own devices for feed. Table 9

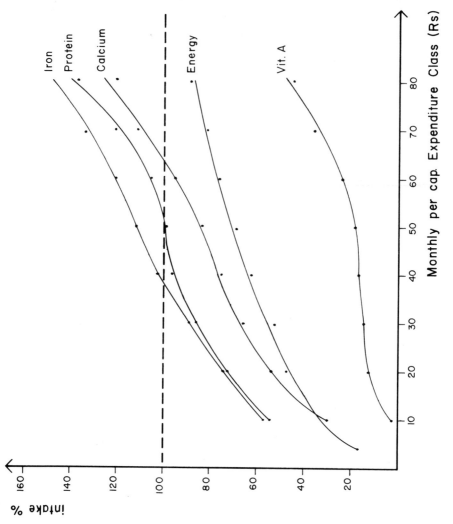

Fig. 3 — Daily per cap. % intake. Gujarat (1961 – 1962)

TABLE 8

LOW-INCOME RURAL POPULATIONS ACCORDING TO LANDHOLDING AND INVOLVEMENT IN DAIRY PRODUCTION

	GUJARAT KAIRA	PUNJAB LUDHIANA	TAMIL NADU MADURAI
	PERCENT OF HOUSEHOLDS		
Landless	26.4	54.7	57.4
Non-farmers	7.3	28.9	41.2
Farm labourers			
With milch animals	18.5	19.6	5.5
Without milch animals	0.6	6.2	10.7
Under 1 ha	22.7	7.1	30.5
Non-farmers	0.4	0.2	1.1
With milch animals	16.5	5.8	12.6
Without milch animals	5.8	1.1	16.8

TABLE 9

ANNUAL MILK PRODUCTION AND NET INCOME IN INDIA

	MILK PRODUCTION (litres)	NET INCOME* (Rs.)
Indigenous cow on current feeding	200	175
Indigenous cow on improved feed	475	700
Crossbred cow on cultivated green fodder and concentrate feed	2,500	1,300

* Excluding labour.

suggests that, in economic terms, it would be very profitable to obtain green fodder and in addition, an adequate amount of concentrates for feed purposes. However, this could have a negative effect on food supply. Producing green fodder will mean that one foregoes producing food grains for other food purposes on an area. Also, the concentrates fed to animals often could be grains, oil seeds, etc., which could be used for various food preparations directly. Thus, when resources put into the feed production or into the feed directly and the food value of the milk is considered one arrives at a calculation like the one given in Table 10.

This table suggests that an uncritical introduction of dairy farming in India could have a catastrophic effect. As an example, India has 230 million heads of cattle; if they were to require each 1/3 of an acre planted with green fodder, they would require 69 million hectares plus some extra concentrate which also would require land. This figure should be compared with the present 75 million hectares with food grains in India.

These calculations clearly indicate that wide introduction of dairying must be based on utilizing dairy cattle's ability as ruminants to use grass, straw, forage and agricultural by-products such as sugar cane tops, rice bran, etc. for feed and convert it into a high grade food, in this case milk.

TABLE 10

NET ANNUAL FOOD CALORIC BALANCE
OF COW MILK PRODUCTION IN INDIA

(kJ)

	INPUT	OUTPUT	BALANCE

Indigenous cow
    fed on agricultural waste in
    addition to wayside grazing
	0	565	565

Indigenous cow
    1 kg concentrate feed/day
    during lactation in addition
    to wayside grazing, etc.
	880	135	745

Crossbred cow
    600 kg concentrate
        feed annually
            and
    crop from 1/6 ha land
	4,800
	11,700
		5,670	−10,830

TABLE 11

AGRICULTURAL DEVELOPMENT IN GUJARAT

(1,000 tons)

YEARS	TOTAL GRAIN CROPS	TOTAL OTHER CROPS	MILK COLLECTION
1955/56			11
1956/57			14
1957/58			21
1958/59	177	83	28
1959/60	145	90	23
1960/61	137	94	24
1961/62	188	77	35
1962/63	184	98	50
1963/64	205	132	62
1964/65	216	73	61
1965/66	215	73	66
1966/67	262	106	72
1967/68	317	141	81
1968/69	233	105	113
1969/70	414	171	124
1970/71			118
1971/72			133
1972/73			148
1973/74			112
1974/75			130

In addition India has many areas with favourable agricultural conditions: Here intercropping of a food crop with a feed crop can be used or one may obtain on irrigated land two food crops and in addition two feed crops in periods between the food crops, thereby obtaining feeds with no sacrifice with regard to food production.

It seems clear that India has been fully aware of these considerations. Cooperative dairying has probably advanced farthest in the state of Gujarat. Table 11 indicates that, combined with very large increases in milk collection, the state of Gujarat had achieved a remarkable increase in both agricultural food crops and other agricultural crops. Thus, there has been no encroachment ot dairing on production of other foods.

CONCLUSION

One interesting conclusion came out as a result of the study of the project. There is a general tendency to evaluate achievements of a development project as compared to the original objectives. In this case it was clear that the original objectives were not achievable. During the course of development, the project has managed to achieve some very remarkable and useful goals. It seems that one should commend the project leaders for having had sufficient flexibility and imagination to pursue useful rather than preset goals.

UTILIZATION OF WASTE FOR FOOD PRODUCTION USING ANIMALS: APPLICATION TO HUMAN NUTRITION

W.J. Pigden

Research Coordinator, Planning and Evaluation Directorate

Research Branch, Agriculture Canada, Ottawa, CANADA

INTRODUCTION

In food production, large quantities of inedible cellulosic material are produced which are of no direct food value to humans. This is readily appreaciated for cereals where for every kg of grain produced one or more kg of straw and chaff is produced (Table 1).

TABLE 1

ESTIMATED CEREAL CROP RESIDUES AVAILABLE
FOR LIVESTOCK FEEDING IN WASHINGTON

CROP	RESIDUES kg/kg GRAIN
Wheat	1.0*
Barley	1.5
Oats	2.0

Source: Anderson, 1977, J. Anim. Sci. 46: 849.

* In wheat harvesting some residue was left on fields for erosion control.

The same is true for fruits and vegetables and many other types of waste. For example in pulping wood about 50 percent of the dry matter is chemically dissolved to isolate the cellulose fiber used for paper production.

Some concept of the huge quantities of waste materials available from cereal production can be gained by examining the data in Table 2. In the U.K. out of a production of about 9 million tons an estimated 3.5 million tons are burned in the fields annually. Likewise, there are reports that 100 million tons are burned annually in the U.S.S.R.

Unlike poultry, pigs and man, ruminants (cattle, sheep, llamas, etc.) can digest available cellulose in their fore-stomach (rumen) by means of their rumen celluloytic microflora and utilize the energy to produce milk, meat, wool, hides, etc. They can also utilize non-protein nitrogen such as urea in place of protein. However, there is a great difference between the availability of the cellulose in good quality forages and in those of poor quality such as cereal straws (Table 3).

These differences are chiefly due to differences in availability of the cellulose to the rumen microorganisms.

Another factor of great importance in animal production is the intake factor. Forages of low digestibility are usually consumed at correspondingly low levels and production of meat or milk is very poor.

However, if the digestibility of cellulosic residues from cereal

TABLE 2

ESTIMATED AVAILABILITY OF
SPECIFIC NON-WOOD FIBROUS PLANT MATERIALS

('000 metric tons)

RAW MATERIAL	WORLD WIDE	CONTINENTAL US
Wheat Straw	550,000	50,000
Barley Straw	40,000	5,000
Oat Straw	50,000	8,000
Rice Straw	180,000	50,000

Source: Atcheson, 1976, Science, 191: 768

TABLE 3

DIGESTIBILITY OF THE DRY MATTER
IN GOOD AND POOR QUALITY FORAGES

FORAGES	DRY MATTER DIGESTIBILITY (%)
Alfalfa (immature)	67
Timothy (immature)	73
Wheat Straw	38
Bagasse	35
Barley Straw	46

and sugar production could be increased to levels approaching those of good quality forages, e.g., alfalfa, enormous amounts of energy would become available for food production. For example, the estimated 50 million tons of cereal straws in Western Canada contain about twice the potential feed energy in the entire Canadian barley crop of 350 million bushels annually.

The major animal products, from milk and meat, are important high quality foods in their own right. In addition, they have a valuable complementary effect when consumed with cereal proteins thereby increasing overall quality and reducing the total protein requirements. Also they contribute valuable minerals, e.g., zinc and iron which are often seriously deficient with high cereal diets. Increasing the supply of animal products from wastes would not only release valuable land for production of food for direct human use, it could greatly increase the total food supply and the quality of the diets.

STRATEGIES FOR UTILIZATION OF WASTES

Types of Wastes and Amounts

The first requirement is to determine the types, amounts and location of waste products. The extent to which they are being used for other purposes is also essential and also if there are any serious contaminants, e.g., pesticide residues. A survey conducted by a team of specialists, economists, nutritionists, etc., is generally the best method for determining this. Where it is not possible to determine the actual amounts of waste but figures are available for the edible product a rough rule of thumb estimate is to consider the amount of the waste to equal the edible product. A useful classification of types of waste is as follows:

Cereal residues	– straws, husks, corn stalks and cobs
Forest products	– wood, liquors, cellulose fibre, waste paper, foliage
Animal wastes	– excreta, paunch manure, whey
Sugar crop residues	– bagasse, sugar cane tops and leaves, rind, pith, beet pulp
Others	– oilseed hulls, treenut hulls, olive residues, vegetable and fruit crop residues, potato wastes, food processing wastes.

Methods of Utilization

When the types, amounts and the location of wastes have been determined within a given region a nutrition-economic strategy for utilization must be worked out. This approach will depend primarily on what other feeds and supplements are available, the type of animal production desired and costs of treated or untreated residues versus alternative feedstuffs. Competition for other uses must always be considered.

Grazing. The simplest and lowest cost per ton of nutrients harvested is to allow ruminants to graze on the residues. This procedure has been utilized successfully in utilizing cereal and corn residues following harvest of the grain for low level production, e.g., maintenance of beef cattle:

The major disadvantages are that unless grazing is carefully controlled the animals frequently trample and waste much of the residue, they graze selectively utilizing the most nutritious material first so the residue becomes progressively of lower quality, it is difficult to provide adequate supplementation and the lower quality residues, e.g., wheat or rice straw, do not lend themselves to this approach. Also heavy snowfall in northern climates reduces the grazing season.

A modification of this approach is to bale the residues in large bales, leave these in the field and allow the animals access to the bales. This controls wastage to a large extent but does not eliminate the supplementation problem nor increase the digestibility.

The grazing — large bale approach can be successfully employed for the more highly digestible crop residues. However, the very low level of extraction of the energy from the lignocellulose makes it unsuitable for the majority of cellulosic crop residues.

Simple Processing and/or Preservation. This involves harvesting and mixing with other ration components to produce a feedstuff. If the materials are wet they may be fed wet, dried or ensiled with other materials. Examples are use of wet or dried brewers grains, wet or dried straw; fresh, ensiled or dry beet pulp. If, as is often the case, the material must be stored, then storages or preservation techniques must be worked out to suit the product and the type of feed. Many fruit and vegetable wastes are in this category. Much of the technology needed has been developed and can be adopted to many situations with little difficulty.

Simple Physical Processing. This generally involves grinding (comminuting) the material to facilitate handling, reduce bulky materials for storage and most important of all, allow the animal sufficient to eat to attain reasonable production. American tub trinders are an efficient method of grinding. Dried and ground bagasse is a good example. Although digestibility is low, (around 37%) it can be mixed with other feeds in quantities of 10-20% and utilized effectively. Some types of cereal straws (e.g. barley, oats) are also in this category. Costs of grinding (assuming no drying is involved) are about $5-6/ton of air dry material. This approach is especially effective where relatively low levels of production are expected, e.g., dry beef cows, or where the ration contains large amounts of highly digestible material, e.g., molasses, grain, etc., and a source of fiber is required to balance the ration. Even hardwood sawdust can be effectively utilized where available. The cost of grinding whole wood for feed is considered to be prohibitively expensive.

Chemical Processing. Processing here involves the application of a chemical to break down the lignocellulose-hemicellulose complex to make the energy available to the rumen microorganisms. It does not usually involve the extraction or removal of the lignin from the feed. In some cases the waste residue must be ground and dried to facilitate treatment and mixing, and in other cases, the material is treated inthe long form either wet or dry.

Sodium hydroxide has been generally the most effective chemical and the most extensively employed in experimental work and in commercialized processes for increasing the digestibility of lignocellulosic residues. Other hydroxides (NH_4OH, $CA(OH)_2$, etc.) are generally less effective but when mixed with NaOH they can be as effective and can contribute useful nutrients, such as calcium and ammonia (Klopfenstein, 1978).

The classic work or Beckmann (1921) must be mentioned. He showed that soaking straw with dilute NaOH would increase its digestibility from around 40 to 75%. The process was developed as the basis for a commercial "on farm" process and was used extensively in Scandinavian countries for many years. Disadvantages of this method include large losses of hemicellulose components and the large amounts of water required for soaking and washing out the alkali. This process is still in use but to a very limited extent.

(a) The "dry" alkali approach - In the early 1960's a dry alkali process was developed in Canada (Wilson and Pigden, 1964), whereby 3-6% of concentrated NaOH was mixed with the ground wheat straw cellulosic waste and neutralized or partially neutralized by the reaction. The procedure was further developed and commercialized in the U.K. and in Denmark to produce a coarsely-ground, pelleted cereal straw product. Equipment for commercial production is being marketed and a number of commercial plants are in operation in both countries. In the U.K., something like eight commercial plants are now in production. The digestibility of the pelleted product is about 58% and it can be readily transported and mixed into dairy and beef rations at levels from 10-30%. The economics ($ Canadian) of this process were as follows, in 1977:

- the total cost of producing one ton of pelleted straw was about $60, which included $42 for straw delivered at the plant;
- the selling price per ton of processed straw pellets was approximately $100;
- cost of barley/ton in England - $160;
- cost of barley/ton in Toronto, Canada - $90.

With the existing price structure for feeds in England, and judging from the expansion in numbers of processing plants, this appears to be a highly profitable operation. But the economics outside of the EEC are something else. Obviously they depend heavily on the costs of straw, labor and chemicals to produce a product for the corresponding industry.

The advantages of this process are that it produces a dense product which can be mechanically handled and transported and can be readily mixed with other feeds for high production rations (dairy, etc.).

The disadvantages are that bulky crop residues must be transported to a centralized processing unit up to twenty miles, the cost of the raw material and processed product is high, it is not suited to on-farm use, and the residual sodium hydroxide places limits on its use. Danish firms are also selling smaller scale equipment for "on-farm" use with the dry alkali process.

(b) Ensiling with alkali - The second alkali treatment method with commercial promise is the application of sodium hydroxide to wastes during the ensiling of straw, corn stalks and/or cobs, bagasse, etc. This method has been developed to a large extent through the work of Klopfenstein and co-workers in Nebraska, U.S.A., and Mowatt at Guelph, Canada. The alkali is added at the time of ensiling along with enough water to increase the moisture to about 65% and the reaction on the lignocellulose occurs over a period of days or weeks. The level of sodium hydroxide recommended is about 3% of the dry matter.

In addition to sodium hydroxide other hydroxides have been tested alone and in combination with sodium hydroxide. Although generally not as effective in increasing digestibility as NaOH, the evidence indicates they can be effective especially in mixtures.

The approximate cost per ton as given by Klopfenstein (1978) is as follows:

	Per metric ton DM
Crop residue	$ 4
Handling	$13
Chemicals	$ 9
TOTAL	$26

He states that when corn grain costs over 8 cents/kg this method would be economically feasible under Nebraska conditions.

The advantages of this method are that it is an "on-farm" process so that bulky residues are transported short distances only and much of the equipment and storage facilities normally available for other crops can be utilized. Also, relatively inexpensive storage, e.g., pit silos, can be constructed.

The disadvantages appear to be mainly the costs of the extra chemical in the feed and the fact that the alkali does not increase digestibility neraly as much as the dry alkali process.

(c) Ammonia treatment - Ammonia, either in the anhydrous liquid form, aqueous ammonia or ammonia derived from urea or ammoinum bicarbonate is effective in increasing the digestibility and intake of cereal straws and corn stover (stalks).

Its advantages are that it is an "on-farm" process, the equipment is simple and relatively inexpensive, it can be used as an emergency method on short notice and would seem to be one of the better methods during times of feed shortages.

The disadvantages are the time that the ammonia requires to act on the straw, particularly in cold weather and increases in digestibility are not as large as those with sodium hydroxide.

FOREST PRODUCT WASTES FOR FEEDSTUFFS

One of the world's greatest renewable resources is its forests with the potential to produce large quantities of animal feeds and thereby release agricultural land for crops for direct human use.

In Canada there are about 600 million acres of merchantable forests with is about twice the acreage of agricultural land. It is estimated that of the approximately 16,000 million tons of timber on this land about half of it can be harvested for conventional forest products and the other half is waste, i.e., foliage, branches, stumps and cull trees (Stone, 1975).

As sources of feed, trees do have some advantages. A forest consists of a mass of 100 tons/acres or more of vegetable matter for large scale harvest operations; there are no annual crop failure problems, no storage problems over winter or during wet seasons and trees grow on land not suited to other crops. The cell walls are a rich source of carbohydrates, about 70%.

Why are trees not used more as a source of feedstuff? Chiefly because of the hard lignified nature of wood which gives it its strength and rigidity and which renders most wood more indigestible than poor cereal straws. Also much of the material is not available where it can be used.

The technology to convert virtually any tree to feed has been available for some time but the economics were poor. However, new, recently developed technology has provided something of a breakthrough in several areas and shows good potential for further development and economic production of feed from wood (Table 4).

Tree Foliage "Muka"

The Russians pionnered the utilization of tree foliage for feeds and currently produce 100,000 to 200,000 tons of a product they call Muka. It includes the needles of conifers and the leaves of broad-leafed species. Heating at a temperature around 210°C for a few minutes to drive off the moisture and essential oils plus grinding is all the processing that is needed.

The feed is similar in composition and feeding value to alfalfa and is used in poultry, cattle and swine rations in Russia. The cost to produce this material in Canada is estimated to around $110/

TABLE 4

MAJOR PROCESSES AND FEED PRODUCTS
FROM FOREST INDUSTRIES WASTE

SOURCE	PROCESSING TREATMENT	FEEDSTUFF PRODUCT
Hardwoods		
Mill waste under-utilized species	Partial hydrolysis Heat + pressure	Low nitrogen high energy roughage
Sulfite pulp mill liquors	(a) Concentrated to a wood molasses	Molasses
	(b) Fermentation to SCP	Yeast or Pekilo protein
Tree foliage	Separation + heat + grinding	Alfalfa substitute
Conifers		
Mill waste under-utilized species	Acid hydrolysis (a) Fermentation to SCP (b) Wood molasses	Yeast, Pekilo protein or molasses

ton. There are pilot studies and feeding trials underway now in Canada at two locations.

Hardwood Utilization

There are certain species of hardwoods, especially poplar, which are very much under utilized by the forest products industry in Canada. Starting with the steaming process developed by Heaney and Bender (1970) a high temperature, high pressure, continuous flow method of processing sawdust or poplar chips into a feed for ruminants has been developed by Stake Technology Limited. Equipment for this process is now commercially available, and can also be used for processing straw or bagasse. The process is a limited hydrolysis method utilizing the acetyl groups to disrupt complex chemical bonds. The product is a brown, dispersed, fibrous product containing about 50% moisture with 1.5 - 2% crude protein, and is 50-65% digestible. It is highly palatable and can be handled and fed much the same as silage. It is acidic (about pH 3.5) and does not appear to mold or deteriorate in open storage. The processing costs are around $15/ton of dry matter and the cost of raw material ranges up to $45/ton of dry matter, depending on the source. Supplementation costs for nitrogen, minerals and vitamins are about $5/ton. Feeding trials with cattle and sheep have shown this product to be a very useful feed for ruminant production.

Waste Sulfite Liquor

Chemical pulping dissolves about 50% of the wood, ligning and carbohydrates. Canada has a large number (over 30) of sulfite mills which generate an estimated 3.8 million tons of dissolved wood annaully (Stone, 1975). Most of this is dumped into the rivers. The spent liquor can be evaporated into a syrup for feeding as a substitute for molasses or it can be used as a substrate to produce torula yeast or Pekilo protein for protein supplements. For the most part the large quantities of plant protein available in Canada has discouraged protein production and we continue to dump our sulfite liquor. In countries such as Finland I understand it is fully utilized for protein production.

Sawdust and Other Wood Waste

Although good progress has been made in developing feed from hardwoods, most northern forests like Canada's contain about 80% softwoods (conifers) and an economical and efficient method has not yet been developed for converting these species to feed. Acid hydrolysis will convert the wood to a form suitable for yeast production but is an expensive way of producing protein. Methods to process softwood to feed are under development. It is worth noting that unlike Canadian forests most of the trees in Latin America are hardwoods.

SUMMARY AND RECOMMENDATIONS

1. Efficient utilization of agricultural crop residues, forestry wastes and food processing waste can substitute for much of the cereal grain now fed to livestock especially to ruminants and can greatly extend the world food supply and improve the quality of human diets.

2. The nutrient composition, palatability and digestible energy of different varieties within cereal (wheat, oats, barley, rice, etc.) should be characterized.

3. Further studies should be carried out on chemical processing and on high pressure, high temperature applications to reduce costs and further increase available energy.

4. Additional information is needed on supplementation.

5. Additional information on storage and preservation is essential.

6. Better equipment for efficient treatment of cereal residues and wood must be developed.

7. Existing information on nutritional value of residues and their use in feeding systems should be made more widely available. Presently it is scattered throughout a large number of publications.

8. Plant breaders in their quest for improved cereal varieties should also consider selecting for quality of the straw as a feed stuff in order to improve the feeding value of "waste" products.

BIBLIOGRAPHY

Anderson, D.C., 1978, Use of cereal residues in beef cattle production systems. J. Anim. Sci., 46: 849.
Atchison, J.E., 1976, Agricultural residues and other nonwood plant fibers. Science, 191: 768.
Garret, W.N., H.G. Walker, G.O. Kohler and M.R. Hart, 1976, Feedlot response of beef steers to diets containing NaOH or NH_3 treated rice straw. 15th California Feeders Day Proc., University of California at Davis, p. 39.
Heaney, D.P. and F. Bender, 1970. Forest Products J., 20(9): 98-102.
Klopfenstein, T., 1978, Chemical treatment of crop residues. J. Anim. Sci. 46: 841.

Klopfenstein, T.J. and W.C. Koers, 1973, Agricultural cellulosic wastes for feed. "Proceedings American Chemical Society. Symposium: Processing Agricultural and Municipal Wastes". AVI Publishing Co. Inc., Westport, CT.

Oji, U.I. and D.N. Mowat, 1978, Nutritive value of steam-treated corn stover. Can. J. Anim. Sci., 58: 177.

Oji, U.I., D.N. Mowat and J.E. Winch, 1977, Alkali treatments of corn stover to increase nutritive value. J. Anim. Sci. 44: 798.

Pigden, W.J., 1977, Nutritional and Economic Aspects of Utilizing Wood-Processing By-Products. New Feed Resources. Proceedings of a Technical Consultation Held in Rome, 22-24 November 1976. FAO, pp. 211-225.

Rexen, F. and F. Vestergaard Thomsen, 1976, The Effect on Digestibility of a new technique for alkali treatment of straw. Anim. Feed Sci. Technol. 1: 73-83.

Stone, John, 1975, Forest Products as Commercial Feed. "Waste Recycling & Canadian Agriculture". Conference Proceedings, Toronto, April 24-24, 1975. Agricultural Economics Research Council of Canada.

REPORT AND RECOMMENDATIONS OF THE WORKSHOP ON MODERN

PRACTICES OF ANIMAL FEEDING: RELATIONSHIP TO ANIMAL NUTRITION

V.A. Oyenuga

University of Ibadan

NIGERIA

It has been estimated that about 462 million individuals in the world have intakes of food below the lower limit. Some 434 million of 94 per cent of these live in the developing regions of the world. Some 301 million persons of them live in the Far East (excluding China) and about 67 million in Africa. An individual with an intake included in this below limit group will be exposed to a high risk of inevitable reduction in activity, adverse effect on growth in the case of a child or of a continuous loss of body weight in adults. There is, therefore, a very high probability that about 462 million individuals in the world, 94 per cent of them in economically poor communities, have food available in quantities insufficient to meet their needs.

In the light of this situation, it will be uneconomic and unrealistic, in modern animal feeding technique, to continue to recycle a substantial part of the available food energy directly utilizable by man through animal organisms whose poor conversion efficiency has been so aptly illustrated in Dr. Rerat's introductory remarks. The important point, which has been made by many of the contributors, and which needs positive emphasis in this symposium and at this Congress, is the need to reduce or completely remove any possible competition with humans in animal feeding. As human populations increase rapidly towards the end of this century and the beginning of the next, so will the competition for the available food resources between man and his beast become increasingly sensitive. At all times and in all cases, it may become increasingly necessary, in view of dwindling world food resources in relation to rapidly increasing human population to carefully weigh the advantages derivable from the unique biochemical values of animal proteins and animal fats in human diets with the vast superior direct utilization efficiency of plant products by man, and thereby carefully allocate resources so as to maximize the production and utilization of energy and protein per hectare of cultivated land.

In this case, however, traditional food habits and tastes of individuals and communities as dictated by purchasing power and rising standards of living, as well as the basic nutritional needs of the weaned infant, the pre-school age and growing child and of the "vulnerable group" generally cannot, in reality, be completely ignored. It may be necessary for the speakers and the audience to comment further on this aspect of the problem.

The ruminant livestock — cattle, buffalo, goats and sheep — is globally becoming a tool for exploiting marginal lands, otherwise less utilizable for arable farming. The use of this land for the production of animal products thus complements rather than competes with the production of energy and protein foods directly utilizable by man. Both Kercher and Perez-Gel Romo have clearly described the modern technique and the problems involved in the highly developed North America and in the somewhat less developed South American conditions. To what extent could existing and more recent technical information be promoted between scientists in the advanced and less advanced countries should form a useful subject for further comments by Dr. Kercher and, further discussion.

Romo mentioned that 446 million hectates (79.9% of agricultural land) were devoted to permanent grassland in South America in 1967. Of the 870 million hectares of total agricultural land in Africa in 1975, 686 million hectares (or 80 per cent of total agricultural land) are permanent pastures (FAO, 1976) for ruminant and herbivores utilized for the production of meat and milk. Since ruminants alone account for practically all the available milk and for over 60 percent of the meat, there appears to be great potential for the production of meat without competing with man. As supplements to grazing, particularly during the drier periods of the year, and for highly productive animals, various other sources of feed wastes including:

- sugar cane by-products;
- citrus pulps; orange peels; pineapple rinds;
- coffee pulp; products from rice-processing;
- meals and cakes from processed oilseeds;
- cassava roots and leaves; peelings from roots and tubers;
- banana by-products; and
- non-protein nitrogenous substances, including urea.

A large number of less known tropical agricultural fruits and nuts, recently detailed investigations which have yielded valuable information; dried poultry manure and many others are emerging, which could be substituted for cereals and legumes that are used for human diet.

Clausen in his paper further reinforced this subject since the modern dairy cow is becoming a more efficient converter and the proportion of protein feed concentrate is diminishing, with indication that these could be effectively replaced in the future by agricultural by-products, non-protein nitrogenous substances or by protein of poor quality.

In his excellent paper, Rerat described the modern pig as less fatty, more of lean, tender meat, better flavour and converting feed more efficiently into lean meat. As on omnivore, it competes with man for grains and protein food. However, the many industrial and agricultural by-products, already mentioned, are becoming increasingly available for this class of domestic animals to convert to animal protein and animal fat for the use of man. In all these a judicious blending of pastures and concentrates as well as by-products and wastes is perhaps the best way out.

We are grateful to Fernando for his elucidation of the possible contaminants that may arise from and outside the use of feed additives to promote rapid and balanced livestock growth and their possible effect on the consumer. It is necessary to bear in mind the information contained in this excellent paper when planning and executing animal feeding experiments and when planning commercial production particularly of poultry and pigs.

BIBLIOGRAPHY

F.A.O (1976), FAO Production Yearbook, v. 30, 1976.

FEED FORMULATION AND FEED TECHNOLOGY

Olaf R. Braekkan

Government Vitamin Institute - Directorate of Fisheries

Bergen, NORWAY

   Feed technology in context with aquaculture is how to prepare the feed and its form and properties. The fish farmer may prepare the feed himself, but for moist and in particular dry pelleted feeds, the feed manufacturer has taken over. In the following the term referred to will be: (1) wet feed, i.e., whole or minced fish, eventually with admixture of vitamins, minerals and binders (water content approx. 70%); (2) moist feed with a higher admixture of dry components such as fish meal (water content approx. 50%); and (3) dry feed with a water content below 20%. Intermediate forms may be applied with regard to the water content.

   Meyers[7] reviewed the formulation of water-stable diets for larval fish. He pointed out the need to understand the requirement at all stages of growth and especially from egg to metamorphosis. The trend is toward a replacement of the traditional larval food to fabricated artificial diets. As the eggs and the fishlarvae may differ much from species to species, the question of the size and shape of the feed particles is important.

   Limborgh reviewed the industrial production of ready to use feeds for mass rearing of fishlarvae, and emphasized the need for very small particles in the first stages. He emphasizes that large scale rearing of fishlarvae can only be achieved provided plenty of suitable larval feeds can be available.

   Ghittino[3] reviewed formulation and technology of moist feed. He points out that with the increased size of the fish farms, fresh or mixed diets very soon became inadequate. The dry feed was introduced and established from 1969 and on. In his opinion a return to moist diets might for several reasons be conceivable in fish farming.

First of all because of economic ones, as the production cost of one kilogram of trout using a moist diet is about half that of trout reared on dry diets. Considering the need of marine fish to obtain water from their diet to compensate for loss osmotically across the gills and other body surfaces, wet diets or diets with added water would find preference.

Solberg[6] reviewed formulation and technology of moist pellets. A moist pellet is made up of four main components: the wet feed, the dry component, brinders and mixtures of minerals, vitamins and eventually preservatives.

Aquacop[1] reports on a study of technology for manufacturing pellets with moistening process in tropical area. In general the methods involves an extrusion without steam through a granulating process followed by drying operation in a sun-drier working an additional heat exchanger. This device seems adaptable in food production for fish in temperate or warm-water.

In fish feed protein constitutes the main and expensive component and studies are frequently reporting on the trial of alternate sources. Several papers reported on single cell proteins. Matty and Smith[6] evaluated a yeast, bacterium and alga as protein source at levels of 20 to 40% in compound feeds fed for 12 and 10 weeks to rainbow trout. At 20% inclusion crude protein digestibility was comparable with conventional protein sources such as fish meal, soybean meal, cotton seed meal and groundnut meal. The growth rates were also comparable. At 40% level the fish fed bacterial protein showed noticeable lowered growth rate, attributable entirely to a lowered food consumption.

Another protein source which in later years has been given special attention is krill of different species. Hilge[5] in studies on young channel catfish found krill meal as total substitute to fish meal to depress weight gain to a considerable degree when it was used as the only protein source.

A key factor in feed formulation and technology is the availability of the different nutrients. Although progress has been made with introduction of more extensive and efficient chemical analysis, there is still a considerable lack of information on the digestibility of the proximate components of feeding stuffs and such data are required for efficient diet formulation. Whereas crude protein and crude fat in most foods are well digested by fish, carbohydrate utilization is still and area of much confusion. As starch in the most common carbohydrate, an improved digestibility is observed upon heating caused by destrination.

REFERENCES

[1] Aquacop (Cuzon, G.M.; Febvre, A.; La Pomelite, Ch., Calves, J., Martin, J.L.; Griessinger, J.M.; Hatt, P.J., Le Bitoux, J.F.; Michel, A.). EIFAC/78/Symp: E/28.

[2] Beck, H.; Gropp, J.; Koops, H. and Tiews, K. Single cell proteins in trout diets. EIFAC/78/Symp: E/67.

[3] Ghittino, P., Formulation and technology of moist feed. EIFAC/78 Simp: R/15.1.

[4] Gulbrandsen, K.E., Experiments with red feed (Calanus finmarchicus) and krill (Meganyctiphanes norvegica) as protein sources in feeds to Coalfish (Pollachius virens). EIFAC/78/Symp: E/32.

[5] Hilge, V., Preliminary results with krill meal and fish meal in diets for Channel catfish (Ictalurus puntactus, Raff.). EIFAC/78/Symp: E/36.

[6] Matty, A.J. and Smith, P., Evaluation of a yeast, bacterium and alga as a protein source for rainbow trout. Effect of protein level on growth, gross conversion efficiency and protein conversion efficiency. EIFAC/78/Symp: E/67.

[7] Meyers, P., 1978, Formulation of water-stable diets for larval fishes. EIFAC/78/Symp: R/13.2.

[8] Solberg, S.O., Formulation and technology of moist feed - Moist Pellets. EIFAC/78/Symp: R/15.2.

[9] Tacon, A.G.J., The sue of activated sludge, single cell protein (ASCEP) derived from the treatment of domestic sewage in trout diets. EIFAC/78/Symp: E/63.

FISH FEEDING TECHNOLOGY

Takeshi Nose

Freshwater Fisheries Research Laboratory

Fisheries Agency, Hino, Tokyo, JAPAN

1. INTRODUCTION

In fish culture, production cost largely depends on the skillful feeding practices as well as on the adequate selection of feed. During the past thirty years, fundamental scientific knowledge of fish nutrition has developed rapidly enough to provide artificial diets for most of freshwater fish cultured in the world. For marine fish, a large portion of their diets still comes from raw materials such as raw fish, putting some special problems in marine fish farms. Modern knowledge regarding feed components and feed manufacturing techniques made it possible to prepare a wide variety of feed forms and compositions. However, the utilization of these feeds seems to be not adequate enough due to the lack of knowledge on feeding behavior, digestion mechanism, energy requirements of each species of fish as well as the influence of abiotic variables such as water quality on the production of fish. Along with the development of fish culture technology, the biomass of fish raised per unit of area tends to increase not only in intensive fish culture but also in extensive fish culture. Thus, more precise technological managements are needed in modern fish culture.

In this paper, present status of feeding technologies will be presented briefly in terms of feed forms, daily feed allowance and feeding equipments.

2. TYPES OF DIET

At present, more than one hundred species of fish including crustacea are cultured in the world, and these fish are raised in raised in many types of culture systems. In case of intensive fish culture, fish must be supplied all the nutrients and energy necessary for performance of growth, maturation and other biological functions, thus, be fed complete diets. In case of extensive culture, fish can forage natural foods and low cost supplemental diets are used to stimulate fish growth replenishing a part of energy and nutrients.

Complete Diet

Wet Fresh Feed - Still nowadays, large amounts of raw of frozen scrap fish are used as the most practical feeds in some countries, due to their high acceptance and low cost. More than 120,000 tons of marine fish were raised on scrap fish in Japan in 1976. Grower feeds for trout cultured in Denmark are mostly minced fresh fish, such as herring, sand eels, whiting and sperling. In Norway, salmon and trout are cultured mainly on frozen trash fish. With a food conversion of 7:1, the feed cost of producing 1 kg of food fish is $0.385 to 0.518 for wet feeds whereas $0.497 to 0.662 for dry feeds with a feed conversion of 1.8 (Brown, 1977).

The trash fish, when it is fresh or stocked in proper freezing conditions before use, can provide all the nutrients required by the cultured fish, especially for carnivorous fish. However, some marine and freshwater fish hold high level of thiaminase in their organs (Harrington, 1954; Fujita, 1954). It has been reported that yellowtail develop thiamine deficiency when were fed only anchovy or saury for long periods. The deficiency can be prevented by the supplement of thiamine coated with hydrogenated beef tallow at a level of more than 1 mg (as thiamine nitrate) per kg of body weight per day (Ishihara et al., 1978). The material for wet fish feeds must be kept as fresh as possible, otherwise, oxidative deterioration proceeds while stocking and oxidative products such as rancid oil cause severe damage on fish. The addition of alpha-tocopherol prevents the ill effects of rancid oil to some extent.

Dropping and small particles of the wet feed disperse in water while feeding, causing heavy load on environmental conditions, not only increasing the demand of oxygen but also activating bacteria in water and bottom mud, resulting in the decrease of carring capacity of water bodies.

Moist pellet - During the 1950's, more economical meat-meal diets were devised for salmonids by blending approximately equal weights of meat and dry mixture. Nutritionally adequate meat-meal diet such as Oregon moist pellet (OMP) has provided good growth performance not only for salmonids but also for other species of fish which are difficult to raise on dry feeds (Snow & Maxwell, 1970; Anderson, 1974; Nagel, 1976). A typical OMP for salmonids is reported to contain 34.4% moisture, 34.8% crude protein, 7.5% fat,

6.8% ash and 16.8% carbohydrate (by difference), and digestible calories of 2,200 Kcal per kg diet. The feed conversion ranged from 1.8 for fall chinook salmon to 2.4 for steel head trout, with an average of 2.0 with the standard OMP (Hublou, 1963).

Moist pellet has an advantage in utilizing wastes generated by the fishing industry and scrap fish unusable for human food or animal feed without any supplemental processing. Recent shortage in the supply of high quality fish meal made it necessary to develop moist pelleys to utilize animal proteins which have thus far been discharged.

During prolonged storage above freezing temperature, hydrolytic and oxidative rancidity develop and reduce feed efficiency and growth rate, though microbial and mold growth were effectively inhibited by inclusion of sorbate (Crawford et al., 1973). Recently, preservation of moist rations has been developed by the aid of fermentation techniques resulting in high quality products with a shelf life of about thirty days at ambient temperature (Webber & Huguenin, 1978).

<u>Dry Pellet</u> - Dry pellets are the most common type of feeds used in fish culture, due to the convenience of preparation, transportation, storage and feeding. Optimum size is the largest pellet that the fish will readily consume. Thus, feed manufacturers provide various size of pellets from the smallest size of crumbs, then granules, to a larger size of about a half inch in diameter. Water stability and hardness are also important characteristics of the pellet. Pellets must remain intact in water for at least 10 minutes even though 15 to 20 minutes stability is much better. For crustaceans much higher stability is needed. High fat ingredients and added fat reduce compression; fat content in the mixture is recommended not to exceed 6%. Rice and oat hulls, alfalfa meals and bones in animal by-products as well as hemicellulose and cellulose derivatives, lignosulfonates, bentonites and others weaken a pellet. Micro-pulverized materials increase pellet compression (Robinette, 1977). The following test was devised to measure the water stability of pellets; place a known weight of pellets on a No. 10 Tyler mesh aluminium screen and immerse in water for 10 minutes, then remove, dry and reweight the pellets. A good stable pellet should not lose more than 10% of its original weight (Robinette, 1977).

Pelleting involves the use of moisture, heat and pressure before the feed mixture passes through holes of die ring. A temperature of about 85º C at a moisture content of about 16% destroys thermolabile nutrients to a certain extent. Vitamin C was reported to be destroyed most readily during the processing. Inclusion of extra amount of heat sensitive vitamins or spraying the vitamins directly after extrusion are practically applied. While stocking, vitamin C decreases rapidly but several B group vitamins such as B_1, B_2, B_6, pantothenic acid, inositol and folic acid were reported to be stable even if pellets were kept for one year in dry conditions at 30º C (Nakagawa, 1964).

Due to high water stability and buoyancy, expanded feeds have become popular recently, especially among catfish farmers in the United States. The feeds are made with a equipment developed for extrusion processing of pet food. A high temperature of 107 to 127º C and a pressure of 440 to 550 x 10^4 kg/m^2 gelatinize starch to high extent (Robinette, 1977) and some nutrients, particularly carbohydrates, become more digestible while thermolabile vitamins will be destroyed to a greater extent. Even though the expanding process costs usually 8 to 15% more than ordinary high density pellets, expanded feeds have many beneficial characteristics which compensate the high price. Fishfarmers can easily control the amount of diet according to the appetite of fish close to maximum rate of consumption without loss of feed and over feeding.

The bulky nature of the pellet increases transportation and storage cost, and extremely expanded diets may satiate stomach volume before enough amount of the diet is consumed by fish. Channel catfish consume 17% more of the high density pellet and gain appreciably more weight than the low density pellet when pellets were offered as much as they were accepted (Lovel, 1977).

Recently, expanded pellet began to be used for yellowtail culture in Japan. Before feeding, the feeds are softened by mixing with water emulsified with fish oil. The softness and fish oil flavour increase the palatability and yellowtail accept the feeds as well as wet fresh feeds. Almost identical growth was reported between the traditional raw fish diet and the softened expanded feed.

<u>Paste Feed</u> - In Japan, eels have been cultured with marine scrap fish for more than 60 years, however, a formulated diet was developed in the middle of the 1960's. The diet was composed mainly of finely pulverized fish meal and gelatinized dried potato starch supplemented with vitamins and minerals. Before feeding, an equal weight of water, and 5 to 10% of fish oil are added to the dry mixture and vigorously blended, mechanically, for 30 seconds. The paste thus prepared is set on a wire-basket hung on the surface of pond water in a dark feeding hut. Proper adhesiveness prevents loss of the diet. Cohesive strength of the paste gradually decreases with lapse of time. A feed conversion of about 2 will be obtained in practical farms. The characteristics of gelatinized potato starch play an important role in the physical consistency of the diet.

The formulated feed is considerably high in price due to high inclusion of gelatinized potato starch. Some trials have been made to feed eel dry pellets or expanded pellets and a high feed efficiency was reported. However, pellets are rarely used in practical farms (Aoe et al., 1966).

Supplemental Diet

A large number of warmwater fish used to be raised in ponds where fish can grow entirely or to a certain extent on natural food, depending on the biomass of fish and natural food. This type of extensive fish culture, especially of herbivorous and omnivorous fish in tropical and semi-tropical zone, became very important for production of animal protein at lesser cost and at higher yield per unit of area when compared with land animals.

The critical standing crop, defined as the biomass beyond which the natural food does not supply enough amount of nutrients required by fish, depends on the ratio between the supply of natural food and nutritional requirements. When the population or biomass of fish increased, available natural food decreased rather rapidly due to over grazing, and these parameters continuously changed along with seasons, amount and type of fertilizer, and many of other biotic and abiotic factors. Availability of natural food, especially of phytoplankton differs among the species of plankton as well as the species of fish. More than 80% of apparent digestibility is reported on Oocystis, Scenedesmus and Euglena when these algae were fed to carp (Hepher, Sandbank & Shelef, 1978). Tilapia nilotica can absorb 70 to 80% of ^{14}C labelled to Microcystis, Anabena and Nitzia but only 50% is absorbed from Chrollera (Moriarty & Moriarty, 1973).

The yield of fish which comes only from natural food is usually very low, and 0.5 to 0.8 ton/ha would be reasonable yield of a pond fertilized with inorganic fertilizer and 2 tons/ha has been reported when a pond is fertilized with chicken manure (Satomi et al., 1973). Thus, the production of extensive culture system depends largely on supplemental diet, the composition of which is simpler and less expensive than the complete diets, and the efficiency of the supplemental diets will depend to large extent on the amount and the composition of the diet.

Natural food which is foraged by fish usually contains about 50 to 60% protein on a dry matter basis. These values are considerably higher than that required by most of warmwater fish. In case of food shortage, ingested food is used to meet the energy requirement at first, and a large portion of dietary protein would be burned for energy production. In these circumstances, supplemental diet, even low in protein content, can afford energy for fish and the protein from natural food can be converted into fish flesh efficiently. Cereal grains (sorghum) have been reported to be as effective as protein rich pellet containing 22.5% protein when standing crop of carp does not exceed more than 800 kg/ha. When the biomass of carp increased up to 1,400 kg/ha, however, a pellet containing 22.5% gave better yield than cereal grains (Hepher, 1978).

There will be no fixed composition of the supplementary diet.

The composition may differ not only between different fish species but also between different standing crop of fish of the same species, and at the same standing crop, at different productivity levels. As a useful practice, "dilution" of a complete or a highly nutritional diet with a low nutritional value diet is recommended in Israel. Up to 800 kg/ha of carp, only cereal grains are used, and up to 1,800kg/ha cereal grains are replaces gradually by 25% protein pellets and only the pellets are used over 1,800 kg/ha (Hepher, 1978).

3. DAILY FEED ALLOWANCE

The amount fed to fish is a fundamental consideration for the highest production at the least cost. Excess feeding may lead to digestive and metabolic inefficiency and uneaten food causes deterioration of the aquatic environment, while under feeding may cause overcompetition for food and result in large variation in size of fish at harvest as well as in poor growth rate.

As most fish eat to satisfy their energy requirement, floating type food is likely to prevent excess feeding. For sinking type food, daily amount of food is based in most cases on feeding tables principally developed by Duel and Brockway (Duel et al., 1952) for trout and salmon hatcheries. Freeman et al. (1967) presented two methods for calculating the amount of daily rations. One is based on estimates of expected percentage of body weight gain and feed conversion, and the other is based on expected fish length increase, energy value of feeds and conversion. Buterbaugh and Willoughby (1967) presented a feeding guide for trout modifying the theory advanced by Haskel (1959), introducing the concept of hatchery constant. These feeding guides for salmon and trout were expanded to other species of fish such as carp raised in net cages at high density (Kurihara, 1966) and channel catfish (Lovell, 1977).

At present, the feeding levels in these tables are expressed as percentage of body weight at different water temperature and fish size. However, energy content differs from diet to diet and it would be preferable for feeding tables to be expressed on energy basis. The energy budget studies developed rapidly in the last ten years. However, no feeding table has been established yet, based on the energy requirement of fish. This fact might be due to the complexity of bioenergetic investigation on fish.

The energy required for growth is the sum of total energy deposited in fish body (ΔB) and total energy of metabolism (R). The latter is composed of energy released in the course of metabolism of unfed and resting fish (standard metabolism (R_s)), in the course of digestion, assimilation and storage of materials consumed (including SDA) (R_d) and of swimming and other activity (R_a) (Braaten, 1978). The sum of energy is equal to the energy assimilated from

FISH FEEDING TECHNOLOGY

diet. For the estimation or assimilated or available energy value of diet, standard physiological fuel values, 4 Kcal/g protein or carbohydrate, 9 Kcal/g lipid, have been often used for carp, catfish and other omnivorous fish. The fuel values proposed by Phillips and Brockway (Phillips, 1972), 3.9 Kcal/g protein, 8.0 Kcal/g lipid and 1.6 Kcal/g carbohydrates, are commonly used for salmonids and other carnivorous fish due to their poor ability to digest dietary carbohydrates. Recently, digestible energy of diets were determined for several species of fish by measuring gross energy of diet and feces (Page & Andrews, 1973; Elliot, 1976). Rainbow trout of 400 g assimilated 4.39, 5.34, 3.37, 1.94 and 7.60 Kcal per g of white fish meal, corn gluten meal, soybean meal, wheat middlings and powdered cod liver oil, respectively, at 16º C (Nakamura et al., 1973). The digestible calorie of a diet is affected by water temperature, amount of diet and size of fish.

Metabolizable energy (ME) is a more absolute estimate of the available energy value to diet. However, very little information is available and the method for ME determination has not been established for fish (Smith, 1978).

The energy of metabolism can be measured by oxygen consumption. For converting the rate of oxygen consumption into ratio of energy production, the energy equivalent (oxicaloric coefficient = Q_{ox} is generally accepted to be 3.36 - 3.42 Kcal/mg of oxygen (Brett, 1973; Warren and Davis, 1967; Braaten, 1978).

4. FEEDING EQUIPMENT

The most common practice is to feed by hand. During hand feeding, fishfarmers can watch the appetite of the fish, which provides important information on the activity and health of the fish and as estimate of the growth and survival of the fish. Careful hand feeding usually results in better food conversion and faster growth than mechanical feeding (Andrews & Page, 1975; Peffer, 1977). Due to the rise in labour cost and expansion of fish farm scale, various mechanical feeders have been devised and introduced in fish farms. Hand operated blowers and "disk-throwers" are useful as mechanical aids to hand feeding. Hand feeding from a slow moving vehicle driven along ponds or from a boat in a large pond are effective methods.

For ad libitum feeding, various types of demand feeder are available. Dry pellet feed is stored in a hopper and a small quantity of the pellet is dropped into water by mechanical movement of a rod actuated by feeding action of fish or by a signal response by an electronic beam. The fish can get feed any time they want. However, a highly sensitive trigger mechanism can release food by inadvertent triggering action by unintentional movement of the fish, resulting in overfeeding and increased BOD load to the pond. On the other hand, the mechanism not sensitive enough results in underfeeding, with poor growth and ununiform size of fish. For automated feeding according to prescribed schedule, feeders equipped with timed-release mechanism are generally in use. Frequency and duration of feeding are adjustable depending on the requirements and behavior of fish. The feeders have the mechanism to disperse pellets widely on the surface of the water by means of centrifugal force of a rotating disk or air from blower or compressed air. A screw conveyor within a tube or belt is also used for sispersing pellets. Several hatcheries have a trolley system travelling on overhead tracks which convey a series of hoppers over the race ways and disperse pellets along the entire length of the race way.

Semi-moist extruded fresh or frozen pellets can be fed by pneumatic tube or mechanical feeders. However, most of the demand or self feeders are designed for pelletized feeds; thus moist rations are still most often fed by hand. Some yellow-tail farms in Japan use a boat equipped with a automatic feeder. A batch of scrap raw fish in a hopper is chopped and dispersed on the surface of water by fast driven belt conveyor, the speed of which is adjustable.

BIBLIOGRAPHY

Anderson, R.J. (1974), Feeding artificial diets to smallmouth bass. Prog. Fish Cult., 36, 145-151.
Andrews, J.W. & J.W. Page (1975), The effects of frequency of feeding on culture of catfish. Transm. Am. Fish. Soc., 104, 317-321.
Aoe, H., Y. Noguchi, I. Masuda & Y. Sugisawa (1966), Eel culture with solid feeds I. A rearing experiment in a laboratory tank. Agriculture (Tokyo), 14, 139-146.
Braaten, B.R. (1978), Bioenergetics - a review on methodology. EIFAC/78/Symp: R/4, 1-70.
Brett, J.R. (1973), Energy expenditure of Sockeye salmon, Oncorhynchus nerka, during sustained performance. J. Fish. Res. Can., 30, 1799-1809.
Brown, E.E. (1977) "World Fish Farming: Cultivation and Economics". The AVI Publishing Co., Westport, Connecticut, pp. 307.
Butergaugh, G.L. & H. Willoughby (1967), A feeding guide for brook, brown and rainbow trout. Prog. Fish Cult., 29, 210-215.
Crawford, D.L., D.K. Law, T.B. McKee & J.W. Westgate (1973), Storage and nutritional characteristics of modified Oregon moist rations as an intermediate-moisture product. Prog. Fish Cult., 35, 33-38.
Duel, C.R., D.C. Haskell & A.V. Tunison (1942), The New York State Fish Hatchery feeding chart. Fish. Res. Bull., No. 3, 1-61.
Elliott, J.M. (1976), Energy losses in the waste products of brown trout (Salmo trutta l.) J. Anim. Ecol., 45, 561-580.
Freeman, R.I., D.C. Haskell, D.L. Longacre & E.W. Stiles (1967), Calculations of amounts to feed in trout hatcheries. Prog.

Fish. Cult. 29, 194-209.
Harrington, R.W. Jr. (1954), Contrasting susceptibilities of two fish species to a diet destructive to vitamin B_1. J. Fish. Res. Board Can., 11, 529-534.
Haskel, D.C. (1956), Trout growth in hatcheries. New York Fish and Game Journal, 6, 204-237.
Hepher, B. (1978), Supplementary diets and related problems in fish culture. EIFAC/78/Symp: r/11.3, 1-8.
Hepher, B., E. Sandbank & G. Shelef (1978), Alternative protein sources for warmwater fish diets. EIFAC/78/Symp: R/11.2, 1-29.
Hublou, W.F. (1963), Oregon pellets. Prog. Fish Cult., 25, 175-180.
Ishihara, T., K. Hara, M. Yagi & M. Yasuda (1978), Studies on thiaminase I in marine fish VIII. Thiamine requirement of yellowtail fed anchovy. Bull. Jap. Soc. Sci. Fish., 44, 659-664.
Kurihara, N. (1966), Amounts of food given to common carp reared in net fish cage placed in lake. Aquiculture (Tokyo), 13, 197-203.
Lovell, R.T. (1977), Feeding practices, in: "Nutrition and feeding of channel catfish", R.R. Stickney & R.T. Lovell, eds., Southern Cooperative Series Bulletin No. 218, Auburn Univ., pp. 44-49.
Moriarty, D.J.W. & C.M. Moriarty (1973), The assimilation of carbon from phytoplankton by two herviborous fishes: Tilapia nilotica and Heplochromis migripinnis. J. Zool., Lond., 171, 41-55.
Nagel, T. (1976), Intensive culture of fingerling walleyes on formulated diets. Prog. Fish Cult., 38, 90-91.
Nakagawa, K. (1964), Kokey shiryo ni tsuite (A comment on pelleted feed). Yoshoku, 1(9), 21-27.
Nakamura, T., M. Hagiwara & T. Higuchi (1973), Determination of nutritive value of fish feeds I. DE and DCP of feeds for rainbow trout. Research Annual, Nihon Nosan Kogyo, 4, 35-42.
Page, J.W. & J.W. Andrews (1973), Interaction of dietary levels of protein and energy of channel catfish (Ictalurus punctatus), J. Nutr., 103, 1339-1346.
Peffer, E. (1977), Studies on the utilization of dietary energy and protein by rainbow trout (Salmo gairdneri) fed either by hand or by an automatic feeder. Aquaculture, 10, 97-106.
Phillips, A.M. Jr. (1972), Calorie and energy requirement, in: "Fish Nutrition", J.E. Halver, ed., Academic Press, N.Y. & London, pp. 1-28.
Robinette, H.R. (1977), Feed manufacture, in: "Nutrition and feeding of channel catfish", R.T. Lovell, ed., Southern Cooperative Series Bull. No. 218, Auburn Univ., pp. 44-49.
Snow, J.R. & J.I. Maxwell (1970), Oregon moist pellet as a production ration for largemouth bass. Prog. Fish Cult., 32, 145-151.
Satomi, Y., J. Toi & T. Ito (1971), Some problems of eutrophication in freshwater fish culture pond. Aquiculture (Tokyo), 19, 141-150.
Smith, R.R. (1978), Methods for determination of digestibility and metabolizable energy of feedstuffs for fish. EIFAC/78/Symp:

R/3, 1-12.
Warren, C.E. & C.E. Davis (1967), Laboratory studies on the feeding, bioenergetics and growth of fishes, in: "The biological basis for freshwater fish production", S.D. Gerking, ed., Oxford, Blackwell Scientific Publications, pp. 175-214.
Webber, H.H. & J.E. Huguenin (1978), Fish feeding technologies. EIFAC/78/Symp: R/10.2, 1-35.

PROTEIN AND ENERGY REQUIREMENTS

C.B. Cowey

Institute of Marine Biochemistry

St. Fittich's Road, Aberdeen, U.K.

PROTEIN REQUIREMENT

The minimal dietary protein level giving optimal weight gain was first studied in Chinook salmon Oncocrhynchus tschawytscha (Delong et al., 1958). This type of experiment has since been repeated on many species. Basically the fish must be given a diet containing graded levels of a high quality protein, a sufficient energy density and adequate balances of essential fatty acids, vitamins and minerals over a prolonged period. From the resulting dose/response curve protein requirement is usually obtained by an Almquist plot. The differences in apparent protein requirement are thought to be due to differences in culture techniques and diet composition (Garling and Wilson, 1976). The relatively high dietary protein levels required for maximal growth of certain fish such as grass carp, Ctenopharyngodon idella, and Brycon sp. are surprising as these fish are omnivorous if not actually herbivorous. In pond culture Brycon sp. are grown on unwanted fruit and other plant material of low protein content (St. Paul, 1977). Under these conditions there is presumably a substantial contribution to their protein intake from a natural food chain.

Zeitoun et al. (1973, 1974) have shown that the protein requirement of euryhaline fish (rainbow trout Salmo gaird neri, Coho salmon, Oncorhynchus kisutch) reared in water of salinity 20 ppt is unchanged from the requirement in freshwater. No data yet appears to be available for the protein requirement of these species in full strength sea water.

Zeitoun et al. (1976) introduced the concept of economic protein requirement. They point out that the parabolic curve relating growth to dietary protein level "does not reflect the practically insignificant differences in percentage gain below and beyond the maximum points".

ESSENTIAL AMINO ACIDS

Quantitative Requirement

For many years the only quantitative data on essential amino acid requirements of fish were those of Halver and his colleagues (Mertz, 1972) relating to Chinook salmon. In recent years data on other species have begun to appear, those for Japanese eel, Anguilla Japonica, (the work of Drs Arai and Nose) being almost complete. The test diets used to quantify amino acids requirements have most frequently contained a "protein" component consisting of a small amount of whole proteins (casein, gelatin) and a larger amount of crystalline amino acids. By varying the composition of the amino acid component, diets containing graded levels of each essential amino acids may be elaborated and fed to obtain dose/response curves. A recent innovation has been the use in test diets of proteins relatively deficient in a given essential amino acid.

Kaushik (1977) determined the arginine requirement of rainbow trout from a conventional dose/response (growth) curve and also by measuring the tissue (blood and muscle) levels of free arginine in groups of trout given increasing amounts of dietary arginine.

The known data suggest that real differences exist between fish species in their requirement for certain amino acids. This leads to difficulties in formulating the protein component of practical diets for those species whose amino acids requirements are not yet known.

Supplementation of Proteins with Amino Acids

One solution to the use in practical diets of proteins that are relatively deficient in one or more amino acids is to supplement the protein with appropriate amounts of the amino acid in question. Unfortunately not all fish appear to utilize free amino acids equally well.

Aoe et al. (1970) found that young carp, Cyprinus carpio, were unable to grow on diets in which the protein component (casein, gelatin) was replaced by a mixture of amino acids similar in overall composition. A trypsin hydrolyzate of casein was equally ineffective.

Channel catfish are also unable to utilize free amino acids given as supplements to deficient proteins. Andrews and Page (1974) showed that when soyabean meal was substituted isonitrogenously for menhaden meal, growth and feed efficiency of channel catfish were substantially reduced. Addition of free methionine, cystine or lysine, the most limiting amino acids, to these soya-substituted diets did not enhance weight gain.

Amino Acid Metabolism

It is too early yet to attempt to explain why certain species (rainbow trout) are able to use supplementary free amino acids while other species (channel catfish) are not. The question will only be answered when there is data on the rate of assimilation of amino acids by different species together with data on (i) tissue levels of sub-strates; (ii) tissue levels of amino acid deaminating enzymes and (iii) kinetic data on these enzymes.

Cowey et al. (1974) did not find any significant changes in total hepatic glutamate dehydrogenase, aspartic aminotransferase or alanine aminotransferase activities in the livers of plaice fed high or low protein diets for several weeks. Nagai and Ikeda (1973) were similarly unable to show any effect of dietary protein levels on the activities of these aminotransferase in carp liver.

It is trite to remark that much more data is required before any conclusions may be drawn on the ability of fish to adapt at the cellular level to variations in protein intake.

Tissue levels of amino acids in fish may vary with species but analyses on channel catfish (Wilson and Poe, 1974) show that in liver, gills and kidney they are greater than 1mM. Levels of amino acids in systemic plasma of rainbow trout given a diet containing 600 g crude protein/kg were in the range 0.1-1.0mM and were much higher than concentrations in the plasma of trout given a diet containing 100 g crude protein/kg (Cowey et al., 1977).

NUTRITIONAL VALUE OF PROTEINS

Biological tests of the nutritional quality of proteins for birds and mammals were designed to compare how different food proteins rate in their ability to meet the amino acid requirements of the animal, how the chemically assessed "score" of a protein compares with its limiting amino acid in food proteins and are useful in determining the availability of amino acids especially in proteins that have been subject to extensive processing.

The tests are non-specific in that they attempt to measure the qualitative and quantitative adequacy of ten different compounds at one time: in addition the limiting amino acid will differ in different compounds at one time: in addition the limiting amino acid will differ in different proteins and even in a given protein at different levels of intake. Measurement of nutritional value ought to be made at or near the maintenance level of protein intake where the amino acid composition of the protein ate the limiting feature.

Many reviews of nutritional quality of proteins in mammals are available and the subject has been discussed with references to fish by Cowey and Sargent (1972). The main methodological innovation recently is that of Ogino et al. (1973), that permits separate collection of faecal and metabolic end products. Water from the fish tank is continuously siphoned through a trap, where ifances are retained and precipitated, and hence to a large column (5.5 x 60cm) containing a strongly basic cation exchange resin (Amberlite IR.120H) where cationic substances are retained. These latter substances are eluted from the column with 5% (w/v) HCl at the end of the experiment. The method has the advantage that nitrogen balance can be measured in fish that are not (obviously) stressed, the limit to flow rate being imposed by the size and capacity of the resin column. The method obviously cannot be applied to marine fish and equally any neutral or anionic substances (urea, taurine) are not retained quantitatively by the resin.

Net protein utilization (NPU) at different levels of protein intake has been measured in a few species (Ogino and Saito, 1970; Cowey et al., 1972; Dabrowski, 1977). In all cases NPU falls with increasing dietary protein level; inspection of the data shows that carp and grass carp given casein at the rate of 400 g/kg retained about 35-40% of it in their tissues, plaice given the same dietary level of freeze dried cod muscle retained 45% of it. These values are much lower than those of omnivorous birds and mammals (efficiency protein utilization approximately 60%) which of course grow maximally at much lower dietary protein levels.

INTERACTION OF PROTEIN
WITH OTHER ENERGY SOURCES

It is long known that fish utilize available energy efficiently. The factors involved have only recently begun to be explored in depth. Metabolic rate influenced by a large number of factors as species, temperature, body size, activity, water chemistry (O_2, CO_2, pH, salinity) and so on.

The optimum temperature for growth for example should be that temperature at which the difference between voluntary energy intake and the energy necessary for maintenance is at a maximum.

The relationship between body size and metabolic functions is of special importance to fish production. It is seldom possible to geed fish ad libitum. Uneaten food is carried away by water or disintegrates and is unavailable. Therefore, feeding rate must be adjusted so that small fish receive enough to meet their needs for rapid growth and large fish are not overfed.

Fish grow from 150 milligrams to 500 grams as move at marketing a 3.000 fold increase.

Smith and his colleagues went on to measure heat increment (specific dynamic action) associated with dietary protein, fat, carbohydrate and complete diets in salmonids. They showed heat increment of protein much lower with fish than mammals amounting to less than 5% of ingested metabolizable energy. That for birds and mammals might be as much as 30%. Including the energy costs of all phases of production including the cost of reproduction and mortality, it is apparent that fish and among the most efficient of all animals in converting feed into high quality protein.

BIBLIOGRAPHY

Andrews, J.W. & J.W. Page, Growth factors in the fishmeal component of catfish diets. J. Nutr., 104: 1091-1096, 1974.
Aoe, H. et al., Nutrition of protein in young carp - I - Nutritive value of free amino acids. Bull. Jap. Soc. Sci. Fish., 36: 407-413, 1970.
Cowey, C.B. & J.R. Sargent, Fish nutrition. Adv. Mar. Biol. 10: 383-492, 1972.
Cowey, C.B. et al., Studies on the nutrition of marine flatfish. The protein requirements of plaice (Pleuronectes platessa), Br. J. Nutr. 28: 447-456, 1972.
Cowey, C.B. et al., Studies on the Nutrition of marine flatfish. The effect of dietary protein content on certain cell components and enzymes in the liver of Pleuronectes platessa. Mar. Biol. 28: 207-213, 1974.
Cowey, C.B. et al., The regulation of gluconeogenesis by diet and insulin in rainbow trout (Salmo gairdneri). Br. J. Nutr. 38: 463-470, 1977.
Dabrowski, K. & T. Wojno. Studies on the utilization by rainbow trout (Salmo gairdneri) of feed mixtures containing soya bean meal and an addition of amino acids. Aquaculture, 10:297-310, 1977.
Delong, D.C., J.E. Halver & E.T. Mertz. Nutrition of salmonid fishes: VI Protein requirements of chinook salmon at two water temperatures. J. Nutr., 65: 589-599.
Halver, J.E., Nutrition of salmonid fishes - IV - An amino acid test for chinook salmon. J. Nutr., 62: 245-254, 1957.
Kaushik, S.J., Influence de la salinité sur le metabolisme azoté et

le besoin en arginine chez la truite arc-en-ciel. Thèse. L'Université de Bretagne Occidentale: 230 p.

Mertz, E.T., The protein and amino acid needs, in: "Fish Nutrition" J.E. Halver, ed. Academic Press, New York, 106-143.

Nose, T., Determination of nutritive value of food protein in fish. Effect of amino acid composition of high protein diets on growth and protein utilization of the rainbow trout. Bull. Freshwater Fish. Res. Lab. Tokyo, 13(1): 41-50, 1963.

Ogino, C., Kanino & M.S. Chen., Protein nutrition of fish - II - Determination of metabolic faecal nitrogen and endogenous nitrogen excretion of carp. Bull. Jap. Soc. Sci. Fish., 39: 519-523, 1973.

Ogino, C. & K. Saito, Protein nutrition in fish - I. The utilization of dietary protein by young carp. Bull. Jap. Soc. Sci. Fish. 36: 250-254, 1970.

Saint-Paul, U. Aspectos generales sobre la piscicultura en Amazonas y resultados preliminares de experimentos de alimentación con raciones peletizados con diferentes composiciones. I Simposio de la Associacion Latino-americana de Acuacultura, Maracay, Venezuela, Nov./1977.

Wilson, R.P. & W.E. Poe. Nitrogen metabolism in channel catfish, Ictalurus punctatus, II - Relative pool sizes of free aminoacids and related compounds in various tissues of the catfish. Comp. Biochem. Physiol. 48B: 545-556.

Zeitoun, I.H. et al., Influence of salinity on protein requirements of rainbow trout (Salmo gairdneri) figerlings. J. Fish. Res. Bd. Can. 30: 1867-1873.

Zeitoun, J.H. et al., Influence of salinity on protein requirements of coho salmon (Oncorhynchus ksutch) smolts. J. Fish. Res. Bd. Can., 31: 1145-1148.

Zeitoun, J.H. et al., Quantifying nutrient requirements of fish. J. Fish. Res. Bd. Can. 33: 167-172. 1976.

VITAMIN, FAT AND MINERAL REQUIREMENTS

John Emil Halver

Associate Professor - College of Fisheries

University of Washington - Seattle - U.S.A.

Vitamin deficiency diseases have been described for many species of fish. Test diets have been developed which can be used to induce specific water soluble or fat soluble vitamin deficiency symptoms in salmonids, intalurids, cyprinids, els and several marine fishes. The water soluble vitamins function in fish metabolism as in homeothermal metabolism. The primary difference is that reaction rates of the enzymes involved differ in poikilotermic animals reared at different water temperatures. The water soluble vitamins are part of coenzyme systems in the same fundamental biochemical pathways as in other animals. Thus, deficiencies induce specific metabolic system failures and resultant deficiency signs. These specific deficiency syndromes have been described and compared with those found in other experimental animals.

Tentative requirements in terms of milligrams of vitamins per kilogram of dry diet have been determined for several species of fish. Sparing effect of one vitamin precursor on other has not been throughly investigated and specific differences in requirements of fish on different carbohydrate or fat intake and different protein intake have not been adequately measured. The requirement for thiamin is related to the carbohydrate intake in most animals studied. This vitamin may be particularly important for certain carbohydrate consuming species of fish such as the grass fishes, the mulleto and many others that eat phytoplankton or other plants. Likewise, the requirement for pyridoxine is related to the level in the ration; consequently pyridoxine is related to the level in the ration; consequently pyridoxine is one of the fish vitamins to be exhausted in carnivorous fish fed low pyridoxine diets.

Coho salmon shows acute symptoms of scoliosis and lordosis when fed deficient vitamin C diets. The ascorbic acid intake required to promote the most rapid growth in long-term experiments had to be increased threefold to fivefold to promote the most rapid wound repair when the fish were stressed with an incision through the skin into the abdominal cavity or into the musculature. Ascorbic-2-sulphate has been isolated as a major metabolite in fish tissues and may represent the chemically stable storage form of vitamin C for fish. Enzyme systems have been isolated which convert ascorbic acid into ascorbic-2-sulphate (vitamin C_2) and to hydrolyze C_2 back to L-dihydro ascorbic acid (vitamin C_1).

Both forms of vitamin C have equimolar vitamin activity to support growth and to prevent the onset of scurvy.

Requirements for that soluble vitamins have been less discretely defined than those for the water soluble vitamins. Tissue storage of fat soluble vitamins is cumulative and hypervitaminosis can be expected when large amounts of fat soluble vitamins are fed to fish for long periods. Typical deficiency symptoms occur when these components are lacking in the diet. Vitamin D is not required in diet supplements when fish are reared in water containing some calcium, but vitamin D_3 is required when fish are reared in water systems containing little or no calcium ions. Hypervitaminosis A and D have been reported. Acute liver toxicity, growth inhibition, poor food conversion and elevated alkaline phosphatase activity in the blood occurs. Therefore, fish, like other experimental animals, have specific qualitative requirements for the fat soluble vitamins. These requirements can be altered by the particular species of fish reared with size, with water temperatures and with environmental and dietary conditions encountered.

FATS AND ESSENTIAL FATTY ACIDS

Fats are a ready source of energy for fish. Most studies indicate that fish can use 20% to 30% of the dry diet ingredients as fat, provided adequate amounts of choline, methionine, and tocopherol are present in the ration. Fats have the distinct advantage of being almost completely digestible. Fish appear to be designed to metabolize fats efficiently as an energy source with a concomitant sparing effect on the protein requirement for maximum growth. The neutral lipid component of the fish rations is therefore a useful component for diet formulation and is especially desirable in feeds of fry or fingerlings which require high energy intake for rapid growth. This asset is not without constraint, however, because some fish such as rainbow trout, salmon, plaice and sea bream do not have sufficient ability to elongate short chain fatty acids and then unsaturate the carbon chain to convert simple animal or vegetable fats into polyunsaturated long chain fatty acids at the levels found in fish tissue lipids.

VITAMIN, FAT AND MINERAL REQUIREMENTS

The polyunsaturated fatty acids which are essential for good fish growth and normal cell functions have concomitant requirements for tocopherols to prevent fatty acid peroxidation, chain rupture, and subsequent toxic reactions in the liver and spleen. Fats form an important part of fish rations for energy and in addition specific polyunsaturated fatty acids are required for specialized cell structure and organ functions.

MINERALS

Very little is known about specific trace mineral requirements of fish for these enzyme activators and metabolic system structural components. Macro-mineral requirements have been studied for a few species under carefully controlled experimental conditions. Calcium can be sequestered from the water through the gill tissues in salmonids. Also, calcium can be absorbed through the gut when adequate amounts of vitamin D_3 are present in the ration.

Phosphorus is absorbed through gut tissues. Calcium:phosphate balance studies have indicated that many fish diets contain inadequate amounts of phosphate.

The particular form of phosphate must be analyzed in diet ingredients because of the unavailability of phytic salts which are present in many agricultural commodities. Sodium:potassium balance is specially important for fish in a sea environment or in fish undergoing changes converting from fresh water to sea water environments. Acceptable ranges are 1:2→2:1. Diets should contain 3-5 g calcium and 3-5 g phosphorous/kg diet.

Sodium and potassium also need to be balanced in diets fed to fish in fresh water. Ranges should be 1-3 g for each kg diet. Higher levels of either elements induce impaired metabolism and poor growth.

Magnesium is essential for proper intermediary metabolism and is especially important for carbohydrate catabolism. Requirements are about 300-500 mg/kg diet.

One of the first recorded trace mineral deficiencies in trout was described over 60 years ago when fish were reared on diets deficient in iodide. Acute proliferation of thyroid tissue, a compensatory reaction occurred, and the fish developed typical goiter. Intermediate stage thyroid proliferation can be reduced with adequate intake of iodide. Requirements are 100-300 µg/kg diets.

Cobalt appears to be present in the form of organically bound vitamin B_{12}. Cobalt may, however, stimulate the production of B_{12} by gut bacteria and consequently small amounts of cobalt are included in fish diets.

Iron, zinc, copper, manganese and many other classical trace mineral elements are also required to activate several enzyme systems in actively growing fish. As a result trace mineral supplements are included in most diet formulations even though specific quantitative requirements for these elements have not been cataloged.

BIBLIOGRAPHY

Halver. J.E., 1972, "Fish Nutrition", Academic Press, New York, 713 p.

VITAMIN REQUIREMENTS OF FINFISH

John E. Halver

Senior Scientist in Nutrition
U.S. Fish and Wildlife Service
Seattle, Washington, U.S.A.

1. HISTORICAL INTRODUCTION

One of the first reports of a specific vitamin deficiency in fish was made by Schneberger at the Thunder River Hatchery in Wisconsin in 1941 when he reported that paralysis in rainbow trout (Salmo gairdneri) which had been fed fresh carp tissue (Cyprinus carpio) was cured when crystalline thiamin was injected into individual fish (Schneberger, 1941). Fish diet disease was reported by Wolf (1942) to be due to thiaminase present in fish tissues which would hydrolyze thiamin in the meat diets fed to fish in hatcheries. Tunison et al. (1942) measured levels of thiamin, riboflavin, and nicotinic acid in the liver, kidney, caeca, and muscle tissue, to establish base lines for requirements of trout for these vitamins. Dietary gill disease could be reduced by incorporating fresh liver or dried yeast in the diet (Wolf, 1945). McCay and Tunison (1934) observed scoliosis and lordosis in brook trout (Salvelinus fontinalis) which had been fed formaldehyde preserved meat, but the symptoms took almost one year to develop and the disease was not correlated with the recently identified vitamin C. Many fish disease reports during the 1940's inferred that dietary deficiencies may have caused or potentiated the symptoms of a specific pathogen.

Typical avitaminosis symptoms of paralysis, convulsions, scoliosis, anemia, cataracts, gill hyperplasia, hemorrhage, anorexia, poor growth and high mortality were reported whenever fish were concentrated and intensive fish husbandry practices used (Halver, 1972). Diets manufactured from agricultural commodities often failed to contain sufficient vitamins to prevent the onset of anemia and diet disease symptoms were common, but specific cause and effect relationships were difficult to define. Hypervitaminosis A and D were reported when large quantities of whale liver or seal oil were fed to young salmon (Burrows et al., 1952), but specific studies to define requirements and acceptable levels for fat soluble vitamins in fish diets were not possible because of the lack of adequate test diets to exercise proper experimental control over the nutrients to be assayed.

2. TEST DIETS FOR VITAMIN REQUIREMENT STUDIES

Barbara McLaren published a vitamin test diet which was used on rainbow trout in 1947 (McLaren et al., 1947a). However, this diet contained crab meal to furnish the anti-anemic factor. As a result, long-term low level vitamin feeding studies with this diet did not establish several specific vitamin deficiency symptoms in trout (McLaren et al., 1947b). Halver, Norris and Donaldson in 1949 showed that the anti-anemic factor H of McCay and Tunison was a combination of vitamin B_{12} and folic acid. Chinook salmon (Oncorhynchus tshawytscha) made anemic by feeding spawned-out salmon carcasses, were cured when a combination of folic acid and vitamin B_{12} was injected into the fish (Halver, 1953a).

Xanthopterin, a colored pigment extracted from insects gave only a partial response; folic acid, injected alone, a partial response; vitamin B_{12}, alone, a partial response; but folic acid plus vitamin B_{12} promoted erythropoiesis and soon a normal blood picture occurred in the fish (Halver, 1953a). This formed the basis for the development of vitamin test diets to measure qualitative and then quantitative vitamin requirements of salmonids. Wolf in 1950, developed a vitamin test diet made of commercial casein, gelatin, starch, vegetable oils, cod liver oil, minerals and crystalline vitamins which raised rainbow trout for 24 weeks. He reported qualitative requirements for 8 of 12 crystalline vitamins included in the test diet components (Wolf, 1951).

A complete vitamin test diet for chinook salmon was developed in 1951. This diet could be used to rear chinook salmon for at least 24 weeks without the appearance of any vitamin deficiency syndrome (Halver, 1957). By deleting one crystalline vitamin at a time from the diet, specific vitamin deficiency syndromes did appear and were described in detail when these deficient rations were fed to chinook salmon, coho salmon (O. Kisutch), sockeye salmon (O. nerka), rainbow trout and cutthroat trout (S. clarkii). The levels of several components were adjusted to conform to protein requirements of young salmonids and the improved vitamin test diet (Table 1) was used successfully to rear several other species of fish (Halver, 1972, 1975).

TABLE 1

WATER-SOLUBLE VITAMIN TEST DIET H-440[a]

COMPLETE TEST DIET (gm)		VITAMIN MIX (mg)		MINERAL MIX (mg)	
Vitamin-free casein	38	Thiamin –HCl	5	USP XII No. 2	plus
Gelatin	12	Riboflavin	20	$AlCl_3$	15
Corn oil	6	Pyridoxine-HCl	5	$ZnSO_4$	300
Cod liver oil	3	Choline chloride	500	CuCl	10
White dextrin	28	Nicotinic acid	75	$MnSO_4$	80
a-Cellulose mixture[b]	9	Calcium pantothenate	50	KI	15
a-Cellulose	8	Inositol	200	$CoCl_2$	100
Vitamins	1	Biotin	0.5	per 100 gm of salt mixture	
	9	Folic acid	1.5		
Mineral Mix	4	L-Ascorbic acid	100		
Water	200	Vitamin B_{12}[d]	0.01		
Total diet as fed	300	Menadione (K)	4		
		a-Tocopherol[c] acetate (E)	40		

[a] Diet preparation: Dissolve gelatin in cold water. Heat with stirring on water bath to 80°C. Remove from heat. Add with stirring – dextrin, casein, minerals, oils, and vitamins as temperature decreases. Mix well to 40°C. Pour into containers; move to refrigerator to harden. Remove from trays and store in sealed containers in refrigerator until used. Consistency of diet adjusted by amount of water in final mix and length and strength of beating.

[b] Delete 2 parts a-cellulose and add 2 parts CMC for preliminary feeding.

[c] Dissolve a-tocopherol in oil mix.

[d] Add vitamin B_{12} in water during final mixing.

3. WATER SOLUBLE VITAMINS

Thiamin

Thiamin deficiency caused neuritis, paralysis and growth of fish slowed or stopped within a few weeks (see Table 2). Violent fits were induced in deficient fish by shocking the container. Recovery was rapid when the missing vitamin was replaced in the ration (Schneberger, 1941; Wolf, 1942; McLaren et al., 1947a; Halver, 1953, 1957; Dupree, 1966; Coates and Halver, 1958; Shellbourne, 1970).

A trunk winding symptom occurred in eels (Anguilla japonica) (Hashimoto et al., 1970). Skin congestion and subcutaneous hemorrhage was reported in carp (Aoe et al., 1969) and in red sea bream (Alvamis, sp.; Yone, 1974).

Riboflavin

Riboflavin deficiency produced cataracts which caused blindness in 12-14 weeks, and the fish grew slowly. The cataracts were irreversible when riboflavin was added to diet and only fish with adequate vision recovered (Halver, 1957; Coates and Halver, 1958; Halver, 1972; NAS/NRC, 1973, 1977).

Pyridoxine

Pyridoxine deficiency caused rapid mortality with most fish exhibiting violent epileptic-type seizures and rapid rigor mortis. Recovery was equally rapid and growth resumed when pyridoxine was added to the diet (Halver, 1972; NAS/NRC, 1973, 1977).

Pantothenic Acid

Pantothenic acid deficiency caused ataxia after fish and continued to exhibit good growth for approximately four weeks. Gill membranes began to proliferate and mat together, and a typical pantothenic acid deficiency symptom occurred in all species of fish tested (Halver, 1972; NAS/NRC 1973, 1977).

Folic Acid

Folic acid deficiency caused macrocytic normochromic anemia, although the growth of the test fish was nearly identical to the control for the first 8-10 weeks on test. Coho salmon seemed to be more sensitive to folic acid deficiency than chinook salmon or rain-

TABLE 2

VITAMIN DEFICIENCY SIGNS IN FISH

NUTRIENT	FISH
Thiamine	Anemia, anorexia, ataxia (terminal), convulsions (when moribund), corneal opacities, degeneration of vestibular nerve nucleus, fatty liver, hemorrhage of midbrain or medulla, loss of equilibrium, melanosis in older fish, muscle atrophy, paralysis of D and P fins, rolling, whirling motion, vascular degeneration, weakness.
Riboflavin	Anorexia, cloudy lens, darkened skin, dim vision, discolored iris, hemorrage in eyes, nares or operculum, incoordination, lens cataract, mortalities, photophobia, xerophthalmia.
Pyrodoxine	Anorexia, ascites, ataxia, convulsions, flexing of opercles, hyperrirritability, indifference to light, microcytic, hypochromic anemia, rapid, jerky breathing, rapid onset of rigor mortis, spasms, weight loss, nervous disorders, increased mortalities.
Folic acid	Anemia, anorexia, ascites, dark coloration, erythropenia, exophthalmia, fragility of caudal fin, lethargy, macrocytic anemia, pale gills, poor growth.
Pantothenic	Anorexia, clubbed gills, flared opercula, gill exudate, general "mumpy" appearance, lethargy, necrosis of jaw, barbels, and fins, prostration, poor weight gain.
Inositol	Anemia, bloated stomach, poor growth, anorexia, skin lesions.
Biotin	Anemia, anorexia, blue slime disease, colonic lesions, contracted caudal fins, dark coloration, erythrocyte fragmentation, mortalities, muscle atrophy, poor growth, spastic convulsions.

TABLE 2 (continued)

VITAMIN DEFICIENCY SIGNS IN FISH

NUTRIENT	FISH
Choline	Anemia, poor food conversion poor growth, vascular stasis and hemorrhage in kidney and intestine.
Nicotaninc acid (niacin)	Anemia, anorexia, colonic lesions, edema of stomach and colon, incoordination, jerky movements, muscle spasms, lethargy, photophobia, swollen gills, tetany, skin hemorrhage, high mortality.
p-Amino-benzoic acid	No significant change in growth, appetite or survival.
Cobalamin (B_{12})	Anorexia, erratic hemoglobin and erythrocyte counts, fragmentation.
Ascorbic acid	Anorexia, impaired collagen production, impaired wound healing, lordosis with dislocated vertebrae and focal hemorrhage, poor growth, scoliosis with hemorrhage in severe cases, twisted deformed hyaline cartilage in gill filaments and sclera of eyes.
Vitamin A	Ascites, edema, exophthalmos, hemorrhagic kidneys.
Vitamin D	Reduced conversion.
Vitamin E (α-tocopherol)	Anemia, ascites, ceroid in liver, spleen, kidney, clubbed gills, epicarditis, exophthalmia, microcytic anemia, mortalities, pericardial edema, poor growth, red blood cell fragility.
Vitamin K	Anemia, coagulation time prolonged.

Adapted from: Ashley, L.E. in FISH NUTRITION p. 442-445, 1972.

bow trout and soon abnormal blood cell morphology could be observed in either blood smears or anterior kidney imprints from test fish. In contrast, salmon receiving a complete test diet or anemic fish placed on a recovery diet containing folic acid recovered rapidly and soon displayed normal blood cells in either smears or anterior kidney imprints (Halver, 1972; NAS/NRC, 1973, 1977).

Ascorbic Acid

No ascorbic acid deficiency symptoms developed in chinook salmon after 16-20 weeks on test, and growth paralleled that of fish fed the complete test diet (Halver, 1957). When these same ascorbic acid deficiency diets were fed to coho salmon for 24-36 weeks, however, acute scoliosis and lordosis appeared (Halver et al., 1969). X-rays of spinal deformities in the affected fish showed extreme dislocation of vertebrae, and Hematoxylin-Eosin stained sections of the spinal column in the area of the fracture dislocation showed displacement of neraly 120° in some fish and indicated atrophy of the spinal cord in the area of the acute deformity. A normal gill arch stained with Mason's stain shows well organized cartilage, bone and support tissue; whereas, distorted and twisted filamentous cartilage in ascorbic acid deficient fish after only 12-14 weeks on test. Rainbow trout and brook trout raised on diets devoid of ascorbic acid developed acute lordosis and scoliosis, similar to the acute deformities observed in young carp. guppies (Poecilia reticulata) and cherry salmon (O. masou) (Kitamura, 1963). A broken back syndrome appeared in groups of channel catfish fed inadequate amount of vitamin C (Lovell, 1973).

Choline

Choline deficiency in chinook salmon and coho salmon was manifest by slow growth and poor diet efficiency. Test fish were slow to recover after choline was added to the ration and increased gastric emptying time was observed in famine. The lesions in the colon reported by others working with trout were only occasionally observed in a few chinook and coho salmon held on choline deficiency for 12-16 weeks. Similar observations were reported in choline deficient lake trout, red sea stream and channel catfish (Halver, 1972; NAS/NRC, 1973, 1977).

Niacin

Niacin deficiency generated nonspecific symptoms in young fish. Growth was inhibited only after 12-14 weeks on experiment and some test fish appeared to suffer from muscle tetany and edema of the stomach and colon (Halver, 1972; NAS/NRC, 1973, 1977).

Biotin

Biotin deficiency in chinook or coho salmon showed no evidence of the "blue slime patch" disease previously reported for brook, brown and rainbow trout by other workers (Halver, 1957). Brook trout appear most prone to "blue slime patch" disease with brown trout and rainbow trout more resistant (Phillips et al., 1949, 1950). Similar symptoms have been reported for other fish (Halver, 1972; NAS/NRC, 1973, 1977).

Vitamin B_{12}

Cyanocobalomin deficiency produced macrocytic anemia in chinook and coho salmon and in rainbow and brook trout raised on vitamin B_{12} deficient diets. Poor growth have been reported in several other species of fish but specific deficiency signs have been lacking (Halver, 1972; NAS/NRC, 1973, 1977).

Inositol

Inositol deficiency inhibited growth of fish, and test fish exhibited some general nonspecific deficiency symptoms after 10-12 weeks (Halver, 1957). Experiments with rainbow trout were inconclusive for specific deficiency symptoms although fish showed inhibition in growth and poor food conversion on inositol deficient diet (McLaren et al., 1947a; Halver, 1972). Other fish tested on inositol deficient diets have responded similarly (NAS/NRC 1973, 1977).

Para-aminobenzoic Acid

Para-aminobenzoic acid deficiency signs have not been induced in young salmonids or in young cyprinids when diets devoid of this vitamin were fed (Halver, 1957; Halver and Coates, 1958; Aoe and Masuda, 1967; NAS/NRC, 1973, 1977).

QUANTITATIVE WATER SOLUBLE VITAMIN REQUIREMENTS

The logical sequence to the qualitative vitamin requirement study was determination of quantitative needs of fish for each of these required vitamins. Replicate groups of fish were fed different levels of each vitamin, with an excess of all other required vitamins in the test ration. Test fish were fed for a long experimental period, and representative random samples were selected from each lot on each diet treatment. Fish were analyzed for vitamins stored in the liver to complement the data on growth, food conversion and mortality (Halver, 1966, 1972).

The diet treatment resulting in maximum liver storage coupled with normal hematology, normal histology of test tissues, low mortality, good growth and good food was selected as the vitamin level meeting the requirement. Tentative vitamin requirements of chinook salmon fingerlings in terms of mg of vitamin per kilogram body weight per day were calculated to be: thiamin, .13-.20; riboflavin, .75-1.00; pyridoxine, .38-.43; choline, 50-60; niacin, 5-7; pantothenic acid, 1.3-2.0; inositol, 18-20; biotin, 0.03-0.04; folic acid, 0.1-0.15; vitamin B_{12}, 0.0002-0.0003; and ascorbic acid, 2-3.

Some of these values for chinook salmon were similar to those reported by Phillips and McLaren in studies with rainbow, brook and brown trout (S. trutta). The sparing effect of one vitamin or vitamin precursor on others was not tested which may account for the relatively high values reported for chinook salmon when compared with those listed for trout. Other possibilities should also be considered, however, such as the more soluble nature of these test diets used compared with the meat-meal mixtures used to develop vitamin requirement studies in trout, and, of course, the obvious implication that Pacific salmon, being of a different genus and having different environmental requirements and growth characteristics, may have a higher specific vitamin requirement than the trout.

This is not always the case observed, however, because in the brown trout study for biotin requirements, the apparent requirement of brown trout was approximately twice that of brook trout and rainbow trout. Likewise, in the ascorbic acid studies between coho salmon and rainbow trout, the rainbow trout appeared to have a requirement for ascorbic acid approximately twice that of the coho salmon.

The growth rate of young coho salmon or rainbow trout fed diets containing intermediate or higher levels of ascorbic acid in the ration were identical. However, when fish were wounded with a 1 cm incision into the musculature, wound repair was slow in trout fed diets containing less than 400 mg of ascorbic acid/kg dry diet and the most rapid wound repair was not observed until fish received at least 1 g of L-ascorbic acid/kg of ration. Coho salmon also showed inhibition in wound repair on diets containing less than 400 mg of ascorbic acid/kg ration and the most rapid repair was observed at higher ascorbic acid intake. The rate of wound repair was directly related to the ascorbic acid content of the ration and varied from no collagen synthesis in trout or salmon fed the diets devoid of vitamin C, slow wound repair at levels of 50 or 100 mg C/kg ration, intermediate repair at 200-400 mg C/kg diet up to rapid repair in fish fed the complete test diet containing 1 g of L-ascorbic acid/kg dry diet ingredients. Blood levels of ascorbic acid did reflect the diet treatment, but ascorbate stores in the anterior kidney tissue proved to be a better indicator of ascorbic acid status of both coho salmon and rainbow trout.

Most of the values listed for the vitamin requirements in salmonides have been calculated based upon the diet level which would result in the maximum liver tissue storage of the vitamin commensurate with good growth and diet conversion. More exacting clinical biochemistry techniques have been applied to fish enzyme assays and the level of a particular vitamin in the dietary which would saturate the enzyme complex and produce maximum activity has been used to estimate the specific vitamin requirement for that species (Cowey et al., 1975; Murai and Andrews, 1978).

These requirements were based on the premise that the dietary intake sufficient to promote maximum liver tissue storage should satisfy tissue demands of the vitamin for maximum growth, resistance to disease, maturation and normal stress encountered in the environment. The growth studies with coho salmon and rainbow trout in the ascorbic acid studies have shown that maximum growth rate can be obtained and maximum tissue storage can be obtained at 1/10-1/15 the dietary intake needed when the animals are exposed to massive stress or trauma. Thus, tentative vitamin requirements for fish must be related to fish size, species, environment and the physiological stress encountered. Fortunately, liver and tissue storage levels for most of the water-soluble vitamins are extensive and should be adequate to support normal growth and physiological function for several weeks. Therefore, the vitamin content of the ration need not be guaranteed on a daily intake basis provided adequate intake for replacement storage is encountered at short periods during the growing period.

Quantitative water soluble vitamin requirements for other species of fish have been more recently investigated. Aoe and Ogino have catalogued the needs of young carp for thiamin (Aoe et al., 1967, 1969), riboflavin (Aoe et al., 1967; Ogino, 1967), niacin (Aoe et al., 1967), pantothenic acid (Aoe et al., 1971; Ogino, 1967), biotin (Ogino, 1970), folic acid (Aoe et al., 1967), pyridoxine (Ogino, 1965), and choline (Ogino et al., 1970). Hashimoto and Nose have been concentrated on the requirements of the eel (Hashimoto, 1975; NAS/NRC, 1977), Yone has worked with sea bream (Yone et al., 1971, 1974) and Cowey with marine flatfish (Cowey et al., 1975). Quantitative requirements for the different species tested are listed in Table 3.

5. FAT SOLUBLE VITAMINS

Fat-soluble vitamin studies have been conducted with chinook and coho salmon, rainbow, brown and brook trout and several other species of fish but the results have been less discrete than in studies with

TABLE 3
VITAMIN REQUIREMENTS FOR GROWTH[a]

VITAMIN (mg/kg dry diet)	RAINBOW TROUT	BROOK TROUT	BROWN TROUT	ATLANTIC SALMON	CHINOOK SALMON	COHO SALMON
Thiamin	10–12	10–12	10–12	10–15	10–15	10–15
Riboflavin	20–30	20–30	20–30	5–10	20–25	20–25
Pyridoxine	10–15	10–15	10–15	10–15	15–20	15–20
Pantothenate	40–50	40–50	40–50	R[b]	40–50	40–50
Niacin	120–150	120–150	120–150	R	150–200	150–200
Folacin	6–10	6–10	6–10	5–10	6–10	6–10
Cyanocobalamin	R	R	R	R	0.015–0.02	0.015–0.02
myo-Inositol	200–300	R	R	R	300–400	300–400
Choline	R	R	R	R	600–800	600–800
Biotin	1–1.5	1–1.5	1.5–2		1–1.5	1–1.5
Ascorbate	100–150	R	R	R	100–150	50–80
Vitamin A	2000–2500 IU	R	R		R	R
Vitamin E[c]	R	R	R		40–50	R
Vitamin K	R	R	R		R	R

[a] Fish fed at reference temperature with diets at about protein requirement.
[b] R = required
[c] Requirement directly affected by amount and type of unsaturated fat fed.

TABLE 3 (continued)

VITAMIN REQUIREMENTS FOR GROWTH[a]

VITAMIN (mg/kg dry diet)	CARP	CHANNEL CATFISH	EEL	SEA BREAM	TURBOT	YELLOW-TAIL
Thiamin	2-3	1-3	2-5	R[b]	2-4	R
Riboflavin	7-10	R	R	R	R	R
Pyridoxine	5-10	R	R	2-5	R	R
Pantothenate	30-40	25-50	R	R	R	R
Niacin	30-50	R		R		
Folacin		R	R	R	R	
Cyanocobalamin		R				
myo-Inositol	200-300	R		300-500		
Choline	500-600	R		R	R	
Biotin	1-15	R	R		R	R
Ascorbate	30-50	30-50		R		R
Vitamin A	1000-2000 IU	R				R
Vitamin E[c]	80-100	R				R
Vitamin K	R	R				

[a] Fish fed at reference temperature with diets at about protein requirement.
[b] R = required
[c] Requirement direct affected by amount and type of unsaturated fat fed

Adapted from Halver, J.E. in FISH NUTRITION, 1972, p. 39; NAS/NRC 1973; 1977.

the water-soluble vitamins. Tissue storage of the fat-soluble vitamins is accumulative; hence, hypervitaminosis symptoms can be expected to be encountered when high intake of dry ingredients containing large amounts of the fat-soluble vitamins are fed for long feeding period to fish. Qualitative fat-soluble vitamin experiments have disclosed that vitamins A, K and E are required for normal growth and maturity of chinook and coho salmon, rainbow and brook trout (see Table 2).

Vitamin A

Vitamin A deficiency symptoms indicate photophobia, dim vision, cataracts of the eyes, poor growth, abnormal cartilage and impaired bone development. Hypervitaminosis A involved abnormal bone and cartilage formation, ankylosis, poor growth and liver toxicity (Halver, 1972; NAS/NRC, 1973, 1977).

Vitamin K

Vitamin K deficiency induces extended blood clotting time, hemorrhage of the tissues and slow tissue repair from contusions. Prothrombin time in fish fed diets devoid of vitamin K was increased 3-5 times over control fish (Halver, 1972; NAS/NRC, 1973, 1977).

Tocopherols

Vitamin E deficiency in salmon and trout involved increased capillary fragility, ceroid in the liver and spleen, hemorrhage in the eye and in acute cases of extended deficiency, macrocytic anemia and liver toxicity (Halver, 1972).

Similar symptoms were described for tocopherol deficient carp (Watanabe et al., 1970; Aoe et al., 1972; Hashimoto et al., 1966), and channel catfish (Murai and Andrews, 1974; Dupree, 1969).

Hypervitaminosis E can be induced but only at very high intake for long time periods and is manifest by poor growth, poor food conversion and liver toxicity (Poston, 1965).

Vitamin D

Vitamin D has not been shown to be required for normal growth and function of salmonids. Podoliak and others (Phillips et al., 1954) have shown that trout and salmon may sequester calcium from the water through the gill membranes into the blood stream to satisfy the calcium requirements for growth and normal physiological function.

Vitamin D requirements in mammals are associated with absorption of calcium in the gut. Therefore, the major requirements for vitamin D in the diet of fish may be circumvented by ability of the fish to sequester calcium ions from the water environment. Consequently, long term studies with very carefully defined environments and vitamin D dietary content must be conducted before intracellular-intercellular vitamin D function and requirement for fish is understood. Hypervitaminosis D has been reported to occur in salmon fed diets containing large amounts of fish liver oils and appears as acute liver toxicity, growth inhibitions, poor food conversion and elevated alkaline phosphatase activity in the blood (Poston, 1968).

Lipoic Acid

Lipoic acid deficiency signs have not been developed in any test group of fish fed vitamin test diets. Metabolism of lipoic acid in fish tissues has not been adequately studied.

6. QUANTITATIVE FAT-SOLUBLE VITAMIN REQUIREMENTS

Amounts of fat-soluble vitamins recommended for complete diets for normal growth of fish are listed in Table 3. These amounts do not allow for processing or storage losses. The interrelationships between vitamins used and other nutrients or feedstuffs in the ration have not been adequately explored.

Vitamins A and A_2 can be interconverted by trout and catfish to supply the vitamin A requirement for the species (Braekkan et al., 1969; Dupree, 1970).

Vitamin D requirements can not be demonstrated except in very low ionic strength waters, when calcium must be absorbed through the intestinal tract rather than through the gill tissues (Poston, 1968).

More work has been completed on the tocopheral requirements of fish, since this vitamin is involved in maintaining the integrity of the long chain polyunsaturated fatty acids incorporated into phospholipid membranes (Halver, 1972). Vitamin E and selenium interrelationships have been tested in Atlantic salmon (S. salar) (Poston et al., 1976), and the effects of oxidized oils and other antioxidants have been explored in catfish (Murai and Andrews, 1974) and carp (Watanabe et al., 1970, 1975a, 1975b; Hashimoto et al., 1966).

Water soluble analogues of menaquinone have been studied for vitamin K activity in fish (Poston, 1976), and results confirm the assumption that several vitamins K may be used to promote synthesis of blood clotting proteins (Poston, 1964; Halver, 1972).

7. SUMMARY AND CONCLUSIONS

Most fish tested have requirements for 11 water-soluble vitamins and for at least 3 of the 4 fat-soluble vitamins. The requirement for para-aminobenzoic acid has not been demonstrated in long term feeding studies for fish. Likewise, the specific role of lipoic acid in salmonid diets has not been demonstrated. These diets used in the most controlled studies contained highly purified ingredients and produced growth in test fish populations at about 90% of that obtained with commercial diets made up of glandular tissues, agricultural products, and industrial food ingredients. Dietary efficiency with less processed ingredients invokes smaller energy demands upon fish metabolism and diet utilization resulting in better growth rates when all the digesting and absorbing systems of the fish are operating; or perhaps several unknown growth-promoting components may still be discovered.

BIBLIOGRAPHY

Aoe, H., I. Abe, T. Saito, H. Fukawa and H. Koyama, 1972, Preventive effects of tocols on muscular dystrophy of young carp. Bull. Jpn. Soc. Sci. Fish. 38(8): 845-851

Aoe, H. and I. Masuda, 1967, Water-soluble vitamin requirements of carp - II, Requirements for P-aminobenzoic acid and inositol. Bull. Jpn. Soc. Sci. Fish., 33(7): 674-680.

Aoe, H., I. Masuda, I. Abe, T. Saito and Y. Tajima, 1971, Water-soluble vitamin requirements of carp - VIII, Some examinations on utility of the reported minimum requirements. Bull. Jpn. Soc. Sci. Fish. 37(2): 124-129.

Aoe, H., I. Masuda, T. Mimura, T. Saito, A. Komo and S. Kitamura, 1969, Water-soluble vitamin requirements of carp - VI, Requirement for thiamine and effects of antith Jpn. Soc. Sci. Fish. 35(5): 459-465.

Aoe, H., I. Masuda, T. Saito and A. Komo, 1967, Water-soluble vitamin requirements of carp - I, Requirement for vitamin B_2. Bull. Jpn. Soc. Sci. Fish. 33(4): 355-360.

Braekkan, O.R., O. Ingebrigtsen and H. Myklestad, 1969, Uptake and storage of vitamin A in rainbow trout (Salmo gairdneri). Int. J. Vit. Research, 39: 123-130, Norway.

Burrows, R.E., D.D. Palmer, H.W. Newman and R. Azevedo, 1952, U.S. Fish and Wildlife Service, Special Sci. Rept. 86: 1-20.

Coates, J.A. and J.E. Halver, 1958, Water soluble vitamin requirements for silver salmon. Spec. Sci. Rep. 281:1-9.

Cowey, C.B., J.W. Adron and D. Knox, 1975, Studies on the nutrition of marine flatfish. The thiamin requirement of turbot (Scophthalmus maximus). Br. J. Nutr. 34: 383-390.

Dupree, H.K., 1966, Vitamins essential for the growth of channel catfish, USDI, BSFW, Techn. Paper No. 7. 12 pp.

Dupree, H.K. (1969). Progress in Sport Fishery Research, 1968, U.S.

Bur. Sport Fish. and Wildlife, Resource Publication 77: 220-221.
Dupree, H.K. (1970), Dietary requirements of vitamin A acetate and beta carotene. Progress Sport Fishery Research, 1969. U.S. Bur. Sport Fish. and Wildlife. Resource Publication 88: 148-150.
Halver, J.E. (1953a), Ph.D. Thesis, University of Washington, Seattle, Washington.
Halver, J.E. (1953b), Fish diseases and nutrition. Trans. Am. Fish. Soc. 83: 254-261.
Halver, J.E., (1957), Nutrition of salmonoid fishes - III, Water-soluble vitamin requirements of chinook salmon. J. Nutr. 62: 225-243.
Halver, J.E. (1966), Vitamin and amino acid requirements of Pacific salmon (Oncorhynchus). EIFAC 66/SC 11-3: 61-68.
Halver, J.E., 1972, The vitamins, in: "Fish Nutrition", Academic Press, New York, p. 29-103.
Halver, J.E., 1975, Nutritional requirements of cold water fish. Proc. 9th Int. Congress. Nutrition 3: 142-157.
Halver, J.E., L.M. Ashley and R.R. Smith, 1969, Ascorbic acid requirement of coho salmon and rainbow trout. Trans. Am. Fish. Soc., 98: 762-771.
Halver, J.E. and J.A. Coates, 1957, A vitamin test diet for long term feeding studies. Prog. Fish. Cult. 19: 112-115.
Hashimoto, Y., 1975, Nutritional requirements of warm water fish. Proceed. 9th Intl. Congr. Nutrition 3: 158-175.
Hashimoto, Y., S. Arai and T. Nose, 1970, Thiamine deficiency symptoms experimentally induced in the eel. Bull. Jpn. Soc. Sci. Fish. 36(8): 791-797.
Hashimoto, Y., T. Okaichi, T. Watanabe, A. Furukawa and T. Umezu, 1966, Muscle dystrophy of carp due to oxidized oil and the preventive effect of vitamin E. Bull. Jpn. Soc. Sci. Fish. 32(1): 64-69.
Kitamura, S. (1963), Personal Communication.
Lovell, R.T., 1973, Essentiality of vitamin C in feeds for intensively fed caged catfish. J. Nutr. 103: 134-138.
McCay, C.M. and A.V. Tunison, 1934, Notes of interest. The nutritional requirements of trout. N.Y. State Conserv. Dept., pp. 31-32.
McLaren, B.A., E. Keller, D.J. O'Donnell, and C.A. Elvehjem, 1947a, The nutrition of rainbow trout - I, Studies of vitamin requirements. Arch. Biochem. 15: 169-178.
McLaren, B.A., E. Keller, D.J. O'Donnell and C.A. Elvehjem, 1947b, The nutrition of rainbow trout - II. Further studies with purified rations. Arch. Biochem. 15: 179-185.
Murai, T. and J.W. Andrews, 1974, Interactions of dietary α-tocopherol, oxidized menhaden oil and ethoxyquin on channel catfish (Ictalurus punctatus). J. Nutr. 104: 1416-1431.
Murai, T. and J.W. Andrews, 1978, Thiamin requirement of channel catfish fingerlings. J. Nutr. 108: 176-180.

National Academy of Sciences/National Research Council, 1973, Nutrient requirements of trout, salmon and catfish. Nutr. Req. Dom. Animals, 11: 1-57.
National Academy of Sciences/National Research Council, 1977, Nutrient requirements of warm water fishes. Req. Dom. Animals, pp. 1-78.
Ogino, C., 1965, B vitamin requirements of carp, Cyprinus carpio. I, Deficiency symptoms and requirements of vitamin B_6. Bull. Jpn. Soc. Sci. Fish. 31(7): 546-551.
Ogino, C., 1967, B vitamin requirements of carp - II, Requirement for riboflavin and pantotheic acid. Bull. Jpn. Soc. Sci. Fish. 33(4): 351-354.
Ogino, C., T. Watanabe, J. Kakino, N. Iwanage and M. Muzuno, 1970, B vitamin requirements of carp - III, Requirement of biotin, Bull. Jpn. Soc. Sci. Fish. 36(7): 734-740.
Phillips, A.M.Jr., D.R. Brockway, A.J.J. Kolb and D.M. Maxwell, 1949, The nutrition of trout. N.Y. State Conserv. Dept. Fish. Res. Bull. 13: 1-12.
Phillips, A.M.Jr., D.R. Brockway and E.O. Rodgers, 1950, Biotin and brown trout: The tale of a vitamin. Prog. Fish-Cult. 12: 67-71.
Phillips, A.M.Jr., F.E. Lovelace, H.A. Podoliak, D.R. Brockway and G.C. Balzer, Jr., 1954, The absorption of calcium by brook trout. N.Y. State Converv. Dept., Fish. Res. Bull. 18: 5-23.
Poston, H.A., 1965, Effect of dietary vitamin E on microhematocrit, mortality and growth of immature brown trout. N.Y. State Conserv. Dept., Fish. Res. Bull. 28: 6-9.
Poston, H.A., Effects of massive doses of vitamin D_3 on fingerling brook trout. N.Y. State Conserv. Dept., Fish. Res. Bull. 32: 48-50.
Poston, H.A., 1976, Optimum level of dietary biotin for growth, feed utilization, and swimming stamina of fingerling lake trout (Salvelinus nomoycush). J. Fish. Res. Bd. Canada 33: 1803-1806.
Poston, H.A., G.F. Combs, Jr. and L. Leibovitz, 1976, Vitamin E and selenium interrelations in the diet of Atlantic salmon (Salmo salar): Gross, histological and biochemical deficiency signs. J. Nutr. 106: 892-904.
Schneberger, E., 1941, Fishery Research in Wisconsin, Prog. Fish-Cult. 56: 14-17.
Shelbourne, J.E., 1970, Marine Fish Cultivation: Priorities and progress in Britain. In: "Marine Aquiculture", W.J. Mcneil, ed., OSU Press, Corvallis, p. 15-36.
Tunison, A.V., D.R. Brockway, J.M. Maxwell, A.L. Dorr and C. M. McCay, 1942, The vitamin B requirement of trout. N.Y. State Conserv. Dept. Fish. Res. Bull. 4: 12-18.
Watanabe, T., F. Takashima, C. Ogino and T. Hibiya, 1970, Effects of alphatocopherol deficiency on carp. Bull. Jpn. Soc. Sci. Fish. 36(6): 623-630.
Watanabe, T., T. Takeuchi and C. Ogino, 1975a, Effect of dietary

methyl linoleate and linolenate on growth of carp - II, Bull. Jpn. Soc. Sci. Fish. 41: 263-269.
Watanabe, T., O. Utsue, I. Kobayashi and C. Ogino, 1975b, Effect of dietary methyl linoleate and linolenate on growth of carp - I, Bull. Jpn. Soc. Sci. Fish. 41(2)257-262.
Wolf, L.E., 1942, Fish diet disease of trout, A vitamin deficiency produced by diets containing raw fish. N.Y. State Conserv. Dept., Fish. Res. Bull. 2: 1-5.
Wolf, L.E., 1945, Dietary gill disease of trout, N.Y. State Conserv. Dept., Fish. Res. Bull., 7: 1-22.
Wolf, L.E., 1951, Diet experiments with trout. Prog. Fish-Cult., 13: 17.
Yone, Y., M. Furuichi and K. Shitanda, 1971, Vitamin requirements of red sea bream - I. Relationship between inositol requirements and glucose levels in the diet. Bull. Jpn. Soc. Sci. Fish. 37(2): 149-155.
Yone, Y. and M. Fujii, 1974, Studies on nutrition of red sea bream. X Qualitative requirements for water-soluble vitamins. Rep. Fish. Res. Lab. Kyushu Univ. 2: 25-32.

STATUS OF FISH FOOD FORMULATION

AND FISH NUTRITION — RESEARCH IN BRAZIL

        Newton Castagnolli

        Assist. Prof. of UNESP - Univ. of São Paulo State

        Campus de Jaboticabal - SP - BRAZIL

This review paper shows that only a few experiments concerning fish feeding have been done in Brazil up to the moment, and most of them are related to diet formulation.

As can be seen, only in the last decade, or even more precisely, within the last five years, this research effort has undergone development, although the first trial was done by Azevedo et al. in 1964.

The latter studies concerning fish nutrition are still at a beginning stage but considering the greater governmental assistance for Institutes and University Research Programs on aquaculture, and also due to the growing private interest on fish farming, it is assumed that fish nutrition experiments will continue to make progress.

The earliest attempts towards fish diet formulation in Brazil were done by Azevedo et al. (1964) who tried different levels of penicillium yeast as supplements to common carp diet. Diets consisted of 80% wheat flour and 20% meat meal supplemented with 0 (control), 2, 4, and 8% penicillium yeast. Best results from an economic viewpoint were obtained with the diet containing 2% of the unicellular bacteria.

Azevedo et al. (1968) also tested different raw forages as food for tilapia (Tilapia rendalli). Among Swannee sp, (gramineae), rami (Boehmeria sp) and the leguminosae perennial soy (Glycine javanica) and cudzu (Pueraria sp) fed ad libitum, the best performance was observed with Boehmeria sp.

Probably due to the absence of governmental assistance and lack of interest in the private sector in fish farming no further studies were undertaken until work was returned through efforts made at the Fish Culture Section of the Faculdade de Ciências Agrárias e Veterinárias (UNESP, Campus of Jaboticabal City, São Paulo State), and at Pentecoste Research Center of Diretoria de Pesca e Piscicultura in Ceará State.

Castagnolli et al. (1974) in a comparative test fed common carp (Cyprinus carpio) and curimbatas (Prochilodus scrofa) the same basal diet consisting of 50% corn meal, 25% meat meal, 15% soybean meal, 8% rice bran supplemented with 2% mineral plus vitamin mixture. The crude protein of the diet was 18.5%. After 6 months average weight gain of carp was around 1.0 kg while Prochilodus presented a poor performance of 0.1 kg weight gain within the experimental period.

Da Silva et al. (1975) in a preliminary trial observed that only male Tilapia nilotica and hybrids crossbred with T. hornorum and T. nilotica presented similar performance in earthern ponds quarterly fertilized with 56.0 kg/Ha simple superphosphate and 56.0 kg/Ha ammonium superphosphate. Fish were also fed daily with rice bran with 14% crude protein content. After 6 months at a stocking rate of 1 fish/m^2 the harvest was around 3.0 ton/Ha in both ponds.

The same authors (da Silva et al., 1975) harvested 2500 kg/Ha hybrid tilapia within 6 months in a consortiation experiment with pigs. Fishes were stocked at the low density rate of 2,400/Ha in a natural pond of 1,000 m^2 with a pig-pen built at the edge of the pond and 70 pigs were reared above the water surface level. The fish were fed strictly on pig manure directly or indirectly on phytoplankton produced through organic pond water fertilization. Recently another experiment has been set-up in order to test different fish and pig stocking rates.

According to Castagnolli and Felicio (1975) sorghum may fully replace corn meal in diets for carp (Cyprinus carpio) and tilapia (Tilapia rendalli) at 70% level in the diet. They observed that at the same experimental conditions carp presented better performance than tilapia since the results have shown a double growth rate and a three-fold weight gain for common carp. Carp showed a 2.5 conversion compared to 8.2 conversion presented by tilapia. Observations were carried on during 4 months in 20 small cement ponds (2.0 x 1.0 x 0.8 m) at the Fish Culture Section of Jaboticabal's Faculdade de Ciências Agrárias e Veterinárias.

In the same ponds, Mendonça (1975) observed that opaque-2 and hybrid corn gave the same performance to common and mirror carp. Corn meal content in the diet was up to the 70% level.

Pereira Filho et al. (1976) have tested diets with the same ingredients but in varying proportions in order to try different crude protein levels such as 20, 25, 30 and 35%. After 180 days observation period, better performance was achieved by fish fed with the diet containing 25% crude protein. Fish showed a weight gain of about 640 g within the experimental period. The estimated crude protein content that could enhance maximum production (25.8%) didn't differ much from the most profitable one (25.5%) in the pelleted diet containing the following ingredients: corn meal, meat meal, soybean meal, Sacharomicies cereviseae yeast, rice bran supplemented with mineral plus vitamin mixture.

Two different crude protein levels (20 and 28%) were tested in diets for Rhamdia hilarii, a Brazilian catfish by Machado and Castagnolli (1976). Fish were fed during 90 days with pelleted diets and best results were obtained with fish fed diet with the higher protein level. The results showed also that Rhamdia is not a fast growing species and might be suitable for polyculture rearing.

Torloni and Campos (1976) have tested the influence of protein level in the performance of rainbow-trout (Salmo irideus Gibbons) fed pelleted diets with 25, 30, 35 and 40% crude protein content at Campos do Jordão Trout Hatchery, in São Paulo State. Best results estimated from collected data correspond to a diet with a crude protein level of 33% within the experimental conditions. One year after the beginning of the experiment it was observed that diet with 35% crude protein content produced a harvest of larger numbers of individuals weighing more than 200 g.

Fuga (1977) has tested the antibiotic BDZ-50 (Zinc Bacitracin) as growth promoting agent for common carp fingerlings at 0 (control), 10 and 20 mg/kg of the diet. After 16 weeks there was no significant difference between treatments.

Scorvo Filho (1977) had observed the performance of tilapia (Tilapia rendalli) fed different raw forages: Salvinia auriculata, a floating water weed, perennial soy (Glycine javanica), confrei (Symphytum pelegrium) and napier grass (Pennisetum purpureum) leaves. He concluded that fish presented similar performance.

Paiva (1978), in Pentecoste, Ceará State, in a preliminary trial tested babaçu (an oleaginous palmacea) cake, cotton (Gossypium malvacearum) seed cake and castor bean seed cake or food for Tilapia nilotica. Later, in cement boxes of 500 liters capacity he tested various proportions of meat meal to castor bean seed cake or protein source for the same species and found that best results were obtained with diets containing 25 and 50% castor bean seed cake.

The same author (Paiva, 1978) in a trial with gradual replacement of meat meal by shrimp wastes, still with the same species (T. nilotica) observed a 20% higher performance in the treatments in which the ingredients were present at 50-50% level.

Da Silva (1978), also in Pentecoste, Ceará State, in a recent experiment observed the growth rate of two characidae amazonic species: tambaqui (Colossoma macropomum) and pirapitinga (C. bidens). Fingerlings of both species were obtained through hormone (pituitary gland extract) injection and the initial average body weight was 25 g. Fish were stocked in 18 earthen ponds at two densities (5,000 and 10,000 fingerlings/ha) and also a polycultur of both species with hybrid tilapia, being 5,000 Colossoma sp and 5,000 all male hybrid fingerlings/ha. The experiment lasted one year and results showed an astonishing score of more than 1.0 kg average body weight gain for both Colossoma species. Fish were fed on a finishing broiler diet with only 17% crude protein content.

Barbieri (1978) studied the feeding dynamics of the cichlid acara (Geophagus brasiliensis) at Lobo Reservoir, near São Carlos City, São Paulo State. It was observed that Geophagus is fatter and has a full stomach during March and April, after the spawning season. Through monthly sampling during an entire year the condition factor and average stomach replenishment rate was also determined.

Macedo et al. (1978) recently started a feeding trial with "tambaqui" Colossoma macropomum fingerlings in 16 aquaria in order to determine the best crude protein level in iso-caloric semi-purified diets containing 14, 18, 22 and 26% crude protein. Best performance ($P < 0.01$) after 2 months observation was presented by the fishes fed on diets containing 22 and 26% crude protein. After the 3rd month the fish will be transferred to 16 cement ponds with approximately 1,800 liters capacity.

Concerning shrimp and fish pelleted diet technology Machado and Eldin (1978) had made attempts to utilize "chitosan" produced from the wastes of shrimp processing. They had tried different viscosities of chitosan (400, 500 and 700 centipoises). It was shown that besides the economic advantages of chitosan when compared with other pellet binders such as alginates or hexane phosphate, it is also a nutrient for growing fingerlings or juvenile shrimps.

The latter papers show that attempts related to fish nutrition in Brazil are still at a beginning stage but due to incressing interests and the importance of fish culture research in such a large tropical country it is assumed that within a few years the scientific level of the contributions will increase.

BIBLIOGRAPHY

Azevedo, P., E. Millen, and H.L. Stempniewski, 1964, Alimentação artificial de Cyprinus carpio. Uso de micélio de penicilina como suplemento de ração. Bolm Ind. Anim., S.P. 22: 69-79.
Azevedo, P., H.L. Stempniewski, and J.M.R. Alckmin, 1968, Alimen-

tação de tilapia melanopleura com diferentes plantas forrageiras. Rev. Med. Vet., S.P. 4(1-2).
Barbieri, G.B., 1978, Estudo sobre a dinâmica alimentar do acará, Geophagus brasiliensis. In preparation.
Castagnolli, N., P. de Andrade, and S. Sobue, 1974, Ensaio de competição alimentar entre carpas, Cyprinus carpio L. e curimbatas, Prochilodus scrofa steind, Científica 1(1): 69-80.
Castagnolli, N. and P.E. Felicio, 1975, Substituição do milho pelo sorgo na alimentação de carpas e tilapias. Ci. e Cult. 27(5): 532-537.
Da Silva, A.B. and L.L. Lovshin, 1975, Diretoria de pesca e piscicultura (D.N.O.C.S., Fortaleza - CE). Annual Report, pp. 1-5.
Da Silva, A.B. and L.L. Lovshin, 1975, Consorciação piscicultura e suinocultura. Ensaio com híbridos de tilapia. Proc. IV Siban - Simpósio Brasileiro de Alimentação e Nutrição. Botucatu, S.P. 26/31/01/1975 (Absts.).
Da Silva, A.B., 1978, Personal communication.
Fuga, J.O., 1977, Influência da bacitracina de zinco (BDZ-50) na alimentação de carpas. Trabalho de Graduação (FCAVJ/UNESP) p. 1-25.
Macedo, E.M., N. Castagnolli, and J.E.P. Cyrino, 1978, Nível proteico ideal na nutrição do tambaqui, Colossoma macropomum (Pisces Characidae). In preparation. Proc. XI ICN - International Congress of Nutrition, p. 486 (Absts.).
Machado, C.R. and N. Castagnolli, 1976, Preliminary observations related to culture of Rhamdia hilarii, a Brazilian catfish. Proc. FIR: AQ/CONF/76/6. 79: 1-9.
Machado, Z.L. and Z. Eldin, 1978, Emprego do chitosan no preparo de dietas peletizadas para camarões. Proc. XI CIN, Rio de Janeiro, Brasil, p. 487.
Paiva, C.M., 1978, Personal communication.
Pereira Filho, M., N. Castagnolli and A.R. Teixeira Filho, 1978, Nível proteico ideal na alimentação de carpas, Cyprinus carpio L., Científica, 6(2): 313-319.
Scorvo Filho, J.D., 1977, Uso de diferentes forrageiras na alimentação de tilapias, Tilapia rendalli. Trabalho de Graduação (FCAVJ/UNESP): 1-28. (XI ICN: 485 Absts.).
Torloni, C.E.C. and C. Campos, 1976, Arracoamento de truta arco-iris (Salmo irideus Gibbons). Influência de diferentes níveis proteicos na dieta. Bolm Ind. An. 34.

II. Food Science and Technology

POTENTIAL UTILIZATION OF UNCONVENTIONAL

FOOD SOURCES AND NEW FOOD PRODUCTS - Introductory Remarks

Dr. Juan Claudio Sanahuja

Fac. de Farmacia y Bioquimica - Univ. de Buenos Aires

Ab. Ing. Huergo 1145 - Buenos Aires - Argentina

The study and development of unconventional sources for animals feeds and direct human use, and the development of new foods maintains its importance for the solution of the food shortage problem that afflicts the major part of the world's population.

It is true that in the past few years these studies have greatly progressed; however, there are still a number of problems common to all unconventional proteins sources that have yet to be solved.

These problems, including the nutritive quality of proteins, their safety functionality, processing technology, as well as acceptability and economic factors, affect their use and consumption.

For the study of these proteins from a nutritive value point of view it is necessary for producers and food processors not only to be acquainted with protein's amino acid pattern, but also to be able to handle rapid and inexpensive protein quality evaluation methods.

Although the value of these foods basically depends of protein quantity and quality, their safety can be affected adversely by several factors: a) antinutritional or toxic factors, as for example in oil seed proteins; b) nutrients that could affect normal metabolism, as for example, the odd-numbered carbon chain fatty acids and the nucleic acids present in single-cell proteins; c) contaminants such as metals, pesticides or mycotoxins or harmful products formed during processing as compounds of lysino-alanine type.

These toxic factors may pose a threat to animals or humans: therefore the studies of toxicity must be well planned and performed and to eliminate the possibilities of adverse reactions.

New technologies are being developed to maintain effectively the desirable properties of proteins for their functional applications. In this area more information is at present required about the correlation between protein molecular properties and their physical and functional characteristics such as hydration, solubility, suspendability, swelling, gel formation, viscosity, elasticity, and absorption.

Organoleptic properties, including low flavor level, or at least, a flavor compatible with the other ingredients, and no "off" flavor or undesirable colors should be studied.

Finally, more information is also needed about processing of these protein sources in order to be able to develop formulated food products.

PLANTS AND OIL-SEED PROTEINS

Most technology for oil-seeds proteins has been developed for soyabeans and their secondary products. Other oil-seeds such as sunflowers, rapeseed, cottonseed are also potential source of good quality proteins. The technology for obtaining "food grade" products from these materials is still under research and more work must be developed to solve problems involving flavor, functionality, color, antinutritional factors, processing technology and many other factors.

LEGUMES

Legumes are a valuable source, not only of proteins but also of energy and of some vitamins and minerals and in many countries constitute a staple food of relatively low price.

But the consumption of legumes is limited by some problems, specially those concerned with their digestibility and, in some of them, with the presence of toxic factors.

LEAF PROTEINS

Leafy plants have tremendous potential for supplying an important part of the protein requirements of animals and humans.

At present, process research is being developed for commercial dehydration facilities and for a leaf protein recovery system.

Assuming the technical problems of objectionable color and taste can be overcome, the major need is for economical refining and production process.

SINGLE CELL PROTEINS

The most common type of non-photosynthetic SCP is from yeast, which seems to have the fewest biological problems; the carbon substrate may vary from gas oil to purified alkanes, methanol, ethanol, natural gas or waste materials (cellulose, citric fruits residues, rice bran, etc.).

At the moment, there are in several countries a number of SCP plants operating for animal feed production.

Improved processes for RNA removal would involve study of RNA levels in cells, the role and activating of endogenous ribonucleases and their control by means of genetic and cell physiological factors.

MARINE PROTEIN FOODS

Antarctic krill seems to be one of the more promising food protein resources and studies are developed for determining the nutritive quality of their protein, and transforming krill into an acceptable food product.

Finally, new food products for improvement of public health should be formulated considering the requirements of all the nutrients that may be lacking in traditional diets.

But the field introducing of these new food products is not an easy matter; the acceptance by the consummers seems to depend on factors that go beyond their nutritional benefits.

IMPROVING PROTEIN QUALITY IN PLANTS BY GENETIC MEANS

In recent years plant breeders succeeded in developing or identifying mutations in corn and other cereals with a higher relative level of lysine and other favorable combinations of amino acids in seed proteins.

This achievement raises the question of obtaining similar mutations in other grain and crop plants.

New technologies need also to be developed for separating upgraded protein from grains or their byproducts — in this way new applications for proteins in formulated food products or in producing meat analogs, extenders or beverages could be obtained.

Some of these topics will be considered in this workshop by speakers that are relevant figures in the science and technology of unconventional food sources; undoubtedly, their papers will make a valuable contribution towards achieving a substantial progress in the production of new food products.

THE NUTRITIONAL VALUE OF WINGED BEAN (Psophocarpus tetragonolobus
L. DC), WITH SPECIAL REFERENCE TO FIVE VARIETIES (CULTIVARS)
GROWN IN SRI LANKA*

N.S. Hettiarachchy[1], H.M.W. Herath[2], and
T.W.W. Wickramanayake[3]

[1-3] - Dept. of Biochemistry, Fac. Medicine, Peradeniya
Campus, Univ. of Sri Lanka, Peradeniya, Sri Lanka
[2] - Dept. of Agric. Biology, Fac. Agriculture, ibid.

INTRODUCTION

The winged bean (Psophocarpus tetragonolobus) plant is a climbing semi-perennial. It is cultivated in the wet zone of tropical and sub-tropical parts of South East Asia and Africa mostly in Papua, New Guinea. In general it is cultivated twice a year, once at the beginning of the wet season between September and November and again during the dry season in April. The wet season planting aims at obtaining tubers. The legume nodulates freely and the bacteria present in the nodules fix nitrogen from the atmosphere enabling the plant to thrive on relatively poor tropical soils. As the plants are climbers, strong stakes are fixed about four feet above the ground for each plant to grow on and then spread over a trellis or fences. The winged bean matures in about 3 to 8 months.

The interesting feature about this legume, unlike other edible legumes, is that every part of the plant can be eaten. In Sri Lanka, winged bean is grown as a home garden vegetable primarily for the immature pods, which may be chopped and cooked in coconut milk or tempered in coconut oil. Leaves are used in the preparation of "mallun" (tempered with coconut scrapings). Tubers and seeds are eaten occasionally.

---

* This paper deals with the following aspects: the protein content, anti-nutritional factor/s (trypsin inhibitor/s) and methods of detoxifying anti-nutritional factor/s.

Detailed studies of winged beans have been reported by Cerny et al.[1]. Much of the presently available information on this plant was compiled by the National Academy of Sciences[2]. A recent review published by Newell & Hymowitz[3] shows that relatively few agronomic, nutritional and plant breeding studies have been conducted on the crop.

The present study was designed to determine the protein content of all the edible parts of the winged bean plant, trypsin inhibitor/s (anti-nutritional factor/s) and methods of detoxifying the inhibitor/s in raw mature seeds. Five cultivars, three indigenous (V_1 - Puttalam, V_3 - Colombo and L_{18} - Polannaruwa) and two introduced (UPS_{61} - Papua - New Guinea and TPT_2 - Nigeria) were studied. The seeds of the five selections were grown in the Central Research Institute of Agriculture, Gannoruwa, from where the samples were collected for the study.

MATERIALS AND METHODS

Preparation of winged bean samples for protein content

Mature seeds (2.5 g) were ground in a Warring Blender to pass through an 80-mesh screen. Portions of this flour were taken for protein analysis.

Samples of leaves, flowers, pods and tubers (5 g) were chopped and suspended in 100 ml. of distilled water and homogenised in a Warring blender. Aliquots of the suspensions were taken for protein estimations.

Protein:

The crude protein content of the winged bean samples were determined by the micro-Kjeldahl's method.

Treatments of winged bean seeds

Roasting: Samples of flour were roasted at $80°$ for 30 min. and the seeds were roasted at $100°$ for one hour.

Boiling: Seeds were soaked for 24 hours, and boiled for one hour.

Autoclaving: Soaked seeds were autoclaved for 30 min. at $130°$.

Germination: Soaked seeds were sown on a moist cotton wool bed for 48 hours.

Trypsin Inhibitor Activity:

Phosphate buffer (o.1M, pH 7.6): 16 ml. of NaH_2PO_4 (0.2M) and 84 ml. of $Na_2HPO_4.12H_2O$ (0.2M) diluted to 200 ml. with distilled water.

Casein solution (2.0%): A suspension of 2 g. casein was made with 80 ml. phosphate buffer, and dissolved by heating on a steam bath for 10 min. The cooled solution was made up to 100 ml. with phosphate buffer and stored at 5°.

Trypsin solution: 0.125 g trypsin was weighed and dissolved in 25 ml. of phosphate buffer (0.1 M). This solution can be stored at 5° for one to two weeks without appreciable loss in activity.

Preparation of winged bean extracts: Crude extracts were prepared by shaking 1 g pulse flour with 50.0 ml. of 0.01N NaOH for one hour in a shaker, the pH being maintained at 9.0. This suspension was diluted (1:30) with distilled water to get about 40 to 60% of trypsin inhibition.

Assay for the enzyme activity:

Method developed by Kakade et al.[4] was used for estimating the trypsin inhibitory activity of the extracts. Stock trypsin solution (0.5 ml.) was pipetted in triplicate into test tubes, followed by 0.4 ml. HCl (0.001N) and the final volume of each tube adjusted to 2.0 ml. with the phosphate buffer. The tubes were kept in a water bath at 37° for 10 min. To one set of tubes, 6.0 ml. of 5% (w/v) trichloro acetic acid was added. This served as a blank. To each tube was then added, 2.0 ml. 1% casein solution previously brought to 37°. The tubes were incubated at 37° for 20 min. after which 6.0 ml. of 5% TCA solution was added to each tube in the second and third sets to stop the reaction. The suspensions were centrifuged and the absorbance of the filtrate measured at 280 mμ in a spectrophotometer (SP 500, model II) against the blank.

Trypsin inhibitor activity was estimated by the procedure as described above, except that the aliquots of winged bean extracts were added to the tubes prior to adjusting the volume to 2.0 ml.

Expression of activity: Enzyme activity is expressed in trypsin units (TU). In this method one trypsin unit was arbitrarily defined as an increase of 0.01 absorbance unit at 280 mμ in 20 min. for 10.0 ml. of reaction misture under the conditions described and the trypsin inhibitory activity as the number of trypsin units inhibited (TUI), Kakade et al.[4]. An inhibition of 50% was obtained by using a volume of 1 ml. of extract (see Table Table III). This volume was therefore used in assessing the trypsin inhibitory activity (see Table IV).

TABLE I

PROTEIN CONTENT OF THE WINGED BEAN SEED
COMPARED WITH THAT OF SOYBEAN

(expressed as g/100 g of fresh seed)

$*+=$ Cultivars	Proteins (%)
$*V_1$	38.
$*V_3$	42.
$*L_{18}$	37.1
$+UPS_{61}$	34.3
$=TPT_2$	34.8
'NAS	29.08 - 37.4
Soybean	35.1

* Indigenous cultivars
+ Seeds from Papua - New Guinea
= Seeds from Nigeria
' National Academy of Sciences
*+= Average of triplicate determinations

RESULTS

The protein content of the seeds and other parts of the plant are given in Tables I and II. These values are compared with those quoted in the report by National Academy of Sciences[3] and the protein content of soybean[5]. The seeds have a protein content of 34.3 - 42.0%, the indigenous cultivars having a slightly higher protein content than the introduced cultivars. These values compare favourably with the range quoted by NAS, and the value for soybean. The

TABLE II

PROTEIN CONTENT OF DIFFERENT PARTS OF THE WINGED BEAN
(composition expressed as g/100g of fresh edible portion)

Average of triplicate determinations

	PROTEIN (g. %)					
	V_1	V_3	L_{18}	UPS_{61}	TPT_2	NAS
Leaves	8.4	5.6	4.7	7.9	6.4	5.7 - 15.0
Flowers	2.3	2.8	2.5	1.8	2.0	5.6
Immature pods	2.5	2.3	2.3	2.8	2.4	1.9 - 2.0
Tubers	7.0	5.8	7.2	5.6	6.4	12.2 - 15.0

TABLE III

DETERMINATION OF THE TRYPSIN INHIBITOR
ACTIVITY OF WINGED BEAN SEED EXTRACT (V_1)

LEVEL OF WINGED BEAN EXTRACT (ml)	*ABSORBANCE AT 280 mµ	+TU	−TUI	PERCENT INHIBITION
−	0.44	44	0	−
0.5	0.338	33.8	10.2	23.2
1.0	0.225	22.5	21.5	48.9
1.5	0.12	12.0	32.0	72.7
2.0	0.098	9.8	34.2	77.7

* Average of duplicate determination.
\+ Trypsin units as defined in the text.
− Trypsin units inhibited.
TUI = TU of control − TU of test.

Percent inhibition = $\dfrac{\text{TUI}}{\text{TU of control}} \times 100$

The assay for tripsin activity and trypsin inhibitor activity were carried out as described in the text.

TABLE IV

DETERMINATION OF THE TRYPSIN INHIBITOR ACTIVITY OF WINGED BEAN SEED EXTRACTS, USING 1 ml OF EXTRACT

WINGED BEAN EXTRACT	*TU	TUI	% INHIBITION	TUI/g OF SAMPLE
-	40.0	-	-	-
V_1	22.0	18	50	2.98×10^4
V_3	20.0	20.0	45	2.69×10^4
L_{18}	11.8	28.2	70.5	4.21×10^4
UPS_{61}	12.7	27.3	68.3	4.07×10^4
TPT_2	18	22.0	55	3.28×10^4

* Average of duplicate determinations.
TU - Trypsin units, as defined in the text.
TUI - Trypsin units inhibited.

TABLE V

COMPARISON OF THE PERCENTAGE OF ANTI-TRYPTIC ACTIVITIES OF UNTREATED AND TREATED WINGED BEAN
(average of triplicate determination)

WINGED BEAN EXTRACT	UNTREATED	ROASTED FLOUR	ROASTED SEEDS	GERMI- NATION	SOAKED & BOILED	SOAKED & AUTOCLAVED
V	50	38.0	40.0	27.0	10	6
V_3	45	36.2	40.4	26.3	11.8	4.9
L_{18}	70.5	60.7	62.0	32.8	14.0	4.6
UPS_{61}	68.3	59.6	52.8	28.2	12.7	6.8
TPT_2	55	40.2	44.2	26.8	8.2	5.8

protein content of the leaves, flowers and tubers, is lower than the values reported by the NAS. Nevertheless, the values for immature pods compare favourably with those quoted by the NAS.

The data and calculation pertaining to the measurement of the trypsin inhibitor activity of a sample of winged bean extract (V_1) is shown in Table III. The original activity of the enzyme without the additions of winged bean extracts in the incubation mixture was treated as control value. Trypsin units inhibited and percentage inhibition were calculated after comparing with the control value. It is seen that about 50% inhibition was obtained when the volume of extract was 1.0 ml. and this volume of extract was used for the assays (see method for details).

The level of trypsin units (TU), trypsin units inhibited (TUI), percent inhibition and TUI per g. of sample for all the five cultivars are given in Table IV. It is seen that V_3 has the least trypsin inhibitor activity and L_{18} the most. These values are two to four times higher than the value reported (1.14×10^4) by Jaffe & Korte[6].

Table V shows the percent inhibition of trypsin present in untreated (raw) and treated winged bean seeds. Germinated seeds appear to have lost about half the anti-tryptic activity. Soaked and boiled or soaked and autoclaved samples have lost an appreciable amount of anti-tryptic activity. The loss of anti-tryptic activity on heating for one hour or autoclaving at 130° for 30 min. appear to show the heat labile nature of the trypsin inhibitor. It is interesting however, that this study found roasting the flour or the seeds unsatisfactory procedure for complete removal of trypsin inhibitor activity.

DISCUSSION

The five cultivars of winged bean studied have higher protein content in seeds than the soybean. The protein values for tubers are about four times more than the root crops of humid tropics — manioc, yams, sweet potato (Newell & Hymowits)[3], probably due to extensive nodulation and nitrogen fixation by this plant. The low content of protein in tuberous roots of the present study (5.6 - 7.2%) in comparison to that reported by NAS value (12.2 - 15.0%) could be due to genetic variation or environmental conditions.

In addition to the protein-rich seed, the tubers, flowers and immature pods are reasonably good sources of protein.

Like soybean and most other legume seeds the winged bean seeds contain trypsin inhibitor/s. This could reduce the dietary effectiveness of the protein; but the inhibitor/s can be inactivated or destroyed by heat treatment. This study found that autoclaving for 10 min. at 130° or boiling for 30 min. (Cerny et al.)[1] insufficient to effect complete inactivation or destruction of the inhibitor. However, autoclaving for 20 min. at $130°$ or boiling for one hour was necessary to inactivate the inhibitor/s effectively. The long period of exposure to heat might be necessary when trypsin inhibitor concentrations are high.

The toxic effects of raw bean have been observed by Claydon[7] and Jaffe & Korte[6]. Besides trypsin inhibiror/s, haemagglutinin activity (Schertz et al.)[8] and the presence of cyanide (Claydon)[7] have been reported in winged bean seeds. These toxic components have been found to be detoxified by heat treatment. More studies to detect the presence of toxic components and methods to detoxify them should however, be carried out with different varieties of winged bean seeds.

It has been shown by Khan[9] that the potential seed production of V_3 cultivar is 2678 kg/ha and tuber production of 535 Ka/ha in comparison to the yield of Papua - New Guinea selection which are 1946 kg/ha and 450 kg/ha respectively. The present study report a protein content of 42% in seeds and 5.8% protein in tubers. The trypsin units inhibited is 2.98×10^4 per g. of sample. The trypsin inhibitor/s could be effectively destroyed or inactivated by boiling or autoclaving the soaked seeds. The cultivar V_3 with a high production of seeds, tubers and a high protein content in seeds and in tubers and with low TUI (in comparison to V_1 and L_{18}) could be a potential cultivar to be cultivated on a large scale.

In developing countries, the winged bean might well be considered in future as a substitute for soybean. In Sri Lanka, it is the tender pod that is eaten as a vegetable. A considerable amount of nutrition education of the rural population will be necessary to make possible an appreciable harvest of the dried seed.

ACKNOWLEDGEMENT

We wish to thank Messers Karalliadde J.B. and Dharmasena, H.A. for the technical assistance. We also wish to thank the National Science Council, Sri Lanka, for the financial assistance made available under "Winged Bean Project" - 2/RG/76/1.

REFERENCES

[1] Cerny, K. Kordylas, M. Pospisil, F. Svabensky, O. & Zajic, B., 1971, Nutritive value of winged bean (Psophocarpus palustris Desv.), Brit. J. Nutr. 26, 293-99.

[2] National Academy of Sciences, 1975, "The winged bean: A high protein crop for the tropics", Washington, D.C.

[3] Newell, C.A. & Hymowitz, T., 1978, The potential of the winged

[4] bean – Psophocarpus tetragonolobus L.DC – as an agricultural crop. Personal communication.

[4] Kakade, M.D., Simons, N.R & Liener, I.E., 1969, An evaluation of natural vs. synthetic substrates for measuring the anti-tryptic activity of soybean samples. Cereal Chem. 46, 518.

[5] FAO, 1972, Food composition table for use in East Asia, Food and Agricultural Organization of the United Nations, Food Policy and Nutrition Division, Rome.

[6] Jaffe, W.G. & Korte, R., 1976, Nutritional Characteristics of the winged bean in rats. Nutr. Reports International, 14, 449-55.

[7] Claydon, A., 1975, A review of the nutritional value of the winged bean Psophocarpus tetragonolobus (L) DC. with special reference to Papua New Guinea. Science in New Guinea 3(2), 103-14.

[8] Schertz, K.F., Boyd, W.L., Jugelsky, W.J.R. & Cabanillas, E., 1960, Seed extracts with agglutinating activity for human blood. Economic Botany, 14(3), 232-40.

[9] Khan, T.N., 1974, Problems and progress in improvement fo winged beans in Papua New Guinea. Meeting on Winged Bean, National Academy of Sciences, D.C., on 24-26 Oct. 1974.

PROSPECTS AND PROBLEMS OF CULTIVATION

OF LUPINS IN SOUTH AMERICA

        R. Gross

        German Agency for Technical Cooperation

        Eschborn, FEDERAL REPUBLIC OF GERMANY

The large-seeded, annual lupins of the varieties Lupinus albus and Lupinus mutabilis can replace soya bean in those regions where the latter cannot be cultivated, due to soil requirements or to frost susceptibility. Under South American conditions, the small-seeded Lupinus luteus and Lupinus angustifolius, with low oil content have not proved better in comparison with other lupin species.

As opposed to other legumes, such as peas and beans, these varieties grow upright and stable, thus facilitating direct mechanical harvest. With regard to yield, and contrary to other legumes, lupin is similar or superior to other cereals, such as wheat and barley, under the same conditions, whereby the fertilizing requirements of this legume are very much inferior. The protein and oil yield of these varieties is, as an average, four times higher than that of other cereals.

LUPINUS ALBUS

The large-seeded Lupinus albus, native of the Mediterranean littoral, is being nowadays successfully cultivated in Brasil[18], Argentina and Chile. Cultivation in Brasil and Argentina is of secondary importance, and chiefly serves in human consumption like salted almonds. In Chile, on the contrary, the introduction of "sweet" lupin in 1949 led to an increasing cultivation in the south of the country.

The species and varieties introduced in Chile came from "sweet" lupins developed by V. Sengbusch[1,2]. These differ from bitter lupins because their alkaloid content reaches a maximum of 0.05% of the whole seed. The bitter varieties contain up to 2% of these alkaloids[11,16]. This low content of bitter factors allows a direct utilization of grains and green plants as a source of food for both humans and animals[1,2,14].

In opposition to Central Europe, where due to hard winters high-yielding lupin varieties are too susceptible to frost for winter cultivation, as well as too unripe for summer cultivation, under South Chilean conditions (37-40 degree of latitude) these lupin species achieve total mature seeds, whereby new selections have successfully endured up to $-10°C$ in the younger stage[9].

At present, the arable acreage reaches 3,000-4,000 hectares and the yield 2,800 kgs., with maximum yields over 5,500 kgs/ha. The arable potential in the IX Area (38-39 degree of latitude) is determined at 73,000 hectares, without displacing other crops[13].

Moreover, lupins can be cultivated to the north, as well as to the south of this region.

The arable acreage depends on the demand of the grain. These are classified according to their degree of importance, as follows:

1. Feeds grain for poultry, swine and milk substitute for calves.
2. Seed for green production of silage and soiling (green-fodder).
3. Grain export for products like salted almonds to Brasil, Venezuela, Spain and Italy for human consumption.
4. Lupin flour as admixture up to 15% for wheat flour, baby-food and other preparations.

Cultivation takes place up to now mainly through the small Mapuche farmers (indian inhabitants), who obtain by their intensive production approx. twice or thrice as much yield in comparison with wheat[13].

Static humidity with the corresponding fungus diseases, weed infestation and late sowing, are factors which mainly reduce yield.

An additional problem has arisen through the bittering of the native "sweet" lupin varieties, when they are cultivated successively for more than 3 years. In practical cultivation it has been observed, that lupin seed shall not be cultivated for more than 3 years, so as to avoid a total alkaloid content of more than 0.05% in the grain. For this reason, a careful breeding preservation of the material is necessary.

The main points of plant breeding lie for the moment in the combination of the characteristics of new selections with the potential yield of the present varieties (Table 1)

Various research studies for the utilization of lupins have been already carried out[9]. In all these studies it was determined that in the case of "sweet" lupins no toxical problems appear, so that in the future beside animal utilization, human utilization as well should be given greater importance.

LUPINUS MUTABILIS

The other genetical variety of this genus is found in South America[7]. Whereby during the pre-Inca period the grain of Lupinus mutabilis was cultivated and used in the Andean region[12]. The proof is found, for example, in the vessel of the Tiahuanaco period (400-1,200 a.D.). In the Andean countries Venezuela, Colombia, Ecuador, Peru and Bolivia, the cultivation and utilization of Lupinus mutabilis free of bitter factors, is still found today[3,5,7,12]

Cultivation takes place between 2,000 and 4,000 meters, having its main production at 3,500 m.a.s. Under such extreme conditions, this variety offers, without fertilizing, average yields of 1,200 kg/ha, whereby 2,500 are not an exception. The principal factor which limits the cultivation of larger surfaces, is the deficient demand, as it consists up to now of the self-supply for the "campesinos"[5,10]. The peasants know that this variety improves the quality of the soil.

Due to a guaranteed demand and technical assistance the average yield could be brought up — within a year — from 1,000 kg/ha to 1,400 kg/ha.

At present, 4,000-6,000 hectares are cultivated in Peru. Production problems occur, especially due to root diseases such as Fusarium sp in the case of too high soil humidity, Antragnose caused by Ascochyta sp and Colletotrichum sp, as well as a Dipterus which perforates the stem of the plant. Moreover, cultivation is limited through high pH values of the soil and frost over 4,000 m.

Under these conditions, the use of ecotypes or local varieties has offered up to now the most sure yields.

Besides the genetical characteristics of the cultivated seed, mainly the early sowing period is decisive for grains and oil content[5]. In this way, it was determined through exact experiments, depending on the opportune sowing period, that the difference in oil content was of 21% at the highest, and 15% at the lowest. In future, the development of suitable varieties will be decisive. Using the selections of the collection in Cuzco, Peru[6] and the breeding material of Campex, Chile, it was possible to find varieties

TABLE 1

BREEDING RESULTS UP TO NOW
OF LUPINUS ALBUS IN CHILE (% DM)

	CULTIVATED VARIETIES	NEW STRAINS
Protein (N x 6,25)	35	40
Oil	12-15	18
Crude fiber	10-14	9

which only need a 3-4 month period of vegetation[8]. Through simple selection, lines with 25-26% oil content were isolated[9] (Table 2). With this oil content, Lupinus mutabilis lies already beyond soya in oil content. Moreover, the protein content was increased to 48%.

These positive phytogenetic varieties are being combined and augmented, and shall serve for production, after cultivation experiments.

TABLE 2

BREEDING RESULTS UP TO NOW
OF LUPINUS MUTABILIS IN CHILE (% DM)

	CULTIVATED VARIETIES	NEW STRAINS
Protein (N x 6,25)	35-45	48
Oil	16-22	25-26
Crude fiber	10	9

In a parallel way, Lupinus mutabilis with low content of bitter factors, were obtained[4]. The many-sided possibilities of utilization are an advantage to the marginal small farmer, who uses the lignified stem as fuel[10]. Beside the usual lupin cultivation for self-supply, approx. 1,000 hectares will be cultivated this year in the Peruvian Sierra Central for oil production. In this way, it is expected, that a constant demand of this species shall be assured in the Peruvian highland. The result shall improve both the socio-economical situation of the highland inhabitants, as well as the food supply of oil and protein for the consumer.

REFERENCES

[1] Baer, E.v., El Lupino Dulce, Simiente XLII 2024 Chile 1972.
[2] Baer, E.v., Suesslupine-Anbau in Suedchile. Landwirt Im Ausland D.20506 F 6.J.H4 Juli/August 1972.
[3] Baer, E.v., Lupinenanbau in den Andenlaendern Entwicklung und Laendlicher Raum D 20506, 9. Jahargang, H4 Juli/August 1975.
[4] Baer, E.v., and Gross, R., Auslese bitterstoffarmer Formen von Lupinus mutabilis, Z. Pflanzenzuechtung 79, 52-58, 1977.
[5] Baer, E.v., Blanco, O. and Gross, R. Z. Acker und Flanzenbau J. Agronomy u. Crop Science, 145, 317-324, 1977.
[6] Blanco, O. Investigaciones agricolas en Tarhui en la Universidad Nacional del Cuzco, in: R. Gross and E.v. Baer 1974. "Proyecto Lupino", Informe N° 2, Instituto de Nutrición, Lima, Peru, 1977.
[7] Bruecher, H., Beitrag zur Domestikation proteinreicher und alkaloidarmer Lupinen in Suedamerika. Angew. Botanik XLIV (1970): 7-27.
[8] Cardenas, B., El Cultivo del lupino dulce en Chile-Caprosem Ltda. Dpto. Técnico, Temuco, Chile, 1977 (13 p.).
[9] Fundacion Chile, Situación, Análises y Perspectivas del lupino en Chile, Reunión de trabajo 1 y 2 de Diciembre 1977, Santiago, Chile.
[10] Gross, R., and Baer, E.v., Die Lupine, ein Beitrag zur Nahrungsversorgung, I. Z. Ernaehrungswissenschaften 14.224-228, 1975.
[11] Hackbarth, J. and Troll, H.J., Lupinen als Koernerleguminosen und Futterpflanzen, In Hand. Pflanzenzuechtung, 2 Aufl. Bd. IV, 1-51. (1959).
[12] Leon, J., Plantas alimenticias andinas, IICA Zona Andina, Boletín Técnico N° 6, 1964, pp. 92-95.
[13] Lopez, Fr., El cultivo del lupino y su incidencia económica y social en la IX Región. Reunión de Trabajo, Situación, Análisis y Perspectivas del Lupino en Chile, Fundación Chile, Santiago, 1 a 2 de Diciembre 1977.

[14] Mangold, E., Sitz-Ber. Dtsch. AKAD. WISS. Berlin Math. naturwiss. kl. Nr. 3, 1-46 (1950).
[15] Martinod M.S., Rev. Ecuatoriana de Med. 11.4, 199-205 (1964).
[16] Sengbusch, R.v., Suesslupinen und Oellupinen ldw. Jb. 91.5 719-880 (1942).
[17] Ufer Max. Erste Versuche mit Suesslupinen in Brasilian. Landwirt im Ausland 7 Jahrg. Heft 1 (1973).

VEGETABLE PROTEIN

J. Mauron

Nestlé Prod. Technical Ass. Co. Ltd.
P.O. Box 88
Ch-1814, La Tou de Peilz - Switzerland

Although it is realized nowadays that, for the time being, there is no real protein gap in the world in the sense that minimal physiological protein requirements cannot be met, it is nevertheless true that there is a protein shortage in the socio-economic sense. Indeed, with increased living standards man everywhere increases his protein intake to levels that apparently exceed his physiological needs and, more important still, he switches from cheap vegetable protein to expensive animal protein consumption. The net result of this progression is that more and more vegetable crops are grown, not to feed man but to feed domestic animals with a very low efficiency. In the USA for instance, 91% of the plant protein produced is fed to live-stock to produce 5.3 MT of meat protein consumed. This craving for meat in the affluent societies puts an enormous strain on cereal production and in this competition the developing countries are on the losing side. I would like to call this craving for animal protein in affluent societies the "hedonistic" protein needs as opposed to the physiological protein needs. Because of this "luxus consumption" of proteins in industrial countries, as well as in the affluent segments of developing countries, there is a constantly increasing demand for protein in the economic sense, leading to the paradoxical situation that the rich get too much protein and the poor less and less. On top of that, the energy crisis made it clear to everybody that the increased conversion of valuable vegetable food into animal products was becoming a nonsense in the present situation.

The only hope, therefore, to feed properly mankind in future will be to produce more food of vegetable origin containing an appropriate amount of protein.

1. THE MAIN VEGETABLE PROTEIN SOURCES

We should be aware that even today the main protein sources of man are of vegetable origin, cereals representing by far the most important one (Table 1). The introduction of high yield cereal strains in the last decade has certainly been the greatest single advance in food production, although some drawbacks are now evident.

The major world oilseed crops are soya, cottonseed, peanut and sunflower. They are rich in oil and protein and after oil extraction a cheap protein rich meal is recovered. Oilseed meals were recognized as a potential new source of protein for human consumption in developing countries in the early 1950's and at the end of that decade special processing conditions necessary to produce oilseed meals fit for human consumption from the major oilseeds were known (Altschul 1958, 1969). In practice, however, progress has been very slow and only a minimal amount of oilseed meals is used in human food.

Pulses have been traditionally the main protein source of the poor people for millenaries. Their production is stationary or even declining in some developing countries like India, because of the increased attractiveness of high yield cereals.

Potatoes are generally considered as a protein source, although they contain 10-12% crude protein on a dry weight basis, 50% of it being in the form of small peptides and free amino acids. Thus a potatoe has a protein concentration comparable with that of cereal grains, but its protein quality is much better and comparable with an animal protein. The potatoe is an excellent and inexpensive source of nutrients for expanding populations and its versatility makes it adaptable to a wide range of environments.

Green leaves are the biggest producers of protein in a plant and supply protein to the other tissues, including the seeds which are subsequently used to nourish humans and animals. However, leafy plants contain high concentrations of structural materials (cellulose, lignine) which are largely indigestible by man. Methods have therefore been developed to produce leaf-protein concentrates for incorporation into human food. Although these leaf-protein concentrates have a certain potential as protein source, their safe production, user acceptance and economics have not yet been sufficiently established. The brightest prospect seems to be for a village level production in some developing countries.

A new promising source of vegetable protein, traditionally grown in New Guinea and in Ghana and now investigated by many agronomists in several tropical countries is the winged beans (Psophocarpus tetragonolobus).

TABLE 1

WORLD PROTEIN PRODUCTION (1974)

	PRODUCTION MILL. TONS	% PROTEIN	PROTEIN PRODUCTION MILL. TONS
Wheat	360	(12)	43.2
Rice	323	(7.5)	24.2
Corn	292	(9)	26.3
Barley	170	(9)	15.3
Millet + Sorghum	93	(11)	10.2
Rye	33	(11)	3.6
Miscellaneous	62	(10)	6.2
CEREALS, TOTAL	1333		129
Potatoes + tubers	427	(2)	8.5
Pulses	44	(24)	10.5
Soya	57	(37)	21.1
Other oilseeds	81	(23)	18.6
TOTAL			58.7
TOTAL PLANT PROTEINS			187.7
Milk	417	(3.4)	14.2
Meat	116	(18)	21
Eggs	23	(12)	2.8
Fish	65	(18)	11.7
TOTAL ANIMAL PROTEINS			49.7

2. ADVANTAGES OF VEGETABLE PROTEIN PRODUCTION

Plants are direct converters of the sun energy, the only renewable energy source, into food. It is known that photosynthesis is a process of high efficiency ($\sim$ 30%) but under practical field conditions only a small part of this potential is used (1-2%). Considerable research is going on to develop crops with higher photosynthetic yield and with more efficient occupation of the land surface. A considerable drawback in tropical countries is the lowering of the photosynthetic yield due to the high night temperature which increases carbohydrate losses through respiration.

During the agricultural evolution of the last 10,000 years, man has selected plant crops adapted to different climates and soils, thus enabling him to survive under very different environments. In this respect the American Indians must be considered agricultural geniuses as they developed such basic food crops as corn, manioc, potatoes and beans. It is interesting to note that not more than about twenty different plants constitute the basic food crops of mankind. Even more interesting is the fact that they were all cultivated by prehistoric man, already. Most of these traditional crops have been greatly improved by modern breeding techniques, but no new basic crop has been developed so far by modern man. The great reservoir of wild plants not yet cultivated represents a great potential for new developments and a challenge to modern plant genetics. Thus, a group of experts selected among 400 wild tropical plants, 36 that could be used as food or raw material. There is no doubt that in future we will have to broaden the genetic basis of our food supply and abandon the present trend towards over selection of a few highly specialized crops.

A basic advantage of vegetable crops is the gain in energy obtained during production, shown by energy ratios always higher than one (Table 2). This ratio can be as high as 60 for manioc produced by subsistence farmers in Africa and is around 30 for cereals produced in the same labour intensive manner. With increasing mechanisation of agriculture the ratio drops rapidly, but does not fall below 1, whereas for animal foodstuffs the ratio is of the order of 0.4 to 0.1, demonstrating again the high inefficiency of this type of production. Even for plant production, however, the energy crisis will put limits to mechanisation and for countries with little fossil energy resources, an energy saving, relatively labour intensive, semi-industrial agriculture must be deviced.

It should also be remembered that the plant kingdom contains all the basic nutrients man needs and that an appropriate mixture of vegetable foodstuffs will provide all essential nutrients in adequate proportions.

TABLE 2

ENERGY REQUIREMENTS FOR PROTEIN PRODUCTION

COUNTRY	AGRICULTURE	CROP	ENERGY/kg PROTEIN (MJ)	ENERGY RATIO
Africa (averages)	subsistence	manioc	15	61.0
Africa (averages)	subsistence	corn	4.2	37.7
Tanzania	subsistence	rice	8.2	23.4
Philippines	semi-industrial	rice	33	5.51
U.S.A.	industrial	rice	143	1.29
U.K.	industrial	wheat	42	3.35
U.K.	industrial	potatoes	96	1.57
U.K.	industrial	milk	208	0.374
U.K.	industrial	broiler-meat	290	0.10
U.K.	industrial	eggs	353	0.14

Finally, it should be noted that man has been familiar with the basic plant crops for millenaries and has developed ways and means to transform these plants into palatable dishes, free of toxic or antinutritional factors. This long experience represents a considerable advantage of vegetable proteins over completely new, biosynthetically produced proteins.

3. THE DISADVANTAGES OF VEGETABLE PROTEINS

Certain drawbacks of vegetable protein should however not be concealed. There is, first, the strict dependence upon nitrogen fertilizers, a dependence that follows the law of diminishing returns. Since after 1973 the price of nitrogen fertilizer has increased about five times, it is becoming completely uneconomical in many industrial countries to use more fertilizer per acre. The situation is different in most developing countries, in which an increase in fertilizer use could still augment the yields considerably but is limited for obvious economic reasons. Nitrogen is vital to increase plant food production. The recent major escalation in the cost of nitrogen fertilizer has stimulated interest in possibilities for increasing the biological means for fixing nitrogen. This requires greater understanding of the processes involved and of the organisms that have this capacity. The genes coding for nitrogenase in bacteria have been successfully transferred from one species to another and the genetic potential exists for extending this nitrogen-fixing ability to a variety of economically important organisms. However, research on this is still rather limited and it will take at least 10 to 15 more years to develop a process of potential agronomic significance.

I do not have to insist here on some nutritional defects of
plant proteins, since they are well known. Vegetable proteins from
a particular source are generally deficient in some essential amino
acids, lysine in cereals, methionine in pulses and so on. Biological
values of most plant proteins are therefore inferior to those of
animal proteins. This fact, however, has been greatly over-empha-
sised in the past, since it is relatively easy to overcome this
drawback by mixing in an appropriate manner different sources of
vegetable proteins. Indeed, man has learned to do exactly this for
millenaries under the most different environment conditions. As an
illustration I show you the relatively balanced aminogramme of the
daily food of a family of substance farmers in central Zaire (Yasa-
Bonga) who live on poor soils where only manioc can provide the bulk
of calories. The balanced aminogramme is obtained by combining the
manioc staple food with protein sources such as green leaves, mush-
rooms and pulses (Table 3).

It is also true that plant proteins have generally a somewhat
reduced digestibility. Proper preparation of the plant material
by trimming, milling, cooking, fermentation etc. is able, however,
to correct for this drawback to a considerable extent. After proper
preparation the difference in digestibility between vegetable and
animal proteins is very small.

A more important point are the numerous toxicants present in
many vegetable protein sources (see Table 4). These are substances
that constitute natural components or natural contaminants of a
plant. The former are very well investigated and can be removed
or inactivated in all known major vegetable protein sources, whereas
the latter are still a subject of considerable concern. Thus the
wide-spread infection of peanuts with Aspergillus flavus leading to
the formation of aflatoxin has ruined the hopes to develop cheap
protein rich food based on this raw material. Top priority must
therefore be given to the prevention of microtoxin formation in
vegetable crops in order to make better use of the huge plant protein
reservoir.

At last, an important practical aspect of plant protein utilization should be mentioned, namely the low appeal of vegetable food
to modern man. This is a great obstacle to the increased use of
vegetable proteins in foods, because the influential people in a
community tend to switch from vegetable to animal food, giving thus
to the latter a high standing in the public image. This has led
the manufacturers of the first generation of textured vegetable
proteins to advocate their products as meat extenders or meat replacers. In my opinion, this is not a very exciting approach. Why
don't we use as starting point the age old traditional dishes prepared from soya in the different countries of the Far-East. These
recipes have evolved slowly by a sort of trial and error process
and are therefore not only very palatable but also highly nutritious

TABLE 3

AMINO ACID COMPOSITION OF THE DAILY DIET OF THE KUMBIMASHI FAMILY

	MANIOC (STAPLE) (42.5% of daily protein)	REST OF DIET (57.2% OF DAILY PROTEIN	TOTAL DIET
Lysine	5.8	5.1	5.1
Histidine	2.1	2.4	2.3
Leucine	4.1	8.0	6.3
Valine	3.1	4.7	4.0
Isoleucine	2.6	3.9	3.4
Phenylalanin	3.3	4.9	4.2
Tyrosine	2.9	3.8	3.4
Threonine	3.1	4.1	3.7
Methionine	1.0	2.4↑	1.8
Cystine	0.9	1.5↑	1.2
Thyptophan	1.2	0.9	1.0
TOTAL EAS	29.3	41.7	36.4
S-AS (limiting)	1.9	3.9↑	3.0

and devoid of toxic factors. Three types of soya preparations from Japan are Tofu, a preparation in which the soya protein is precipitated by calcium salts; Yuba, a dish in which the protein is coagulated by heat and soya sauce, a condiment in which the protein protein is hydrolyzed by enzymes. This illustrates the traditional artesanal practices that should become the basis for new technologies. Along these lines our Research Laboratory developed a new process for texturing soya protein, deriving from the Japanese Yuba process, but using Western equipment such as roller-driers. We are just at the beginning of a new era where food technology and tradition will meet to give the new impetus to the development of vegetable protein foods to help to feed mankind in a future of dwindling energy resource, increasing costs and permanent population pressure.

TABLE 4

TOXIC SUBSTANCES IN FOODS

DESIGNATION	CHEMICAL NATURE	SOURCE
Trypsin inhibitors	Protein	Leguminosae (soya, etc.) cereal germs
Haemagglutinins	Protein	Leguminosae (soya, etc.) Euphosbiaceae
Goitrogens	Thioglucosides	Cruciferae (Brassica, Sinapis, etc.)
Cyanogens	Glucoside	Manioc, Phaseolus, almonds
Saponins	Glucoside	Leguminosae (soya) spinach, Asparagus
Gossypol	Phenol	Cottonseed
Subst. inducing Lathyrism	β-Aminoproprionitril + others	Lathyrus sativus, odoratus (pea)
Subst. inducing Favism	Divicine (?)	Vicia faba (Metabolic disease + toxic subst.)
Allergens	Protein, macromolecules	Cereals (wheat) Leguminosae, fruits, nuts, etc.
Mycotoxins	Aflatoxine + others	Peanuts, Cereals
Metalbinding subst. (Zn, Mn, Cu)	Phytic acid	Soya, Pea, sesam

BIBLIOGRAPHY

Altschul, A.M. (1958), "Processed Plant Protein Foodstuffs", Academic Press, New York.
Altschul, A.M. (1969), Food Protein for Humans. Chem. Eng. News 24, 68-81.
Leach, G. (1975), "Energy and Food Production", Intern. Institute for Environment and Development, London and Washington, D.C.
Mauron, J. (1975), Future Trends in the application of new sources of protein. Bibl. Nutr. et Dieta, No 21, pp. 147-162, Karger, Basel.
Mauron, J. (1976), Future Trends in the use of protein resources, in: "People and Food Tomorrow", D. Hollingsworth and E. Morse (ed.), Applied Science Publishers Ltd., 1976.
Mauron, J. (1978), Vor- und Nachteile neuartiger Eiweisstoffe, in: Symposium "Lösung von Ernährungsproblemen: ber Beitrag des Produzenten, des Verteilers und des Ernährungswissenschaftlers Bibliotheca Nutritio et Dieta, in press.

STUDIES DONE IN ECUADOR ON THE POTENTIAL UTILIZATION

OF NON-CONVENTIONAL SOURCES OF FOODS

Ligia P. de Benitez

Institute for Technological Research

Aptdo. 2759 - Quito - Ecuador

In Ecuador several research programs have been carried out, and are still being carried out, with the purpose of getting better use from national raw materials, in particular those with no industrial application and which, for this reason, lead to instability in the production and price mechanisms.

The National Institute of Nutrition has provided valuable support to these investigations through studies on the composition of traditional foods, diets and formulas for the more vulnerable groups of the population, nutritional and anthropometric measures, etc., some of which have already been tested and accepted. The Institute for Technological Research of the National Polytechnic School initiated studies about non conventional foods several years ago, emphasizing the economical and technological aspects which might provide feasible solutions in the shortest time possible.

The investigations have been oriented towards 3 sets of programs:

1. Utilization of the vegetable resources of local origin for the elaboration of flours or protein concentrates which could be utilized to improve the nutritional quality of widely-consumed but low protein content traditional foods;

2. Replacement of the raw materials which the country imports at elevated prices for others of national origin, of easy cultivation and low price, or with surpluses which are not yet utilized.

In both cases, efforts have been oriented towards trying not to alter the physical and organoleptic characteristics of the new product, in order to avoid consumer acceptability problems, considering that this would be the most economical, easy and feasible way of combatting malnutrition problems; and

3. Elaboration of substitutes and new products which could serve as substitutes of certain foods of animal origin such as meat, milk, etc.

Within the first set of programs the following projects have been developed:

JORGE DAVILA T.
"STUDY AND POSSIBLE INDUSTRIALIZATION OF LUPINO OR CHOCHO"
INSTITUTE FOR TECHNOLOGICAL RESEARCH, QUITO, 1964

Objective

The objective of this study is the industrialization of the seeds of the "lupino tricolor Sodiro", popularly known as chocho, for the obtention of an oil and a flour with high protein content.

Project Reach

- extraction of lupinine, spartein and lupanine, which are the main alkaloids contained in the seed and which, aside from making it toxic also give it a bitter taste,

- drying of the residual product after the extraction of the alkaloids, and extraction of the oil,

- obtention of the flour from the residual cake.

Summary of results

The seed is ground and the alkaloids extracted through percolation with cold and hot water ($10°$ F or $50°C$) in a proportion of seed to water of 1/75 which, although not eliminating totally the spartein, removes the bitter taste of the seed; a posterior extraction by vapor hauling of the ground and slightly alkaline seed totally eliminated the alkaloids. The solvent-extracted oil has a yield of 18%. The resulting flour contains 55% protein with an average biological value of 55%.

EDWIN ORSKA
"PROTEIN EXTRACTION FROM BANANA LEAVES"
INSTITUTE FOR TECHNOLOGICAL RESEARCH, QUITO, 1971

Objective

The objective is the study and design of a pilot plant for the obtention of a powdered protein concentrate, from banana leaves, which could be easily added to foods of popular consumption.

Project Reach

- physical-chemical characterization of the raw material
- previous treatments
- obtention of the protein concentrate
- characteristics of the product obtained
- design of the pilot plant
- industrial pre-feasibility test

Summary of results

In any of the Ecuadorian zones there is sufficient availability of raw material to supply the needs of the plant which is the purpose of the study, 50 tons/day (20 hours), and 14,000 tons annually, which will produce 247,000 kg of refined protein concentrate.

The product has the following characteristics: light coffee-colored powder, with a slightly balsamic odor and slight herby taste, with a humidity content of 5.3%, 56.6% protein, 1.9% fiber, 0.8% ether extract, and 14.1% ashes.

The estimated sale price at the year of the presentation of this study is of 90 cents of the dollar per kg.

Tests of inclusion of this concentrate in 2 types of popular foods have been carried out: in wheat cookies and in "pinol" (a mixture of toasted barley flour, brown sugar and spices such as cinnamon and cloves), with substitution levels of 25% and 50%. At 25% substitution, acceptability levels of 90% and 95% were found for cookies and "pinol", respectively.

RAUL GANGOTENA
"PROTEIN EXTRACTION FROM MANIOC LEAVES"
INSTITUTE FOR TECHNOLOGICAL RESEARCH, QUITO, 1972

Objectives

This work is oriented towards the obtention of concentrates with high protein content from manioc leaves, which are abundant throughout the country, for mixing with other foods of popular consumption.

Project Reach

- botanical and agronomical characteristics of the raw material
- physical-chemical characteristics of the manioc leaves
- extraction of the juice from the manioc leaves
- obtention of the protein concentrate
- characteristics of the product obtained.

Summary of Results

The amount of leaves which can be produced in a hectare planted with manioc is estimated to be of 12 to 18 tons per year.

For the extraction of the leaf protein the following process is proposed:

The fresh leaf is made into a pulp in a mill, with a water/leaves proportion of 2/1; the pH is regulated to 8.5 with NaOH; the juice is extracted in a channeled press, in which 40% of the total leaf protein is recuperated.

For the obtention of the protein concentrate two methods are proposed:

- acidify the juice to a pH of 4.5, heat with direct vapor to $26.67°F$ ($80°C$), decant and dry the residual "mud" in an atomization dryer using an entrance air of $85°F$ ($185°C$). A green-brown powder is obtained, with 6.2 humidity, 30.5% protein, 8.5% ether extract, 2.9% crude fiber and 8.7% ashes.

- dry the juice directly by atomization, using an entrance air of $85°F$ ($185°C$), obtaining a slightly dark green-brown powder with 6.2% humidity, 31.9% protein, 6% ether extract, 1.1% crude fiber and 10.6% ashes.

For the refinement, the resulting powder is extracted by percolation using 95% ethanol as the solvent. The resulting product is cinnamon-colored and has a toasted-grain smell, with 6.5% humidity, 42.4% protein, 1.2% ether extract, 2.3% crude fiber, and 12.2% ashes.

The inclusion tests of this concentrate in foods was done with bread and pinol. Good results were not obtained with bread because the percentages used were too high, but with pinol a substitution of up to 30% is feasible, increasing the protein content from 3.6% to 14.9%.

EDUARDO MORAN
"PROTEIN EXTRACTION FROM EXTRACTION RESIDUES (SESAME OIL CAKE)"
INSTITUTE FOR TECHNOLOGICAL RESEARCH, QUITO, 1972

Objectives

The objective is to study the obtention of a protein concentrate from the residues of sesame seed (Sesamun oriental Lin) oil extraction, and its possible uses in human feeding.

Project Reach

- studies about the availability of the cake
- physical-chemical characteristics of the residual cake
- protein extraction
- obtention of the protein concentrate
- characteristics of the product obtained
- potential production and outline for a processing plant

Summary of results

Presently aobut 8,000 tons of sesame seed are processed, leaving approximately 4,000 tons of residual cake, which constitutes the raw material for the protein extraction. This study utilized the cake from an oil plant which utilizes solvent extraction systems.

Centrifugation is the more adequate method for the separation of the protein extract. The resulting product is a light brown fine powder, with a smell similar to cereal flour, with a solubility of 75% in reference to nitrogen, and the following composition: 6% humidity, 65% protein, 1.5% crude fiber, and 0.9% ashes.

The substitution tests for this concentrate are done with pinol and cookies. The pinol with 25% sesame seed flour added to it has the same acceptability as the original product. Cookies with the same percentage of sesame seed flour have a slightly darker dough, but there is practically no variation in the taste.

PABLO POLIT
"OBTENTION OF PROTEIN ISOLATES FROM DEFATTED FLOUR OF HIGUERILLA"
INSTITUTE FOR TECHNOLOGICAL RESEARCH, QUITO, 1975

Objective

The purpose of this study is to obtain a protein isolate from defatted higuerilla flour, to be used for animal feeding.

Project Reach

- physical-chemical characterization of the raw material
- oil extraction and obtention of the flour
- obtention of the protein isolate
- determination of the biological value

Summary of Results

Some 20,000 tons of higuerilla are processed annually in Ecuador, leaving a residual cake, after the oil extraction, containing 36% protein.

In this work two different processes are utilized to prepare the cake:

- trituration of the seed and extraction with ethylic ether;
- trituration of the seed, cooking of the paste to 37.78°F (100°C) and extraction with hot hexane.

In both cases the protein is extracted by suspension of the cake in an aqueous medium, agitation and elimination of the insoluble solids through centrifugation.

The solubilities of the proteins vary according to the pH of the solvent: it is also possible to increase the solubility by adding salts, as long as the pH of the solvent is equal to or greater than 11.

The temperature for greatest extraction is 7.22°F (45°C), and the time needed to reach equilibrium (maximum solubility) is 10 minutes, when the proportion solvent/cake is 5/1.

The levels of extraction and of recuperation of the protein

obtained in the 2 methods of cake preparation are:

In the first case:

a. when extraction is done at pH 11, extraction is 86% and recuperation is 63.8%;

b. when extraction is done with 0.5% NaOH, extraction is 88.7% and recuperation is 72.3%.

In the second case:

a. when extraction is done at pH 11, extraction is 52.8% and recuperation is 38%;

b. when extraction is done with 0.5% NaOH, extraction is 74.7% and recuperation is 57.8%.

Determination of the aminoacids of the isolated protein demonstrates that the thermal treatment and the increase in alkalinity cause a greater destruction of certain aminoacids such as lysine, threonine, tyrosine, cystine and methionine.

Feeding tests with mice prove that the protein isolates are not toxic and that the net protein utilization (NPU) is of 36% (similar to that of wheat gluten), reaching 46% in the case of the isolate obtained from the cake extracted with ether and 0.5% NaOH. The addition of 3% L-lysine and 1.54% DL-methionine increases the NPU from 37.5 to 49%, which is 73% of the value found for casein.

HERNAN AYALA MEJIA
"UTILIZATION OF THE QUINOA FOR PROTEIN FORTIFICATION
OF WIDELY-CONSUMED FOOD PRODUCTS" - (1977).

Objective

This work deals with the utilization of varieties of bitter quinoa cultivated in the country and the possibility of industrialization in order to obtain flour which could be added to bread for protein fortification.

Project Reach

- agricultural aspects
- physical and chemical aspects of the raw material
- elimination of saponines
- obtention of the flour and physical-chemical characterization

- production levels
- utilization of the quinoa flour in bread
  . tests at laboratory level
  . tests at pilot level
- acceptability tests

Summary of Results

The quinoa for this study was collected in different regions of the country, and the flour was prepared in the following manner:

First the saponine was eliminated by a peeling process with 3% soda solution in ebullition, during two minutes, with a volume proportion of one to two for raw material and solution. Then the skin is promptly removed with a water stream and agitation. The washing time lasts 15 minutes for 5 kg of material, with a water stream of 500 cc/sec. Then the grain is dried and ground.

Tests with bread showed that it was possible to substitute up to 15% quinoa flour without altering the physical and organoleptic characteristics of the bread. In addition, bread with quinoa has a considerably longer conservation bread than wheat bread.

Seeing as how quinoa today, in Ecuador, is a non-technified cultivation, production is scarce.

Within this same set of programs, work has also been carried out on new oilseeds which grow wild in the Ecuadorian Orient, and which have arisen great interest because of the quality of their oils and because of their high protein content, as well as by their potential use in the future as new sources of protein.

the tree peanut (orinocense karat). It is a tropical tree; the seeds have the follwing composition on a dry basis: ether extract, 54.3%, protein (Nx6.25) 19.7%, non-nitrogenated material, 26%. The residual cake, after the oil is extracted, has a 43% protein content, 6% crude fiber and 6.7% ashes.

- Waki (schium Sp.). It is a climbing vine; its fruit has the following composition: 5% humidity, 64% ether extract, 26% protein, 1.3% total carbohydrates, 1.7% crude fiber and 1.9% ashes. The oil is of good quality, with a high content of fatty acids: oleic and linoleic. The cake, after the oil is extracted, has 4% humidity, 80% protein (Nx6.25), 5% crude fiber and 6% ashes. This cake can be consumed directly; if fried in oil it produces a product similar in taste to crisply fried pork's rind.

In the second set of programs a number of projects have been developed as part of the Program of Composite Flours:

"COMPOSITE FLOURS"
INSTITUTE FOR TECHNOLOGICAL RESEARCH, QUITO, 1972-1978

Objectives

- partial substitution of imported wheat flour by flour obtained from national sources such as corn, potato, rice, and other starchy products;
- improvement of the nutritional quality of breads including flours rich in protein.

The work under this program has been oriented towards the elaboration of breads and pastas.

Project Reach

- agricultural and market diagnosis at the national level of all raw materials which are the object of the study, in order to determine the most apt to be utilized, in terms of production, productivity and costs.

- physical-chemical and nutritional characterization of the varieties selected in the chapter above and re-selection of the same.

- elaboration of raw and pre-cooked flours, depending on the case, and physical-chemical and nutritional characterization of the same.

- elaboration of breads and spaghetti and physical-sensory evalution of the same. In the investigations, corn, rice, potato and manioc flours were used as substitutes and soy flour as enricher for the bread.

- market surveys and marketing studies for wheat, flour and bread.

- acceptability studies of breads made with composite flours.

Summary of Results

It coincided that for both potato and corn the least attractive varieties for direct consumption, and thus the ones obtained at the least cost are the ones that offer the best industrialization characteristics, not only because of their higher productivity and low cost, but also because of their higher starch content.

Among the varieties of rice and manioc no significant variations were found, and thus it is not possible to recommend, in the case of industrialization, the most convenient one from the economical and production points of view.

The potato flour must be pre-cooked. The one with best results in bread is that with an 89% level of starch modification.

Corn, rice and manioc flour are elaborated from the raw product.

In all cases a 10% substitution with these flours in the bread improved the physical and organoleptic characteristics of the bread, with a better volume and higher level of acceptability on the part of the consumer. With 20% substitution a bread with characteristics similar to wheat bread was obtained.

As the substitution level increases, the noted characteristics decrease.

From the nutritional point of view we could mention a slight improvement with regards to the wheat bread when substitution was made with potato, rice and corn flour, and a significant impoverishment when manioc flour was used.

In spaghetti, studies were carried out with tertiary mixtures — wheat-soy and pre-cooked corn flour, wheat-soy and potato, wheat-soy and manioc. The best results were obtained with mixtures of 25% soy, 50% corn and 25% wheat semolina.

Acceptability Tests of Bread with Composite Flours

The objectives of this study are:

- to investigate the degree of preference for the composite flours breads as compared to wheat bread
- to investigate the preference at different substitution levels
- to compare the bread prepared in the pilot bakery with the bread from commercial bakeries.

The consumer aspects analysed were the following:

- reactions and gestures when faced with the bread characteristics;
- comments made at the time of purchase;
- type of bread selected and volume of purchase.

In the third set of programs, the following project has been developed:

"PRODUCTION OF SOY MILK"
INSTITUTE FOR TECHNOLOGICAL RESEARCH, QUITO, 1975

Objectives

Produce a milk substitute at a considerably lower cost.

Project Reach

- selection of the raw material
- aqueous extraction of the soy milk
- addition of flavoring agents
- acceptability tests

Summary of Results

The product obtained can be stored for a period of 6 months without refrigeration, and without the product suffering any alterations. The product can be compared to partially de-fatted cow's milk.

Acceptability tests were done with 50 children aged 4 to 6 years, with an acceptability level of 97%.

PROJECTS BEING DEVELOPED

- continuation of the work on composite flours for the bread industry, working with tertiary and multiple mixtures which include soy;

- production of intermediate foods of high nutritional value and low cost, dehydrated in roller dryers and formulated from cereals, tubers, fruits, oilseed cakes and whole soy, destined to serve as the basis for the preparation of sweet dishes (snacks and pastries) and salty dishes (soups, patties), not only in the home but also in food assistance programs).

B.P.C. - A FISH PROTEIN CONCENTRATE

FOR HUMAN CONSUMPTION

A.H. Delfino, M.D.

Coordinator of Research, Clinic at Medical School

Montevideo, URUGUAY

B.P.C. is obtained by a method of biological hydrolysis applied to fresh fish by means of a yeast (Hansenula montevideo sp nova) (Bertullo, 1970).

It is a hypoproteic, hypofat, acid food. It is 72% protein, formed by 60% polipeptides and 40% free amino acids, with fats in a 0.2-6% ratio, but retaining the phospholipids; pH is between 5.35 and 5.55.

Bromatological analysis: amino acid content is detailed in Table 1, and mineral content in Table 2.

Biochemical properties: all the amino acids are "L" form. The present of L-Dopa was estimated qualitatively.

Bacteriology: it is sterile as regards pathogenic bacteria.

Toxicology: no toxicity during the three years of experiments on albino rats (Westlar).

Research on protein quality shows the following:

- Score: the score of B.P.C. in relation to the Protein Standard (FAO/WHO) is 99%, with valine being the limiting amino acid (see Table 3).
- PER (Protein Efficiency Ratio) and N.P.U. (Net Protein Utilization): the average PER for B.P.C. is 2.51 as compared to casein with adjusted reference to 2.50; N.P.U. is 68.67 (Table 4).
- Digestibility: by the A.O.A.C. Method, it is 98-99%. Appar-

TABLE 1

AMINO-ACID CONTENT IN B.P.C.

AMINO ACID	PROTEIN (mg/gr)
Isoleucine	47.7
Leucine	70
Lysine	126
Methionine + Cistine	40
Phenylalanine + Tyrosine	36.7
Threonine	41.8
Tryptophan	11.5
Valine	49.5
Histidine	18.9
Alanine	65.7
Aspartic Acid	103.2
Arginine	113.8
Glutamic Acid	135.3
Glycine	75.4
Proline	38.4
Serine	35.5
Tyrosine	31.7
γ-Aminobutiric Acid	8
γ-Amino-N-Butiric Acid	traces
Taurine	3.5
Sarcosine	2.7
Phosphoserine	6
Beta Alanine	traces
Ornitine	1.9

TABLE 2

MINERAL CONTENT OF B.P.C.

Ca	2.30%
P	1.80%
Na	0.48%
K	1.07%
Mg	0.93%
Fe	10 ppm
Cu	6.7 ppm
Zn	21 ppm
Mn	21 ppm
Co	6.5 ppm
Br	12.1 ppm
Ba	21.5 ppm
Sr	17 ppm
Al	51 ppm
Fl	0.1-0.2 ppm

TABLE 3

SCORE OF B.P.C.

AMINO ACID (AA)	PROTEIN STANDARD* AA mg/gm PROT.	B.P.C.** AA mg/gm PROT.	%
Isoleucine	40	47.7	119.2
Leucine	70	70.7	101
Lysine	55	125.6	228.4
A.A.S.T.	35	40	114.3
A.A.A.T.	60	80	133.3
Threonine	40	41.8	104.5
Tryptophan	10	11	110
Valine	50	49.5	99
Histidine	14	19	135.9

* FAO/WHO - 1973
** Bertullo et al., 1974

TABLE 4

PROTEIN EFFICIENCY RATIO* AND NET
PROTEIN UTILIZATION OF B.P.C.**

SAMPLE Nº	PER VALUE[1]	SAMPLE Nº	PER VALUE[1]
H.1	2.38	H.11	2.50
H.2	2.39	H.12	2.50
H.3	2.64	H.13	2.58
H.4	2.51	H.14	2.44
H.5	2.44	H.15	2.62
H.6	2.66	H.16	2.38
H.7	2.42	H.17	2.40
H.8	2.62	H.18	2.53
H.9	2.63	H.19	2.57
H.10	2.41	H.20	2.61

SAMPLE Nº	N.P.U. VALUE[2]	SAMPLE Nº	N.P.U. VALUE[2]
1	67.50	6	74.74
2	61.85	7	64.42
3	64.32	8	70.01
4	72.44	9	68.32
5	70.01	10	72.10

* Bertullo et al (1974).
** All the samples were Hake, prepared with fish caught all year round.
[1] Reference casein adjusted to 2.50.
[2] Reference casein was 66.66.

ent digestibility (Ap. D.) is 91 (Fomon, 1975).
- Biological Value: the apparent biological value for B.P.C. is of 84, having started at 42.
- Protein Evaluation: the comparison made between the sum of essential amino acids of B.P.C. and casein in different types of Fish Protein Concentrates (F.P.C.) shows a figure of 40.49 gms/100 gms of protein, according to the Animal and Plant Health Inspection Service (1976) (Table 5).

This F.P.C. has been used for human beings during 14 years. It has been taken by more than 100,000 people with no toxic or pathological reaction; it was controlled in supplementary feeding by more than 200 doctors in Uruguay.

In this presentation we aim to present:

1) the study of substituing cos's milk by B.P.C., and
2) the latest research regarding pregnant mothers and B.P.C.

THE STUDY OF SUBSTITUTING
COW'S MILK BY B.P.C.

We wish to prove that children who cannot tolerate cow's milk can substitute it by the B.P.C. preparation shown in Table 6 and grow and develop within normal standards.

Intolerance to cow's milk was accompanied in 45% of the cases by digestive symptoms (vomiting and diarrhea), 35% broncheolitis or other respiratory troubles, the rest had skin or optic symptoms. The diagnosis was confirmed by controlled repetition tests (Latin American Congress of Gastroenterology and Nutrition, April 1978), 90% showing immediate response and 10% a little latter.

Method

The child population was all of the same socio-economic level, the same geographic area and race (Table 7). Systematic medical assistance was given by a team that worked in Family Benefits Hospital Nº 1 (Sanatorio Nº 1, Asignaciones Familiares), IMPASA Hospital; Policlinics of Gastroenterology and Neumology of Pedro Visca Hospital and CASMU Hospital, all with the same system of control and progressive feeding and the same child care. The mothers showed interest in the growth control of the children. Two references were taken as control elements:

1. Weight in kilograms by age and standard deviation (LACP- Latin American Center of Perinatology and Human Development, 1976) (Fig. 1); and

TABLE 5

PROTEIN EVALUATION* - MINIMUM ESSENTIAL AMINO-ACIDS** PER 100

	gms MIN. ESSENTIAL AMINO ACIDS PER 100 gms PROTEIN***
F.P.C. (Herring, Sweden)	40.15
F.P.C. (Red Hake, USA)	38.02
F.P.C. (Sardine, Morocco)	38.13
F.P.C. (Hake, Chile)	38.03
F.P.C. (Hake, Uruguay - B.P.C.)	40.49
Cow's milk casein	27.69

* Animal and Plant Health Inspection Service (1976).
** Phenylalanine + isoleucine + leucine + lysine + methionine + valine + threonine.
***Minimum should be over 25%.

TABLE 6

B.P.C. PREPARATION

B.P.C. (Powder)	5 %
Rice Mucilage (according to age)	5 %
Corn Oil	1 %
Glucose	7 %
Cl-Na	1 %
Cl-K	2 %
Boiled Water	100 cc

TABLE 7

SUBSTITUTION OF COW'S MILK BY B.P.C. IN CASES OF INTOLERANCE

Name	Age	Institution/Doctor	Weight gain	Condition
Eliz.	13 months	H.P. Visca / Dr. R. Maggi	At the end of 45 days, went up 200 gms.	Intolerance to cow's milk all beef protein. Dystrophy 2º (500 gr. in 1 month)
J. Car.	14 months	H.P. Visca / Dr. R. Maggi	In 1 month, went up 800 gms.	Intolerance to cow's milk and soy (Preparation 4). Dystrophy 2º.
Dario	6 months	H.P. Visca / Dr. R. Maggi	In 15 days went up 600 gms.	Allergy to milk. Muscoviscidosis. Bronqueolitis Dystrophy 2º
Serg. G.	8 months	H.P. Visca / Dr. Nairac	In 20 days went up 450 gms.	Allergy to cow and mare's milk. Dystrophy 2º.
Pablo	24 months 1/2	H.P. Visca / Dr. Nairac	Went up 200 gms a month for 10 months, height 92 to 96.	Allergy to milk and gluten. Dystrophy 1st. degree. Fed B.P.C. without gluten.
Nelson	11 months	H.P. Visca / Dr. Nairac	Went up 400 gms in 15 days. Height 80 to 83.	Allergy to cow's milk. Did not tolerate Preparation 4.
Luis	12 months	H.P. Visca / Dr. Marzolino	Went up 500 gms in 15 days.	Intolerance to cow's milk and gluten.
Viviana	7 months	H.P. Visca / Dr. J. Pereira S.	Went up 1000 gms in one month.	Intolerance to cow's milk. Diarrhea and lung trouble.
Ana L.	17 months	H.P. Visca / Dr. J. Pereira S.	Went up 200 gms in 15 days. Hypogam became normal in 2 months.	Intolerance to cow's milk. Hipogam globulinemia.
Ana P.	10 months	H.P. Visca / Dr. J. Pereira S.	Went up 60 gms in 15 days.	Intolerance to protein of beef origin
Raúl A.	5 months	Military H. / Dr. R. Maggi	Went up 500 gms in 15 days.	Vomiting and bronchial trouble, improved after 24 hs.
Madelón	3 months	CASMU / Dr. F. de los Santos	Went up 500 gms in 15 days.	Intolerance to cow's milk and soymilk.
Mariana	1 month	IMPASA / Dr. A.H. Delfino	Went up 550 gms in 15 days.	Intolerance to cow's milk. Hospitalized twice: 1) diarrhea; 2) bronchitis.
Oscar	2 months	Asig. Familiares / Dr. A.H. Delfino	Went up 700 gms in 1 month's treatment.	Intolerance to cow's milk. Diarrhea - distrophy 2nd degree
Sebastián	1 month	Asig. Familiares / Dr. A.H. Delfino	Went up 800 gms in 1 month's treatment.	Intolerance to cow's milk. Diarrhea - Dystrophy 2nd degree.
Silvana	3 months 1/2	Asig. Familiares / Dr. A.H. Delfino	Went up 480 gms in 10 days.	Intolerance to cow's milk. Vomiting and diarrhea. Dystrophy 2nd degree.
Nelson	3 months	Asig. Familiares / Dr. A.H. Delfino	Went up 600 gms in one month.	Intolerance to cow's milk. Vomiting and diarrhea. Dystrophy 3rd degree. Was operated anal imperforation.

2. Average speed of growth per unit of weight over weight of previous examination ((ASG/U (VMC/U) published in June 1976 and April 1978) (Fig. 2).

Dr. Anibal Capano was responsible for the processed data. Specific examples are commented below.

In the case of Sebastian C. (Fig. 3), we see: normal weight at birth, 3,400 gms; 45 days later he weighs 400 gms less. Intolerance is proven, the B.P.C. preparation substitutes cow's milk, and six months later his S.D. is normal; at 18 months his weight is above average.

Maura C. had normal weight at birth; with a dystrophy at seven months she weighs 5 kg. The B.P.C. preparation is administered and two months later her weight is average; 12 months later it is S.D. (Fig. 3).

In Fig. 4 we see that Maura C's weight is at first below P10, then rapidly it goes over P90 until she attains higher weight. Sebastian C., who had negative growth, soon reaches P50, once he began taking the B.P.C. preparation and remains above this limit.

Mariana O. was hospitalized twice before she was 45 days old (Fig. 5). Intolerance was proven and the B.P.C. preparation administered; S.D. remained above average for weight.

Veronica M. had dermatitis problems after breast-feeding stopped; cow's milk was suspended and the B.P.C. preparation given (Fig. 5). At about her fifth month her weight becomes stationary (hospitalized); at age one year her weight is above S.D.

Results

Growth Curves - Development of Body Weight - they all, after different spaces of time, attained normal 50 percentile and almost 25% went above standard deviations.

Speed of Growth - A.S.G./U. with regards to previous weight: they all are within normal limits, with the same significance as the previous one, but the variables in growth are much more evident.

Height - within normal limits for the 25 cases.

Neurological development - normal in 20, abnormal in 5 (important dystrophy in first months of life).

B. P. C. — A FISH PROTEIN CONCENTRATE

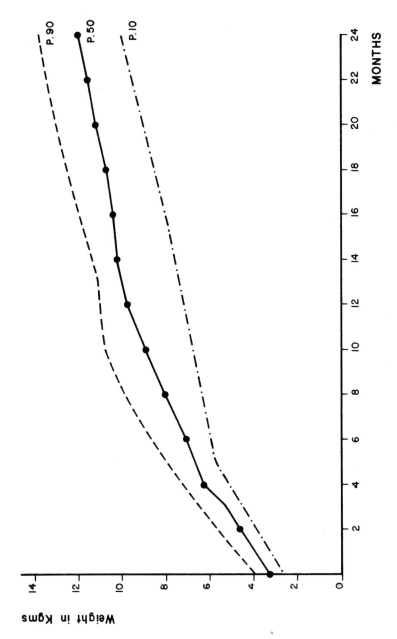

Fig. 1 — Growth Curve

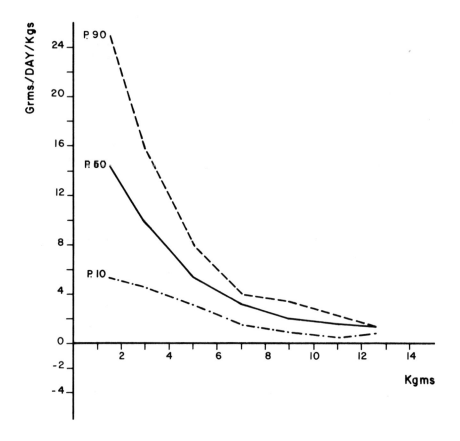

Fig. 2 - Average Speed of growth per unit of weight of previous examination.

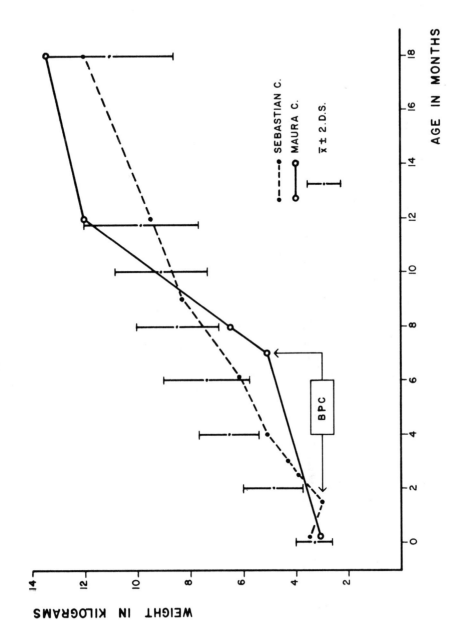

Fig. 3 — Weight of two children, from birth until 18 months of age, showing time when B.P.C. is introduced in the diet.

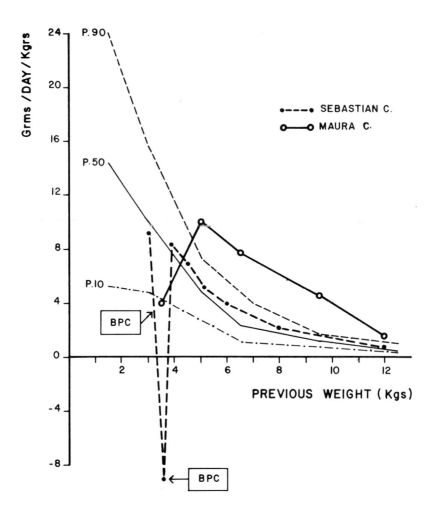

Fig. 4 - Average speed of growth per unit of weight over weight of previous examination of two children showing time when B.P.C. is introduced in the diet.

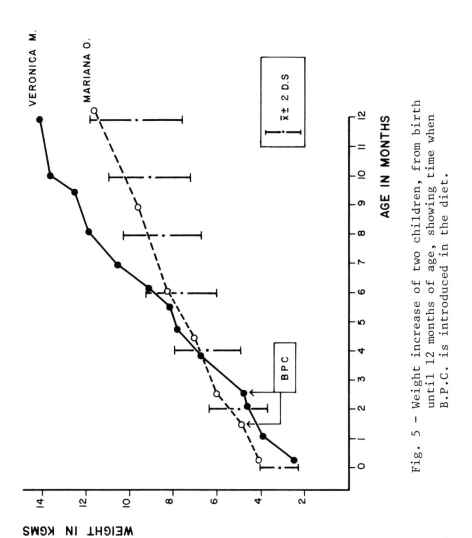

Fig. 5 — Weight increase of two children, from birth until 12 months of age, showing time when B.P.C. is introduced in the diet.

E.E.G. - there were some abnormal cases (4) in their first six months; after the sixth month they became normal.

The Beginning of Walking - this is expressed in an arithmetical average and the standard deviation of ages for the beginning of walking. x in months 12.9; S.D. 1.7.

Ocular Examination: normal.

Gessell Test - within normal variables.

Conclusions

The 25 cases of children with intolerance to cow's milk proved that by giving them the B.P.C. preparation their growth and development became normal, and their weights, in kg, above normal, according to age and S.D. and according to Average Speed of Growth per Unit (A.S.G./U.) of weight over the previous examination (LACP).

RESEARCH ON PREGNANT MOTHERS AND B.P.C.

Dr. Poseiro, in the High Risk Policlinc of the Family Social Services Hospital (Asignaciones Familiares) has made a study of the intrauterine growth by uterine height during gestation with 2.S.D. (Fig. 6).

In this study carried out by Dr. Poseiro, we see 22 mothers who had uterine height below normal average and would give birth to babies with S.G.A. (Small for Pregnancy Time); 12 of these mothers were given 9 gms of B.P.C. every day besides the usual treatment (Fig. 7 and 8). Within 4-5 weeks those heights had reached 90% of S.D. and the women gave birth to babies with adequate weight for their age. The other 10 mothers continued with the usual treatment (without B.P.C.), and their babies were born S.G.A. We are continuing to study these very promising results.

SUMMARY

The B.P.C. is a fish protein concentrate obtained by a biological hydrolysis, by means of a proteolitic yeast, producing excellent results in the studies mentioned (intolerance and pregnancy). There are experiences in other fields of Medicine that have already been verified and others in the process of research, such as the one presented here on uterine height and nutrition with B.P.C.

This food marks the opening of a new field in the search for solutions to the world problem of nutrition.

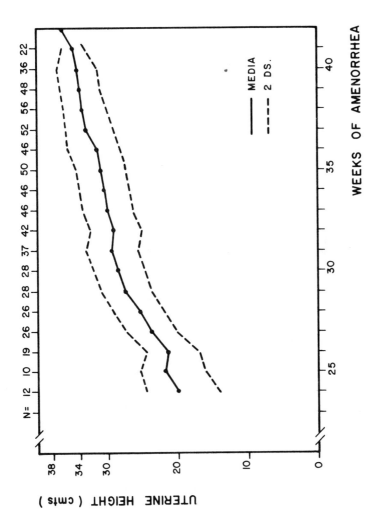

Fig. 6 — Uterine height (back-synphisis) according to the development of the pregnancy.

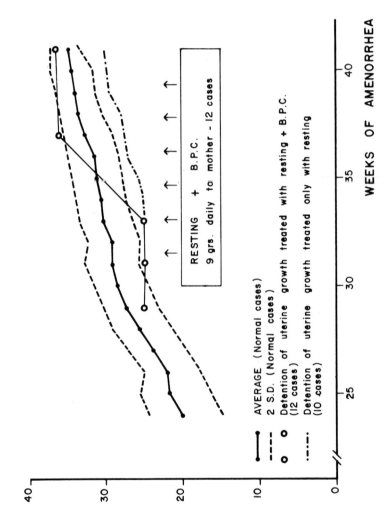

Fig. 7 – Uterine height (back-synphisis) according to the development of the pregnancy

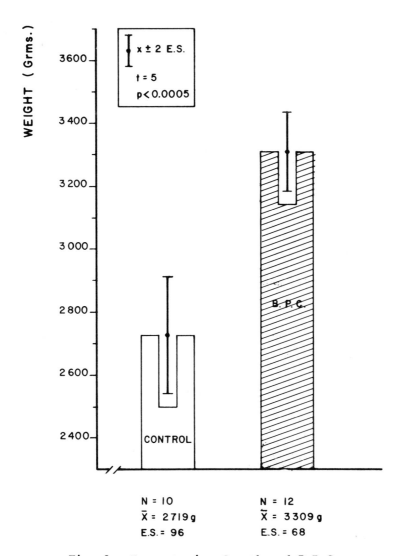

Fig. 8 — Intrauterine Growth and B.P.C.

BIBLIOGRAPHY

Animal and Plant Health Inspection Service (1976). Federal Register 1976, 41-N⁰ 82 - U.S.A.
A.O.A.C. (1960) Official Methods of Analysis, 9th ed.
Bertullo, V.H. (1962), Hydrolysis of Proteins of Animal Origin based on proteolitic micro-organisms. Publ. Fish Inst. Montevideo 1(2).
Bertullo, V.H. (1962), The B.P.C. (Catenulated Bio-Proteose) for human use (Technology of the process, characteristics and bromatological analysis), Publ. Fish Inst. Montevideo 1(2).
Bertullo, V.H. (1964), The B.P.C. (Catenulated Bio-Proteose) in Human Food Publ. Fish Inst. Montevideo 1(3).
Bertullo, V.H. (1970), Patent n⁰ 3516349.
Bertullo, V.H. (1978), Hydrolysis of Proteinic Substances. Uruguay Patent N⁰ 11525.
Bertullo, V.H. and Delfino, S.H. (1968), Present Outlook of Nutrition in Uruguay. Contribution of B.P.C. to hipoprotein diets. Publ. Fish Inst. Montevideo 2(2).
Bolla, M., Chiessa, M., and Doassans, E. (1977), "C.B.P. A Fish Protein Concentrate. A Food of High Biological Value". II Assembly of Dietitians and Nutritionists, Córdoba, Argentina.
Campbell, R.M. and Kosterlitz, H.W. (1972), Am. Journal Physiol. 72-395 U.S.A.
Delfino, A.H., Caillabet, E., Bidegain, S., Gomez Rincon, M., Bertullo, V.H., Alvarez, C., Moris, A., and Corbo, M. (1968). The Catenulated Bio-Proteose (B.P.C.) of fish in the recovery of the undernourished children. Publ. Fish Inst. Montevideo 2(2).
Delfino, A.H., Pecarovich, R., Bidegain, S., and Bertullo, V.H. (1974), Considerations on the answer to immunization in children that supplement their diet with B.P.C. Publ. Fish Inst. Montevideo 2(3).
Delfino, A.H. (1977), The B.P.C. a Protein Concentrate for Human Use. Nitrogenated Balance and Clinical Aspects. III Congress of Gastroenterology and Nutrition, São Paulo, Brazil, and at the XIII Pediatric Assembly of Uruguay 1977, Fray Bentos, Uruguay.
FAO/WHO, The Need for Energy and Proteins (1973), Report of a Special Combined Committee of FAO/WHO Experts. Report N⁰ 522. Rome, Italy.
FAO/WHO, The Need for Proteins (1976), Report of a combined group of FAO/WHO experts. Report n⁰ 37, Rome, Italy.
Fomon, S.J. "Child Nutrition", 2nd ed., 1975, p. 138.
Kacevas, D. and Bertullo, V.H. (1974), The Use of B.P.C. in the Research on Ateriosclerosis. Publ. Fish Inst., Montevideo 2(3).
Magmessan, J.H. (1945), Pediatric Record (Uppsala) 32, 599.
Martell, M. and Falkner, F. (1977), Early Post-natal Growth Eval-

uation in Full-Term, Preterm and Small-for-Dates Infants. 1/4, 313-323, U.S.A.

Martell, M. (1976), LAPC (C.L.A.P.) Growth and Development in the two first years of post-natal life. Publ. Cient. Nº 672-PAHO/WHO.

National Academy of Sciences, National Research Council, Report of; and International Conference, Committee on Protein Malnutrition, Food and Nutritional Board (1963), U.S.A.

Whipple, G.H. and Robscheit-Robins, T.S. (1971). Amer. J. Physiol. 72-395, U.S.A.

UTILIZATION OF NON-CONVENTIONAL FOODS IN BRAZIL

Ronaldo A. Salum

Organizações Bittencourt de Alimentos S.A.
R. Esteves Junior 34
Florianopolis, SC - Brazil

Humanity's growing need for food is demanding great efforts on the part of scientists and investigators throughout the world, in terms of research for producing better and more economical items.

Genetics has demonstrated to be an area with enormous potential, and has already made valuable contributions towards increased productivity levels. Such is the case of the production of grange poultry, where genetics was responsible for the advent of industrial aviculture, in the models which are universally practiced.

The advancement of nutritional research has further permitted the equation of the nutritional requirements of each animal species with the utilization of foods which can best satisfy such needs.

The world demand for meats is greatly unsatisfied, since yearly demographic growth has been placing substantial demands in increased food production, especially in terms of protein. Rabbit-raising can contribute to and represent an important area of support to the supply of meat, in the combat against food crises, due to some of the positive factors of rabbits.

Origin: mammal belonging to the order of the rodents, family of the Leporideos, genus ORYCTOLAGUS - Oryctolagus cuniculus.

It is still discussed whether they originated in Southern Europe, Asia or Africa, having crossed the Straight of Gibraltar and spreading throughout the Iberic Peninsula.

Raising: Rabbit-raising well deserves the attention of governments.

Aside from being easy, it is low cost, under all aspects, from small-scale raising (family use) to industrial exploration.

The big advantages offered by rabbit raising lie in their easy feeding, their extraordinary fecundity, and in the speedy results.

Rabbits are essentially herbivores; the females, when in the wild, have a gestation period of 28-30 days, being capable of at least 5-6 deliveries per year. In captivity they can deliver 8-9 times per year, with 6-10 offsprings each time. They procreate exponentially. In one year, a rabbit can produce at least 25 times its weight, or 2,500% of its own weight, as compared to a cow, for example, which weighs about 300 kg, takes 1 to 1.5 years to produce a 100 kg calf, which is the equivalent of only one third of its weight. (With feeding being quite expensive if the animal is in a stable).

Some countries have such a high level of rabbit-raising development that there are governmental and producers' organizations that protect them.

In Italy there is the National Rabbit-Raising Institute; in France there is INRA, which devotes a large portion of its research to rabbit-raising; in the USA there is ARBA. In Brazil, although there is an ANC (National Rabbit-Raising Association), rabbit-raising has not yet received the attention it deserves, even though there is a small number of producers who advertise through agricultural magazines. There is also ACILAC (Latin American Rabbit-Raising Scientific Association), funded in 1977.

Breeds: There are several types of rabbit breeds. In Europe, however, the raising in captivity is done mainly with the cross of two breeds, the New Zealand and the California. These breeds adapt themselves throughout the years, and produce good conversion meat and skin in Europe. In the USA other breeds are also raised.

In Santa Catarina, Brazil, a breed was developed in the municipality of Imarui which has good skin and meat conversion; in São Paulo, in the Selecta Grange, a breed was developed, a cross initiated with the Norfolk breed; in Rio Grande do Sul there are no breed specifications, although there is a predominance of New Zealand. There are producers, however, who own Angora rabbits, for wool. If given the opportunities for expansion these could be raised in large-scale.

Demand and supply: Whereas in Brazil the consumption of rabbit meat is only now awakening, that is not true of the international market, where rabbit meat is a conventional food, as important as poultry and other meats. China has an annual production of 800 million rabbits; the USA, 200 million; Italy, 120 million; France,

100 million; and Spain, 70 millions. All, however, in small farms or cooperatives.

European countries such as Italy, France, and Spain are the main consumers of rabbit meat. Italy consumes 160,000 tons/year, this supply being provided by family productions of 1,260,000 rabbit-raisers. The consumption in Italy is growing, since 50,000 tons were produced in 1960 and 100,000 tons in 1970. France, aside from its own production also requires approximately 100,000 tons/year from the external market (China). Spain has a deficit of 55,000 tons/year.

With each passing year there is a growing interest in rabbit-raising.

With respect to demand and supply in the Brazilian market, consumption may be estimated at 1 million animals/year. The markets in São Paulo, Rio de Janeiro and Porto Alegre are presently the biggest consumers of rabbit meat.

In all our contacts with producers, technicians and enterprises linked to the production and marketing, there was unanimity in the opinion that the supply is smaller than the demand. The lack of greater divulgation of the qualities of rabbit meat and the small supply of the product have not conditioned the consumer to look for it.

I believe some factors were preventing rabbit-raising propaganda in Brazil:

- the abundance of other types of meats;
- the cheap price of other meats;
- ignorance about rabbit meat;
- non-existence of a group which really believed in rabbit-raising and was really decided to promote it.

Today these factors are disappearing:

- cow meat, which would be the most abundant in Brazil, is becoming scarce; prices are increasing;
- there are already two groups in Brazil which have decided to seriously consider rabbit-raising, and are putting it in practice.
  These groups own and are constructing their own slaughter-houses.

We have no doubt about affirming that now there will be a great impulse in the exploration of rabbits, and soon we will see the Brazilian population enjoying a preference for rabbit meat.

A large number of producers and their rabbit farms have been appearing daily throughout the Brazilian territory. However, the experimental industrialization of rabbit exploration has demonstrated several gaps in our knowledge of its nutritional requirements in relation to its meat.

In São Paulo there is the Selecta Grange in Itu, with approximately 4,500 breeders, with 6 deliveries/year and an average of 6 offsprings/delivery.

In Santa Catarina there is O.B.A. (Bittencourt Food Organization S.A.), a family organization which, in the municipality of Imarui, decided to act on what they believed and established a grange, which presently counts with 15,000 breeders, with a procreation rate of 8 deliveries/year and 6-7 offsprings/delivery.

In Rio Grande do Sul some small farmers have formed rabbit-rasing cooperatives.

Rabbit Meat: It is irrefutable and unquestionable that rabbit meat is excellent in all aspects. It is composed of white muscle, thus containing a greater percentage of soluble protein than red muscle. According to research, it has greater digestibility, with a higher percentage of glycogen and less fat.

Based on studies done on the percentual composition of different species such as cows, pigs and poultry, animals which belong in the daily diet in Brazil, and rabbit, the supremacy of the latter over the others is unquestionable.

Below we transcribe a table on the amino acid content of foods (Table 1), from the Handbook of Diet Therapy by Dorothea Turnes of the American Dietetic Association, which has been translated by Lais B. Camara and is used in the Hospital dos Servidores do Estado do Rio de Janeiro.

Meat Acceptability: We know that rabbit meat is a conventional food in Europe and in the USA, but the concept is not prevalent in Brazil. However, the conventionalization of a product depends, aside from other factors, on the existence of a supply of that product. Studies carried out in the Brazilian markets, consumption adaptations, and the fact that the consumption of a specific product can provide subsistence and possibly large-scale production at a reasonable cost, are three factors which led us to study the hypothesis that "although rabbit meat may not be a conventional food in Brazil it could become so if conditions were granted to those who are becoming interested in its production".

TABLE 1

SPECIAL FOOD CONSTITUENTS PER 100 gm OF EDIBLE PORTION

FOOD GROUPS	ESSENTIAL AMINO ACIDS										OTHER CONSTITUENTS
	Arginine	Phenylalanine	Histidine	Isoleucine	Leucine	Lysine	Methionine	Threonine	Tryptophan	Valine	Cholesterol
	mg	mg	mg	mg	mg	mg	mg	mg	mg	mg	mg
Cow											
lean	1.212	0.773	0.653	0.984	1.540	1.642	0.466	0.830	0.220	1.044	37.0
medium	1.128	0.720	0.608	0.916	1.434	1.529	0.434	0.773	0.204	0.972	76.0
med. fat	1.051	0.670	0.566	0.853	1.335	1.424	0.404	0.720	0.190	0.905	
Pig											
fat	0.601	0.386	0.339	0.503	0.721	0.804	0.245	0.455	0.127	0.510	
lean	0.864	0.555	0.487	0.724	1.038	1.157	0.352	0.654	0.183	0.733	
medium	0.729	0.428	0.411	0.611	0.876	0.977	0.297	0.552	0.154	0.610	70.0
Rabbit											
cooked											
raw	1.435	0.374	0.229		3.120	1.997	0.666		0.374	0.624	71.0
Chicken											
white	1.608	0.885	0.536			1.957	0.792	1.095	0.303		
dark	1.491	0.966	0.483			1.764	0.756	0.966	0.252		
whole	1.302	0.811	0.593	1.088	1.490	1.810	0.537	0.877	0.250	1.012	113.0
Hen											
white											
dark											
whole											
cooked											
raw	1.346	0.838	0.613	1.125	1.540	1.971	1.556	0.907	0.259	1.047	101.0

TABLE 2

COSTS OF DIFFERENT TYPES OF MEATS, ACCORDING TO PLACE OF PURCHASE
(São Paulo, Nov. 1977)

SPECIES AND ANIMALS	BUTCHER'S SHOP	SUPER-MARKETS	DISTRICT MARKET	AVERAGE COST (kg)	QUANTITY	AVERAGE COST (Cr$)
Rabbit	—	34	38	36.4	180	6.61
Pig						
- meat cut 1	38	40	32	36.7	180	6.61
- meat cut 2	14	68	70	70.70	150	10.60
- meat cut 3	48	44	40	44	180	7.92
Cow						
- meat cut 1	66	35	55	52.20	120	6.26
- meat cut 2	42	24.5	37	33.50	120	4.02
- meat cut 3	40	23	37	33.30	140	4.66
- meat cut 4	38	24	35	32.30	120	3.88
Chicken	24	20	20	21.30	180	3.83
Turkey	—	30	35	32.50	180	5.85

TABLE 3

COSTS OF DIFFERENT TYPES OF MEATS, ACCORDING TO PLACE OF PURCHASE
(Florianópolis, July 3, 1978)

SPECIES AND ANIMALS	BUTCHER'S SHOP	SUPERMARKETS AVERAGE COST	MUNICIPAL MARKET	AVERAGE COST /kg
Rabbit	35	30	35	33.30
Pig				
- meat cut 1	45	45	45	45.00
- meat cut 2	65	45	68	66.00
- meat cut 3	32	35	35	34.00
Cow				
- meat cut 1	80	78	78	78.60
- meat cut 2	60	60	58	59.30
- meat cut 3	58	—	—	—
- meat cut 4	60	60	58	59.30
- meat cut 5	50	55	58	57.00

TABLE 4

CHEMICAL COMPOSITION PER 100 gm OF MEATS OF DIFFERENT ANIMAL SPECIES

ANIMAL SPECIES	ENERGY VALUE (Cal)	HUMIDITY (%)	PROTEIN (gm)	FAT (gm)	THIAMINE (mg)	RIBO-FLAVINE (mg)	NIACIN (mg)	ASHES (gm)	P (mg)	Ca (mg)	Fe (mg)
Cow	244	62.1	18.7	18.2	0.06	0.17	4.3	1.0	207	4	3.2
(semi-fat)	225	63.9	19.4	15.8	0.08	0.17	4.7	0.9	180	11	2.9
	146		20.5	6.5	0.06	0.17	4.3				
	232	63.4	18.7	16.9	0.06	0.17	4.3	1.0	210	6	3.1
Pig	216	66.9	15.5	16.6	0.83	0.20	4.4	1.0	204	5	0
(semi-fat)	185	68.0	17.3	12.3	0.84	0.20	4.5	2.4	197	10	2.6
					0.83	0.20	4.4				
	248	62.8	16.5	19.7	0.76	0.22	2.4	1.0	180	5	2.0
Chicken	170	69.6	18.2	10.2	0.08	0.16	9.0	1.0	200	14	1.5
	111		19.7	3.1	0.08	0.16	9.0		200	2	1.9
	178	69.6	20.2	10.2	0.08	0.16	9.0	1.0	200	14	1.5
Rabbit	159	68.0	20.4	8.0	0.04	0.18	10.0	1.2	210	18	2.4
	162	68.0	21.0	8.0	0.08	0.06	12.8	1.0	352	20	1.3
	175		20.8	10.2	0.04	0.18	9.0		199	13	1.9
	140	68.0	20.0	6.0	0.04	0.18	10.0	1.2	220	22	2.8

In Florianopolis, Santa Catarina, studies were carried out in the Hospital dos Servidores do Estado, in the Hospital Santa Casa Menino de Deus, in supermarkets, and even in public markets, to verify the acceptability of rabbit meat.

The results, in spite of the market prices being slightly higher than that of poultry, were satisfactory, since it is logical to expect that habits will not change radically. However, no statement can be made if the product is not easily available, and everywhere in Brazil there is a lack of rabbit meat. There is a need to place it within popular reach. In this case, we think the relationship cost-nutrition could become accessible to the popular reach.

Tables 2 and 3 detail the costs of different types of meats found during a study done in November 1977 in the capital of São Paulo by nutrition professors of USP (University of São Paulo), and in a study done in Florianopolis, Santa Catarina, in 1978.

Table 4 details the percentual chemical composition of different types of meats, according to studies carried out by nutrition professors of USP (published in the Revista nº 43, Março 1978 of SBCTA - Sociedade Brasileira de Ciencias e Tecnologia de Alimentos).

The analytical data always vary from place to place, but the average is maintained.

THE EXPERIENCE IN MEXICO ON THE UTILIZATION

OF NON-CONVENTIONAL PROTEIN SOURCES

H. Bourges and J.C. Morales de León

Inst. Nac. de la Nutrición, Div. de Nutrición

Viaducto Tlalpan y Av. S. Fernando, México 22, D.F.

Before trying to talk about the experience in Mexico with reference to the utilization of non-conventional protein sources, two reflections seem to be necessary: first, in spite of a rather inefficient agriculture, food production in Mexico has been grossly sufficient in the last two years, utilizing of course, the conventional systems. For example the total protein availability amounts to 73.5 g. per person per day, 34.5% (26 g) being of animal origin.

However, availability and consumption are low for most of the population, due to the very low income, inefficiency of food utilization at home level and at the rural areas, the sparcity of 85,000 communities whose food exchange is minimum.

The second reflection is regarding the relativity of the terms conventional and non-conventional, which vary from region to region from culture to culture, as well as with time. So as to fix what will be meant by non-conventional, the discussion will include protein sources which are not widely used in the country at the present time.

When a very active search for new protein sources was in fashion in the late fifties and during the sixties, several research groups began studies of non-conventional proteins which have continued until the present moment.

Although research has been rather active, little application of results has taken place. Briefly, studies have been conducted in the following areas: 1) leguminous seeds; 2) oil seeds; 3) single cell protein; 4) leaf protein; 5) insects; 6) Whey; 7) Bovine blood.

TABLE 1

STUDIES OF NON-CONVENTIONAL PROTEIN SOURCES
IN MEXICO — LEGUMINOUS SEEDS

GROUP	INSTITUTION	TYPE OF STUDY
Calderón, R., et al.	IMIT	Development of an infant formula based on chick-pea protein concentrate
Sotelo, A., et al.	IMSS	Composition, antinutritional factors of 33 species and varieties consumed in some regions in México.
Pérez-Gil, F., et al.	INN	Composition, nutritive value, antinutritional factors and incorporation into products of: - guaje (Leucaena leucocephala; L. esculenta); - mezquite (Prosopis laevigata); - guamuchil (Pithecollobium dulce); - alegría (Amaranthus hypocondriacus).

Source: DFNTA/(INN), Méx. 1978.

In Table 1, a summary is made of the studies on leguminous seeds. It's worthwhile commenting that many of the seeds studied are used for human or animal feeding in small regions thus suggesting the possibility of promoting a wider use in the future. So far, composition, nutritive value, content of antinutritive value, content of antinutritional factors and incorporation into different products have been the aspects more actively investigated[1,2,3].

In relation to oil seeds, the most outstanding studies are shown in Table 2. Soybean has been by far the best explored source of this group and in general, of all the non-conventional proteins, both by academic institutions and the industry. Since industrial research is not made public, it is not included in the table. Soybean has been used for enrichment and complementation of cereals but extension of animal products such as meat, eggs, milk and fish have obtained considerable success.

TABLE 2

STUDIES OF NON-CONVENTIONAL PROTEIN SOURCES IN MEXICO — OIL SEEDS

GROUP	INSTITUTION	TYPE OF STUDY
Camacho, J.L. et al.	INN	Adaptation of traditional Asian techniques. Germination and fermentation; "soymilk".
Morales, J.C. et al.	INN	Direct use by soaking and cooking.
Tovar, R. et al.	UNAM	Adaptation of traditional Asian techniques. Germination and Fermentation.
Del Valle, F. et al.	ITESM	Limecooking of corn soybean mixtures to prepare "tortillas".
Perez-Villaseñor, J. et al.	UAMI	
Bourges, H. et al.	INN	

Adaptation of oriental techniques for the direct domestic use of soybeans is receiving special attention as well as the introduction of the soybean grain in traditional Mexican recipes, particularly "tortilla" and "frijoles". Sesame sunflower and peanuts have also been investigated[4-14].

Table 3 shows the studies conducted on single cell proteins. Casas Campillo and his group have gained considerable experience in relation to the production of several microbiological proteins, particularly yeast[15,16].

Several years ago, Monroy[17] studied "xastle" — which is the residue of the fermentation of agave juice — to obtain "pulque", which is one of the main fermented beverages utilized in Mexico, of which millions of liters are produced daily in the rural areas. Xastle is a heterogeneous mixture of microorganisms which are currently wasted or utilized for animal feeding.

Finally pionner studies were carried out in Mexico in relation to blue-green algae of the "spirulina" genus, growing wildely in alkaline lakes which are very common in the country[18,19].

TABLE 3

STUDIES OF NON-CONVENTIONAL PROTEIN SOURCES IN MEXICO – SINGLE CELL PROTEIN

GROUP	INSTITUTION	TYPE OF STUDY
Casas Campillo, C. et al.	IPN	– Composition, nutritive value and production technology of several species of yeast, bacteria, fungii and algae. – Study of different sustrates (hydrocarbons, molasses and alcohols).
Viniegra, G. et al.	UNAM	– Fermentation processes and utilization of SCP for animal feeding.
Monroy, R. et al.	INN	– Composition nutritive value and use of "Xastle".
Bourges, H. et al.	INN	– Composition, nutritive value of spirulina and its incorporation in mixtures. – Composition and nutritive value of saccharomyces cererisae.

Source: DFNTA/INN (Méx.), 1978.

Table 4 summarizes the studies on leaf proteins, insects, whey and bovine blood.

With reference to leaf protein, Parada and his group at the Instituto Politécnico has studies several plants which are abundant in Mexico, as well as the extraction of the protein and its incorporation into different products[20-24].

Several species of insects which are consumed at regional level were explored as to their protein content and quality[25-26].

Whey is included here because, at least in Mexico, in 1973 it was hardly considered a conventional food source for humans; that year, two products were developed at the Institute and introduced into market for nutritional programs: CONLAC, a modified milk for infants and "biscuits" or "waffers" made from a pressed mixture of skim milk, whey and soybean flour[27]. Racotta[28] developed a simple procedure for the obtention of whey proteins concentrates at rural level.

Finally, Lara et al[29] at the National University found a procedure to concentrate whole bovine blood protein which unfortunately has a low protein quality.

This compilation has by no means been completed nor is it in a chronological order. We have included all the studies we could find published and those we know about but probably more has been done and not published or is on the way.

It may seem surprising to realize how little of the research results has been utilized. It may be due to the fact that some of the projects have not been practicaly oriented, that most of the protein sources mentioned are too expensive and their sensory characteristics are poor. Since Mexico is a country whose conventional protein production is not fully consumed due to the low purchasing power of great sectors of the population it can not compete with export and animal feeding demands. For a new protein source to succeed it should have: a very low price, a high availability and be sensorial and culturaly acceptable.

Only soybaen and to a certain extent whey, have these caracteristics in Mexico and these are the reasons why they are the ones utilized at the moment.

Although soybean was introduced in Mexico 40 years ago, only during the present decade has cultivation been significant. Except for the last two years, production has increased up to half a ton per year. The total consumption is one million tons, most of which is used as a source of oil, the protein cake being utilized for animal feeding. Several companies have built plants for production

TABLE 4

STUDIES OF NON-CONVENTIONAL PROTEIN SOURCES IN MEXICO

GROUP	INSTITUTION	TYPE OF STUDY
Leaf Proteins		
Carviotto, R.O. et al.	IN	– Composition of several plants common in Mexico.
Parada, E. et al.	IPN	– Composition nutritive value and obtention of LPC from: alfalfa (Mendicago sativa); Watercress (Nasturtium aquaticum) and water lily (Eichornia crassipes). Incorporation of alfalfa LPC into tortillas and adaptation of extraction process for rural use.
Insects		
Conconi, J. et al.	UNAM	– Composition and nutritive value of species consumed in Mexico: Atta mexicana; Atizies taxcoensis; Hypota agavis and Lasius eskamole.
Bourges, H. et al.	INN	
Del Valle, F. et al.	ITESM	– Composition nutritive value and obtention of flour of cricket.
Whey		
Bourges, H. et al.	INN	– Composition nutritive value and incorporation into several products.
Racotta, V. et al.	IPN	– Obtention of whey protein concentrates.
Bovine Blood		
Lara, F. et al.	UNAM	– Composition nutritive value and obtention of the concentrate.

of whole defatted and extruded soybean for human consumption; the production amounts to only 20,000 tons per year which is barely 2%.

Most of the flour for human consumption is used by bakeries as partial substitute of eggs and milk and most of the extruded soybean is utilized by meat processors to extend products, a very small amount is used by chocolate and icecream manufacturers.

One company has specialized in selling soybean based products to the public, such as: powdered beverages, breakfast cereals, meat analogs, special products for lactose-intolerant babies, as well as plain flour for domestic use. All these products, which are for direct human consumption, represent a very small amount within the 20,000 ton that have already been mentioned.

Since the possible benefit of using soybean would be its low price and since unfortunately most of the industrial products mentioned in the last paragraph are very expensive, their little success is not surprising, for example, extruded soy is sold for a price over ten times that of soybean seed and quite similar to the price of meat.

If the price of the soybean seed is considered, it becomes the cheapest of all the protein sources in Mexico, including ceverment to use it in nutritional programs.

Among the first was the enrichment of "tortilla" flour; this project has not had relevance since: a) soy-bean tends to alter the typical elasticity and flavor of "tortilla"; b) population in need of nutritional help does not consume industrialized corn flour and c) the price of "tortilla" is controlled and any change becomes a delicate political issue.

An alternative to give soybean a role in social benefit has been the extension of animal protein. At least in theory, this procedure should reduce the price and increase the weight of the final mixture of soybean with the animal protein source.

However extension should not lower the nutritive quality of the product if it is done correctly. Furthermore, experience of the Institute shows a very high potential demand of animal protein by the population not expressed normally due to their very high prices.

The first product of this kind was a beverage based on skim milk extended with soluble soybean flour and dextrins called SOYACIT, which was commercialized on the free market for a private firm at prices below those of milk. Sales reached approximately 10 tons a month, equivalent to 80,000 liters of the reconstituted beverage. Today this product has virtually disapaered due to circumstancial disagreement between the government and the manufacturing firm.

Nevertheless, many similar formulations have been developed which may be produced in the future.

Another interesting product is a compressed "waffer" or "pellet" based on a dry mixture of skim milk, whey, soybean flour and vegetable oil, which, with the name NUTRIMPI and LACTODIF has been utilized during the past year for the school breakfast program with a production of almost 1,000,000 million units per day at a price of 0.03 US Cy per unit[30].

In relation to meat extension, two products were developed in 1974 for IDA, a company dependent of the Mexico city slaughterhouse. One of them, PROTE-IDA, consisted of 30% lean beef meat, 30% extruded soybean, 20% pork leftovers, 16% wheat flour and 4% condiments, presented as meat balls or patties. Since the marketing of this product was difficult, a second product MOL-IDA was developed, which consisted simply of lean beef mixed with 20% texturized soybean and commercialized by a system of trailler shops which every day visited the poor areas of the city, selling the kilogram of product at a price equivalent to one dollar.

The demand for this product was very high, so procution had to be increased in a few months to 20 tons a day and plans exist for a plant to produce 50 tons/day. A considerable number of families who could not normally buy meat were able to buy this product[30, 31].

With reference to whey, as has been mentioned previously, before 1974 it was not used in our country for human consumption. The local production was either given to animals or thrown away.

The same year, a special formula for child feeding was developed for regions where milk is not available in sufficient qualities. It consisted of dry skim milk, dry-whey, vegetable oil sacarose, dextrins and was supplemented with a mixture of vitamins and iron. The price of this mixture, in a hard paper package, was about 12 pesos/kg, equivalent to 1.50 pesos/liter compared to 4 pesos for fresh milk and 6 pesos for commercial powdered infant formulas.

This product is being sold with the name CONLAC. Unfortunately, the original goals were some what defeated. Circumstances have demanded the product for competition with other modified milks in the urban market and the high and medium economical strate. The inexpensive paper package has also been substituted by a tin can, too sophisticated to maintain a low price, reaching 24 pesos per pound at the present time.

However it is still cheaper than the first class fresh milk (6.50 to 7.00 pesos) and of the infant formulas (9 pesos per liter).

In summary, CONLAC still has an advantage since it represents a saving for the people, but returning to the original concept, it would certainly help the sectors in need. Present production is 10 ton a day.

As mentioned previously, while discussing about soybeans, NUTRIMPI and LACTODIF include whey in their formulation. Actually these products are a soybean extended pressed CONLAC.

To conclude this short presentation, we may comment that at the project level there are many products and formulations intended to help our population in acute need of inexpensive foods, particularly for child feeding; many of these products which include soybean and whey, could not be considered as non-conventional protein sources, nowadays.

LIST OF ABREVIATIONS

IMIT   — Instituto Mexicano de Investigaciones Tecnológicas

IMSS   — Instituto Mexicano del Seguro Social

INN   — Instituto Nacional de la Nutrición

UNAM   — Universidad Nacional Autónoma de México

ITESM   — Instituto Tecnológico y de Estudios Superiores de Monterrey

UAMI   — Universidad Autónoma Metropolitana Ixtapalapa

IPN   — Instituto Politénico Nacional

REFERENCES

[1] De Sandi, J.O., Uribe, M.G., and De la Vega, J.A., Evaluación Nutritiva de Leguminosas Silvestres. Tesis UNAM Fac. de Química (1976).
[2] Calderón, R., IMIT Personal Communication (1978).
[3] Pérez-Gil, F., INN Personal Communication (1978).
[4] Camacho, J.L., INN Personal Communication (1978).
[5] Morales de León, J.C. and y Zardain, M.I., Evaluation of different treatments on the soybean to inhibit the activity of the anti-nutritional factors and increase its direct consumption in the rural area. Rev. Technol. Aliment. (Méx.) XII (4-5): 133, (1977).
[6] Tovar, R. UNAM Personal Communication. Trabajos presentados en el V Congreso Nacional de ATAM, México (1974).
[7] Berra, R. Efecto del remojo en algunas propiedades físicas, bio-

químicas y organolépticas de la soya. Rev. Tecnol. Aliment. (Méx.) IX(2): 76 (1974).

[8] Pérez Villaseñor, J., Del Valle, F. & Saleme, M.M., Enriquecimiento de las tortillas con proteínas de soya por medio de la nixtamalización de mezclas de maíz y frijol de soya. Rev. Tecnol. Aliment. (Méx.) IX(1): 24 (1974).

[9] Bourges, H. Personal communication (1978).

[10] Marcos Báez, F. Elaboración de una harina de ajonjolí - evaluación biológica y su posible uso como alimento humano. Tésis UNAM - Fac. de Química (1975).

[11] Morales de León, J.C. Desarrollo de productos alimenticios a base de proteína de origen vegetal para consumo humano. Rev. Tecnol. Aliment. (Méx.). In press.

[12] Rocío Hernández. Obtención de harina y concentrado proteínico a partir de semillas de girasol (Helianthus Annuus L.) Tesis Fac. Química UNAM (1978).

[13] Trejo, A. et al., Incorporación de concentrados y aislados proteínicos de semillas de girasol y coco a refrescos y bebidas con pulpa de frutas. Trabajo presentado en el Premio Nacional en Ciencia y Tecnología de Alimentos (1977).

[14] Tufiño, S. Unpublished data.

[15] Casas Campillo C. Proteínas alimenticias de origen unicelular. Rev. Tecnol. Aliment. (Méx.) XI, (3): 124 (1976).

[16] Viniegra, G. et al., UNAM, Personal communication (1977).

[17] Monroy, R., Chávez, A. & Hope, P. El valor nutritivo del sedimento del pulque. VII Reunión Anual Soc. Mex. de Nutr. y Endocr. Guadalajara, Jal. (1977).

[18] Bourges, H., Sotomayor, A., Mendoza, E. & Chávez, A., Utilization of the alga spirulina as a protein source. Nutrition Reports International 4, (1): 31-43 (1971).

[19] Bourges, H. et al. INN, Personal communication (1978).

[20] Cravioto, R.O., Masieu, J. & Guzmán, J. Composición de Alimentos Mexicanos. Ciencia 11:129-155, (1951).

[21] Parada, E. & Hope, P., La Alfalfa - alimento como fuente de proteínas para alimento humano. Rev. Tecnol. Aliment. (Méx.) V, 1 y 2, 36 y 4 (1970).

[22] Parada, E. & Alcantar-González, E. Estudio Preliminar del lirio acuático como alimento humano. Rev. Tecnol. Aliment. (Méx.) X(2): 68, (1975).

[23] Lara García, E. & Parada, E. Separación térmica de las fracciones proteínicas de hojas. Rev. Tecnol. Aliment. (Méx.) XI(6): 269 (1976).

[24] Parada, E. Elaboración rural y industrial de concentrados proteícos de hojas. Rev. Tecnol. Aliment. (Méx.) XI, (6): 269 (1976).

[25] Conconi, J. & Bourges, H., UNAM-INN Personal communication. (1977).

[26] Del Valle, F. I.T.E.S.M. Personal communications. (1977).

[27] Bourges, H. INN Personal Communication. (1978).

[28] Racotta, V. Posibilidades de aprovechamiento del suero lacteo.

Rev. Tecnol. Aliment. (Méx.) In press.

[29] Lara, F., Ramírez, F. & Sánchez, N. Estudio para la obtención de proteínas de sangre de bovino. Rev. Tecnol. Aliment. (Méx.) XII (4 y 5): 141 (1977).

[30] Bourges, H. Experiencia del Programa Nacional de Alimentación (México) con el uso de la Soya. En Memorias de la Primera Conferencia Latinoamericana sobre la Proteína de Soya, p. 33 (1975).

[31] Camacho, J.L., Bernal, E. & Bourges, H. Diseño y Evaluación nutriológica de productos a base de carne extendidos con soya. Rev. Tecnol. Aliment. XI(1): 5 (1976).

AGRICULTURAL PRODUCTION OF SOYBEANS:

RELATION BETWEEN VARIETIES AND NUTRITIVE VALUES

A. Lam-Sánchez

School of Agric. & Vet. Sci., State Univ. of São Paulo

Jaboticabal, São Paulo - BRAZIL

Soybeans have been pointed out as one of the solutions for the nutrition problem in the tropics, because, besides being a good source of protein and oil, they have a good yield potential in this area, where yields above 3.000 kg/ha have been obtained in experimental results[1]. There are still some problems that need to be solved. One of them is related to the obtaining of adapted varieties and consequently related to plant breeding.

A good variety is a result of three main factors: yield, nutritive value is concerned with the total amount of protein, based on a good balance of the essential amino acids, and protein digestibility.

In general soybeans present 40% protein, 20% oil and around 34% carbohydrates[2]. A wide range of variability for protein and oil has been found and it is conditioned by genetic and environmental factors. Piper & Morse[3] found 30% to 46% protein in 500 samples; Dies[4] screening 128 varieties found a range of 32.4 to 50.2% protein; Commercial varieties planted in the United States present as average of 39 to 41%[5]. Results obtained in São Paulo show a variability of 31.8 to 38.0%[6] with brazilian material and introduced lines from USA; and 25.8 to 40.6% in material planted in Guatemala[7].

Protein and oil are negatively correlated, r: -0.26 to -0.74[8] and also protein is correlated with yield in the same way, r -0.08 to -0.33[9,10]. Lately, some varieties were released in the United States such as 'Protana' and 'Bonus' that produce 43% protein without a significant reduction in yield and oil.

The nutritional quality of soybean protein is limited by a defficiency in sulfur amino acids, methionine being the limiting one. Cystine will supplement methionine to a limited extent[11,12]. On the other hand soybean protein presents a good lysine content which makes it suitable for use in the cereal-legume system of nutrition.

Variability of methionine content has been found; it also is conditioned by genetic and environmental factors. Alderks[13] in 1949 found 1.28 to 1.53 g/16gN in 20 varieties; Krober[14] in 1956 found 1.3 to 1.7 g/16gN this range was conditioned by locality and planting season; Krober & Cartter[15] found 1.0 to 1.7 g/16gN related to protein content; Kakade et al[16] found values of 1.0 to 1.9 g/17gN Our results showed a range of 1.01 to 1.82 g/10gN in Brazilian material and 0.41 to 2.34 g/16gN in the Guatemala material[6,7].

There is a positive correlation value between methionine content and protein content in the seed (0.56 to 0.58)[15]. However, if breeding for protein is being carried out, this has to be done in such a way to increase two protein fraction that can be easily changed genetically and has at the same time, a high content of the limiting amino acids; as indicated by Roberts & Briggs[17], the 7S fraction of the globulins presents only 0.19% of menthionine.

Trypsin inhibitors have been extensively studied since they decrease 50% of the soybean nutritive value if they are not eliminated[18]. Generally this elimination is achieved by heat treatment at 15 lb/in^2 for 10 to 15 minutes. Variability for trypsin inhibitor was found by Kakade et al[16] in 108 varieties with the values of 66 to 233 TIU/mg of protein. Our results show a range of 22.4 to 37.2 TIU/ml in the Brazilian material and 25.4 to 45.5 TIU/ml in the Guatemalan material[6,7].

There is some uncertainty with relation to the real effects of these toxic factors. In experimental results extruding 'Pelican' variety by the Brady Crop Cooker, trypsin inhibitor values were not decreased significantly, but the results obtained with biological assays were good. The same material was also extruded at Colorado, obtaining the same results.

CONCLUSIONS

Nutritive value of soybean protein is the result of proper content based on a good balance of essential amino acids and digestibility. Breeding soybeans for nutritional quality has to be based on the following considerations:

Due to the correlations found between protein and yield, as well as between protein and oil, as correct amount of protein has to be achieved without decresing yield. Breeding for protein must be done in increasing the protein fractions that can be easily changed genetically and at the same time have a high content of the limiting amino acid. However, care must be taken since possible changes in the content of certain amino acids could occur if the first and second limiting ones are increased.

As shown by the results obtained, more evidence is needed to understand the effect of trypsin inhibitors on protein utilization in soybeans.

REFERENCES

[1] Lam-Sánchez, A., 1975, Alguns aspectos de produção da cultura da Soja, in: IV Simpósio Brasileiro de Alimentação e Nutrição (SIBAN), Botucatu, São Paulo, January 1975.
[2] Nakamura, S., 1967, Review of PL 480 work on soybean carbohidrates, in: "Proceedings of International Conference on Soybean Protein Foods", ARS-71-35, May 1967, ARS, U.S. Dep. Agr. Peoria, Illinois. pp. 249-254.
[3] Piper, L.W., & W.J. Morse, "The soybean", Mc Graw-Hill Book Co., New York, 1923.
[4] Dies, E.J., "Soybeans - Gold from the soil", Mac Millan Co., New York, 1942.
[5] Hartwig, E.E., 1973, Varietal development, in: "Soybeans: improvement production and uses", B.E. Caldmell (ed.), Amer. Soc. Agron. Inc. Madison, Wisconsin, pp. 187-210.
[6] Lam-Sánchez, A., & J.E. Dutra de Oliveira, 1973, Protein and oil content in several soybean varieties (unpublished data).
[7] Lam-Sánchez, A., L.G. Elias, R. Gomez Brenes & R. Bressani, 1977, Estudo de variedades de soja (unpublished data).
[8] Weiss, M.G., C.R. Weber, L.F. Willians & A.H. Probst, 1952, Correlation of agronomic characters and temperature with seed compositional characteres in soybeans, as influenced by variety and time of planting. Agron. J. 44: 289-297.
[9] Johnson, H.W., H.F. Robinson & R.E. Comstock, 1955, Genotipic and phenotypic correlations in soybeans and their implications in selection. Agron. J. 47: 477-483.
[10] Byth, D.E., C.R. Weber & B.E. Caldwell, 1969, Correlated truncation selection for yield in soybeans. Crop. Sci. 9: 699-702.
[11] FAO, Amino acid contents of foods and biological data on proteins, Rome, 1968.
[12] Cravens, W.W., & E. Sipes, 1958, Soybean oil meal in: "Processed Plant Protein Foodstuffs", A.M. Altschul (ed.), Academic Pres Inc. Publishers, New York, pp. 333-397.
[13] Alderks, O.H., 1949, The study of 20 varieties of Soybeans with respect to quantity of oil. isolated protein, and the nutritional value. J. Amer. Oil. Chem. Soc. 26: 126-132.
[14] Krober, O.A., 1956, Methionine content of soybeans as influence by location and season. J. Agric. Food. Chem. 4: 254-257.

[15] Krober, O.A., & J.L. Cartter, 1966, Relation of methionine content to protein levels in soybeans. Cereal Chem. 43: 320-325.

[16] Kakade, M.L., N.R. Simons, I.E. Liener & J.W. Lambert, 1972, Biochemical and nutritional assesment of differenties of soybeans. J. Agric. Food. Chem. 20: 87-90.

TYPES OF SOY PROTEIN PRODUCTS

J. Wapinski

Manager, Intern. Tech. Serv. Ralston Purina

Checkerboard Square, St. Louis, Missouri, U.S.A.

Soybeans were grown in Eastern Asia and China long before writeen records. However, soybeans have only been produced on a commercial scale in the United States since the 1920's.

The total world production of soybeans since 1945 has had a dramatic rise from 503 million bushels to 2,400 million bushels in 1975. While the United States is the leading producer of soybeans in the world, Brazil's crop has shown a dramatic increase over the last 10 years. The expanding market for soybeans has been principally for high-protein supplements for animals, and soybean oil for food uses. Soybean meal is principally used in high-protein supplements for poultry, hogs, cattle and livestock. The demand for soybean protein products in food applicantions has been growing rapidly; however, less than 5% of the total supply of soybean protein is used directly in food applications.

The common field variety of soybean used in food applications are shperical and yellow in color. There are approximately ten varieties grown in the U.S., bassically in the corn belt.

The major structural parts of the soybean are the hull and the cotyledon. The surface of the cotyledon is covered with an epidermis and the interior is filled with numerous elongated palisade-type cells. The majority of the proteins occur as protein bodies in these cells. The oil is present in these cells in the form of spherosomes. The bulk of the protein and oil are found in these cells.

The raw material for protein products derived from soybeans for human consumption is nº 1 and nº 2 U.S. grade soybeans. Proper

handling and cleaning of soybeans is important in maintaining high-
-quality raw material. Soybeans are stored at a moisture content
of less than 12%, which helps maintain them is a stable condition
because they may be in storage up to one year. At the time of processing, the beans are cracked and dehulled. The cotyledons are
then broken into 6 to 8 parts and flaked to produce thin particles
so that the oil can be extracted more easily. The most common oil
extraction method in commercial operations utilizes hexane as the
solvent. Products from this oil extraction system are crude soybean oil and defatted soybean flakes. The defatted soybean flakes
require desolventization and can then be further processed to food
ingredients.

   The crude oil is refined into a number of products for food use.
Soybean oil has been a principal product from soybeans for many
years. Major food uses of soybean oil are shortening, margarine,
salad and cooking oils, and prepared dressings. Soybean oil is present in 82% of margarine, 75% salad and cooking oils, and 80% of prepared dressing markets.

TYPYCAL COMPOSITION OF SOYBEANS
AND SOYBEAN PRODUCTS

   Soybeans contain approximately 42% protein, 20% oil, 35% carbohydrates, 5.0% ash and 5.5% crude fiber. Defatted soy flakes are
used as the starting material for nearly all food protein products
and they contain at least 50% protein. Soy protein concentrate
contains at least 70% protein and approximately 24% insoluble carbohydrates. Isolated soy protein contains a minimum of 90% protein
and small amounts of carbohydrates and curde fiber. The ash content
is approximately 5.0%, but can vary depending on the process employed
in production. The residue oil content is less than 1%.

   The sulfur containing amino acids are usually considered the
limiting amino acids in animal feeding. Soybean protein is considered a high-quality protein and is used widely. The methionine
and cystine content of isolated soy protein is slightly lower than
soy flour and soy protein concentrate.

   Soybean contain soluble and insoluble carbohydrates. The soluble carbohydrates are principally raffinose, stachyose, and sucrose.
Raffinose and stachyose have been implicated as the flatulence factor
of soybean products. These sugars are not digested in the stomach
and small intestine and are thought to be fermented in the lower
intestinal track.

   Insoluble carbohydrates are principally the cellular structure
of the cotyledon cells. Hemicellulose and cellulose are the major
components with some lignin and pectin present.

TYPES OF SOY PROTEIN PRODUCTS

SOY FLOUR

Various types of soy flour can be produced. The primary differences are the types of desolventizer used to remove the solvent from the defatted soybean flakes. Soy flour and grits are generally classified to specific particle sizes, the particle size of soy flour will be less than 200 mesh.

The various types of soy flour products are classified as dry powders and textured products. Dry powders can range in particle size from flakes to less than 300 mesh. Solubitily and trypsin inhibitor activity can vary from high levels in an uncooked soy flour to low levels in a toasted soy flour. A number of products can be produced from soy flour as: defatted, low-fat, and lecithinated products. These products vary in their fat content.

Textured Soy Flour

The two most common methods of texturizing soy flour are the steam expansion and the thermo-plastic extrusion processes. The textured soy flour can be produced to desired shape and particle size. Colors and flavors can be added to produce a wide variety of products for a number of food applications.

SOY PROTEIN CONCENTRATE

Soy protein concentrate is a product prepared from high-quality, sound, clean, dehulled soybeans by removing most of the oil and water soluble non-protein constituents, and shall contain not less than 70% protein on a moisture-free basis.

There are three generally recognized commercial processes for producing soy protein concentrates. They are alcohol leach, dilute acid leach, and the moist heat, water leach process. Concentrates are prepared from defatted soybean flakes by immobilizing or insolubilizing the major protein fraction. The soluble sugars, ash, and protein are removed from the insolubilized protein fraction and insoluble polysaccharides.

Soy protein concentrate products can be of two types: Dry powders and textured products. The dry powders can vary in particle size from flakes to spray-dried particles. The aqueous alcohol and moist heat process usually produces products with low solubility, whereas, soy protein concentrates produced using the acid leach process can be of higher protein solubility. The fat and water absorption of these products can be varied to some extent based on the pH and solubility of the concentrates produced.

The dry powders are available in various particle sized and pH. The textured products can be produced in various particle sizes and shapes. They are processed similar to textured soy flours. Colors and flavors can also be added to fit the food application.

ISOLATED SOY PROTEIN

Isolated soy protein is the major protein fraction of soybeans prepared from high-quality, sound, clean, dehulled soybeans by removing a preponderance of non-protein components and shall contain not less than 90% protein on a moisture-free basis.

The general process for producing isolated soy protein is as follows. High DPI defatted soy flakes are aqueous extracted. The proteins and soluble carbohydrates are solubilized and then the residue, which is primarily the insoluble polusaccharide fraction of the defatted flakes, is removed by centrifugation. The extract is then adjusted to pH 4.5 with food grade acid. This causes the major protein fraction to precipitate into a product known as protein curd. Protein curd is then washed and dried to produce an isoelectric form of isolated soy protein or can be adjusted in pH with a number of different cations to produce neutralized products, and spray dried.

EXTRACTABILITY OF PROTEINS IN DEFATTED SOYBEAN MEAL AS A FUNCTION OF pH

The principle of isolating the major protein fraction of soy is based on the pH extraction curve. The protein is solubilized at a pH above 7 and below pH 9.5 under controlled time and temperature. The major high molecular weight proteins are precipitated at pH 4.5, thus allowing for separation. The primary protein is the high molecular weight globulin fraction, of the protein present in soybeans. The soluble sugars are separated in the soy whey along with low molecular weight proteins and soluble minerals.

TYPES OF ISOLATED SOY PROTEIN PRODUCTS

Isolated soy proteins are available in dry powdered form and textured form. The dry powder can be produced to have varying dispersibility, emulsification properties, gelation, solubility, and viscosity properties. The functionality of the isolated soy protein can be altered by mechanical, physical and enzymatic processing methods. These products are generally designed to perform specific functions in food systems.

Three forms of isolated soy proteins are available. The spray--dried powder, spun protein and structured isolated soy protein. The

TYPES OF SOY PROTEIN PRODUCTS

textured products are usually sold in a hydrated state either refrigerated or frozen. They can be produced with varying textural properties, color, and flavor.

Soy protein products are used in a wide variety of food products. The successful incorporation of soy proteins into these food products requires that the protein exhibit the functional properties similar to traditional protein. Therefore, the protein products are designed and processed to exhibit desired functional properties. The major uses of soy protein products are in emulsified meat products, bakery products, dairy-type products, infant formulas, protein supplements, meat analog products, and fish products.

There are many examples of emulsified meats in which soy protein products are used. The products are used to emulsify fat and water and give texture and stabilize the emulsions.

The type of soy protein product employed depends on the functional requirement of each specific application. The protein products are hydrated with various amounts of water, depending on the product employed. Emulsified meat products include the balogna, sausages, hot dogs, and the wide variety of related products. The course ground products such as hamburger-type products utilize different soy products. Protein products are also used in poultry products such as turkey rolls and emulsified poultry products.

A new technique of injection of isolated soy protein into ham, fish, and poultry products without griding or chopping the meat products has been developed. This allows for increased yield of products such as ham, roast beef, corn beef, fish and turkey breasts.

A more recent development has been the use of isolated soy protein to replace the functional properties of surimi, which is a fish paste product used in a wide variety of traditional Japanese food products. An example of these products are Kamaboko, Chikuwa, Agekama, and fish sausage. A portion of the fish paste is replaced with isolated soy protein and water. The protein is required to function in the product by supplying the desired textural and organological properties similar to the traditional product.

Soy protein products have been used in baked goods for several years. Soy protein products are used, in general, to replace non--fat dry milk, to control textural properties, fat absorption, water absorption, and volume of these products. Also, certain high protein foods can be produced such as high protein bread and cookies. Soy protein products are used in varying amounts in products such as cake, donuts, and bread.

Isolated soy protein is the principal protein source used in hypoallergenic infant formulas. There are many examples of commercially marketed infant formulas in which the protein source is isolated soy protein. The protein emulsifies and stabilizes the fat and water, as well as provides the protein requirement.

Soy protein products are also used in a wide variety of protein supplements. The soy protein is used primarily as a protein nutritional source for these products.

Various types of soy protein products are also used in dairy--type products. The functional requirement in these products is very specific. The products include whipped toppings, coffee whiteners, and yogurt.

Meat-analog type products can be formulated to meet various dietary requirements and to supply nutrition to populations that are vegetarian. The soy proteins are utilized with various other vegetable protein sources in combination with the appropriate vitamins, minerals, carbohydrates, and fats.

In conclusion, soybeans and soybean products have a long history of use as food for humans. The soy protein products are used in a wide variety of food products to meet specific nutritional, functional, and organological requirements. It is important to recognize that there are different types of soy protein products and they vary in chemical, physical, and functional properties. It is important that the appropriate type of product is used in each application, so that the ultimate value and quality of food products be maintenained or enhanched.

CONVERSION OF RAW SOYBEANS INTO HIGH-QUALITY PROTEIN PRODUCTS

J.J. Rackis

Research, Science & Educ. Adm. U.S. Dept. Agric.

Peoria, Illinois 61604

INTRODUCTION

Use of soybeans for food in the Orient goes back to ancient times; their history elsewhere covers a more recent period. Commercial processing, which began in Europe in 1908, accelerated rapidly in the United States. The invention of solvent extraction in 1934 and subsequent development of desolventizer-toasters made it possible to inactivate antinutritional factors and to monogastric animals. Although nutritionally soy is also suitable for another monograstric species (man), its introduction into the Western diet was greatly limited by its poor flavor qualities. Today, emerging technology has resulted in the production of a wide array of soy protein products with improved organoleptic properties, ranging from simple flours to sophisticated spun food analogs.

Among agricultural products offering the most promise for human nutrition is the soybean, which by virtue of its high protein content (ca.40%) and its desirable amino acid composition can very nearly satisfy the amino acid requirements of humans. The sulfur amino acids are first limiting in soybean protein. It appears that methionine supplementation may not be required for soy proteins to successfully replace animal protein in a mixed diet to maintain nitrogen balance in young children and adults (Keystone Conferencem in press). Soy protein has had a long history of use, primarily as fermented foods in the Oriental countries; since 1928 it has been used in this country as a replacement for milk protein in infants allergic to cow's milk and intolerant to lactose (Sarett, 1976), as a substitute for meat protein in the school lunch program (Food and Nutrition Service, 1971a, 1971b), and as products available to the consumer in the form of textured meat analogs.

The U.S. production of 4 million bushels of soybeans in 1922 has increased to an estimated figure of more than 1.7 billion bushels in 1978. The Foreign Agricultural Service has estimated 1977 world production of soybeans at 72.2 million tons; U.S. and Brazil crops account for 62 and 17% of the world crop, respectively. Thus, soybeans emerging from the Far East have become a primary international protein resource.

In view of the predictions that the use of soy protein will propably double in the next 5 to 10 years, it becomes increasingly important to assess, on a continuing basis, the significance of the biologically active constituents in soybeans, particularly as new processes are developed to manufacture products with improved organoleptic and functional properties.

The trypsin inhibitors (TI) are a major factor and require special considerations. The chief physiological response to the TI's is the development of pancreatic hypertrophy, which is believed to be the underlying cause of growth inhibition that accompanies the ingestion of raw soybeans. Although conventional processing of soybeans reduces TI activity to levels that appear innocuous to animals in short-term experiments, the question of whether the ingestion of low levels of TI over a long time would cause pancreatic hypertrophy is just begining to be answered. Human tests on soy protein quality are of short duration and as a consequence, we must turn to animal experiments for answers. Results of numerous chemical analyses and animal experiments indicate that properly processed soy protein products are unlikely to cause health problems in man (Rackis, 1974).

This report reviews the evidence that the simultaneous destructions of TI's and transformation of raw protein into forms more readily digested are primarily responsible for the conversion of raw soybeans into products of excellent protein quality. This article also provides information on the factors that affect mineral bioavailability and the goitrogenic, estrogenic, and allergenic constituents in soy proteins.

NUTRITIONAL ASSESSMENT OF SOYBEAN TRYPSIN INHIBITORS

Soy TI's are proteins that inhibit proteolytic enzymes. Of specific nutritional interest is the inhibition of trypsin and chymotrypsin in the intestinal tract of animals. TI's are present in most animal tissue. They are widely distributed in the plant kingdom and are present in large amounts in Leguminoseae (peas, beans), Gramineae (cereal grains) and Solonaceae (potatoes, eggplant, and various other vegetables). Large numbers of species within these families are important sources of foods that are eaten directly or after cooking. The multiple forms of TI's present in soybeans are genetically controlled variants. Frequency of distribution of the

TABLE 1

NUTRITIVE VALUE AND TRYPSIN INHIBITOR CONTENT
OF DIFFERENT VARIETIES AND STRAINS OF SOYBEANS[a]

SAMPLE	PROTEIN (%)	RAW FLOUR TIU[b]/mg PROTEIN	PER[c,d]	
			Raw	Toasted[e]
Disoy	39.6	100	1.47	2.66
Provar	41.2	106	1.60	2.20
PI 153319	36.6	168	0.88	2.46
Hark	39.1	100	1.21	2.32
PI 153206	37.5	139	1.21	1.95

[a] Kakade et al. (1972).

[b] Trypsin inhibitor units

[c] Protein efficiency ratio (gram weight gain per gram protein consumed). Data adjusted to a basis of PER = 2.50 for casein.

[d] Pancreas weights were normal in rats fed toasted soy and were negatively correlated with PER in rats fed raw soy.

[e] Autoclaved at 15 psi (120°C) for 30 min.

TI variants differs in Japanese (Kaizuma and Hymowitz, in press), USDA (Hymowitz, 1973), and European soybean germ plasm (Skorupska and Hymowitz, in press). Kakade et al. (1972) analyzed 108 soybean varieties and strains for TI activity and found values ranging from 66 to 233 TI units per mg protein. Raw flours prepared from these soybeans inhibited rat growth and enlarged pancreas. Data for five of these flours are given in Table 1. Toasting was required to obtain maximum nutritional value and to eliminate pancreatic hypertrophy.

Biochemical Effects of Raw Soybeans

A listing of the biological and physiological effects that occur in animals fed raw soybean meal is given in Table 2. Hemo-

TABLE 2

PROPERTIES OF HEAT-LABILE FACTORS IN SOYBEANS

Inhibit growth	Stimulate pancreatic enzyme secretion
Reduce protein digestibility	Stimulate gallbladder activity
Increase sulful amino acid requirements	Reduce metabolizable energy
Enlarge pancreas	Inhibit proteolysis

glutinating activity is no longer considered to be a significant physiological factor since Turner and Liener (1975) demonstrated that soybean hemoglutinin selectively removed from crude raw soy extracts by affinity chromatography had no effect on growth rate of rats; but as expected, growth was greatly improved after the extract was heated to improve digestibility.

There are qualitative differences in responses of animal species fed raw soybeans (Table 3). Rats and chickens are the most sensitive. No adverse effects were noted in dogs. The only report on man indicates that a diet containing 180 g raw soy flour supports positive nitrogen balance, but at a level 80% of that for toasted soy flour. The heat-labile effects shown in Tables 2 and 3 are physiologically interrelated. The following discussion presents evidence that the simultaneous elimination of these effects by toasting results from the destruction of TI activity and transformation of raw protein into more readily digestible forms. The salient effects of TI, i.e., pancreatic hypertrophy and growth inhibition, represent an animal's inability to digest protein and utilize the amino acids in the most efficient manner rather than an irreversible response to a toxic substance. Growth inhibition and pancreatic hypertrophy occuring in rats fed raw soy diets can be readily reversed by replacing raw soy flour with toasted soy flour or casein. After continuous feeding of raw soybean meal for ca. one-fourth of their life-span, adult rats maintened body weight and except for the enlarged pancreas, all other organs were normal (Booth et al., 1964).

No deleterious effects were noted in pigs fed toasted soy rations continuously through four generations. A diet of soy milk (plus vitamins and iron supplement) permitted excellent growth, reproduction, and lactation through at least three generations of rats (Rackis, 1974).

TABLE 3

BIOLOGICAL EFFECTS OF RAW SOYBEAN MEAL
IN VARIOUS ANIMALS[a]

SPECIES	GROWTH INHIBITION	PANCREAS	
		Size	Enzyme Secretion
Rat[b]	+[c]	+	+
Chicken[b]	+	+	+
Pig[b]	+	−	±
Calf	+	−	±
Dog	−	−	d
Human	e	e	e

[a] Rackis (1974).

[b] Although adult animals maintain body weight, pancreas effects still occur.

[c] + = growth inhibition and pancreatic hypertrophy and hypersecretion;
− = no effect; ± = hyposecretion.

[d] Hyposecretion initially; normal after continued feeding.

[e] Unknown. However, two adults, in a 9-day feeding trial had positive nitrogen balance for both raw and autoclaved soy flour.

Growth Inhibition and Pancreatic Hypertrophy Proposed Mechanism

Substantial evidence now indicates that the feeding of raw soybeans and purified soy TI's accelerates protein synthesis in the pancreas and stimulates hypersecretion of pancreatic enzymes into the intestinal tract (Figure 1). Continuous secretion of the pancreas causes pancreatic hypertrophy and an excessive fecal loss of protein secreted by the pancreas. In the young animal, growth inhibition

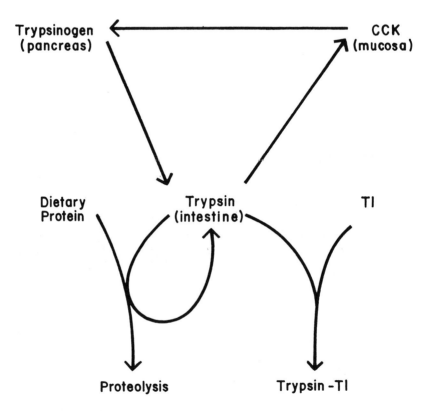

Fig. 1 - Regulation of trypsin secretion.
CCK, cholecystokinin;
TI, Trypsin inhibitor.

occurs because dietary protein could not compesate for the loss of essential amino acids in the endogenous protein. In the adult animal, because of a lower requirement for amino acids, there is no loss of weight, but pancreatic hypertrophy does occur. TI's from lima beans, navy beans, eggs, potatoes, and other foods elicit the same effects (Neiss et al., 1972).

The secretory response of the pancreas to dietary TI is an indirect one that is initiated in the intestinal tract and not in the blood (Olds-Schneeman and Lyman, 1975; Olds-Scheemam et al., 1977). Experiments with rats have demonstrated that pancreatic enzyme secretion is normally suppressed by negative feedback inhibition resulting from the presence of trypsin and chymotrypsin in the intestinal tract (Green and -Lyman, 1972; Green et al., 1973; Olds-Schneeman and Lyman, 1975).

The TI's evoke increased pancreatic enzyme secretion by forming inactive trypsin-TI complexes, thereby decreasing the suppression exerted by free trypsin. Feedback inhibition also occurs in humans (Ihse et al., 1977) and pigs (Corring, 1974) but not in dogs (Olds--Schneeman et al., 1977). A high casein diet (18%) is also a stimulant of pancreatic secretion in the rat, but addition of soy TI to such a diet increases secretion over that of casein aline (Olds--Schneeman et al., 1977). Casein also forms a complex with trypsin during digestion, thereby counteracting feedback inhibition and resulting in secretion of additional trypsin into the small intestine.

In the presence of protein and TI's in the duodenum, cholecystokinin (CCK) is released from biding sites in the mucosa. CCK is one of the hormones involved in regulation of the pancreas and bile secretion from the spleen. CCK is a TI, and repeated injections will cause pancreatic hypertrophy and inhibit rat growth (Rackis, 1974).

It would appear that TI's and dietary protein stimulate pancreatic activity by a common mechanism — it is only the extent of stimulation that differs. This mechanistic concept reinforces other evidence to be discussed that residual TI activity in toasted soy protein products may have no nutritional significance, since the TI content of these products is below the thereshold level at which pancreatic hypertrophy occurs.

Quantitative measurements are based on the fact that the major TI in soybeans has a specific activity of 1.0 (μg bovine trypsin inhibited per μg inhibitor) (Rackis et al., 1962).

Bovine trypsin is generally used for assays of TI activity. Only 30% of human trypsin is inhibited by an equivalent weight of soy TI (Mallory and Travis, 1973). This has prompted speculation that soy TI's may have less significance in human nutrition, because higher TI levels may be needed to counteract the suppression of pancreatic enzyme secretion exerted by trypsin in the intestinal tract. To extrapole TI threshold levels needed to cause pancreatic hypertrophy in rats to humans, it will be necessary to determine the relative specific activity of soy TI with rat and human trypsins. Preliminary data indicate that rat trypsins are inhibited by soy TI, but the extent of inhibition has yet to be quantitated (Liner, personal communication).

BIOLOGICAL THRESHOLD LEVELS OF TI

A series of experiments was conducted to determine quantitatively the relative ability of crystalline soy TY to inhibit growth and cause pancreatic hypertrophy in ratas compared to that of a raw soybean meal diet containing equivalent TI activity (Table 4). Maximum pancreatic hypertrophy occurs with raw meal diets containing 10% raw meal, which supplies 0.4% TI (diet 2). Compared with the casein control diet, about 17% raw meal (diet 3) was required to produce the greatest reduction in weight gain and PE. The TI content of the diet was 0.68%. Higher levels of raw meal (diet 4) begin to reverse growth inhibition but not the pancreatic effect.

In diet 5, crystalline TI at about the same level greatly reduced weight gain, lowered protein efficiency, and caused maximum pancreatic hypertrophy.

In comparing diet 6 containing purified TI with diet 3 containing comparable levels of TI as raw meal, the decrease in weight gain of rats on diet 6 was 66% of that obtained with diet 3. Also, a dietary level of 0.63% TI accounts for 56% of the reduction in PE associated with raw meal. On the other hand, 0.45 and 0.63% TI in diets 5 and 6, respectively, account for nearly all of the pancreatic hypertrophic effect of raw meal diets at an equivalent level of TI activity (diets 2, 3, 4).

The relationship between destruction of TI activity and the increase in PER of defatted soy flakes as a function of toasting is illustrated in Figure 2. Maximum PER corresponds to a destruction of only 80% of the TI activity of raw meal (Rackis et al., 1975).

The relative capacity of defatted soy flours containing graded levels of TI activity to inhibit growth, reduce PER, lower nitrogen digestibility, and enlarge the pancreas is summarized in Table 5. Pancreatic hypertrophy no longer occurs in rats fed soy flour in which only 54% of the TI activity was destroyed (diet 10). The diet contained 464 mg TI per 100 g diet. The further increase in growth and PER values with continued toasting is atributed to an increase in nitrogen digestibility values (Rackis, 1975; Kakade et al., 1973). Maximum PER corresponds to a destruction of only 80% of the TI in raw soy flour (diet 12), at a residual level of 212 mg TI per 100 g diet.

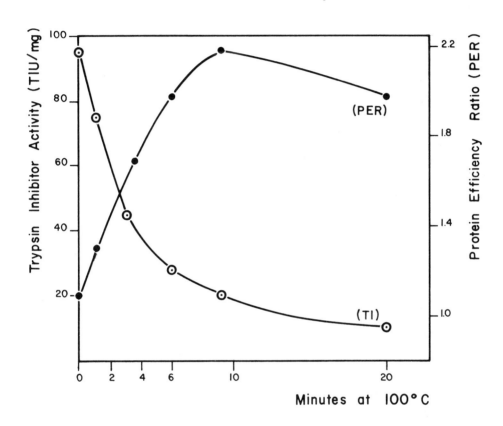

Fig. 2 - Effect of steaming on trypsin inhibitor activity (TI) and protein efficiency ratio (PER) of defatted soy flour (Rackis et al., 1975).

TABLE 4

RELATIONSHIP BETWEEN RAW SOYBEAN MEAL AND TRYPSIN INHIBITOR ON RAT GROWTH, PROTEIN EFFICIENCY, AND PANCREAS[a]

DIET No.	CASEIN DIET[b]	% INHIBITOR IN DIET	MEAN WEIGHT GAIN ± SE[c], g	PE	MEAN PANCREAS WEIGHT ± SE, g/100 g BW
1	Control, 10%	---	65.4 ± 7.04	2.48	0.60 ± 0.22
2	Raw Meal, 10%	0.40	35.6 ± 3.97[d]	1.54	0.79 ± 0.04[d]
3	Raw meal, 17%	0.68	24.6 ± 5.44[d]	1.13	0.77 ± 0.03[d]
4	Raw meal, 23%	0.97	35.8 ± 3.22[d]	1.24	0.75 ± 0.03[d]
5	STI[e]	0.45	40.2 ± 5.50[f]	1.76	0.74 ± 0.02[d]
6	STI	0.63	38.4 ± 4.00[d]	1.72	0.74 ± 0.04[d]

[a] Rackis (1965).

[b] All diets contain 10% protein, except diet 11 which is 12.3% protein; 35-day assay.

[c] SE = Standard error; PE = protein efficiency; BW = body weight

[d] $p < 0.01$

[e] Crystalline soybean trypsin inhibitor

[f] $p < 0.05$

TABLE 5

NUTRITIVE VALUE OF DEFATTED SOY FLOUR CONTAINING GRADED LEVELS OF TRYPSIN INHIBITOR (TI) WHEN FED TO RATS[a]

DIET NO.	DIETARY PROTEIN[b]	TI CONTENT mg/100 g DIET	MEAN BODY WEIGHT (g) ± STANDARD DEVIATION	PER[c]	NITROGEN DIGESTIBILITY[d]	PANCREAS WEIGHT ± STANDARD DEVIATION g/ GBW[e]
7	Casein (0)	0	157 ± 16ab[f]	2.50	92	0.48 ± 0.03c[f]
8	Soy (0)	1001	84 ± 4f	1.13	74	0.68 ± 0.11a
9	Soy (1)	774	94 ± 8ef	1.35	78	0.58 ± 0.01b
10	Soy (3)	464	123 ± 5d	1.75	77	0.51 ± 0.06c
11	Soy (6)	288	141 ± 12c	2.07	83	0.52 ± 0.04c
12	Soy (9)	212	146 ± 11bc	2.19	84	0.48 ± 0.06c
13	Soy (20)	104	139 ± 13c	2.08	83	0.49 ± 0.05c
LSD[g]			11	0.17		0.06

[a] Rackis et al. (1975).

[b] Time (min) of heat treatment at 100°C is given in prentheses.

[c] PER = protein efficiency ratio corrected on a basis of a PER = 2.50 for casein.

[d] Digestibility = intake-fecal nitrogen/intage X 100.

[e] GBW = grams body weight.

[f] Letters not in common denote statistical significance (P<0.05).

[g] LSD = Least significant difference at the 95% confidence level.

TABLE 6

EFFECT OF TRYPSIN INHIBITOR REMOVAL ON THE PROTEIN EFFICIENCY
RATIO (PER) AND RAT PANCREAS WEIGHTS[a]

SOY EXTRACT	TIU/mg PROTEIN	PER[b]	GRAMS PANCREAS PER 100 g BODY WEIGHT[b]
Original	125.1	1.4	0.74
Heated original	13.2	2.7	0.52
Inhibitor-free[c]	12.9	1.9	0.65

[a] Kakade et al. (1973).

[b] Data significantly different at $P<0.03$.

[c] Trypsin inhibitor removed by affinity chromatography.

Affinity chromatography with Sepharose^R-bound trypsin was used to remove TI from a raw soy flour extract (Kakade et al., 1973). After comparison of the TI-free and original extracts in rat feeding experiments (Table 6 and Figure 3), they concluded that approximately 40% of both the growth-inhibiting and hypertrophic pancreas effects of the original extract was attributable to the TI. The remaining 60% was due to poor digestibility of the raw protein.

Rats can tolerate rather high levels of TI in short-term feeding trials of only 4 weeks (see Table 5), and pancreatic hypertrophy is readily reversible (Booth et al., 1964). More recently, it was found that pancreatic hypertrophy in rats fed 30% raw soy flour (950 mg TI/100 g diet) can be reversed by replacing raw soy flour with toasted soy flour containing 99 mg TI/100 g diet (Rackis et al., unpublished data). Of special importance was the fact that chronic ingestion of TI as food-grade soy flour, concentrate or isolate for as long as 300 days did not cause pancreatic hypertrophy in rats. TI content of the diets ranged from 170 to 420 mg/100 g diet. All organs were normal in size and appearance.

TI Content of Soy Products

A modified procedure, particularly suitable for determining insoluble froms of TI in heat-processed and alcohol-treated products, was used to determine the removal of TI during manufacture of soy

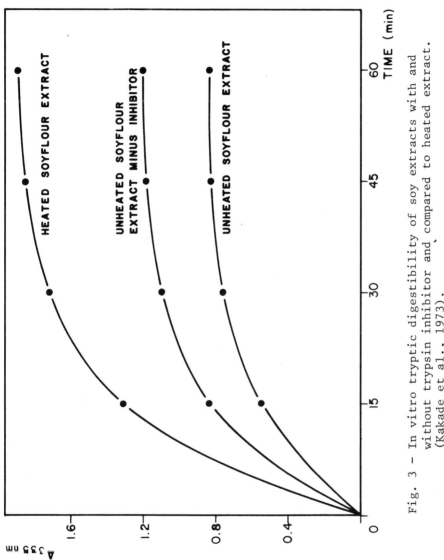

Fig. 3 — In vitro tryptic digestibility of soy extracts with and without trypsin inhibitor and compared to heated extract. (Kakade et al., 1973).

TABLE 7

COMPARISON OF TRYPSIN INHIBITOR ACTIVITY OF
COMMERCIAL SOY PRODUCTS DETERMINED IN TWO DIFFERENT LABORATORIES

PRODUCT	TIU/mg SAMPLE[a]		
	USDA[b]	Soy Processor A	Soy Processor B
Isoelectric soy protein isolate (80 PDI)[c]	31.0	35.1	---
Neutralized soy protein isolate (80 PDI)	26.2	23.2	---
Soy protein isolate-III	9.3	9.2	---
Raw defatted soy flour (85 PDI)	72.5	68.1	---
Heated defatted soy flour (25-30 PDI)	4.8	4.3	---
Defatted soy flour (75 PDI)	57.5	---	55.0
Defatted soy flour (11.3 PDI)	6.3	---	6.4

[a] TIU = Trypsin inhibitor units as defined by Kakade et al. (1969)

[b] Northern Center, Peoria, Illinois.

[c] PDI = Protein dispersibility index

protein products. As shown in Table 7, there was good agreement between the original collaborators involved in developing the assay (Kakade et al., 1974) and analysts in the quality control laboratories of two major soybean processors. The TI assay has been approved for publication in the revised "Approved and Tentative Methods of the American Oil Chemists' Society and American Association of Cereal Chemists", designated methods Ba 11-74 and 71-10, respectively.

TI content of several commercial heat-processed soy protein products is given in Table 8. Residual TI activity in these products is well within the tolerance levels established in rat bioassay. A number of soy-based infant formulas also have low levels of

TABLE 8

TRYPSIN INHIBITOR (TI) ACTIVITY OF
VARIOUS COMMERCIALLY MANUFACTURED SOY PROTEIN PRODUCTS

PRODUCT	TRYPSIN INHIBITOR ACTIVITY			
	TIU/mg[a]	Mg TI/g sample[b]	% of raw soy flours	Reference
Raw soy flour[c]	99.0	52.1	100	d
Toasted soy flour[c]	6-15	3.2-7.9	6-15	e
Soybean concentrate	12.0	6.3	12	d
Soybean concentrate	26.5	13.7	27	e
Soybean isolate	8.5	4.4	9	d
Soybean isolate	19.8	10.4	20	d
Soybean isolate	20.9	11.0	21	e
Soy food fiber	12.3	6.5	12	f
Chicken analog	6.9	3.6	7	f
Ham analog	10.2	5.4	10	f
Beef analog	6.5	3.4	7	f
Textured soy flour	9.8	5.2	10	f

[a] TIU = Trypsin Inhibitor Units as defined by Kakade et al. (1969).

[b] Calculated on the basis that 1.90 TIU is equivalent to 1 ug of TI, Kakade et al. (1969).

[c] Several lots were analyzed.

[d] Kakade et al. (1974).

[e] Rackis, J.J. and J.E. McGhee, unpublished data.

[f] Liener (1975).

TABLE 9

TRYPSIN INHIBITOR CONTENT OF FOOD LEGUMES[a]

LEGUMES	TRYPSIN INHIBITOR ACTIVITY X 10^{-4} UNITS/g
Kidney beans (Phaseolus vulgaris)	4.25
Hyacinth beans (Dolichos lablab)	4.38
Soybeans (Glycine max)	4.15
Lima beans (Phaseolus lunatus)	4.04
Pigeon peas (Cajanus cajan)	2.77
Cow peas (Vigna sinensis)	1.91
Lentils (Lens esculenta)	1.78

[a] Liner (1976).

residual TI activity, and these products likewise do not cause pancreatic hypertrophy in rata (Churella et al., 1976).

The relative TI activity in soybeans and other food legumes is shown in Table 9. TI content of many fruits and vegetables (Chen and Mitchell, 1973) is relatively high, but these values cannot be compared directly with soybeans because activity units are not equivalent.

MINERAL BIOAVAILABILITY IN SOYBEANS

Reports suggesting that "soy protein may inhibit the body's ability to utilize essential minerals, particularly zinc" have stirred great controversy and produced questions of whether mineral supplementation of soy protein diets is required (Rackis and Anderson, 1977; Erdman and Forbes, 1977). The first focus of this controversy involves the effect of soy phytic acid on bioavailability of minerals from other dietary ingredients, i.e., does soy phytic acid in extended ground meat depress mineral absoprtion from the meat protein? Erdman and Forbes (1977) indicates that soy protein products do not affect absorption of zinc as carbonate or zinc from other dietary sources. Also, food products based on soy protein isolates, specifically manufactured for use in infant formulas, have been highly effective in furnishing high levels of absorbable iron form iron salts to these formulas (Rackis, 1974).

The second focus concerns the bioavailability of the minerals in soy protein products themselves. A review of the literature reveals great ambiguity (Rackis, 1974; Rackis and Anderson, 1977). Magnesium availability in soybean meal is as high as in skin milk and eggs (Guenther and Sell, 1974). Whole soybeans and soy flour contain high levels of iron that are higly available. Data on soy protein isolates conflict, ranging from very poor iron avialability to absorbable iron neraly equivalent to animal protein. When fed as the sole source of protein in the diet, some soy products may require supplemental levels of zinc ranging from 0 to 100 ppm.

These differences in bioavailability occur because the many different types of soy products are manufactured for either nautritional or functional purposes (Smith and Circle, 1972). Often, investigators in the past did not identify the type of product used or supply available information concerning conditions of manufacture.

Phytic acid (myo-inositol 1,2,3,4,5,6-hexakis dihydrogen phosphate) is present in most plant foodstuffs, but not in animal protein, either as a phytate salt or complexed with protein. Phytic acid content of various plant foodstuffs is given in Table 10. However, as shown in Table 11, phytic acid content per se has little relationship to zinc availability. For example, zinc availability os peas is comparable to that of nonfat milk, even though peas contain 1.23% phytic acid. Compared to milk protein, zinc availability of defatted soybean meal is moderate (67%), whereas the soy protein isolate had a value of only 44%. These differences in zinc bioavailability occur even though both soy flour and soy protein isolates contain 1.53% phytic acid (be Boland et al., 1975). The relative biological availability of endogenous and supplemental zinc in a cow's milk-based formula was about 0.86 and that of a soy-based formula was 0.69 (with zinc sulfate equal to 1.00) (Momcilovic et al., 1976). Another report (Churella and Vivian, 1976) revealed no differences in bone or carcass ash, calcium, phosphorus, or zinc between rats fed soy isolate containing reduced levels of phytic acid and those fed an isolate with normal levels of phytic acid ca. 1.52%. In the reports of Momcilovic et al. (1976) and Churella and Vivian (1976), the isoelectric from of soy protein isolate was tested. Such isolates contain concentrations of zinc comparable to casein, and the zinc is readily available.

Evidence suggests that the formulation of protein-phytate-mine-

TABLE 10

PHYTIC ACID CONTENT OF VARIOUS PLANT FOODSTUFFS

SOURCE	PHYTIC ACID CONTENT[a], %
Sesame seed	5.18
Corn germ	4.97
Soy products[b]	1.15 to 2.17
Cereal grains	0.6 to 1.30
Food legumes[b]	0.6 to 1.58

[a] Data from several sources.

[b] Range for flours, concentrates, isolates.

TABLE 11

ZINC AVAILABILITY IN VARIOUS FOODS[a]

FOOD PRODUCT	AVAILABILITY, %	
	Chick Assay	Rat Assay
Peas, immature	--	95
Peas, mature	--	75
Nonfat milk	82	79
Fish meal	75	84
Soybean meal, defatted	67	--
Corn	63	57
Rice	62	39
Wheat	59	38
Casein	--	84
Soy protein isolate	--	44

[a] Data from several sources

ral complexes during manufacture of soy isolates, rather than phytic acid per se, may be a primary factor responsible for the reduced mineral availability in some soy isolates (Rackis and Anderson, 1977). Phytic acid reacts with proteins in the water extract of soybeans during acidification of protein at pH 4.4. Upon neutralization of the precipatated protein, insoluble protein-phytate complexes are formed (Smith and Rackis, 1957).

Whether phytic acid and other components can affect bioavailability of essential minerals in soy and other plant foodstuffs depends upon several factors (Rackis and Anderson, 1977): ability of endogenous carriers in the intestinal mucosa to assimilate essential minerals bound to phytic acid and other constituents; hydrolysis of phytate phosphorus by phytase enzymes in the intestine· phytase inhibition by additives in bread. Dietary substances, such as certain dietary fiber, oxalates, phenolic compounds, and other unknown naturally occuring chelating ingredients can reduce mineral absorption. Current research underway should make it possible to evaluate these factors and lead to processing changes that may increase mineral availability in plant proteins.

OTHER BIOLOGICAL-PHYSIOLOGICAL FACTORS

Goitrogens

That soybeans can cause goiter was first observed in 1933. Since that time several reports have shown that soybean goiter in ratas could be reserved completely by the administration of small amounts of iodide. That soy milk not fortified by iodide can produce goiter in humans has been confirmed by several investigators. Today all infant formulas in the United States contain iodide. New technology and nutritional knowledge ensure that infant formulas, based on soy protein isolate, contain all nutrients known to be essential for the infant (Sarett, 1976).

The greatest problem with goiter in humans is associated with vegetables other than soybeans. Glucosinolates are the responsible factors and are present in large amounts in plants of the Cruciferae family such as cabbage, turnips, radishes, rutabaga, and of the oilseeds rape, mustard, and crambe (VanEtten and Wolff, 1973). These glucosinolates, during enzyme hydrolysis, are transformed into substances referred to as goitrins. Unlike most goitrogenic substances, administration of large amounts of oidide cannot countreract the goitrogenic activity of the goitrins derived from glucosinolates.

Soybeans do not contain glucosinolates. The goitrogenic constituent in soybeans appears to be an oligopeptide composed of two or three amino acids or a glycopeptide (Konijn et al., 1973). Only small amounts of iodide are needed to counteract its goitrogenic

TABLE 12

GOITROGENICITY OF SOY PRODUCTS[a]

PRODUCT	IODINE CONTENT, µg/100 g DIET	THYROID wt. mg/100 g BODY wt.
Raw soy flou	1.0-2.3	37
Toasted soy flour	0.7	19
Soy isolate	0.9	16
Soy infant food, iodized	40	7
Raw soy flour + 10 µg I_2/g protein	196	8
Casein	30	7

[a] Block et al. (1961).

activity. In a review of the action of the soybean goitrogen, Konijn et al. (1973) indicate that the soybean goitrogen inhibits idodine uptake by the thyroid and decreases incorporation into organic constituents in the gland. Other workers believe that the goiter resulting from the feeding of raw soybeans to rats results from a diminishde circulation of iodine or an increased fecal loss of thyroxine. In a detailed report, Block et al. (1961) suggest that goitrogenicity results from a simple iodine deficiency.

Although the lack of iodine is most likely the principal cause of soybean goiter, raw soybeans, which contain slightly more iodine than toasted, defatted soy flour, produce greater thyroid enlargement. Results are given in Table 12. Heat processing of raw soy flour greatly reduces but does not completely eliminate the goitrogenic activity. Nordsiek (1962) prevented the goitrogenic effect of raw soybean meal fed to rats by adding casein to the diet. The casein effect was not attributed to its iodine content, but rather that part of the soybean goitrogenicity may be related to the poor digestibility of the undenatured proteins in raw soybeans.

The lack of iodine is also the principal factor that accounts for the goitrogenic activity of soy protein isolates. This conclusion was derived from the fact that only small amounts of iodine are required to prevent goiter. Also, based on the characteristics of the goitrogenically active substance isolated by Konijn et al. (1973), the goitrogen would be removed during the preparation of soy isolates.

Excessive intake of iodine can also lead to thyroid hypertrophy. Landau et al. (1972) reported that some soy milk preparations contained up to 1,400 µg iodine per liter. An iodine level of 34-160 µg per liter is adequate to abolish the goitrogenic effect of soy milks. Recommendations pertaining to protein quality and nutrient fortification of milk-based and milk-substituted infant formulas have been reviewed (Sarett, 1976).

Although there are indications that American diets contain iodide to satisfy Recommended Daily Allowances, there is still a need to continually reassess the nutrient adequacy of new foods because of the rapicly changing dietary patterns of foos intake. There are about 300 compounds known to be goitrogenic, and undoubtedly others will be discovered.

Estrogens

The isoflavone glucosides, genistin, daidzin, and glycitein--7-0-B-glucoside, are the major phenolic compounds in soybeans (Naim et al., 1974). The concentration of each glucoside and its relative estrogenicity are shown in Table 13. Most of the isoflavones are removed during the manufacture of soy protein concentrates and isolates. Soybeans contain 1.2 ppm coumesterol, two-thirds of which is extracted with the oil (Knuckles et al., 1976). The coumesterol content in defatted soybean meal, soy concentrate, and isolate ranges from 0.2-0.6 ppm. Fresh soybean sprouts contain about 71 ppm coumesterol. Although the estrogenicity of glycitein-7-B-glucoside is not known, estrogenic activity of present soybean varieties appears to be below threshold level for estrogenic effect.

Allergens

Since 1928, soy protein has been used as a replacement for milk protein in infants allergic to cow's milk or intolerant to lactose (Sarett, 1976). A weak allergen has been isolated in crude form. Nevertheless, soybeans can be regarded as hypoallergenic in the sense that less of the population display a sensitivity to soy than to either other legumes or cow's milk (Rackis, 1974). On the other hand, four children were found to be intolerant to soy formula but not to cow's milk (Halpin et al., 1977). However, the number of recently reported clinical cases concerning adverse reactions to soybeans are very few. Five newborn infants developed eczema on first postnatal contact with soybeans (Kuroume et al., 1976), and four infants developed diarrhea, vomiting, hypertension, lethargy, and fever following oral challenge with soy protein isolate (Whitington and Gibson, 1977). The infants were also sensitive to cow's milk protein.

TABLE 13

SOY MEAL ESTROGENICITY

ESTROGEN	CONCENTRATION IN SOY MEAL, ppm	RELATIVE ESTROGENICITY[a]	PRODUCT[b]
Diethylstilbestrol[c]	---	1×10^5	---
Genistin	1644[d]	1.00	1644
Daidzin	581[d]	0.75	436
Glycitein 7-0-B-glucoside	338[d]	---[e]	---
Coumestrol	0.4	35	14

[a] Bickoff et al. (1962).

[b] Concentration X estrogenicity.

[c] Included for comparion purposes.

[d] Naim et al. (1974).

[e] Estrogenicity unknown.

REFERENCES

Bickoff, E.M., Livingston, A.L. Hendrickson, A.P., and Booth, A.N. 1962. Relative potencies of several estrogen-like compounds found in forages. J. Agric. Food Chem. 10: 410-413.

Block, R.J., Madl, R.H., Howard, W., Bauer, C., and Anderson, D.W. 1961. The curative action of iodide on soybean goiter and the changes in the distribution of iodoamino acids in the serum and in thyroid digests. Arch. Biochem. Biophys. 93: 15-24

Booth, A.N., Robbins, D.J., Ribelin, W.E., DeEds, F., Smith, A.K., and Rackis, J.J. 1964. Prolonged pancreatic hypertrophy and reversibility in rats fed raw soybean meal. Proc. Exp. Bull. Med. 116: 1067-1069

Chen, I., and Mitchell, H.L. 1973. Trypsin inhibitors in plants. Phytochemistry 12: 327-330.

Churella, H.R., and Vivian, V. 1976. The effect of phytic acid in soy infant formulas on the availability of mineral for the rat.

Fed. Proc. 35: Abstr. No. 2972.

Churella, H.R., Yao, B.C., and Thomson, W.A.B. 1976. Soybean trypsin inhibitor activity of soy infant formulas and its nutritional significance for the rat. J. Acric. Food Chem. 24: 393-397

Corring, T. 1974. Regulation de la secretion pancreatic par retro-inhibition negative Chez le porc. Ann. Biol. Biochem. Biophys. 14: 487-498

de Boland, A.R., Garner, G.B., and O'Dell, B.L. 1975. Identification and properties of phytate in cereal grains and oilseed products. J. Agric. Food Chem. 23: 1186-1189.

Erdam, J.W., Jr., and Forber, R.M. 1977. Mineral bioavailability from phytate-containing foods. Food Prod. Dev. 11 (10): 46,48

Food and Nutrition Service. 1971A. USDA, Washington, D.C. FNS Notice 218, February 22.

Food and Nutrition Service. 1971B. USDA, Washington, D.C. FNS Notice 219, February 22.

Green, G.M., and Lyman, R.L. 1972. Feed-back regulation of pancreatic enzyme secretion as a mechanism for trypsin inhibitor--induced hypersecretion in rats. Proc. Soc. Exp. Biol. Med. 140: 6-12.

Green, G.M., Olds, B.A., Matthews, G., and Lyman, R.L. 1973. Protein as a regulator of pancreatic enzyme secretion in the rat. Proc. Soc. Exp. Biol. Med. 142: 1162-1167

Guenter, W., and Sell, J.L. 1974. A method for determining true availability of magnesium from foodstuffs using chickens. J. Nutr. 104: 1446-1457

Halpin, T.C., Byrne, W.J., and Ament, M.E. 1977. Colitis, persistent diarrhea, and soy protein intolerance. J. Pediatr. 91: 404-407.

Hymowitz, T. 1973. Electrophoretic analysis of SBTI-A_2 in the USDA soybean germplasm collection. Crop. Sci. 13: 420-421.

Ihse, I., Lilja, P., and Lundquist, I. 1977. Feedback regulation of gestion 15: 303-308.

Kaizuma, N., and Hymowitz, T. In press. On the frequency distribution of two seed proteins in Japanese soybean cultivars: Implications on paths of dissemination of the soybean from mainland Asia to Japan. Jpn. J. Breed.

Kakade, M.L., Hoffa, D.E., and Liener, I.E. 1973. Contribution of trypsin inhibitors to the deleterious effects of unheated soybeans fed to rats. J. Nutr. 103: 1772-1778

Kakade, M.L., Rackis, J.J. McGhee, J.E. and Puski, G. 1974. Determination of trypsin inhibitor activity of soy products: A collaborative analysis of an improved procedure. Cereal Che. 51: 376-382

Kakade, M.L., Simons, N.R., Liener, I.E., and Lambert, J.W. 1972. Biochemical and nutritional evaluation of different varieties of soybeans. J. Agric. Food. Chem. 20: 87-90.

Kakade, M.L., Simonson, N., and Liener, I.G. 1969. An evaluation of natural vs. synthetic substrates for measuring the anti-

tryptic activity of soybean samples. Cereal Chem. 46: 518-526
Keystone Conference, in press. Proceedings of Conference on Soy Protein and Human Nutrition, Keystone, Colorado, May 22-25, 1978.
Knuckles, B.E., de Fremery, D., and Kohler, G.O. 1976. Coumesterol content of fractions obtained during wet processing of alfalfa. J. Agric. Food Chem. 24: 1177-1180
Konijn, A.M., Gershon, B., and Guggenheim, K. 1973. Further purification and mode of action of a goitrogenic material from soybean flour. J. Nutr. 103: 378-383
Kuroume, T., Oguri, M., Matsuma, T., Swasaki, I., Kanbe. Y., Yamada, T., Kawabe, S., and Negishi, K. 1976. Milk sensitivity and soybean sensitivity in the production of eczematous manifestations in breast-fed infants with particular reference to intra-uterine senzitization. Ann. Allergy 37: 41-46
Landau, H., Rabinowitz, D., and Freier, S. 1972. An uncommon cause of high serum protein-bound iodine. Isr. J. Med. Sci. 8: 1749-1751
Liener, I.E. 1975. in: "Protein Nutritional Quality of Foods and Feeds", M. Friedman, editor, Dekker, Inc., New York.
Liener, I.E. 1976. Legume toxins in relation to protein digestibility: A review. J. Food Sci. 41: 1076-1081
Mallory, P.A., and Travis, J. 1973. Human pancreatic enzymes, characterization of anionic humar trypsin. Biochemistry 12: 2847-2851.
Momcilovic, B., Belonji, B., Giroux, A., and Shah, B.G. 1976. Bioavailability of zinc in milk and soy protein-based infant formulas. J. Nutr. 106: 913-917
Naim, M., Gesterner, B., Zilkah, S., and Birk, Y. 1974. Soybean isoflavones. Characterization, determination, and antifungal activity. J. Agric. Food Chem. 22: 806-810
Neiss, E., Ivey, C.A., and Nesheim, M.C. 1972. Stimulation of gallbladder emptying and pancreatic stimulation in chicks by soybean whey protein. Proc. Soc. Exp. Biol. Med. 140: 291-296.
Nordsiek, F.W. 1962. Effects of added casein on the goitrogenic action of different dietary levels of soybeans. Proc. Soc. Exp. Biol. Med. 110: 417-420
Olds-Schneeman, B., Chang, L., Smith, L.B., and Lyman, R.L. 1977. Effect of dietary amino acids, casein, and soybean trypsin inhibitor on pancreatic protein secretion in rats. J. Nutr. 107: 281-288.
Olds- Scheeman, B., and Lyman, R.L. 1975. Factor involved in the intestinal feedback regulation of pancreatic enzyme secretion in the rat. Proc. Soc. Exp. Biol. Med. 148: 897-903
Rackis, J.J. 1965. Physiological properties of soybean trypsin inhibitors and their relationship to pancreatic and growth inhibition of rats. Fed. Proc. 24: 1488-1493
Rackis, J.J. 1974. Biological and physiological factors in soybeans. J. Am. Oil Chem. Soc. 51: 161A-174A.
Rackis, J.J., and Anderson, R.L. 1977. Mineral availability of soy

protein products. Food. Prod. Dev. 11 (10): 38, 40, 44.

Rackis, J.J., McGhee, J.E., and Booth, A.N. 1975. Biological threshold levels of soybean trypsin inhibitors by rat bioassay. Cereal Chem., 52: 85-92

Rackis, J.J., Sasame, H.A., Mann, R.K., Anderson, R.L., and Smith, A.K. 1962. Soybean trypsin inhibitors: Isolation, purification, nd physical properties. Arch. Biochem. Biophys. 98: 471-478.

Sarett, H.P. 1976. Soy-based infant formulas. in: "World Soybean Research, L.D. Hill (editor), Proceedings of World Soybean Research Conference", The Interstate Printers and Publishers, Inc., Danville, Illinois, pp. 840-849.

Skorupska, H., and Hymowitz, T. In press. On the frequency distribution of alleles of two seed proteins in European soybean {Glycine max (L.) Merrill} germplasm. Implications on the origin of European soybean germplasm. Genet. Pol.

Smith, A.K., and Circle, S.J. (editors) 1972. "Soybeans: Chemistry and Technology", Vol. I. Proteins, Avi Publishing Co., Westport, Connecticut.

Smith, A.K., and Rackis, J.J. 1957. Phytin elimination in soybean protein isolation. J. Am. Chem. Soc. 79: 633-637

Turner, R.H., and Liener, I.E. 1975. The effects of selective removal of hemogglutinins on the nutritive value of soybeans. J. Agric. Food. Chem. 23: 484-487

VanEtten, C.H., and Wolff, I.A. 1973. Natural sulfur compounds. in: "Toxicants Occuring Naturally in Foods", 2nd Ed., National Academy of Science-National Research Council, Washington, D.C.

Whitington, P.F., and Gibson, R. 1977. Soy protein intolerance. Four patients with concomitant cow's milk intolerance. Pediatrics 59: 730-732.

STUDIES ON NUTRITIVE VALUES OF MIXED PROTEIN
(50% SOY AND 50% FISH)
AND SOY PROTEIN ISOLATE IN JAPANESE ADULTS

G. Inoue

Dept. of Nutrit., School of Med., Tokushima Univ.,

Tokushima, JAPAN

In Japan, soybean and its products, e.g. soybean curd and Miso, have been used from old as the important and familiar dietary foods. In 1977, protein intake from soybean was estimated statistically to be 7.6 gm per head per day, corresponding to be about 10% of total daily protein intake. At present, the consumption of soy protein is increasing with popularization of many kinds of foods mixed soy protein isolate. This tendency will become more and more necessary, indicating that the human studies on evaluation of soy protein and its mixed foods are urgent in our country. Requirements and NPUs of a soybean protein isolate (Supro 620) and when mixed with 50% fish protein were compared with those of fish protein in 21 university students.

EXPERIMENTAL DESIGN

Each subject was given consecutively 4 levels of low protein diets. Protein sources were 100% codfish (as a reference) for 8 subjects, 100% Supro 620 for 5 subjects and the mixed protein (50% Supro plus 50% codfish) for 8 subjects, where the same 5 subjects were studied for the fish and the mixed protein. These protein foods were made homogenously into paste products (Kamaboko) and were given an allocated amount equally three times a day. A combination of 1-day protein free diet and following 10-day low protein diet was a unit of experiment. This unit study (one period) was repeated 4 times with 3-day interruption of ad libitum feeding between each period, where the level of protein intake varied in every period under a random planning among 0.35, 0.45, 0.55 and 0.65 gm protein/kg of body weight (Table 1). Energy intake was kept as constant as possible for an individual to maintain body weight, being average

TABLE 1

EXPERIMENTAL DESIGN

| | | FISH PROTEIN | | | | | SOY + FISH | | | |
| | | Period | | | | | Period | | | |
	(Subj.)	I	II	III	IV	(Subj.)	I	II	III	IV
STUDY 1 (Oct.-Dec., 1977)	A	1	2	3	4	E	4	3	2	1
	B	2	3	4	1	F	3	2	1	4
	C	3	4	1	2	G	2	1	4	3
	D	4	1	2	3	H	1	4	3	2
STUDY 2 (Jan.-Mar., 1978)	E	4	3	2	1	A	1	2	3	4
	F	3	2	1	4	B	2	3	4	1
	G	2	1	4	3	C	3	4	1	2
	H	1	4	3	2	D	4	1	2	3

Energy intake level : 45 ± 2 (38-50) kcal/kg
Protein intake levels : 0.35, 0.45, 0.55 & 0.65 g/kg
 in 1, 2, 3 & 4, respectively.

of 44.6±2.4 kcal/kg (varied from 38 to 50) for 21 subjects (Table 2). Apparent N balance was calculated using the mean urinary and fecal N for the last 5 and 8 days, respectively, where the dermal N loss was not taken into account. Basal metabolism, urine and blood analysis and anthropometry were also measured.

RESULTS OBTAINED

Fig. 1 shows the relationship between N intake and N balance which was studied in the mixed protein. 32 balance data for 8 subjects were plotted against respective N intakes. The individual variations were fairly scattered, but the correlation between the two was statistically significant in either the individual data or the pooled data. The similar findings were obtained in the other two studies for the fish and the Supro. Three regression lines estimated from the pooled data were shown together in Fig. 2. The slope for the mixed protein was somewhat higher than that for the fish protein, while the intersect was lower in the fish than the mixed. In Supro, the slope was smallest and the intersect was highest. As the results, the maintenance requirements and NPUs estimated from the N balance based on both of pooled and individual data are shown in Tab. 3.

In both estimation, the difference between requirements for the fish and the mixed was insignificant, while those among the Supro and the other two were significant ($P<0.01$). Considering together with the data on the slope ratio, requirements and NPUs, the relative NPU of soybean protein isolate may be evaluated to be approximately 75% against fish protein. While in the mixed protein, the relative value raised up to about 95%.

The figures obtained for the fish in this study would be comparable with those for egg protein measured previously by ours and others, although the coefficient of variation in this study was larger than that in our egg studies. In human adult studies by Dr. Scrimshaw et al. and Dr. Calloway et al., addition of methionine or sulfer containing substances to soy protein isolate resulted in a significant increase in N retention. Our results suggest also that the lower content of S-amino acid in soybean protein may be compensated by adding sufficiently with codfish protein, being similar with the data with soy plus beef protein measured by Dr. Young et al.

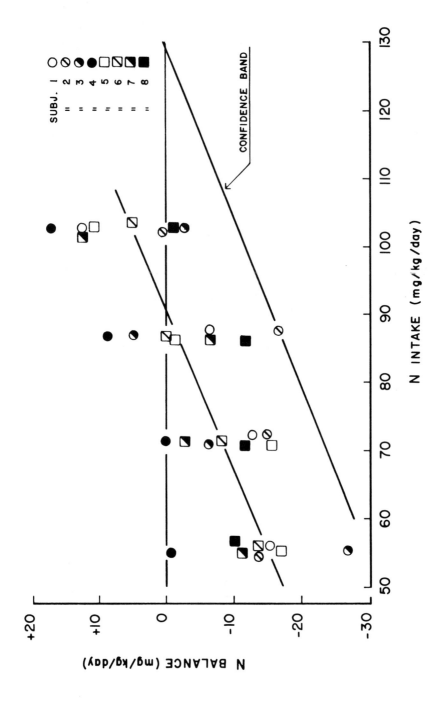

Fig. 1 – Relationship between N intake and N balance for the mixed protein of Fish and Supro 620.

TABLE 2

DIET COMPOSITION

(an example)

Subj.: 65 kg, Mixed protein 0.55 g/kg,
Energy 45 kcal/kg.

MATERIALS	g/day
Cod and Supro	31.9 + 19.4
Sugar	323
Cornstarch	300
Margarine	32
Corn oil	30
Agar	3
Baking powder	5
Salt	2.5
Mineral mixture	6
Vitamin mixture	3

1) N contents of dried cold and Supro are 89.6 and 147.2 mg/g.

2) Total energy intake is 2925 kcal/day (45 kcal/kg).

3) N content was 2.86 g equally in Cod & Supro

TABLE 3

REQUIREMENTS AND NPUs SUMMARIZED

	SAMPLE NO	MAINTENANCE REQUIREMENT (mgN/kg)	COEF. OF VARIATION (%)	NPU
(Pooled data)				
Fish	26	87.1	19.7	53
Fish + Supro	32	90.5	18.9	51
Supro 620	20	118.1*	13.0	39
(Individual data)				
Fish	7	87.7	16.5	54
Fish + Supro	7	95.1	22.3	51
Supro 620	5	119.4*	12.4	39

* The difference from the Fish was significant ($p<0.01$)

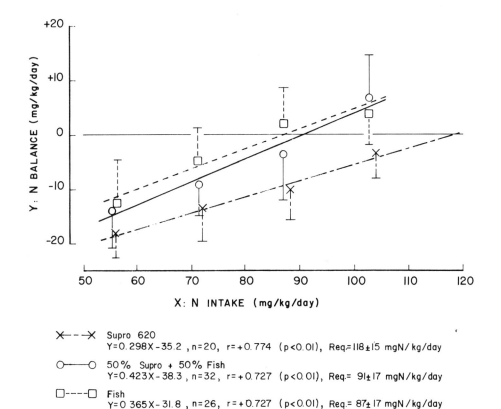

Fig. 2 - Nitrogen balance relating to nitrogen intake

THE TECHNOLOGY OF IRON FORTIFICATION

Dudley S. Titus, Ph. D.

Mallinckrodt Inc. - 8343 Ardsley Dr.

St. Louis, MO. 63121 - USA

Of all the nutrients used for fortification purposes, iron is probably the most difficult to successfully incorporate into foods. Early in the development of the food processing industry it was found that even traces of free iron or copper caused rapid deterioration in the quality of many foods. These problems have been minimized over the years by avoiding any contact with iron equipment and careful control of ingredients. The success fo these efforts is evidenced by the high quality and excellent storage stability of today's processed foods. Fortification and enrichment programs require that we now reverse the process and deliberately add iron to foods and do it in such a way that product quality is not impaired.

Iron compounds vary widely in bioavailability. Soluble iron salts are readily assimilated by the body. Some studies indicate that ferrous salts are more readily absorbed than are ferric salts, but these differences appear to be relatively minor. Highly insoluble compounds such as the ferric oxides are utilized to a very limited extent. Partially soluble products such as the phosphates and iron powders have intermediate bioavailabilities.

In order to have good bioavailability, an iron compound must be solubilized before or during digestion so that the iron can enter the non-heme iron pool in the gut and be absorbed. Unfortunately, those iron sources which meet this criterion are usually those which are most likely to have an adverse affect on the quality of the foods to which they are added.

WHO[1] and INACG[2] have outlined the problems that may be encountered in fortifying a food with iron. Some of the chemical changes that may occur are:

1. ferrous salts may auto-oxidize to form yellow, green, or black ferric oxides;

2. iron salts may react with phenolic compounds such as tannins and propyl gallate to form blue-black colors;

3. iron compounds may react with sulfur compounds and form a black color;

4. iron compounds may increase the activity of oxidative enzymes that cause off-colors, -odors, and -flavors;

5. iron is a known pro-oxidant and may catalyze the development of oxidative reactions.

Chemical reactions such as these not only have undesirable affects on color and flavor, but in some cases functional properties may be affected. For example, after prolonged storage at elevated temperatures, wheat flour fortified with ferrous sulfate loses some of its gluten-forming ability and does not leaven properly. The assimilability of the iron is undoubtedly impaired also as a result of many of these reactions.

Of particular importance with regard to iron's pro-oxidant characteristics is its ability to catalyze the oxidation of Vitamin C. Precautions should be taken in adding iron to a food which is an important source of Vitamin C or in fortifying a food with both iron and ascorbic acid to be sure that Vitamin C activity is not lost.

All of the chemical reactions described above require water and problems associated with added iron can often be minimized or even eliminated by maintaining low moisture levels in the fortified foods. In products such as salt, levels must be kept below 1% to prevent the oxidation of soluble ferrous salts. However, in blended wheat flour products containing about 5% water, soluble iron compounds such as ferrous sulfate are stable for extended periods. Even in wheat flour containing 13% moisture, ferrous sulfate is stable for a few months. Oxidation of ferrous salts can also be prevented by maintaining a pH of 3 or below, but this is often impractical. Keeping the pH as low as possible, however, will often increase stability, and in combination with low moisture levels can be quite effective in maintaining the quality of the fortified food. Low temperature storage will also help slow the rate of chemical reactions.

A second method of avoiding the chemical side reactions of iron salts is the use of complexing agents. This can be done by selecting an iron compound in which the complexing agent is a part of the compound or by adding it separately. The iron in compounds wuch as ferrous citrate, ferrous gluconate, ferric ammonium citrate, and sodium ferric EDTA is complexed sufficiently to greatly retard its

TECHNOLOGY OF IRON FORTIFICATION

chemical reactivity and yet it still becomes part of the non-heme iron pool and is readily assimilated. The main objection to these compounds is that they are usually quite expensive and often have limited commercial availability.

A third alternative to prevent chemical reactions of iron is to physically isolate the iron source from the food. Several encapsulated or coated iron salts have been developed which effectively prevent chemical reactions in foods and release the iron in the stomach or gut. These products are also quite expensive and have achieved limited commercial use. It must be emphasized that whatever steps are taken to prevent chemical reactions in the food, the conditions must be maintained not only during commercial storage and distribution, but also through conditions and usages that normally occur in the home. For example, sugar can be fortified with ferrous sulfate and because of the high purity and low moisture content of sugar, the fortified product is stable and is suitable for a variety of uses. However, when used to sweeten tea, the ferrous sulfate dissolves and reacts with the tannin in the tea, turning it a dark black.

The physical problems WHO[1] and INAGG[2] listed that may be encountered in iron fortification are:

1. it may be difficult to achieve and maintain uniform distribution of high density iron sources such as reduced iron powders. Distribution characteristics can sometimes be improved by selecting particle sizes or by milling the iron to a surface/weight ratio similar to that of the food being fortified. Segregation during storage can be minimized in some products by adding low levels (2-3%) of propylene glycol which acts as a water insoluble adhesive.

2. the color and/or flavor of the iron compound may be undesirable and carry through into the food product. Some iron compounds such as reduced iron and ferrous fumarate are highly colored and are unsuitable for use in a very light colored product.

3. the solubility characteristics of the iron may not be compatible with the product to be fortified. A highly insoluble iron source is obviously unsuitable for liquid products.

Problems resulting from physical characteristics can usually be resolved only by selecting an iron source that is compatible with the food to be fortified and by careful attention to good manufacruting techniques to assure proper levels and uniform distribution.

SUPPLEMENTATION

The incorporation of iron into pharmaceutical products presents fewer problems than does the fortification of food products. The chemical characteristics of the iron source are less important in pharmaceutical products because the other ingredients in the preparation can be carefully selected and controlled. In most iron-containing pharmaceutical preparations there are relatively few other ingredients and no problems would be anticipated if there are no other active ingredients (such as ascorbic acid) which might be affected by iron's pro-oxidant characteristics. It is also important that none of the fillers and diluents in the products will react with iron to make it unassimilable.

Some iron compounds may be difficult to make into tablets, particularly if the dosage is large and the tablet size restricts the amount of excipients that can be used. Under these conditions the tablets are often very hard and do not desintegrate well. The use of special excipients with very high water absorption and swelling properties are sometimes helpful in alleviating this problem.

REFERENCES

[1] WHO, Control of Nutritional Anemia with Special Reference to Iron Deficiency. Technical Report Series 580. World Health Organization, 1975.
[2] INACG, Guidelines for the Eradication of Iron Deficiency Anemia. A Report of the International Nutritional Anemia Consultative Group (INACG).

THE TECHNOLOGY OF VITAMIN A

J.C. Bauernfeind

Nutrition Research Coordinator - Roche Research Center
Hoffmann-La Roche Inc.
Nutley, New Jersey 07110 - U.S.A.

INTRODUCTION

Last year, 1977 was the 30th anniversary of the chemical synthesis[1,2] of vitamin A, one of the outstanding accomplishments in improving world-wide public health. Large-scale commercial manufacture and process efficiency improvements in the immediate years following synthesis have enabled this vital biochemical to be produced in virtually unlimited quantities at a fraction of its initial cost (Figure 1). Yet, today significant segments of the world's population suffer because of vitamin A dietary insufficiency. Problems of inertia, unsuccessful or inadequate education programs, resistance to the concept of vitamin nutrification of foods as a public health approach, and particularly, inadequate food distribution systems need to be overcome to eradicate vitamin A deficiency and the subsequent blindness which frequently follows.

Vitamin A, retinol ($C_{20}H_{29}OH$), is the isoprenoid polyene alcohol, 3,7-dimethyl-9-)2'6'6'-trimethyl-1'-cyclohexen-1'-yl) 2,4,6,8,-non-atetraene-1-ol, also known as axerophthol, the antixerophthalmic vitamin and the anti-infective vitamin. With a trimethylcyclohexenyl ring and a methylated side chain containing four conjugated double bonds (Figure 2), it can exist in different isomeric forms, each with different biological activities. The all-trans isomer possessing the highest biological activity is primarily produced commercially by chemical synthesis. Vitamin A complexed with protein (retinol binding protein) is the principal form of transport of the vitamin in blood from the liver. Vitamin A esters are the liver storage form and vitamin A aldehyde (retinal) in the retina combines with protein to form the essential photosensitive pigment, rhodopsin, in the visual process of man and ani-

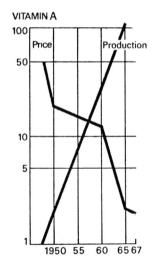

Fig. 1 - Logarithmic production - Price pattern of vitamin A with time

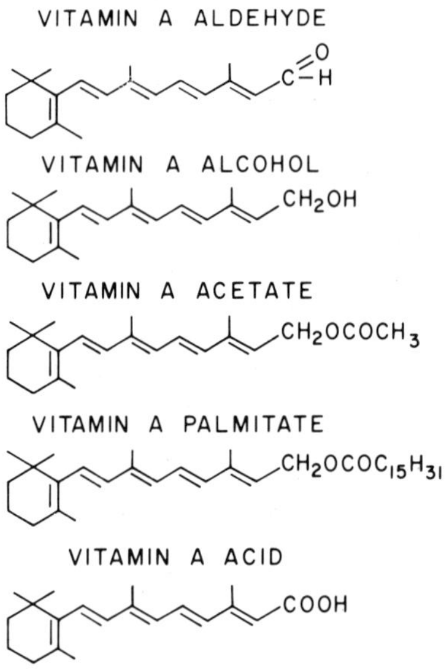

VITAMIN A ALDEHYDE

VITAMIN A ALCOHOL

VITAMIN A ACETATE

VITAMIN A PALMITATE

VITAMIN A ACID

Fig. 2 - Structural formulas of vitamin A compounds

mals. Vitamin A palmitate, propionate, or acetate, also of the all-trans type manufactured by chemical synthesis, are employed most frequently in food applications. The chemical properties of vitamin A compounds are fairly well known (Table 1) and a technology of application has been well developed[3] as has occurred with the other vitamins[4].

SYNTHESIS

All industrial syntheses[2] of vitamin A are based on β-ionone (Figure 3). In 1947, Roche researchers and others announced the synthesis of vitamin A. In 1950, full scale commercial production of vitamin A was begun, the world's first commercial synthetic vitamin A production. The Roche total synthesis of vitamin A from acetone proceeds via methylheptenone, dehydrolinalool, and pseudoionone, and twice makes use of a condensation with isopropenyl ether. The vitamin A procedures of BASF and Roche lead directly to vitamin A acetate, which crystallizes easily. The synthesis by Philips first gives vitamin A aldehyde, which is reduced to the alcohol and only then esterified to the acetate or palmitate ester. In the BASF process, β-ionone is converted to β-C_{15}-vinyl carbinol by acetylene addition and partial hydrogenation; this gives in a Wittig reaction with 2-acetoxytiglic aldehyde, vitamin A acetate. The present Roche process is, in principal, identical with the original vitamin A synthesis. It is regarded as an economical one and this view finds support in the fact that more than two-thirds of the world's production has been so produced[2]. Vitamin A is produced in the USA, Switzerland, Scotland, India, France, Germany, Holland and the USSR.

LABILITY AND STABILITY

The stability characteristics of the vitamin A (a labile structure) and its derivatives are influenced by a number of factors such as:

 a. Oxygen or ozone. In the absence of antioxidants vitamin A is very unstable toward oxygen.

 b. Other oxidizing agents. Chemical agents having oxidizing potential are a threat to unprotected vitamin A structures. This can be demonstrated with exposure of vitamin A to oxides, dioxides and peroxides of inorganic and organic structures. Peroxidases influence these reactions.

 c. Acids. Vitamin A is sensitive toward acids, which can cause rearrangement of the double bonds and dehydrations, eventually followed by cis-trans isomerization.

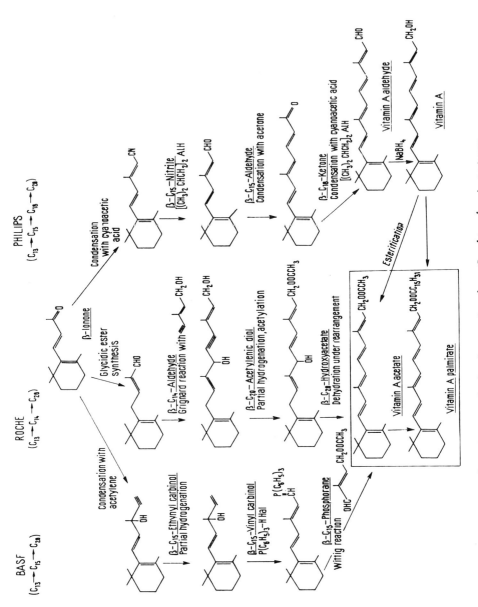

Fig. 3 - Technical synthesis of vitamin A esters

TABLE 1

PHYSICO-CHEMICAL DATA ON VITAMIN A

	VITAMIN A PALMITATE	VITAMIN A ACETATE	VITAMIN A ALCOHOL
Color of crystals (pure)	Semi-solid, yellow	light yellow	light yellow
Melting point (°C)	28-29	57-58	62-64
Molecular weight	524.9	328.5	286.4
Solubility (%), 20°C			
Fats, oils) ether, chloroform)	> 50	> 50	> 50
95% ethanol,) isopropanol)	< 10	< 10	< 10
Spectrophotometric data max. at 325 nm $E_{1cm}^{1\%}$ (ethanol)	975	1550	1835
Biological Activity Vitamin A value (IU/g)	1.82×10^6	2.9×10^6	3.3×10^6

d. Heat. Thermal application to solutions and dispersions of all-trans vitamin A will result in the formation and eventual equilibrium of several cis isomers with trans vitamin A.

e. Moisture and trace elements. Trace metals such as copper, iron, manganese and vanadium accelerate oxidative changes initiated by heat, oxygen, and moisture. Moisture and pH are somewhat more critical for the acetate ester versus the palmitate ester of vitamin A in respect to maximum stability in a water-base product.

Vitamin A is fairly stable when heated to moderate temperatures in an inert atmosphere in the absence of light, but it is unstable in the presence of oxygen or air or when exposed to ultraviolet light. In the presence of alkali, vitamin A is quite stable so that alkaline saponifications of vitamin A containing materials may be carried out without serious loss of vitamin A. The presence of trace metals may also accelerate the oxidation of vitamin A. Therefore, in handling vitamin A, some precautions are: (a) an inert atmosphere; (b) subdued light; (c) avoidance of trace metals and a strongly acid environment. Stability performance in food products[5] and other products is difficult to predict in advance because insufficient details are usually known on pH, moisture content, packaging, storage conditions and other significant variables.

APPLICATION FORMS

Vitamin A is offered in a number of application forms[6,7]. The principal methods of stabilizing vitamin A involve (a) sealing under vacuum or inert gas, (b) storage at low temperatures, (c) addition of antioxidants, (d) absorbing and/or sealing with a protective matrix such as a gelatin or vegetable gums or wax composition, and (e) chemically complexing with other compounds. The first two methods are common methods of storing any unstable compound, but they are not always convenient and practical. The use of antioxidants to conserve vitamin A is quite common practice. One is confined in this respect, however, to those antioxidants that are completely acceptable in a given country for food or pharmaceutical use. This list may include tocopherols, ascorbic acid and derivatives, butylated hydroxyanisole (BHA), butylatedhydroxytoluene (BHT), propylgallate, etc. When preparing vitamin A in an acceptable dry particle form, the selected matrix or absorbant must be nontoxic, edible, digestible, impervious to oxygen, relatively resistant to moisture, easy to handle, relatively economical and available. Much effort[6] has gone into the development of dry, stable, fat-soluble vitamin compositions as detailed by Klaüi et al. (Table 2).

Basically the application forms of vitamin A fall into two categories, liquid products and dry products. Where a maximum concentration of vitamin A in liquid form is desired, vitamin A palmitate at a label potency of 1,650,000 to 1,820,000 per gram as well as vitamin A esters diluted with a vegetable oil to a standardized potency of 1,000,000 IU per gram. Specialty liquid vitamin A formulated products in emulsion form are produced where water-dispersibility or spray-on characteristics are desired in food application use. Dry vitamin A products prepared from the various esters, at potencies of 250,000 IU or more per gram, have been developed in various particle sizes and shapes with digestible coatings. In some instances the product is consumed in dry form; in others, the dry form must become water-dispersible at a given temperature before it is consumed, thus special characteristics are required for the intended end use. In addition to physical characteristics the application forms of micronutrients must have certain chemical and physiological properties which contribute to, for example, adequate performance when used for the intended purpose[8,9]. When liquid or dry vitamin forms are used in the nutrification of foods and intended for human use, they should not influence significantly the color, appearance and flavor of the nutrified food, should be fully available biologically, should be reasonably stable and should be relatively economical to use[10-13,49]. To monitor the content in foods[14] and tissues[15], qualitative and quantitative methods of analysis for vitamin A have been published. Most foods to be successfully nutrified require centralized processing locations, followed by food distribution programs which will insure regular access to and consumption of the nutrified foods.

TABLE 2

STABILIZED DRY FORMS OF FAT-SOLUBLE VITAMINS AND PROCESSES

PRODUCTS	FORM OR TYPE	PROCESSES
1. Adsorbates and mixtures	Powders Granulates	Mixing, air-mixing, spray-mixing Granulating Crushing
2. Fat-based powders	Microspheres ("Fat-beadlets") Flakes Powders	Spray-congealing Drum-cooling Spraying or dropping into liquids Comminuting, (crushing of chilled mass), deep freezing and grinding Dispersion in aqueous liquids Mixing, air-mixing, spray-mixing.
3. Dried emulsions and suspensions	Spherical or spheroidal beadlets Flakes Powders	Spray-drying (spraying into "hot gas") Spraying into liquids ("liquid catch") Spraying into powders ("powder catch") Spraying onto moving or stationary surface Double dispersion process (dual emulsion) Drum drying Air suspension coating Comminuting sheets of dried emulsion
4. Microcapsules	Spherical or spheroidal beadlets Agglomerates	Coacervation Air suspension coating (Wurster/WARF, USA) Multiorifice encapsulation (Southwest Research Institute, USA) Vacuum encapsulation (National Research Corporation, USA)
5. Inclusion compounds and various products (including chemical derivatives, precipitates and coagulates)	Crystals Powders Granules	Crystallization Precipitation and coagulation Chemical reaction

APPLICATION FORM USE

Over the past thirty years a continuous effort (Figure 4) has been made to meet the demands[16] of the food, feed and pharmaceutical industries for suitable vitamin A formulations. In practice, vitamin A can be given as an oral supplement or it can be administered parenterally. It can also be added to a variety of foods and feeds by nutrification. Nutrification, defined, is the incorporation of micronutrients into foods for nutritional improvement. In addition to vitamin A products, variations include combination with vitamins D and/or E in stabilized form. Premixes of vitamin A may include the water-soluble vitamins (e.g. folic acid, ascorbic acid, etc.) and minerals (e.g., iron, iodine, etc.) as well. Thus, many application-market forms are possible with their respective spefications and uses[3, 7, 17, 19].

Vitamin A can either (a) be consumed daily in foods in amounts to meet physiological needs, or (b) be consumed intermittently in amounts greatly beyond daily needs and the excess vitamin stored in the liver for future needs. Those delivery systems dealing with vitamin A of chemical synthetic origin can be divided into two principal categories:

I. Prevention of Vitamin A Deficiency
    A. Short-term measures:
        1. Intermittent massive oral dosing
    B. Longer-term measures:
        1. Nutrification of regularly consumed indigenous foods
        2. Regular ingestion of a daily dietary supplement

II. Treatment of Vitamin A Deficiency
    A. Emergency measure
        1. Initial intramuscular injection of a water-dispersible preparation
        2. Subsequent oral ingestion of an oil-soluble preparation.

INTERMITTENT DOSING[10, 19-24]

For intermittent massive oral dosing of young children (or adults) a technology exists for the manufacture of sealed soft gelatin capsules of vitamin A ester, usually palmitate, (with vitamin E) which can be swallowed directly, or the contents squeezed out onto the tongue of the child not yet old enough to swallow the capsule. Soft gelatin capsules are made and filled continuously on the rotary die process machines by skilled operators from ribbons of gelatin forming the capsule which is simultaneously filled. Content variation among capsules is very small. The vitamin A palmitate (200,000 IU), α-tocopherol (40 IU) in oil formulation is filled into a six minim, one-piece, clear, soft gelatin capsule (photograph)

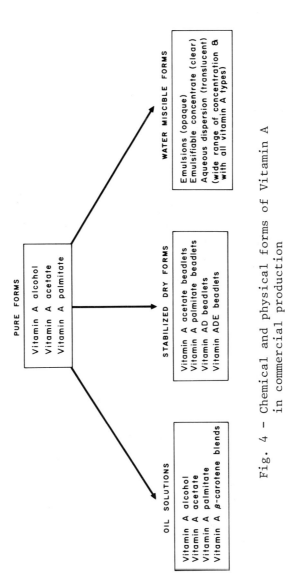

Fig. 4 – Chemical and physical forms of Vitamin A in commercial production

with a nipple or nozzle tip. The tip may be cut off if the contents are to be squeezed out. After swallowing the capsule, dissolution occurs within 10 minutes. Capsules containing less or more vitamin A can be prepared; however, due to the lengthy preparatory steps in machine adjustments and its operations, commercial runs of less than one half million capsules are difficult to arrange. Stability of vitamin A in the capsules is excellent as determined by assay data (Table 3). Field use studies[19,20,22,24] indicate that one capsule should be given every six months, as a minimum program, to young children as xerophthalmia preventing measure. If cost and distribution systems permit, one given every three to four months would give additional assurance, particularly for the older children whose vitamin A requirements are higher and to catch young children missed in the previous dosing schedule. Objections to the capsule approach are (a) the skipped child, (b) the possibility of regurgitations of contents after swallowing, or not swallowing the capsule and (c) the association that one has to take "medicine" to prevent blindness, not really an educational approach, as faulty diet is the real cause. Intermittent massive oral dosing has been used intensively in India and Bangladesh and to some extent in other countries, such as Haiti, Indonesia, Philippines, El Salvador, etc.

TABLE 3

RETENTION OF VITAMIN IN STORED SOFT GELATIN CAPSULES

VITAMIN A VALUE (IU/CAPSULE)

Capsulation Run	Initial	After Storage	
A	214,000[a]	---	211,000 (20 mo)
B	205,000	---	219,000 (25 mo)
C	209,000	---	206,000 (29 mo)
D	214,000		207,000 (31 mo)
E	303,000	294,000 (17 mo)[b]	296,000 (42 mo)
F	295,000	313,000 (17 mo)	300,000 (42 mo)
G	300,000	293,000 (17 mo)	292,000 (42 mo)
H	290,000	280,000 (17 mo)	297,000 (42 mo)

[a] All values by duplicate chemical assay.
[b] Months of storage, 23°C.

PARENTERAL DOSING[19, 22-25]

When a patient (child or adult) is recognized as having signs of corneal ulceration with xerosis and there is danger of keratomalacia and irreversible loss of sight, one wants to apply vitamin A most effectively. Intramuscularly (IM) injected oil solutions of vitamins have been found to be relatively ineffective compared to water-dispersible vitamin A palmitate parenterals which are quick-acting. A water-base formulation of vitamin A palmitate (100,000 IU) containing appropriate emulsifiers, antioxidants and preservatives, heat sterilized in one or two ml glass ampules for hospital use, may be administered intramuscularly in a sterile syringe. IM administered vitamin A from this type of formulation moves quickly into the blood stream and the excess is stored in the liver[24, 25] as can be shown by animal experiments (Figure 5). Liver stores will maintain blood levels as long as such stores remain and protein foods are consumed as part of the diet. Properly formulated vitamin A parenterals are quite stable in respect to vitamin A potency as monitored by storage assay data (Table 4).

Current recomended schedule of treatment by WHO[22] and IVACG[23] of severe xerophthalmia is as follows:

Treatment Schedule	Preparation
Immediately upon diagnosis	Inject IM, 100,000 IU water-dispersible vitamin A
Second Day	Give orally, 100,000 IU oil solution, vitamin A
Prior to discharge	Give orally, 100,000-200,000 IU oil solution vitamin A

All hospitals in countries with recognized cases of vitamin A deficiency coming to the hospital should keep in stock at all times, ampules of water-dispersible vitamin A and oral capsules of oil solutions of vitamin A.

NUTRIFIED FOODS[3, 7, 9, 13, 17, 18, 26-30]

Vitamin A has been successfully added[7] to salad oils, margarine, peanut butter, liquid skim milk, canned evaporated or condensed milk, nonfat dry milk, ice cream, mellorine, parevine, butter, cheese, white flour, bread, cakes, cookies, corn grits, corn meal, white rice, breakfast cereals, snacks, potato chips, potato flakes, potato granules, fruit juice beverages and powders, infant and geriatric

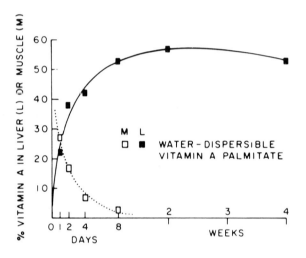

Fig. 5 - Translocation pattern of an effective water-dispersible vitamin A formulation injected IM in the muscle to the liver

TABLE 4

RETENTION OF VITAMIN A IN STORED WATER-DISPERSABLE
VITAMIN A PARENTERALS (INITIAL ASSAY 120,000 IU/ml)

STORAGE CONDITIONS	PER CENT RETENTION
6 wks., 45°C	91
3 mo., 37°C	94
6 mo., 23°C	98
12 mo., 23°C	94
18 mo., 23°C	90
12 mo., 5°C	100
24 mo., 5°C	97

liquid and dry formulas, confectionaries, tea dust, tea leaves, refined crystalline sugar, and seasonings such as refined white fine crystalline salt and monosodium glutamate.

Oils and Fats[31-36]

Oils and fats (shortening, butter, margarine, ghee, vanaspati, etc.) are dietary ingredients of most populations. Liquid vegetable oils (cotton-seed, maize, olive, peanut, soybean, etc.) used in light cooking or consumed with leafy vegetables or other foods can be nutrified with added vitamin A. After the liquid vegetable oil in a bulk container during processing has been clarified, a liquid vitamin A ester concentrate in a proper amount is added and is agitated for uniformity before placing the nutrified oil product in sealed containers. While nutrification of vegetable oils is not practiced to a large extent in commerce, it is feasible if the oils are packaged in sealed opaque or dark colored glass containers to eliminate the effect of light, a stimulant to oxidative processes. Edible antioxidants may also be added to protect both the vegetable oil and the added vitamin A ester from oxidation. Inert gas replacement for the air in the headspace of the container is advisable at the time of initial packaging, if practical to do so. Currently there is some interest in the Philippines for vitamin nutrified edible oils.

Margarine, commonly used as a bread spread interchangeable with butter, is generally nutrified with vitamin A ester (20,000 to 50,000 IU per kg) to equal or exceed the average levels of this nutrient in butter. Nutrification of margarine with vitamin A is now carried on in Australia, Austria, Belgium, Brazil, Canada, Chile,

Colombia, Costa Rica, Denmark, Finland, Germany, Greece, Israel, Italy, Japan, Mexico, the Netherlands, Newfoundland, Norway, Philippines, Peru, Portugal, South Africa, Sweden, Switzerland, Turkey, Venezuela, UK, USA and USSR. Nutrification of margarine with vitamin A or vitamins A, D and E, is relatively simple. Vitamin A, premeasured in amounts of the vitamin for the batch size of the margarine oil tanks used in a particular margarine factory, is added to one tank of oil, a simple act and an easy factory control operation. If synthetic β-carotene is also to be used, both color and vitamin A can be added simultaneously. A safe color, β-carotene provides color and much of the required vitamin A activity. Available data (Table 5) indicate added vitamin A to be quite stable[33-35] during the margarine manufacturing operations and during use of the nutrified product in the home. Furhtermore test results demonstrate, in biscuits, cakes and breads prepared with vitamin A nutrified fats, that 80 to 100 percent of the added vitamin survives the baking process. Lard, shortening, ghee, vanaspati and other fats, likewise can be easily nutrified with added vitamin A.

Beverages

When carriers of supplementary micronutrients such as vitamin A are considered, foods which can be served as beverages might be considered to have special merit since in illness when solid foods may be refused, there is a greater tendency to drink, if only to meet one's need for the water in the beverage.

Milk[37-42]

When fat is removed from whole fluid milk all fat-soluble vitamins including vitamin A are likewise removed. That the resulting skim milk can serve as a vehicle for added fat-soluble vitamins has been well documented. All skim milk or low-fat liquid or dry milk products should have vitamin A added and, if given to infants, should probably have vitamins D and E and possibly C added as well.

Two techniques have been developed for the nutrification of nonfat dry milk (skim milk powder).

a. Wet Technique - A liquid form pure vitamin A palmitate (preferably with added antioxidants) with or without vitamin D is diluted in hydrogenated coconut oil. This warm oil dilution is homogenized into liquid condensed skim milk to make a high-fat premix. The required amount of the premix is blended into a known quantity of condensed skim milk and mixed thoroughly. The mixture is spray dried and then, if desired, instantized in the usual manner. The use of coconut oil has been found desirable for good stability of the added vitamins. The amount used increases the total fat content of nonfat

TABLE 5
RETENTION OF VITAMIN A AND β-CAROTENE IN COMMERCIAL MARGARINE

MANUFAC-TURER	FORM OF VITAMIN A	INITIAL ASSAY	AFTER 2 MONTHS' STORAGE			AFTER 6 MONTHS' STORAGE		
			40°F.	40°F. & 75°F.	75°F.	40°F.	40°F. & 75°F.	75°F.
			(U.S.P. Units of Vitamin A Activity per Gram)					
J	Vitamin A	13,900	12,000 (86)	12,000 (86)	11,300 (81)	14,700 (106)	13,400 (97)	13,600 (98)
	Carotene	6,400	6,700 (105)	6,300 (98)	6,000 (94)	6,100 (95)	5,900 (92)	5,500 (86)
	Vitamin A	13,300	13,100 (99)	13,300 (100)	13,100 (99)	---	---	---
	Carotene	6,100	5,900 (97)	5,900 (97)	5,900 (97)	---	---	---
K	Vitamin A	13,800	14,400 (104)	13,500 (98)	13,800 (100)	---	---	---
	Carotene	5,300	5,200 (98)	5,100 (96)	4,900 (93)	---	---	---
	Vitamin A	12,300	13,100 (107)	12,700 (103)	12,300 (100)	---	---	---
	Carotene	5,300	5,100 (96)	5,100 (96)	4,900 (93)	---	---	---
L	Carotene	6,100	---	---	---	6,100 (100)	5,700 (93)	5,200 (85)
M	Vitamin A	11,000	---	---	---	10,600 (96)	---	10,100 (92)
	Carotene	5,200	---	---	---	4,800 (92)	---	4,800 (92)
	Vitamin A	10,300	---	---	---	10,600 (103)	---	10,400 (101)
	Carotene	5,200	---	---	---	5,000 (96)	---	---
	Vitamin A	13,400	13,400 (100)	13,400 (100)	13,800 (103)	13,100 (98)	12,800 (94)	5,300 (102)
	Carotene	5,200	5,050 (96)	5,200 (100)	4,850 (92)	5,300 (102)	4,700 (91)	---
N	Vitamin A	13,350	---	---	---	11,900 (89)	12,000 (90)	11,100 (83)
	Carotene	5,950	---	---	---	5,400 (91)	5,400 (91)	5,300 (89)
	Vitamin A	12,700	13,000 (102)	12,500 (98)	11,800 (93)	13,600 (107)	12,800 (101)	11,800 (93)
	Carotene	6,400	6,300 (98)	5,800 (91)	5,800 (91)	6,700 (105)	6,300 (98)	5,300 (83)
O	Vitamin A	14,200	13,900 (98)	12,800 (90)	13,200 (93)	13,400 (94)	13,700 (97)	12,700 (89)
	Carotene	5,850	5,900 (101)	6,100 (104)	5,800 (99)	5,900 (101)	5,800 (99)	5,500 (94)
	Vitamin A	13,500	13,100 (97)	12,300 (91)	12,600 (93)	14,200 (103)	13,500 (100)	12,800 (95)
	Carotene	6,050	6,200 (103)	6,200 (103)	5,900 (98)	5,700 (94)	5,100 (84)	5,200 (86)
	Vitamin A	11,800	11,600 (98)	11,000 (93)	10,500 (89)	12,000 (102)	12,800 (108)	10,900 (92)
P	Carotene	5,750	5,700 (99)	5,600 (97)	5,400 (94)	5,600 (97)	5,500 (96)	4,900 (85)
	Vitamin A	13,500	13,000 (96)	13,800 (102)	13,800 (102)	12,400 (92)	12,800 (95)	11,500 (85)
	Carotene	6,000	5,700 (95)	5,700 (95)	5,400 (90)	6,100 (102)	5,900 (98)	---
	Vitamin A	11,800	11,400 (97)	12,000 (102)	11,600 (98)	12,000 (102)	11,800 (100)	---
	Carotene	5,550	5,300 (96)	5,200 (94)	5,100 (92)	5,700 (103)	5,500 (99)	5,200 (94)
	Vitamin A	12,300	12,000 (98)	13,100 (107)	12,300 (100)	12,100 (98)	12,100 (98)	12,300 (100)
	Carotene	5,050	4,900 (97)	4,900 (97)	4,800 (95)	5,400 (107)	5,050 (100)	4,500 (89)
Q	Vitamin A.	12,400	12,500 (101)	12,600 (102)	13,100 (106)	12,100 (98)	11,700 (95)	10,900 (88)
	Carotene	5,500	5,600 (102)	5,600 (102)	5,400 (98)	5,300 (96)	5,200 (95)	4,700 (86)
	Vitamin A	13,000	12,900 (99)	14,200 (109)	11,400 (87)	11,500 (88)	12,600 (97)	11,300 (87)
	Carotene	6,500	6,400 (99)	6,300 (97)	5,900 (91)	6,300 (97)	6,200 (95)	5,400 (83)
	Average							
	Vitamin A		(99)	(99)	(96)	(98)	(98)	(92)
	Carotene		(99)	(98)	(94)	(99)	(95)	(89)

Note: Figures in parentheses are % retention based on initial assay value.

dry milk (about 0.1 to 0.2%). Blending of the premix into the condensed skim milk may be done on a batch basis or on a continuous basis, the latter requiring controlled metering of the introduced vitamin A premix with the flow rate of the condensed milk.

b. <u>Dry Tecnique</u> - A dry, stable, water-dispersible form of vitamin A ester (Type 250-S or 250-CWS) is blended with a weighed quantity of nonfat dry milk to form a premix. An aliquot or proper portion of the premix is added to a given quantity of nonfat dry milk and mixed in, to produce the final nutrified product. Blending may be done batch style in a suitable dry blender, such as a ribbon mixer, or continuous blending may be employed, metering in the dry vitamin A or the prepared premix in proportion to the flow of the dry milk product.

Both the wet and dry methods are practical and proven. The dry technique provides slightly better stability but is slightly more costly than the wet technique because the dry forms of vitamin A are somewhat more costly than the liquid vitamin concentrates and also because of the extra labor involved in the dry mixing operation. Problems to be aware of in the nutrification of dry skim milk are (a) uniformity of introduction of vitamin A in the continuous process, (b) good blending in both the continuous and batch systems, and (c) avoidance over processing which leads to off-flavors. Stability of synthetic vitamin A added to nonfat dry milk is good to excellent (Table 6). Data also exists to show that synthetic vitamin A added to dry milk is fully available biologically[40]. Proper packaging does influence the vitamin A stability of the dry product; packages with high moisture resistance are preferred. When reconstituted[39,41], nutrified nonfat dry milk retains its added vitamin A content (Table 7).

Nonfat dry milk exported from the USA under the Food for Peace Programs has been nutrified with vitamin A and vitamin D over the past decade. After adequate experiments had demonstrated vitamin A to be stable in nonfat dry milk, the product was nutrified with 5,000 IU of vitamin A and 500 IU of vitamin D/100 g for the P.L. 480 overseas distribution in May 1965. Some years thereafter, these levels were reduced to 2200 IU of vitamin A/100 g and 440 IU of vitamin D/100 g, to coincide with the same levels in nutrified nonfat dry milk initiated for the USA in 1968. World Health Organization (WHO) still recommends nutrification for developing countries at 5000 IU of vitamin A and 500 IU vitamin D/100 g, which has been adopted by the EC in Brussels. The vitamin A nutrification of nonfat dry milk has been long favored by WHO. UNICEF, over a decade ago[40] encouraged studies resulting in the developed technology for incorporating vitamin A in the product. Once again in 1977 WHO[43,44] stressed the desirability of nutrifying nonfat dry milk with vitamin A as one means of improving vitamin A intake and hence preventing xerophthalmia blindness in children. To assure consumers of vita-

TABLE 6

RETENTION OF ADDED VITAMIN A IN STORED NONFAT DRY MILK

% RETENTION

	IU/lb		Polyethilene bags						Kraft-foil bags						Tin cans					
SAMPLE Nº	VITAMIN A ADDED	INITIAL ESSAY	Wk/98°F.			Wk/113°F.			Wk/98°F.			Wk/113°F.			Wk/98°F.			Wk/113°F.		
			3	6	12	3	6	12	3	6	12	3	6	12	3	6	12	3	6	12
1	10,000	10,100	93	97	100	101	88	89	104	99	100	111	100	86	99	100	95	92	88	85
2	10,000	10,400	100	91	95	97	88	86	99	100	98	109	100	87	90	88	94	82	91	82
			Mo/75°F.			Mo/40°F.			Mo/75°F.			Mo/40°F.			Mo/75°F.			Mo/40°F.		
			3	6	12	3	6	12	3	6	12	3	6	12	3	6	12	3	6	12
1	10,000	10,100	108	87	69	100	76	85	104	79	78	109	74	84	111	82	85	115	85	96
2	10,000	10,400	93	82	85	106	80	84	103	78	77	121	79	107	108	85	89	115	78	94

TABLE 7

STABILITY OF VITAMIN A FOLLOWING RECONSTITUTION
OF VITAMIN A FORTIFIED NONFAT DRY MILK

SAMPLE[a]	RECONSTITUTION STORAGE (hr at 40°F.)	% VITAMIN RETENTION	
		Wet[b]	Dry[b]
A	24	102	99
	72	101	96
B	24	96	95
	72	86	89
C	24	99	97
	72	98	95
D	24	92	98
	72	92	96
E	24	96	99
	72	92	97

[a] Samples chosen for reconstitution studies already had a variable history of storage in the dry state, for example, sample A, 1 month at 75°F; sample C, 3 months at 98°F; and sample E, 12 months at 75°F. Values represent duplicate vitamin A assays.

[b] Technique of nutrification.

TECHNOLOGY OF VITAMIN A

min A nutrified nonfat dry milk that vitamin A is actually present, a simple qualitative test has been devised which can be used in the field[45].

Tea[46]

Tea is widely consumed in some parts of the world. A food habits survey conducted in the Western India states by the Protein Food Association (1969) revealed that 87% of the villagers gave tea to their children, including 84% of those in the lowest income families. The idea of using tea as a vehicle for conveying meaningful quantities of vitamin A to an entire population was conceived at the New Delhi Mission of the Agency for International Development as early as 1967.

It has been determined at Roche that tea dust could be nutrified by dry mixing the dust with a fine powdered vitamin A palmitate product (Type 250-SD). Tea dust, so nutrified, retained 85% of the vitamin A activity after storage for one year at room temperature (23°C). Tea leaves required another approach. Successful trials were made with emulsions of vitamin A palmitate sprayed onto the leaves. The emulsions were made by homogenizing the vitamin A ester into thick acacia or dextrin solutions at a concentration of 500,000 IU of vitamin A per gram. When diluted with 50% sucrose solution and sprayed on tea leaves, satisfactory stability resulted as judged by storage trials for as long as one year.

Since added vitamin A must survive brewing, the retention of vitamin A was measured after brewing the nutrified tea products. The initial target was a five minute boiling time period, since this does produce a drinkable tea. However, a review of tea making practices revealed that tea is often left on the fire for as long as an hour. Trials indicate that boiling can be detrimental to some of the vitamin A ester emulsion products. However, vitamin A palmitate applied in both the powdered form to the dust and an emulsion form sprayed onto leaves showed 100% retention even after one hour of boiling (Table 8).

Tea dust or tea leaves can be readily nutrified with properly selected form of vitamin A. The initial studies were aimed at nutrifying dry tea to a level of 125 IU of vitamin A per gram. The usual adult cup holding 150 ml of tea brewed from 3 g of dry nutrified tea would supply 375 IU of vitamin A, however, other vitamin A levels are possible. Trials on a commercial scale in a tea factory in South India have demonstrated the feasibility of nutrifying large quantities of tea during its processing, either in individual factories or in tea blending plants. Presumably tea leaves would be sprayed after the fermentation stage near the final drying stage. Some interest in nutrifying tea has been shown in the past in India

TABLE 8
RETENTION OF VITAMIN A IN BREWED TEA

TEA TYPE	FORM OF VITAMIN A	HOW APPLIED TO TEA	PERCENT RECOVERY IN BREWED TEA	
			5-min Cooking	1-min Cooking
Dust	Palmitate 250-SD powder	Dry mixing	100	100
Leaves	Palmitate special water-soluble oil[a]	Sprayed	45	30
Leaves	Palmitate emulsion[a]	Sprayed	100	100
Leaves	Acetate emulsion[a]	Sprayed	50	4

[a]Diluted in 50% sucrose solution.

and Pakistan. By a similar technology it is probably possible to nutrify coffee, if merit exists for such a product.

Cereal-Grain Products[18, 47-55]

Cereal-grain products usually constitute an important calorie source of many populations. In some countries, Mexico, India, Africa, they make up 55-65% of the daily caloric intake[55]. In the USA it is estimated that about 26% of the daily caloric intake[51] comes from products based on cereal-grains (wheat, corn, rice, etc.) and their by-products. In 1974 a proposed nutrification policy[51] for USA cereal-grain products was set forth by the Panel of the Committee on Food Standards and Fortification Policy of the Food and Nutrition Board, National Research Council, National Academy of Science. Since cereal-grain products are suitable carriers technically for the addition of added nutrients and since the Panel has evidence of potential risk of deficiency of vitamin A, thiamin, riboflavin, niacin, pyridoxine, folacin, iron, calcium, magnesium and zinc among significant segments of the population, the Panel suggested the nutrification of cereal-grain products with these ten nutrients. The recent studies of Anderson et al.[52] Cort et al.[53], Rubin et al.[54] and other attest to the technical feasibility of cereal-grain product nutrification. In the past, nutrification of cereal-grain products with added thiamin, riboflavin, niacin and iron has been fairly widely adopted in the USA and has spread to other countries[56] since that time.

Weaning food, principally with a cereal-grain base, usually mixed with a vegetable protein or milk and nutrified with micronutrients, have been tried or considered in Egypt, Algeria, India, Iran, Morocco, Tunisia, Turkey, Guyana, Jamaica, etc., as a nutrient delivery system for infants. Varying degrees of success have been observed. Extruded cereal-grain, vegetable protein product mixtures, are of current interest in Sri Lanka, Bolivia, Costa Rica, Honduras, etc.

Flour; Maize Meal

The development in the last decade of stable dry forms of vitamin A ester of smaller ranges of particle size[50] has greatly broadened the type of food products amenable to vitamin A nutrification. Foremost among these food products are cereal-grain products, such as white flour, corn or maize meal, white rice, etc. The need for the nutrification of cereal-grain products with vitamin A for the developing countries was first suggested by UNICEF personnel. The small particle, dry vitamin A ester product (Type 250-SD), introduced into cereal-grain products in premix form in laboratory trial runs, has shown good stability in corn meal (15% moisture), namely, 86% retention of initial values after six months at room temperature (23°C), and in wheat flour (11% moisture), 90% after the same time interval. Furthermore, when incorporated into nutrified cereal-grain products and baked into chappaties, breads, and tortillas, vitamin A retention values between 87 and 95% were observed. Vitamin A premix for white wheat flour is designed to be metered into the continuous flow of flour with the aid of a mechanical feeder (Figure 6) at the rate of 1/4 oz (7.09 g) per 100 lb (45.5 kg) of flour without problems of bridging, compression, lumping or segregation. The fine particle size of the dry vitamin A form developed for flour permits screening, if necessary, of the flour with a coarse screen without excessive loss or physical separation of the vitamin A particles from the flour.

Recently it has become desirable to retest the stability of vitamin A when incorporated into white flour, corn meal and white rice as a multi-micronutrient premix as part of the feasibility studies of the 1974 proposed NAS-NRC-FNB nutrification program for cereal-grain product. Two premixes were prepared, Premix A containing vitamin A plus five other vitamins (thiamin, riboflavin, pyridoxine, niacin and folacin) and four minerals (iron, calcium, magnesium and zinc) and Premix B like A but without two minerals (calcium and magnesium). Nutrified flour was prepared in commercial bakery facilities with Premix A and B, and breads were baked from each nutrified flour. Nutrified flour samples were stored and assayed for nutrient content, likewise the baked breads[53, 54]. Tabular data (Tables 9 and 10) show excellent retention of the added vitamin A in the multi-nutrient nutrified flour and good to excellent retention in bread baked with such flours. Bread characteristics such as color, grain score, and flavor, although slightly modified

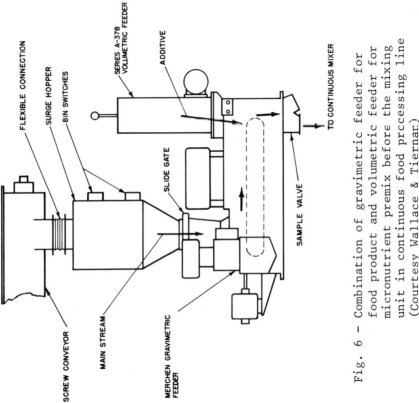

Fig. 6 — Combination of gravimetric feeder for food product and volumetric feeder for micronutrient premix before the mixing unit in continuous food processing line (Courtesy Wallace & Tiernar.)

TABLE 9

STABILITY OF THE VITAMIN-MINERAL PREMIX ON FLOUR (12% H_2O)

SUPPLEMENT	IU PER mg PER 100g			% RETENTION 6 mo. at 23°C	
	Label Claim	Initial Assay Premix A[a]	Premix B[b]	Premix A[a]	Premix B[b]
Vitamin A[c], IU	1600	1730	1800	101	100
Vitamin B_1, mg	0.64	0.76	0.71	99	101
Vitamin B_2, mg	0.40	0.39	0.40	100	100
Vitamin B_6, mg	0.44	0.48	0.56	100	100
Niacin, mg	5.29	5.60	5.60	100	100
Folic acid, mg	0.07	0.063	0.07	100	96
Iron, mg	8.81	9.34	9.46	97	100
Calcium, mg	198.2	209.0	NR[d]	NR	---
Magnesium, mg	44.1	47.0	NR	NR	---
Zinc, mg	2.2	2.3	NR	NR	NR
Moisture, %		11.6	12.1	11.8	12.5

[a] Fortified with the complete vitamin-mineral supplement.
[b] Vitamin-mineral supplement without calcium and magnesium.
[c] Type 250-SD
[d] NR = Not run

TABLE 10

RETENTION OF ADDED VITAMINS-MINERALS IN BAKED BREAD (38% H_2O)

NUTRIENT	LEVEL ADDED IU or mg/1b	% RETENTION			
		After baking		5 Days at RJ	
		Premix A[a]	Premix B[b]	Premix A[a]	Premix B[b]
Vitamin A, IU	992	83	95	83	95
Vitamin B_1, mg	1.8	101	101	101	100
Vitamin B_2, mg	1.1	105	101	108	101
Vitamin B_6, mg	1.2	100	105	100	105
Niacin, mg	15.0	100	102	100	106
Folic acid, mg	0.19	94	105	80	102
Iron, mg	24.8	104	105	NR[c]	NR
Calcium, mg	558.0	NR	---	NR	---
Magnesium, mg	124.0	107	---	NR	---
Zinc, mg	6.2	NR	NR	NR	NR

[a]Fortified with the complete vitamin-mineral supplement.
[b]Vitamin-mineral supplement without calcium and magnesium.
[c]NR = Not run.

when closely compared with unnutrified controls, were suitable and with slight modification of the magnesium and/or calcium levels were judged to be very acceptable products with all ten added nutrients present. In some of these studies, vitamin E (as α-tocopheryl acetate) was added in the micronutrient premix and subsequent flour and bread studies demonstrated excellent stability of added vitamin E[53]. Yellow corn (maize) meal and corn grits were likewise nutrified with a multi-nutrient premix and the nutrified products observed for nutrient retention after storage and when cooked to make ready for consumption (Tables 11 and 12). While the added vitamin A showed some losses, 80% or more was retained in the storage tests and 70% or more in the cooking tests of the nutrified corn products[53, 54].

Vitamin A nutrification of corn meal became a part of P.L. 480 purchase specifications for export food from the USA December 1, 1968 and of flour May 1, 1969. The range specified was 4,000 to 6,000 IU/pound. Currently, additional nutrified foods sent abroad from the USA under the USA-AID or Food for Peace Program have the following specifications: soy-fortified corn meal, 2-3 mg thiamin, 1.2-1.8 mg riboflavin, 16-24 mg niacin, 4000-6000 IU vitamin A, 13-26 mg iron and 500-750 mg calcium per lb; soy-wheat flour blend, 2-2.5 mg thiamin, 1.2-1.5 mg riboflavin, 16-20 mg niacin, 4000-6000 IU vitamin A and 500-1107 mg (for 6% soy) or 750-1364 mg (for 12% soy) calcium per lb; other foods have other specifications.

In December 1967 fortification of white bread with vitamin A, the B vitamins, and iron was instituted in a new government-owned bakery in Bombay[48]. Vitamin A was an ingredient of the fortification mixture at a level to provide 8,000 IU of vitamin A per kg of bread made from both dark and light wheat flour. This program was started in the largest flour mills in major population centers, and was to be extended to nearly 200 flour mills. Programs have been considered for nutritified flour in countries such as Brazil, Chile, India, Tunisia, Pakistan, Jordan and Iran. In the past Guatemala and South Africa have shown an interest in nutrified maize meal. It can be said that the technology exists for the incorporation of vitamin A into wheat and maize products in amounts capable of making substantial improvement in the diets of consumers without impairing significantly the acceptance of the nutrified products. Other cereal grain products made from barley, sorghum, rye, etc., can be nutrified.

Rice[53, 54, 57]

Rice is consumed as a whole kernel product which makes nutrification more difficult than for a meal or flour product. Nutrification is compounded further by the custom of housewives washing rice before cooking. The added nutrients must be placed on rice in a rinse-resistant coating, yet in one which dissolves during the cooking process releasing the nutrients. White rice nutrification

TABLE 11

STABILITY OF VITAMIN-MINERAL PREMIX IN YELLOW CORN MEAL
(6.5% H$_2$O) AND IN CORN GRITS (11.4% H$_2$O) AT ROOM TEMPERATURE (23°C)

NUTRIENT (IU or mg/lb)	CORN MEAL			CORN GRITS		
	Initial	3 mo.	6 mo.	Initial	3 mo.	6 mo.
Vitamin A[c], IU	6000 IU	5820 IU	5880 IU	6000 IU	5700 IU	4850 IU
Thiamin, mg	3.17	3.25	3.07	3.83	3.81	3.86
Niacin, mg	26.0	25.7	NR[b]	36.0	36.0	NR
Pyridoxine HCl, mg	4.5	4.0	4.5	4.4	4.3	4.5
Folic acid, mg	0.6	0.5	0.5	0.49	0.45	0.5
Riboflavin, mg	2.02	1.81	1.96	2.82	2.64	2.58
Iron, mg	41.0	39.0	40.0	55.0	52.1	56.0

[a]Type 250-SD
[b]NR = Not run

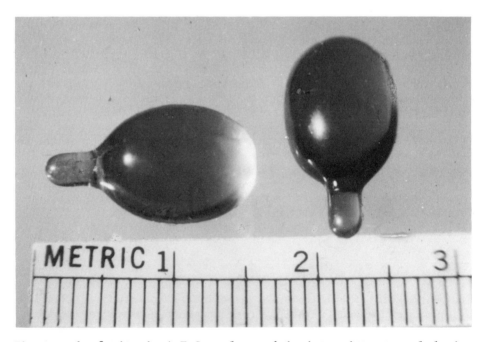

Photograph of vitamin A-E Capsule used in intermittent oral dosing

TABLE 12

RETENTION OF VITAMIN A IN COOKED GRITS
AND COOKED YELLOW CORN MEAL

	PERCENT RETENTION AFTER COOKING TIME (minutes)				
	4	5	6	10	30
Quick grits	80	---	75	70	---
Regular grits	---	---	---	---	66-75
Yellow corn meal	---	87	---	---	---

requires, therefore, an entirely different approach from that of flour or meal. Rice nutrification is accomplished by mixing one nutrified grain or kernel with 199 kernels of regular white rice. The nutrified kernel can be made by either coating white rice kernels with the nutrients or by producing simulated kernels made from a cereal dough composition containing the nutrients. Commercially, only the former is being carried out at the present time. There are two procedures for manufacturing the nutrified rice grains. One of these, referred to as the HLR-Mickus procedure, was developed at Hoffmann-La Roche many years ago and described by Mickus[57]. This procedure entails essentially soaking the rice in an acidified mixture of water-soluble vitamins, then adding various coatings, followed by a powdered iron source, vitamin A and calcium in a Trumbol mixer, finally whitening and polishing the nutrified grains. A second variation is the Wright process in which all the vitamins and minerals are introduced in powder form with suitable coating agents. Both procedures yield a stable coated product, with good vitamin A retention (Table 13), rinse resistance and the nutrients biologically available in the cooked rice. With any nutrified rice product, excess water should not be used in cooking as water poured off after cooking will mean some loss of nutrients.

Sugar (Sucrose)[10, 11, 58-64]

When selecting a vehicle food carrier to be nutrified, it must be one or several which are consumed on a regular basis over the seasons of the year by the population of the involved country. For the Central American countries crystalline white sucrose was found to be such a food carrier for added micronutrients. The average consumption of sugar by the population of Central American countries was determined from data collected during the regional nutrition

TABLE 13

STABILITY OF HLR-MICKUS NUTRIFIED RICE PREMIX
AND RETENTION AFTER COOKING OF NUTRIFIED RICE

NUTRIENT	PER 1b OF PREMIX			PER 1b OF RICE	
	Initial	6 mo. room temp. (23°C)	4 wk (45°C)	Uncooked rice with premix diluted 1:200	Rice cooked nutrified (1380g)[a]
Vitamin A[b], IU	1,200,000	1,050,000	1,050,000	6,000	5,800
Pyridoxine, mg	400	400	390	2.1	2.5
Folic acid, mg	46	44	41	0.23	0.27
Vitamin E, IU	3,240	3,210	3,200	16	15,9
Thiamin, mg	550	530	NR	2.6	NR

[a] 1 lb of uncooked rice = 1,380 g of cooked rice.
[b] Type 250-SD.
NR = Not run.

surveys of eating habits carried out in 1965-67 by the INCAP-OIR-National Governments cooperative effort. The chosen amount of vitamin A to be added to sugar would provide the FAO-WHO Recommended Daily Allowance for an adult man, namely, 750 µg or 2500 IU of vitamin A.

A project was initiated at INCAP with the cooperation of Roche in 1969 to test and implement the nutrification of sugar with a suitable form of vitamin A. Of the vitamin A types on the market, vitamin A palmitate, Type 250-SD or Type 250-CWS, were found to be adaptable to the project. Product Type 250-CWS was finally chosen. It is a pale yellow beadlet product with particles approximately 100 to 450 microns in size. Separation of the vitamin A particles from the sugar crystals is prevented by providing an edible bonding agent when blending the dry vitamin A with the crystalline sugar. In the nutrification of sugar, a vitamin A-sugar premix is first prepared. This premix is added in a set ratio, by means of a mechanical feeder, to the flow of nutrified sugar, followed by adequate mixing to insure uniformity of the final vitamin A nutrified product. At the concentration of vitamin A in nutrified sugar (50 IU per g), no significant color or flavor problems were encountered when consumed in hot coffee, tea, or orange, pineapple and lemon beverages and other food products. Biological effectiveness of the added vitamin A in sugar was demonstrated by animal assay and by increased vitamin A levels of blood serum of mothers, babies and children, and in the milk of nursing mothers after the consumption of vitamin A nutrified sugar. Stability data (Table 14) demonstrate adequate retention of added vitamin A in the product for time periods involved in manufacturing, marketing and consumer use. Long storage of vitamin A nutrified sugar is not necessary since sugar and vitamin A are regularly available for the nutrifying operations. Nutrification of sugar is commercially practiced in Costa Rica, Guatemala, Panama and Honduras and is expected to occur in El Salvador and Nicaragua. ·Interest has been shown in Brazil, Haiti and Chile.

Seasonings

Seasonings in the past have been considered carriers of nutrients. Objections to seasoning as nutrient carriers are that (a) there is no uniform volume or weight of daily consumption, and (b) use is by individual preference. Salt is universally consumed and in some population units it may be one of the few processed items secured outside their agricultural economy. To consider it a vehicle or carrier for missing or inadequate nutrients consumed by these populations is a natural expectation.

TABLE 14

STABILITY OF VITAMIN A IN NUTRIFIED SUGAR

			PERCENT RETENTION OF VITAMIN A						
			45°			23-25°			
PRODUCT	TRIAL	TYPE VITAMIN A	1 mo.	2 mo.	3 mo.	1 mo.	2 mo.	3 mo.	6 mo.
(Premix)[a]									
A	1	250 - CWS	92	83	--	99	98	--	99
B	1	250 - SD	87	79	--	97	95	--	96
C	2	250 - CWS	--	--	86	--	--	92	94
D	2	250 - SD	--	--	77	--	--	93	89
E	3	250 - CWS	93	--	81	97	--	90	90
F	3	250 - SD	92	--	74	100	--	95	93
(Nutrified Sugar)[b]									
G	4	250 - CWS	91	--	76	96	--	96	92
H	4	250 - SD	90	--	73	100	--	88	85

[a] Vitamin A - sugar premix (50,000 IU/g)
[b] Vitamin nutrified sugar (50-70 IU/g)

Salt (NaCl)

Technical problems exist in the nutrification of salt. If the salt is of a pure grade, of low moisture content, of a uniform small particle size and will be packaged in moisture resistant containers, this type of salt can be successfully nutrified with a dry form of vitamin A ester as is evidenced by data (Table 15) secured on white fine crystalline (Diamond) salt. This salt was nutrified with a premix containing vitamin A palmitate (Type 250-SD), thiamin, riboflavin, pyridoxine, niacin and ascorbic acid. Over 90% of the vitamin A was retained after one year of storage. In the USA vitamin nutrified uniform crystal salt (Morton) is employed by some bakers for the nutrification of bread, buns and rolls with thiamin, riboflavin, niacin and iron. Another nutrified salt has been developed for snack toppings such as potato chips to provide supplementary vitamin A, thiamin, riboflavin, niacin and ascorbic acid. The general experience with nutrified high grade, uniform crystal salt therefore is similar to that with nutrified sucrose. Both must be centrally produced, have some degree of standardized quality, and be subjected to proper controls in the nutrification process and in subsequent packaging and distribution programs.

Salt from several countries, India, Java, Etc., received for the purpose of nutrification trials has been damp, of high moisture content (as high as 4%), of highly variable crystal size (up to 12 mm or more), and containing dissolved trace mineral impurities. As mentioned earlier in this presentation, vitamin A is a labile com-

TABLE 15

RETENTION OF ADDED VITAMIN A[a] IN NUTRIFIED SALT[b]

	IU/15g	% RETENTION
Initial value	4300	---
3 mo., 26°C	4200	98
6 mo., 26°C	4110	96
9 mo., 26°C	4100	95
12 mo., 26°C	4030	94

[a] Fine crystalline Diamond salt.
[b] Data of Cort, Roche Product Development Dept.

pound subject to these stress factors found in impure salt. These stress conditions overcome the stabilizing characteristics built into the dry vitamin A products. There is no known technology today that can bring about the successful uniform nutrification of these cruder salt products. Salt, it must be remembered, is a hygroscopic compound and unless it contains a dessicant and/or is packaged in relatively moisture resistant packages it will absorb moisture eventually.

MSG[12, 65]

There are situations, with some seasoning items wherein small volume packaging is practiced and where daily or uniform use of such packages is employed. Such is the case in the Philippines where surveys sponsored by the Cebu Institute of Medicine indicated that 49% of the children consumed monosodium glutamate (MSG) in seasoned food prepared in the home, that little variation existed in the per capita consumption and that MSG was centrally produced and marketed by two local firms. MSG is produced as a pure crystalline substance of uniform physical characteristics and of low moisture content. The survey also revealed that one or two 2.4 g packets of MSG were added to the daily series of meals depending on the size of the family. With the cooperation of Philippine personnel and Cornell consultants, a study was initiated at Roche to nutrify MSG as all indications showed MSG to be a promising vehicle for conveying vitamin A to the Philippine population. A technology was developed to bring about uniformity of particle distribution of a vitamin A nutrified MSG. Packets (15,000 IU) were prepared and stored to determine vitamin A stability (Table 16) which was found to be good to excellent over a

TABLE 16

RETENTION OF ADDED VITAMIN A IN NUTRIFIED MSG
WITH Fe AND A (STORAGE DATA)

TIME TEMPERATURE (°C)	PERCENT RETENTION		
	Vitamin A[1] only	VITAMIN A PLUS F[2]	
		Low[3]	High[4]
1 mo., 45°C	82	92	89
2 mo., 45°C	61	--	--
3 mo., 37°C	78	95	89
6 mo., 37°C	75	81	73
12 mo., 37°C	--	62	60
3 mo., 23°C	--	100	100
6 mo., 23°C	89	89	83
12 mo., 23°C	79	87	84

[1] Vitamin A palmitate type 250-SD (60 mg or 15,000 IU/packet).

[2] Storage trial started 8-1/2 months after A nutrification.

[3] Ferric orthophosphate (13.5 mg Fe/packet).

[4] Ferric orthophosphate (27 mg Fe/packet).

six month period. Since MSG is locally available in the Philippines there is no need for long storage requirements. Biochemical and clinical studies have shown vitamin A nutrified MSG to be an effective delivery system of getting vitamin A to the Philippine population. Current studies in progress shows that MSG can serve both as a carrier of added vitamin A and iron[66].

REFERENCES

[1] Isler, O., 1950. The chemistry of Vitamin A. Chimia 4:103-118.
[2] Isler, O., 1970. Developments in the Field of Vitamins. Experientia, 26:225-240.
[3] Bauernfeind, J.C., 1973. Vitamin A Technology. Volume III of Series. Vitamin A, Xerophthalmia and Blindness. Office of Nutrition, Technical Assistance Bureau Agency for International Development, U.S. Dept. State, Washington, D.C.
[4] Bauernfeind, J.C., 1978. Vitamins: Essential Micronutrients for Man,"Encyclopedia of Food Science", M.S. Peterson and A.H. Johnson, editors, AVI Publishing Co., Westport, Conn., pp. 788-843.
[5] De Ritter, E., 1976. Stability Characteristics of Vitamins in Processed Foods. J. Food. Technol., 30(1):48-54.
[6] Klaüi, H., Hausheer, W. and Huske, G., 1970. Technological Aspects of the Use of Fat-soluble Vitamins and Carotenoids and the Development of Stabilized Marketable Forms, International Encyclopedia of Food and Nutrition, (Pergamon Press) 9:113-158.
[7] Bauernfeind, J.C. and Cort, W.M., 1974. Nutrification of Foods With Added Vitamin A, Crit. Rev. Food Technol., 4:337-375.
[8] NAS-NRC, 1974. General Policies in Regard to the Improvement of Nutritive Quality of Foods. Food and Nutrition Board. National Research Council, National Academy Science, Washington, D.C.
[9] Borenstein, B., 1971. Rationale and Technology of Food Fortification with Vitamins, Minerals and Amino Acids. CRC Rev. Food Technol. 2(2):171-186.
[10] Arroyave, G., Aguiler, J.R. and Flores, M., 1977. Evaluation of Programs to Control Vitamin A Deficiency. Presentation: Western Hemisphere Nutr. Congr. V, Quebec City, Canada.
[11] Arroyave, G., Beghin, I., Flores, M., Stoto de Guido, C., y Ticas, J.M., 1974. Efectos del consumo de azucar fortificada con retinol, por la madre embarazada y lactante cuya dieta habitual es baja en vitamina A: Estudio de la madre y del niño. Arch. Latinoamer. Nutr., 24:485-512.
[12] Solon, F.S., Fernandez, T.L., Latham, M.C. and Popkin, B.M., 1977, Research to Determine the Cost and Effectiveness of Alternate Means of Controlling Vitamin A Deficiency (Xerophthalmia): Final Report. Cebu Institute Medicine, Cebu, Philippines.
[13] Crowley, P.R., 1974. Current Approaches for the Prevention of

Vitamin A Deficiencies: Food Fortification. Presentation: Joint WHO/USAID Meeting on the Control of Vitamin A Deficiencies. Jakarta, Indonesia, World Health Organization, Geneva, Switzerland.

14 Parrish, D.B., 1977. Determination of Vitamin A in Foods: A Review. CRC Crit. Rev. Food Sci. & Nutr., 9(4):375-394.

15 Underwood, B.A., 1974. The Determination of Vitamin A and Some Aspects of its Distribution, Mobilization and Transport in Health and Disease, World Rev. Nutr. & Disease, (Karger, Basle), 19:123-172.

16 Bauernfeind, J.C., Rubin, S.H., Surmatis, J.D. and Ofner, A., 1970. Carotenoids and Fat-Soluble Vitamins: Contribution to Food, Feed and Pharmaceuticals, Int. Z. Vitaminforsch., 40(3):391-416.

17 Klaüi, H., 1974. The Technological Aspects of the Addition of Nutrients to Foods, Proc. 4th Int. Congr. Food Sci. Technol., 1: 740-762.

18 Harris, R.S., 1959. Supplementation of Foods with Vitamins. J. Agric. Food Chem., 7(2):88-102.

19 Bauernfeind, J.C., Newmark, H., and Brin, M., 1974. Vitamins A and E Nutrition via Intramuscular or Oral Route. Amer. J. Clin. Nutr., 27: 234-253.

20 Reddy, V., 1977. Preventive Programs of Vitamin A Deficiency in India. Presentation: Internat. Vitamin A Consultative Group, Geneva, Switzerland, May 16-20.

21 McLaren, D.S., 1964. Xerophthalmia: A Neglacted Problem. Nutr. Rev., 22:289-291.

22 WHO. Vitamin A Deficiency and Xerophthalmia, 1976. Report of a Joint WHO/USAID Meeting. Technical Report Series 590, World Health Organization, Geneva, Switzerland.

23 Internat. Vitamin A Consultative Group (IVACG), 1976. Guidelines for the Eradication of Vitamin A Deficiency and Xerophthalmia. A Report. The Nutrition Foundations Inc., New York, N.Y.

24 Olson, J.A., 1972. The Prevention of Childhood Blindness by the Administration of Massive Doses of Vitamin A. Israel J. Med. Sciences, 8(8-9):1199-1206.

25 Bauernfeind, J.C., Marusich, W.L., and Roncalli, R.A., 1972. The Intramuscular Administration of Vitamin A to Livestock. Presentation: Canadian Veterinary Medical Association Meeting, Quebec, Canada, July 1.

26 Bauernfeind, J.C., Rokosny, A.D., and Seimers, G.F., 1953. Synthetic Vitamin A Aids Food Fortification. Food Eng., 25(6): 81, 82, 85, 87.

27 Bauernfeind, J.C., 1970. Vitamin Fortification and Nutrified Foods. Proc. 3rd Internat. Congr. Food Sci. Technol., Washington, D.C., 217-232.

28 Bauernfeind, J.C. and Brooke, C.L., 1973. Guidelines for Nutrifying Processed Foods. Food Eng., 45(6):91-97, 100.

29 Menden, E. and Cremer, H.D., 1958-59. The Problem of Improving Nutritive Value Food. Manufacture, July, Aug., Sept., Oct.,

Nov. Issues 1958; Feb. Mar. Issues, 1959.
30. NAS-NRC, 1975. Technology of Fortification of Foods. Food and Nutrition Board. National Research Council, National Academy Science, Washington, D.C.
31. Wodsak, W., 1952. Vitaminierung and Farbung der Margarine, Dt. Lebm. Rdsch., 48:208.
32. Wagner, K.H., 1954. Die Vitaminisierung von Milch, Milchprodukten, Margarine und Fetten mit den Vitamen A und D. Fette, Seifen, 56:581.
33. Marusich, W., DeRitter, E., and Bauernfeind, J.C., 1957. Pro-Vitamin A Activity and Stability of β-Carotene in Margarine. J. Amer. Oil Chem. Soc., 34:217-221.
34. Melnick, D., Luckmann, F.H., and Vahlteich, H.W., 1953. Retention of Preformed Vitamin A and Carotene and Margarine Based on Physic-Chemical Assays. J. Food Res., 18: 504-510.
35. Deuel, H.J. and Greenberg, S.M., 1953. A Comparison of the Retention of Vitamin A in Margarine and in Butter Based upon Bioassays. J. Food Res., 18:497-503.
36. Morton, R.A., 1970. The Vitaminization of Margarine. Roy. Soc. Health J., 90(11), 21-28.
37. Wagner, K.H., 1952. Die Vitaminisierung der Sterilmilch. Milchwissenschaft. 7:250.
38. Conochie, J. and Wilkinson, R.A., 1956. The Fortification of Nonfat Milk Solid with Vitamin A. 14th Internat. Dairy Congr., Proc. I (2):357.
39. Wilkinson, R.A., and Conochie, J., 1958. The Stability of Vitamin A in Reconstituted Fortified Nonfat Milk Solids. Australian J. Dairy Technol., 13:29.
40. Bauernfeind, J.C. and Allen, L.E., 1963. Vitamin A and E Enrichment of Nonfat Dry Milk. J. Dairy Sic., 46(3):245-254.
41. Bauernfeind, J.C. and Parman, G.K., 1964. Restoration of Nonfat Dry Milk with Vitamins A and D. J. Food Technol., 18(2):52-57.
42. Coulter, S.T. and Thomas, E.L., 1968. Enrichment and Fortification of Dairy Products and Margarine. J. Agric. Food Chem., 16:158-162.
43. WHO, 1976. Vitamin A Enrichment of Donated Cords: with Special Reference to Dry Skim Milk Powder. PAG Bulletin, WHO, Geneva, 6(4): 1-7.
44. FAO, 1977. Enrichment of Dried Skim Milk: with Special Reference to Vitamin A. Food and Nutrition, 3(1): 1-7.
45. Dustin, J.P., 1977. A Simple Field Test for Checking Vitamin A in Dried Skim Milk. Food and Nutrition, 3(1): 7.
46. Brooke, C.L. and Cort, W.M., 1972. Vitamin A Fortification of Tea. J. Food Technol., 26(6): 50-52, 58.
47. Harris, R.S., 1963. Attitudes and Approaches of Supplementation of Food Nutrients. J. Agric. Food Chem., 16(2): 149-152.
48. Brooke, C.L., 1968. Fortification of Food Products with Vitamin A, with Special Reference to Fortification of Cereals and Cereal Products in India. Proc. West Hemispehre Nutr., Congr. II (Puerto Rico), 137-139.

[49] Brooke, C.L., 1968. Enrichment and Fortification of Cereals and Cereal Products with Vitamins and Minerals. J. Agric. Food Chem., 16:163-167.

[50] Borenstein, B., 1969. Vitamin A Fortification of Flour and Corn Meal: A Technical Discussion. Northwest Miller, 276(2):18-19.

[51] NAS-NRC, 1974. Proposed Fortification Policy for Cereal-Grain Products. Food & Nutrition Board, Nat. Res. Council, Nat. Acad. Sci., Washington, D.C.

[52] Anderson, R.H., Maxwell, D.L., Mulley, A.E. and Fritsch, C.H., 1976. Effects of Processing and Storage on Micronutrients in Breakfast Cereals. J. Food Technol., 30(5):110-114.

[53] Cort, W.M., Borenstein, B., Harley, J.N., Osadca, M. and Scheiner, J., 1976. Nutrient Stability of Fortified Cereal Products. J. Food Technol., 30(4):52-62.

[54] Rubin, S.H., Emodi, A. and Scialpi, L., 1977. Micronutrient Additions to Cereal Grain Products. Cereal Chem., 54(4):895-904.

[55] Austin, J.E., 1978. Cereal Fortification Reconsidered. Cereal Foods World, 23(5):229-233, 265.

[56] Bauernfeind, J.C., Benevenga, N.J., and Mertz, W., 1978. Nutrification of Foods with Micronutrients, Vitamins, Minerals and Amino Acids (In press).

[57] Mickus, P.R., 1955. Seals Enriching Additives on White Rice. Food Eng., 27:91-94.

[58] Arroyave, G., 1971. Distribution of Vitamin A to Population Groups. Proc. West. Hemisphere Nut. Congr. III (Miami, Fla.), 68-79.

[59] INCAP, 1974. Fortification of Sugar with Vitamin A in Central America and Panama. Institute of Nutrition of Central America and Panama, INCAP, V-36 (English translation, Jan. 1975).

[60] Arroyave, G., Aguilar, J.R. and Portela, I.E., 1975. Manual de operaciones para la fortificacion de azucar con vitamina A. Instituto de Nutricion de Centro America Y Panama. Guatemala Publication. INCAP E-853.

[61] Toro, O., de Pablo, S., Aguayo, M., Gattas, V., Contreras, I., and Monckeberg, F., 1975. Prevention of Vitamin A Deficiency by Enrichment of Sugar: A Field Study Report. Dept. Nutr. & Food Technol. Univ. Chile, Santiago, Chile.

[62] Aguilar, J.R., Arroyave, G., y Gallardo, C., 1977. Manual de supervision y control del programa de fortificacion de azucar con vitamina A. Instituto de Nutricion de Centro America y Panama. Publication INCAP E-913, Guatemala.

[63] Arroyave, G., 1977. Control of hypovitaminosis A in Central America and Panama: Nutrification of Sugar with Retinol Palmitate. Presentation: Internat. Vitamin A Consultative Group, Geneva, Switzerland, May 16-20.

[64] Araujo, R.L., Souza, M.S.L., Mata-Machado, A.J., Mata-Machado, L.T., Lourdes Mello, M., Costa Cruz, T.A., Vieira, E.C., Souza, D.W.C., Palhares, R.D. and Borges, E.L., 1978. Response of Retinol Serum Level to the Intake of Vitamin A-fortified Sugar by Pre-School Children. Nutr. Report Internat.,

17(3):307-314.
[65] Solon, F.S., Popkin, B.M., Fernandez, T.L. and Latham, M.C., 1978. Vitamin A Deficiency in the Philippines: A Study of Xerophthalmia in Cebu. Amer. J. Clin. Nutr., 31:360-368.
[66] Bauernfeind, J.C. and Timreck, A., 1978. Monosodium Glutamate, a food carrier for added vitamin A and iron (in progress).

BIOLOGICAL AND CHEMICAL PRODUCTS APPLIED IN AGRICULTURE:

THEIR INFLUENCE ON FOOD QUALITY

Hermann Schmidt-Hebbel

Casilla 3968

Santiago, CHILE

Substances of chemical or biological origin utilized by agriculture can reach our foods mainly through the application of fertilizers to the soil or by means of contamination of the food, whether of chemical or biological character.

Man, as the last link of the food chain, ingests foods originating from vegetables and animals which have been exposed to the environment immediately aroung them. For this reason the foods constitute — together with the ari we breathe, the water we drink and bathe in, and the soil we step on — ecological parameters, and the control of their quality constitutes, in turn, an ecological problem.

In our society of advanced technology, man remains exposed to the ingestion of a number of chemical products of various natures through its foods, and although these may present great advantages towards improving the foods from a technological or nutritional point of view, their selection and purity must be continuously supervised by sanitary authorities, both national and international, in order to avoid their becoming a risk to human health.

In this context, the Codex Alimentarius Commission of FAO/WHO has clearly defined what constitutes a food additive, a nutritional supplement and a contaminant.

Food Additive is a substance of known composition, usually of non-nutritional character, which is incorporated into a food in small and carefully controlled amounts in order to accomplish a technological purpose, usually with the view of improving its presentation (organoleptic characteristics) or preservation conditions.

With the exception of a few additives (ascorbic acid, calcium propanate, gelatin), it is the lack of nutritive value that differentiates it from the Nutritional Supplement, which is added to a food in order to enrich its nutritional value, such as vitamins, minerals, protein and amino-acids (these last ones with the purpose of completing the aminoacidgram).

On the other hand, Food Contaminant is defined as the foreign substance which is not intentionally added to accomplish a determined role; on the contrary, its presence is the result of food "handling" during its harvesting, preparation, elaboration or processing, packaging, transportation and storing. Its origin is, thus, an environmental "contamination" of the food, its presence not being necessary, nor even convenient in the finished product, in which normally it is not destined to subsist. Vegetables may have absorbed it through the soil, water or air, through excesses of fertilizer, residues of pesticides or toxic metals, or even radioactive elements, all of which may reach man in this manner, through his contaminated foods. From the point of view of the etymological origin of the word, "contamination" originates from the latin "contaminare", which means to touch, penetrate, being, thus, the penetration of an impurity in a body. This definition is sufficiently broad to take in the different circumstances united under the present concept of "contaminated food". Thus, this concept includes contamination of biological origin, caused by rodents, insects, worms, parasites, microorganisms and their toxins, capable of transmitting diseases to man and animal, as well as those produced by chemical impurities, and those caused by radioactive contamination.

Even though the X International Congress of Hygiene, in 1900, in Paris, had approved the total prohibition of chemical agents for food preservation, this resolution did not find legal sanction. Presently one can justify, in determinde cases, the correct technological application ("good manufacturing practice") of chemical additives which have been proved inoffensive, as long as applied in concentrations such that insure at the same time their efficacy and the absence of a health risk. One example of this may be the use of an antisceptic in a determined food product, which previously obeyed strictly economic reasons, whereas today it may constitute an efficient protection against the formation of toxins generated by microorganisms, especially mycotoxins which, once produced, cannot usually be eliminated from the infected food of fodder.

In contrast to the utility a determined technological additive can serve, in the conditions indicated above, the penetration of chemical or biological products as contaminants, as recently defined, naturally deserves special supervision, and the regulations, both at the national and international level, are very zealous with respect to restricting their presence to widely acceptable limits so as not to constitute a risk to human health.

We shall now proceed to analyze the different categories of substances which, under the tittle of Contaminants, can become present in foods.

In this respect we should indicate the different pathogenic microorganisms which, alone or through their toxins, constitute the biological contaminants of foods, with the highest incidences in our environment being those by Staphylococcus aureus, Salmonellas, Escherichia coli and, due to its severity, although not its frequency, Clostridium botulinum.

Among the chemical contaminants one must mention in particular the residues of pesticides, metallic contaminants, those resulting from excessive fertilization and those cause by the migration of components from bottles and containers towards the foods or drinks contained in them.

I. PESTICIDES RESIDUES

By "residues" it is understood the amounts of a pesticide which have remained in a food or in the environment as immediate consequence of chemical measures for the protection of vegetables. As it is known, after the application of a pesticide to a vegetable, an active deposit is formed, which can remain in the surface of the plant or be incorporated into the vegetable tissue. In most pesticides, this deposit decreases in a considerably fast manner, both due to physical action (wind, rain, vegetable transpiration) as well as chemical and enzymatic action (hydrolisis, oxidation, reduction, interchange reaction, isomerization and conjugation). The "average life" of a residue, as a numerical expression for the speed of its decomposition is usually only of hours or days, which can be easily carried out with the completion of the necessary waiting time from the last treatment until harvest or consumption. One exception is constituted by the chlorinated hydrocarbons, such as DDT and others, which have residues with rather retarded decomposition and car persist in the vegetable until the next vegetation period. If the pesticide is applied again, a dangerous accumulation of residues can occur, factor which has led to the restriction of its quantity in the form of maximum limits or "tolerances", or even the prohibition of some of these pesticides.

II. METALLIC CONTAMINANTS

The natural presence of trace metals in foods, along with their occasional presence due to chemical contamination, have attracted attention from different points of view.

The biologist becomes interested in the role played by the natural presence of different metals in the metabolic processes; the nutritionist for the importance of the contribution of some of these metals to the organism; the food technologist worries about the influence they may exert on the organoleptic characteristics in some foods and, finally, the toxicologist studies the possible noxious actions of metallic elements in the foods.

In many cases, the natural metallic content in the different foods and drinks is insignificant, in comparison to the amounts which can result from chemical contamination from machinery, pipes, containers and bottles which have been in contact with the foods.

In this respect it is interesting to note that the toxicity of a chemical element does not depend only of its concentration, but also of its combination state. Thus, magnesium, silica and oxygen alone do not cause any worry, but when they form the anhydrous magnesium silicate, they form abestos, of known cancerous action.

Amont the different trace elements present in foods, lead, mercury, cadmium and arsenic have been qualified as "public poisons", due to their great incidence as toxic contaminants of ecological origin in foods and drinks; whereas copper, zinc and tin, although they may also cause noxious effects when absorbed in excess are, together with iron, more of indicators of technological contamination during the processing or storing of foods and drinks.

Excessive fertilization may also generate toxicity phenomena in vegetables, which are capable of absorbing it and thus transmitting it to such foods as garden vegetables. One example of this is the high content of nitrates which vegetables such as spinach, chard, carrots and representatives of the Brassica family (i.e., cauliflower) may present through this mechanism. As it is known, the nitrates can easily undergo a reduction reaction, whether chemical with metals such as iron or manganese in the absence of air, or by microbiological contamination of the vegetables itself, or by coliform bacteria of the intestines, especially in young children. The result of this reduction reaction is the formation of the dangerous nitrites. Due to their metahemoglobinizing and cyanotic action, and especially due to their interaction with amino groups with the resulting formation of nitrosamines — many of which are cancerous, mutagenic and occasionally teratogenic — the nitrates are the object of much discussion at the world level, and this not only because of their possible presence as food contaminants but also because of their use as technological additives in meat products, where they have the double function of increasing the red color and aroma and inhibitting the growth of Clostridium-type germs, of such well known pathogenicity as Clostridium perfringens and Cl. botulinum. In all cases, it has been verified that the simultaneous addition of ascorbic acid, and especially of its sodium composite permits the lowering of the necessary amount of nitrite without reducint its anti-microbial action.

Within the possibilities of chemical contamination of foods and drinks one should also mention those substances which may originate from the migration of the components of the container where the food is stored. In this context one must take into account that the attack on the materials of the container can be exerted by the food or drink in an aqueous media, especially if this contains acid (as happens in fruit juices, wines and non-alcoholic beverages), in a fatty media (fatty foods), and also in dried products with hard or rough texture which then can exert an abrasive action (scraping) over the material of the container. Seeing as how plastic containers are made using macromolecular materials, resulting from the polymerization of derivatives of vinyl, vinylidene, ethylene, propylene, styrene, and others, the macromolecule in itself is not toxic, but its chemical and solubility properties will depend on its macromolecular skeleton and its degree of polymerization. The danger of toxicity lies, then, in the existence of monomers, whether present as impurities or resulting from the decomposition of the molecule. Simulation experiments done in aqueous and fatty media have shown that the migration of toxic monomers is possible, if these are present in the material of the container. In the face of this danger, then, the tendency is to accept for food containers, only those plastic materials based on copolymers or mixed polymers, with a high degree of polymerization and not susceptible to decomposing into its monomers.

With respect to the plastic materials for food containers and for drinks, special attention has also gone to the research of the possible toxicity of the numerous coadjuvants which are incorporated to the plastics to improve their flexibility and rigidity properties and their resistance to physical and chemical agents. The evident toxicity of some of these compounds has been demonstrated, such as for tri-o-cresyl phosphate and chlorinated diphenyls, which is why they should not be used in food containers. For this same reason one should use as coating materials, for both plastic and metallic containers, only sanitary varnishes, shellacs and enamels which offer guarantees of toxicological safety.

In 1942 I had the opportunity to participate in the organization of a Latin American Seminar of Unesco, in Lima, which had the suggestive theme: "Education, Agriculture, Food and Man". One of the main objectives of this event consisted in calling the attention of the participants to the fundamental emphasis which should be given to the teaching of nutrition at the agronomic and veterinarian level, with the final goal being, however, the benefit of man, that is, without detaining such teaching in the stage of vegetable and animal nutrition. This thought coincides, also, with that already manifested by the German agricultural chemist, Joseph König, who, upon beginning, on his own initiative, the first food analysis, over 100 years ago; protested that, until then, there was more preoccupation with vegetable and animal nutrition, through broad analysis of fertilizers and fodder, than with the chemical composition of the foods consumed by man.

It seems to me that, similarly, aiming always at the health of man, the multi-disciplinary scientists who collaborate in the area of toxicoloby are the ones who should intervene to evaluate the problematic between the blessing and the curse, that is, between the valuable effect of the different chemical substances which may be present in the foods and their possible damaging effects, with the objective of maintaining the ecological risk within acceptable limits.

ENVIRONMENTAL CONTAMINANTS

Ian C. Munro

Toxicology Research Division - Health Protection Branch

Tunney's Pasture, Ottawa, Ontario K1A OL2 CANADA

Of the numerous chemical substances to which man is exposed, including those naturally found in the food supply, the group of chemicals loosely classified as the "environmental contaminants" probably presents the greatest potential threat to humam health. Outbreaks of human disease due to ingestion of certain of the chemicals, on some occasions, involving unprecedented numbers of individuals, serves to illustrate the notoriety of these substances. During the course of this presentation, it will not be possible to review all facets of knowledge regarding environmental contaminants in food. Thus, comments will be restricted to those contaminants of greatest interest and public health concern. Also, some principles relating to the safety evaluation of contaminants will be discussed.

Environmental contaminants are comprised of a large group of substances of diverse chemical structure but which, interestingly can be divided into two broad chemical classes. The first of these are the inorganic metals, organo-metals and metalloids. The second class is the organic compounds the most significant of which are the aromatic halogenated hydrocarbons.

It is of more than passing interest that we should be beset with similar problems from chemicals with such divergent sturctures. However, a closer look at the physical propertires of these two groups provides an interesting lesson for those of us who are engaged in the study of structure/activity relationships. Although not every environmental contaminant possess all of the same properties there is a surprising commonality among these substances to the extent that most workers closely aligned with this field of toxicology will view with justifiable suspicion any aromatic halogenated compound or organometal substance detected in environmental media. Let us review these properties. First, we have persistence in the environment. Contaminants resist environmental degradation and are extremely stable in many environmental compartments. Secondly, they tend to biomagnify in the food supply, especially in fish and it is this biomagnification that renders them a potential hazard to man. The characteristic which makes them an actual hazard to man is their slow rate of elimination and/or metabolism which results in their accumulation in tissues. Finally, their toxicity is usually greater to higher order mammals in which they biomagnify than in species of lower phylogenetic order. For exemple, fish, seals and crustaceans can tolerate much higher tissue levels of mercury and arsenic than can man. Chemicals released into the environment that have one or more of these properties can be predicted to have a potential for human harm.

The source and occurrence of these materials in environmental media warrants attention at this point. Contamination of the environment by complex organic substances is largely due to man's activity. For example, the presence of PCB's and Mirex in fish is a result of insufficient care being taken to avoid the release of these materials into watersheds. The situation with the inorganic substances and organometals is somewhat different. Except for instances of gross contamination of the environment due to direct release of these substances, man's activity plays a somewhat reduced role. This fact is supported by the observation that the levels of mercury in museum specimens of marine species taken from the seas 95 years ago are similar to those found in these species today[10]. Thus, it can be concluded that man's industrial activities of the past 100 years or so have not substantially influenced the levels of mercury in marine fish. Where human activity has influenced the levels of mercury in fish the effects have been profound. Fish taken from Pelican Pouch Lake contain levels of mercury considerably below the 0.5 ppm guideline. These data are considered to reflect the low end of the natural background levels of mercury in inland fish of the species shown. On the other hand, the identical species of fish taken from Ball Lake, a body of water contaminated due to industrial use of mercury adjacent to the Lake contain levels of mercury an order of magnitude higher than those found in fish from Pelican Pouch Lake. On the other hand, the mercury levels of fish taken from Lac Seul, where there is no known industrial source of mercury, show that certain fish still contain levels of mercury above the 0.5 guideline. These data are considered to reflect a high natural background of mercury. The elevated levels of mercury in fish from this lake are due to the presence of geological formations in this area from which mercury is leached and assimilated by the food chain.

I mentioned earlier that most contaminants possess the ability to concentrate in the food chain. This ability is, for the most part, selective and limited to certain food products. Arsenic, for example, appears only to be a potential problem in the meat, fish and poultry group of foods, the levels in all other food categories being low and reasonably consistent among the food groups[8]. Although fish is a major source of dietary arsenic, if we look into this a little further, we find that the high levels reported for fish products is due entirely to its ability to concentrate in certain marine animals notably the bottom feeding species such as grey sole and shrimp. The levels in these species considerably exceed those found in most other edible marine products. Pelagic marine species and fresh water fish do not appear to concentrate arsenic to the same extent. This difference in ability to concentrate may be due in part to the feeding habits though other factors such as selective metabolism and renal clearance appear to play some role.

I want to spend some time discussing the health significance of arsenic in fish, but before doing so, it would be instructive to briefly review the transformation of arsenic in the environment. It has been established that Methanobacterium can methylate inorganic arsenic to an organic form. In vitro studies by McBride and Wolfe[9] suggest that in this reaction arsenic is reductively methylated to dimethylarsenic acid by methylcobalamin. The conversion of arsenic to methylated derivatives by algae, sea urchin and other marine biota, including fish has been conclusively demonstrated[7,12]. The chemical form of arsenic in Canadian marine fish is not known and attempts to identify its structure have been unsuccessful to date. However, recent studies from Australia indicate that the chemical form of arsenic in Australia rock lobster is arsenobetaine[5] but arsenobetaine does not occur in Canadian fish in measurable quantities. The gut flora of fish will convert inorganic arsenic to organic forms that are structurally dissimilar to those that occur naturally in fish muscle[12]. Clearly, this is a complex problem and we are only beginning to understand the molecular events involved in the metabolic transformation of arsenic in environmental biota. Well what can we conclude about the significance to health of arsenic in marine species. Studies in humans have demonstrated that the majority of arsenic present in shrimp, crab and Australian lobster appears to be excreted in the urine within a few days following consumption[3,4,14]. Although on the surface those studies appear conclusive, analytical difficulties in these investigations coupled with the lack of adequate balance studies led us to repeat these studies in animals.

The results of our research, although preliminary at this stage, demonstrate quantitative relationships between consumption and excretion. Non-human adult and adolescent primates were given a single oral dose of 1 mg As/kg from a fish slurry of grey sole that contained high levels of arsenic. Whereas adult monkeys excreted approximately 77% of the dietary arsenic over a seven day period, adolescent monkeys excreted a somewhat lesser amount. The data clearly indicate that with fish arsenic about one-third of the dose was retained. The significance of this retention is currently being investigated in multiple dose studies using high arsenic fish. However, more definitive studies on the pharmacodynamics and mechanism of toxicity of fish arsenic in mammalian species must await the isolation and structural identification of this form of arsenic.

The significance of biological methylation on the toxicity of other trace elements has not been investigated adequately. Except for methylmercury which has been extensively studied, there is a relative paucity of information on the toxicity (particularly long-term effects) of other methylated metals and metalloids. On the basis of the electrochemical characteristics of the trace elements, Wood[15] has suggested that other metals such as thallium, tin, palladium and platinum be closely scrutinized for the possibility of biological methylation. Available data on methylmercury would support the need for further studies in this regard. Methylmercury provides a classic example fo the effect environmental transformations may have on the toxicity and pharmacodynamics of the metals. If we look, for example, at the elimination rates for inorganic and organic mercury from mammals we find surprising differences. The t1/2 of elimination of mercuric chloride from the whole body of humans is 40 days compared to 72 days for methylmercury. Also, the t1/2 of elimination of methylmercury appears to vary considerably between individuals. The longer t1/2 of elimination of methylmercury means that a greater amount of methylmercury will accumulate in the body than mercuric chloride administered at comparable dose levels for extended periods. This fact, coupled with the high susceptability of the central nervous system to methylmercury accounts, in a large part, for the increased sensitivity of humans and animals to the toxic effects of methylmercury compared to mercuric chloride.

The methylated form of mercury, because of its greater lipid solubility, can also cross biological membranes much more readily than inorganic mercury. Consequently, methylmercury more readily crosses the placenta resulting in higher exposure of the developing embryo and fetus. In fact, methylmercury concentrations in blood samples of newborn infants shortly after birth are consistently higher than those in their mothers. The ratio of methylmercury concentration in newborn blood compared to maternal blood ranges from 1.3 to 2.1 in cases of long-term medium level exposure as in Japan, long-term low level exposure as in Canada and short-term high level exposure as in Iraq. The ratio of 15 observed in the single case in New Mexico is probably erroneously high because exposure of the mother to methylmercury ceased about 2 months prior to parturition. Clarkson et al. have demonstrated that the elimination rate of methylmercury from infants is slower than from adults; therefore, the mother would loose methylmercury more rapidly than the fetus and an abnormally high infant/maternal blood ratio could be expected in this particular case. In all case, the higher levels of methylmercury in the blood of newborns is probably due to the greater binding of methylmercury to fetal red blood cells compared to the adult (Scanlon, adult (Scanlon, J., 1972).

The higher blood levels of methylmercury in the fetus coupled with the vulnerability of the developing nervous system to toxic insult, suggests that the fetus may be at greater risk from methylmercury exposure in utero than the newborn after birth or the adult. This possibility has recently attracted considerable interest. Japanese studies (Hirada, 1976) have demonstrated that infantile Minimate disease can occur in infants exposed to methylmercury in utero while their mothers do not show symptoms of methylmercury intoxications. Similar data have been collected from accidental exposure of pregnant women to methylmercury in New Mexico (Snyder, 1971) and Iraq (Amin-Zaki et al., 1976).

These data all indicate that in utero to methylmercury may be much more critical than exposure during the postnatal period or adulthood.

The toxicity of other environmental contaminants to the neonate is also currently a matter of great concern in public health. This is exemplified by the concerns of rebulatory agencies regarding the exposure of infants to lead. Abundant evidence supports the fact that during early life, human infants as well as the young of other animals are uniquely susceptible to lead exposure. This increased sensitivity of the young to lead intoxication relates to differences in the pharmacokinetics of lead in infants as well as the greater susceptibility of developing organ systems to the toxic effects of this metal. For example, it is now well established that infants retain more of an oral lead dose than adults. Studies in our laboratories indicate that infant monkeys retained 60 to 70 percent of an oral lead dose while adult monkeys only retained about 4%. Similar data are available for other animal species (Kostial et al., 1973, and Forbes and Reina, 1974) and humans (Kehoe, 1971, and Alexander et al., 1974). This decrease in lead retention with increasing age, is reflected in lower bloos lead levels. The blood lead concentration of infant monkeys on milk diets was 25 to 35 ug Pb/dl whole blood with a lead exposure of 100 ug Pb/kg body weight/day. The infant monkeys were weaned onto an adult primate diet at 28 to 29 weeks of age and the blood lead concentrations declined to 15 to 20 ug Pb/dl even through the lead dose remained constant at 100 ug Pb/kg body weight/day. Therefore, the change in lead retention appears to be related to a change in diet. A wide variety of dietary constituents such as proteins, fats and the essential elements calcium, phosphorus, iron, zinc and magnesium have been shown to alter lead absorption from the G.I. tract (Barltrop et al; Kostial, 1972, 1973). The change in lead retention in infant monkeys at weaning may be related to changes in some of these dietary constituents. However, it has been demonstrated that lead was more toxic to rats fed milk diets whether lead was administered orally or intraperitoneally. The data suggest that lead absorption from the

G.I. tract may not be higher on milk diets since the intraperitoneal dose of lead was more toxic to the rats even though it did not have to be absorbed from the gut. Other factors, such as differences in lead distribution within the body or changes in the excretion of lead from the body due to changes in diet are currently under investigation in an attempt to explain the toxicological significance of the dietary effect.

In addition to differences in lead retention between adults and infants, lead is distributed differently within the body of adults than infants. The ratio of tissue to blood lead was much greater for bone and brain in infant monkeys than adults. These data indicate that a greater portion of the retained lead is distributed to bone and brain in infants than adults. This is consistent with the fact that the major symptoms of lead intoxication in infants is of central nervous system origin whereas in adults primary effects occur on the kidney and liver (Lin fu, 1972 and Waldron, 1974). For example, acute lead poisoning in children is characterized by encephalopathy, the sequelae of which is very discouraging even after lead exposure is stopped. At lower levels of lead exposure, more subtle behavioral effects may occur, such as increased motor activity, impaired fine motor coordination and decreased behavioral adaptability (Brown, 1975; Needleman, 1976). These behavioral alterations have been well documented in rodents but are less well established in human infants. Some epidemiological studies have shown positive correlations between lead exposure and behavioral anomalies in children (David, 1972; Needleman, 1976) while other studies have not. Thus, possible health hazards, such as behavioral anomalies, cannot be assessed at this time.

In 1972, the World Health Organization convened an Expert Committee to evaluate the lead problem. The Committee was able to establish a provisional tolerable weekly intake of 3 mg lead per person or about 7 ug lead per kilogram body weight per day[6] for adults. This was based on toxicity data in adult humans and the assumption that only 10% of orally ingested lead was absorbed. The Committee noted that this figure did not include infants and young children due to uncertainties regarding the degree and nature fo increased risk in this population subset. For these groups no guidelines could be established. This has posed a difficult problem for those of us concerned with regulations relating to acceptable levels of lead in infants foods, particularly whole milk and canned milk products. To develop a further appreciation of the degree of risk for infants and young children, a study was undertaken by the Health Protection Branch in 1975 using data from the 1972 Nutrition Canada Survey of 257 children. The aim of the study was to determine: a) the proportion of infants in Canada getting all or part of their milk from canned products; b) estimates of the lead intake of infants in Canada consuming milk products; and, c) estimates of the percentages of infants with daily lead intakes above the WHO guidelines for

adults. A large percentage of the children had as their source of
mild either powdered milk preparations or 2% fluid milk. Daily consumption of milk in milliliters per kg was plotted against lead intake from milk products. It was readily apparent that a substantial
proportion of infants could exceed the WHO recommended intake for
adults and that infants fed canned milk are a somewhat higher risk
than those fed whole milk due to the presence of higher levels of
lead in the canned products. Lead in canned milk products comes
almost entirely from lead solder use in can seams. On the basis of
this data, one would conclude that every effort should be made to
reduce the lead levels in infant foods to the minimum technologically
feasible. We at the Health Protection Branch in Canada are currently
working with industry in attempts to reduce the levels of lead in
infant food products.

I would like to turn for a moment to the organic environment
contaminants. Due to the immensity of this subject we will deal here
only with one or two special problems of current interest. In this
connection remarks will be restricted primarily to the aromatic
halogenated substances since this group of materials are considered
to pose the greatest risk to human health. Unlike the inorganic substances many of the persistent organic contaminant show an affinity
for fatty tissue. The extent to which this may be expected to occur
can be predicted from the partition coefficient of these materials
in n-octanol/water mixture. Work by Neely[11] demonstrates that the
short chained aliphatic halogenated substances are less likely to
persist for long periods in body fat than the more complex aromatic
substances. This should not be interpreted to mean that the former
are less toxic, for many of the short chained substances in addition
to several of the aromatic substances have been demonstrated to be
carcinogenic in animals.

Appart from occupational exposure, humans are exposed to very
low levels of the short chained halogenated substances and this minor
degree of exposure is not considered to be significant in comparison
with widespread, low level, continuous exposure to the more complex
aromatic substances which tend to persist in the environment and have
slow rates of elimination from human adipose tissue. At the present
time there is a growing concern over the presence of these substances
in human breast milk. It is well known that lactation results in a
mobilization of body fat with a concomitant release of these substances into the breast milk. The intake data on the chemicals assumes a daily consumption of 150 ml whole milk per kg body weight.
It can be noted that in many instances the intake is very close to
the ADI and for some substances exceeds it considerably. Following
the marked restrictions placed on the use of the halogeneted hydrocarbon pesticides the levels of most of these substances in breast
milk appear to be declining. This is particularly true for DDT and
its major metabolites, the levels of which have decreased dramatically from 1967 to the present. Of more pressing concern at the moment

is the presence of PCB's in human breast milk. PCB use in industry, and thus release to the environment was severely restricted following their discovery as major environmental contaminant. Currently, they are permitted only for use in electrical transformers and capacitors but may also occur as a contaminant in recycled paper. However, their persistence in the environment, particularly in certain edible species of inland fish, coupled with new knowledge regarding their toxicity has heightened concern regarding their potential for adverse effects on the human population notably infants. In 1968 a disease known as Yusho was reported in Japan. The symptoms of this disease included chloracne, pigmentation of the skin and nails, swelling of the eyelids with loss of eyelashes and eye discharge. The disease was traced to the consumption of rice oil contaminated with PCB/s from a leaking heat exchanger. It should be noted that the PCB's contained other contaminants including the chlorinated dibensofurans which may have accounted, at least in part, for the observed symptoms. The smallest dose of PCB causing toxic symptoms in the Yusho patients was estimated to be about 500 mg consumed over a period of 120 days. Thus, the daily dose was in the range of 0.2 mg/kg/day. The PCB levels in the blood plasma of babies born to Yusho patients were considerably higher than those born to mothers without a history of PCB exposure. This data coupled with the symptomology in the infants indicates that PCB's readily cross the placenta and accumulates in the fetus. Studies conducted by Allen et al[1], in which groups of monkeys were exposed prior to mating, produced clinical signs of toxicity in the monkeys, almost identical to those observed in the Yusho incident. In Allen's studies a commercial PCB formulation (Aroclar 1248) was used. The daily doses, estimated from intake from the diet were 0.1 or 0.2 mg/kg/body weights corresponding to the 2.5 and 5.0 ppm levels in the diet. Thus, the highest dose, 0.2 mg/kg, corresponded to the dose received by the Yusho patients. The Arochlor 1248 contained only trace quantities of chlorinated dibenzofurans. Thus, it was concluded that these impurities did not play a role in the observed clinical signs. Following 6 months of exposure, the monkeys were bred. The conception rate for the 2.5 and 5.0 ppm dose groups was 8/8 and 6/8 respectively. However, only 5 monkeys from the 2.5 ppm group and 1 from the 5 ppm were born alive. These infant monkeys had a decreased body weight and showed focal areas of hyperpigmentation of the skin. Within two months of birth during which time the infants suckled their mothers, the clinical signs worsened and three of the six infants died including the one borne to the dam receiving 5 ppm PCB. Analysis of milk samples from the lactating monkeys revealed levels ranging from 154-397 ppb PCB. The estimated intake of PCB was 30.8 - 74.9 ug/kg/body weight/day.

After 4 months the remaining infants were placed on a synthetic milk replacer and their condition improved decidely over the next four months. It was reported, however, that these monkeys displayed learning difficulties and hyperkinesis following cessation of exposure[2].

The Yusho incident coupled with the data on monkeys bu Allen et al. demonstrate clearly the transplacental toxicity of PCB's and raise concern over the health of human infants born to mothers with high PEB exposure particularly if the mother desires to breast feed the infant. In 1975 the Health Protection Branch undertook a survey of PCB's in human breast milk in Canada. In this survey of 100 samples the maximum PCB level was found to be 68 ppb and 98% of the samples had residues equal to or greater than 1 ug/kg. The meal leval of PCB was 12 ug/kg. For the human infant this would provide an average intake of 1.8 ug/kg or roughly 10 to 80 times lower than the level known to produce toxic effects in the Fla of the monkey study by Allen et al.

Although the margin of safety for human infants exposed to PCB through their mothers milk is not known definitively at present there have been no reports in Canada of the symptoms resembling those seen in PCB poisoning. The Health Protection Branch has initiated a program to analyze breast milk for PCB in instances where the family physician feels, on the basis of detailed examination of the infant and mother, including a history of her PCB exposure, that this would be desirable. Considering the innumerable advantages of breast feeding it is considered unlikely that Canadian mothers having elevated levels of PCB in their breast milk would be recommended to discontinue breast feeding except in a very few instances, where in the view of the attending physician, this would be considered desirable.

Public health concern regarding contaminants in the food supply stems from the fact for the major contaminants in food the margin of safety between the actual exposure level and levels that will produce adverse effects is quite narrow. For the problem contaminants safety factors rarely exceed one order of magnitude and in some instances may not exist at all. Furthermore, it is only recently that legislation has been enacted to control these materials in environmental media.

Regulatory alternatives in dealing with contaminants are limited. Banning foodstuffs containing potentially hazardous levels of environmental contaminants is not usually an acceptable alternative since this will restrict the availability of otherwise nutritious food. Removing the offending substance from commerce may be acceptable in some instances but has the disadvantage that it deprives the economy of potential useful chemical substances and will not, in any event, produce the desired results in the short-term because of the persistance of most contaminants in environmental media. Additionally, in many instances natural background levels contribute significantly to the total exposure and obviously this cannot be controlled through regulatory actions. The regulatory option selected in most cases has been that of establishing some sort of tolerance or guideline for acceptable levels of contaminants in food commodities. For example, guidelines have been established in Canada prohibiting the sale of fishery products containing in excess of 0.5 ppm mercury or 2 ppm PCB. In developing acceptable guideline levels, regulators must balance the risk from consumption of contaminated products against the economic and other losses to the affected industries.

Further perusal of this problem leads one to the conclusion that the concept of establishing guidelines on acceptable levels of contaminants in food requires some re-examination. The guideline approach has several disadvantages that warranted consideration. Toxicity is a function of dose and with environmental contaminants the dose is the product of the exposure level and the amount of food consumed per day. Thus, it is easily reckoned that without information on acceptable consumption levels, guidelines do not provide the ultimate in the way of health protection. On the contrary. anouncements by governments of guidelines on levels of contaminants in food is usually interpreted by the public to mean that the product is now safe for consumption. The fallacy in this approach becomes readily apparent when one considers that consumption of as little as 60 grams of fish containing 0.5 ppm mercury would result in an individual reaching the World Health Organization provisional tolerable daily intake of 30 ug. Admittedly, the average person would not consume that much fish containing 0.5 ppm mercury on a daily basis but it has been well established that certain heavy fish consumers will easily exceed this quantity. Secondly, regulatory use of guidelines implies that all individuals in the population are of equal sensitivity to contaminants. In other works, the special sensitivity of pregnant mothers, infants and the very old are not reflected in the guideline approach. Thirdly, establishement of across the board guidelines, say for all fish products, have had marked economic impact on certain segments of the food industry, for example, in Canada commercial fishing for swordfish essentially ceased following pronouncement of the 0.5 ppm guidelines.

In addition to economic hardship the guideline has decreased the availability of important sources of dietary protein not only from swordfish but from several other marine and fresh water species many of which exceed the guideline depending on their size, food supply and geographical location.

It will become increasingly important in the future for regulatory agencies to deal effectively with this problem. One means of accomplishing this would be to couple the guideline approach with programs aimed at increasing consumer awareness of toxicological concerns much the same as we are doing through nutrition labelling. The activist position that the consumer has the "right to know" must be met through education and labelling programs designed to educate the consumer regarding the hazards of environmental contaminants. It would be useful to provide the consumer with information on suggested maximum daily or weekly consumption of contaminated products containing various amounts of contaminants. In this way the consumer has the ultimate decision and the special sensitivity of certain population subsets would be taken care of. This approach has the additional advantage of reducing economic hardship an affected segments of the food industry and at the same time optimizing the availability of otherwise nutritious food.

During the course of this presentation, I have attempted to deal with some of the difficult and interesting problems retaling to environmental contaminants in food. Obviously, we have much to learn about these substances and will always be faced with the problem of discovering new ones. Needless to say, for those of us involved in this endeavour, particularly regulators, the task of dealing with these problems has only begun, but we are at least warning to understand the complexities of the problem and, I trust, will develop means of handling this problem more effectively in the future.

REFERENCES

[1] Allen, J.R. and D.A. Barsotti. The effect of transplacental and mamary movement of PCBs on infant monkeys. Toxicology (in press), 1977.

[2] Bowman, R.B., M.P. Heironimus and J.R. Allen. Behavioral effects of low levels of polychlorinated biphenyls (PCBs) in infant monkeys submitted to Science.

[3] Coulson, B.J., R.E. Remington and K.M. Lynch. Metabolism in the rat of the naturally ocurring arsenic of shrimp as compared with arsenic trioxide. J. Nutrition, 10: 255-270, 1935.

[4] Crecelius, B.A. Changes in the chemical speciation of arsenic following ingestion by man. Envir. Hlth. Presp. 19: 147-150, 1977.

[5] Edmonds, J.S., K.A. Francesconi, J.R. Cannon, O.L. Rston, B.W. Skelton and A.H. White. Isolation, crystal structure and synthesis of arsenobetaine, the arsenical constituent of the western rock lobster Panulinus longipes cygnus Bore. Tetra. Letters, 18: 1543-1546, 1977.

[6] FAO/WHO Expert Committee on Food Additives. 16 Report of the Joint FAO/WHO Expert Committee on Food Additives. Wld. Hlth. Org. Techn. Rep. Ser., 505 FAO Nutrition Meetings Report Series, 51, 1972.

[7] Irogolic, K.J., E.A. Woolson, R.A. Stockton, R.D. Newman, N.R. Bottino, R.A. Zingaro, P.C. Kearney, R.A. Pyles, S. Maeda, W.J. McShane and B.H. Cox. Characterization of arsenic compounds formed by Daphnia magna and Tetraselmis chuii from inorganic arsenate. Environ. Hlth. Prespect. 19: 66-66, 1977.

[8] Jelinek, C.P. and P.E. Corneliussen. Levels of arsenic in the

United States food supply. Environ. Hlth. Perspect. 19: 88-87, 1977.
9 McBride, B.C. and R.S. Wolfe. Biosynthesis of dimethylamine by Methanobacterium. Biochem. 10: 4312-4317, 1971.
[10] Miller, G.E., P.M. Grant, R. Kishore, P.J. Steinkruger, P.S. Rowland and V.P. Guinn. Mercury concentration in museum specimens of tuna and swordfish. Science 175: 1121-1121, 1972.
[11] Neely, W.B., D.R. Branson and G.E. Blau. Partition coefficient to measure bioconcentration potential of organic chemicals in fish. Environ. Sci. Technol. 8: 1113 (1974).
[12] Penrose, W.R. Biosynthesis of organic arsenic compounds in brown trout (Salmo trutta). J. Fish Res. Board Can. 32: 2385-2390, 1975.
[13] Smith, D.B., E. Sandi and R. Leduc. Pesticide residues in the total diet in Canada II - 1970. Pestic. Sci. 3: 207-210, 1972.
[14] Westoo, G. and M. Rydalv. Arsenikhalter i vissa livsmedel. Var Poda. 24: 21-40, 1972.
[15] Wood, J.M. Biological cycles for toxic elements in the environment. Science 183: 1049-1052, 1974.

CONTAMINANTS AND ADDITIVES IN ANIMAL FEEDING:

RELATIONSHIP TO HUMAN NUTRITION

Prof. Dr. R. Ferrando

Laboratoire de Nutrition-Alimentation
Ecole Nationale Vétérinaire d'Alfort
7, Avenue du Général de Gaulle
94704 Maisons-Alfort Cédex - FRANCE

During the last 25 years the conventional aspects of animal husbandry have been changed by the increased trend of intensive production due to industrialization, urbanization, a rising standard of living and, at the same time, a decrease of lands available to be farmed. In order to improve both growth rate and feed utilization as well as prevention of microbism[7] or diseases such as coccidiosis, minute doses of various substances, termed feed additives, were mixed in the farm animals' diets. Also, natural toxic substances in food and food contaminants, such as mycotoxins, were discovered and studied.

The most important problems now are not contaminants but the use of additives in animal feeding. On one hand, industrial animal management has been improved by that use. On the other hand, hygienists have become concerned about their use, even though the quantities produced — especially those of antibiotics used as growth factors — are still below the volume produced and sold for both human and animal chemotherapy.

CONTAMINANTS

We shall consider first the problem of contaminants. In this context, as for additives used in animal feeding, it is advisable to overlook the important part that is to be played by farm animals. The consumers are not directly exposed to the toxic. The contaminants and the additives do not reach man directly but after there has been a run through such relays as vegetables and animals for contaminants, animals for feed additives. Such relays should not

be overlooked. With Truhaut[8-20] we have devoted research activities
to this specific aspect of toxicology and human nutrition and developed the socalled relay toxicity methodology.

Milkaflatoxin extracted from milk is toxic for ducklings but
not lyophilised milk containing milkaflatoxin given directly to
ducklings[9]. There are other examples: ochratoxin[11], Senecio Jacobaea[6] etc. Therefore milk from dairy cows fed with Brassica oleacea var. acephala is goitrogenic.

There are foods of plant and animal origin that may induce a
pharmacodynamic action because they contain pharmaco-active compounds as normal or natural components. Sapeika[18] has recently reviewed this question. The interpretation of contaminants' problem
in animal feeding and the relationships with human nutrition will
require thinking and a critical approach. In this connection, one
should take, for experiments to be carried out, the environmental
factors into account. Among these factors, the diet of man. Its
balance and quality will indeed determine different metabolic processes according to the physiological and/or pathological conditions,
likely to be further changed by various environmental conditions
including diet. The same thing is to be considered in toxicology
experiments. We have studied these aspects with Truhaut[21].

The presence of pesticides in trace amounts in some foodstuffs
is largely adventitious. We have already underlined the importance
of animal relay in this question.

More generally speaking W.R. Boon[3-4] has written: "It is important to realise that agriculture itself is a disturbance of the
balance of nature. Ideally, the parasites and pests of foodcrops
are kept under control by the animals, birds and insects found in
the hedgerows and countryside surrounding agricultural land. However, the present escalating rate of world population growth necessitates the intensive cultivation of available land. The wildlife
population is not large enough to cope with the quantities of pests
which thrive in areas of intense cultivation. Leaving the control
of pests to nature under these circumstances would means that millions of people would die of starvation because of unnecessary
crop destruction".

We do not forget that to ensure a safe and effective employment
of pesticides all governments have improved regulations governing
their use.

Analytical techniques of a great sophisticarion and sensitivity
show that residues on growing plants or in animal products, mainly
milk, are usually minute, never more than a few ppm and sometimes
ppb In other respects, culinary practices most often destroy them.

Many substances having a pesticide-like action (antitrypsin, gossypol, flavonoïd compounds, etc.) occur naturally in plants, animals or micro-organisms. Many of these constituents are unknown and the argument that natural substances must be safe is illusory.

We stressed some years ago than DDT can protect duckling against aflatoxicosis[9]. It is ever important to consider antagonism or synergy between residues and natural contaminants. Errors in or complete misknowledge of this may entail misinterpretation of findings which will be the cause of unfortunate confusions.

ADDITIVES IN ANIMAL FEEDING

Main additives used in animal husbandry are:

- Growth - promoting agents acting on breeding performances;
- Anabolizing agents and endocrine modifiers as hormones, antihormones and related compounds;
- Low-dosage specific prophylactic compounds (e.g. Coccidiostats-anti-histomoniasis etc.);
- Miscellaneous substances including antioxidants, stabilizing agents, emulsifiers, aromatic, antimottants, fungicides, etc.

Generally no objections have ever been raised against miscellaneous substances. All such additives have also been approved for use in human foodstuffs.

Coccidiostats or other agents used for the prevention of some specific diseases in the breeding populations have never worried consumers. Neither their names nor their properties awake any interest or even fear in the mind of the public, and calling them by their names will not strike the imagination of the main-in-the-street. Coccidiosis or histomoniasis mean nothing at all to a town dweller or even to most medical men. In the eyes of public and medical opinion, growth factors, mainly those described as antibiotics and hormones are alarming. They are part of the magic of modern life which eludes their comprehension, so that they fear and respect them at the same time. People and sometimes scientists are neglecting the complexity of metabolic pathways. They often mask biological simplicity.

The intestinal flora, re-fashioned by the various components of the diet, happens to be one of these action relays for adjuvants. Does not the carbohydrate-protein balance act similarly, or at least partly?

We will leave it up to the reader to select the right term: additive, food adjuvant, or growth factor. The three expressions are equivalent to our minds, and we might define these compounds as follows: they are substances which, when present in small amounts in the diet of animals reinforce and/or alter directly or indirectly some metabolic process through some known or unknown action, and help in protecting the animals against specific and non-specific diseases, thus promoting, under economic conditions, their general health and performance, without endangering the human consumers.

This definition may be applied to all dietary elements, from essential aminoacids to vitamins and trace elements. It is also an EEC concept.

Vitamin A acts at the epithelial level. It protects against infections, makes the synthesis of steroid hormones possible, of which some are anabolic, and facilitates some detoxication mechanisms. It also promotes growth and food conversion, although the correct mode of action cannot be explained in each specific case. At a high dosage, however, vitamin A becomes harmful. Vitamin D protects the body against bone dystrophies through its metabolites which act via endocrine relays. At a high dosage it becomes dangerous.

Selenium is the active metal of an enzyme (glutathione peroxidase) that destroys peroxides; it is, therefore, essential for the body in very low amounts. At a high dosage selenium becomes toxic. American poultry farming has generally suffered from incorrect ideas about this trace element and from misapplication of the Delaney clause. The discovery of glutathione peroxidase and its role should change the inconsistent stand taken by some authors.

Such examples could be repeated. The position, in about 1947-1948, of vitamin B_{12} and the APF (Animal Protein Factor) which were often confused, illustrates the paradoxical situation of the additives. At a given time in their history, they were on the border line between a necessary drug and an essential foodstuff and, according to the time in question, they would be now on this side and then on the other of this line.

Why should we refuse to accept what is accepted for others? We do not know entirely the mode of action of such additives as antibiotics. Influenced as we are by the concepts of therapy, we confine it to their bacteriostatic or bactericidal power. Antibiotics, as stressed by Riedel[17], have other actions that seem to occur at the metabolic level. They act on the structure of the intestinal epithelium and promote absorption[10,14]. It would be desirable for us to free ourselves from concepts imbued with ideas belonging to the old type of medical principles. The very small dosages of additives used for nutritional purposes are quite independent of the overdoses commonly used in medicine, and the abolition of their use as additives for animal feed would lead to an increase in their rate of use and to conditions impossible to supervise.

TABLE 1

RESIDUE LEVELS OF SOME FEED ADDITIVES AUTHORIZED IN FRANCE AND/OR THE EUROPEAN COMMUNITY

PRODUCTS	DOSES ppm	ABSORBED IN THE INTESTINE	THERMO-LABILITY	RESIDUES	
				Muscle	Liver
Bacitracin Zn	5-20 5-80*	−		< 0,02 µg/g (for 500 ppm)	< 0.01 µg/g (for 500 ppm)
Carbadox**	10-50	+	±	0-NM	0-NM
Olaquindox***	50	+	±	0-NM	0-NM
Erythromycin	5-20	+		0.78 µg/g	1.14 µg/g
Flavomycin	0.5-25	−	−	0	0
Hygromycin B	12	−	−	0-NM	0-NM
Oleandomycin	5-15	+	+	0-NM	0-NM
Payzone-Nitrovin	5-15 20-80*	+	?	0.04 ppm	0.06 ppm
Spiramycin	5-20 5-80*	+	−	0.02 ppm (for a 35 ppm supplementation)	0.15 ppm
Tylosine	5-40	+	−	0	0
Virginiamycin	5-20	+	−	0.12 µg/g	0.12 µg/g
Amprolium	62-125	+		0.36 ppm (whole carcass)	
Amprolium plus ethopabate	62+8 125+8	+		0.5 ppm	1.5 ppm
Buquinolate	82,5	+		traces	traces
Decoquinate	10-40	+	±	< 0.05 ppm	0.05-0.16 ppm
Zoalene-DOT	75-125	+		2.1-2.3 ppm	6.3 ppm
Monensin	100-125	±		(time 0) 0-NM	0-NM
Dimetridazole	150-175	+		traces	traces
Robenidine	30-36	±		0-NM	0-NM

* For milk substitute only.
** Withdrawn when hog reaches 70 kg (153 1g), i.e., 4-5 weeks before slaughter.
   This additive is authorized in swine mainly in France, Belgium, Italy, Spain, Ireland, Switzerland, Luxemburg, Holland, Canada, USA and other countries.
   0-NM = not measurable.
*** Withdrawn in same conditions than Carbadox.

What happens in real and potential dangers? Residues are very low as shown in Table 1 for some feed additives used in France and/or in the EEC.

With respect to anabolizing agents of natural steroids group (oestradiol - progesterone - testosterone) or derived compounds, residues are no more important than in meat of control animals. These residues are innocuous for the consumers. Our work on diethystilbestrol (DES) has forced us to admit this compound as very dangerous[10] for men and the environment. A large number of our investigations were confirmed by Prokowskii et al.[15].

Residues have also been blamed as potential allergens. While this is mainly true of penicillin and streptomycin, now withdrawn in EEC countries, we have always held that such a risk, which is often attached to the use of antibiotics as animal feed additives, was more theoritical than real. The findings of famous allergists, both in France and elsewhere, have strengthened our opinion. There may, of course, be individual idiosyncrasis. Malten[12] writing already in 1968, could draw no conclusions as to the absence or the existence of the risks. In a Symposium on "Antibiotic and Agriculture" (cf. ref. Malten) Frazer voiced the right opinion that the observations made concerned isolated individuals. He added: "Many people are allergic to various foodstuffs and it is, therefore, necessary to establish, by carefully controlled studies carried out under scientific conditions, whether the allergic effect is indeed due to an antibiotic residue and not to some substance in the food".

Such carefully controled studies have not yet been made, and their standardization seems a delicate matter.

Another public health hazard raised against the use of additives, mainly antibiotics, in animal feedstuffs referes to possible bacterial resistances and to their transfer due to RTF. It is more serious and worth gerater consideration. They have given rise to reports published in USA and UK[16-19]. This bacterial resistance to various antimicrobial agents has been down before advent of sulfonamides and antibiotics. In Institut Pasteur de Paris there are strains dating from 1926 and carrying RTF. We must underline that some heavy metals (Cd, Hg, Cu, Co) are inducing RTF[24]. The various processes by which bacterial resistance may be transmitted include transformation, transduction and conjugation or sexual mating[1-2-5-22-23].

The pros and cons of R factor transfer in E. Coli from animal to man are listed in Table 2 adapted from Smith, 1973.

Therefore it is important not to forget that RTF comes also, and perhaps mainly, from use of antibiotics in human and veterinary medicine. The withdrawal of an antibiotic from the list of authoriz-

TABLE 2

EVIDENCE ON THE TRANSFER OF R-FACTORS
IN E. COLI FROM ANIMALS TO MAN

FOR	AGAINST
Animals usually receive antibiotics in a manner more likely to produce R + E. coli (as feed additives)	More antibiotics are used in the human population than in the animal population
Actual amount of R + E. coli is higher in the animal population than in the human population	There is a high incidence of R + E. coli in the human population
Salmonellae are transmitted naturally from animals to man, on meat; why no E. coli? R + E. coli are frequently found on meat	Their numbers are small, especially on cooked meat
R factors have been transmitted experimentally from animal E. coli to resistant E. coli in the human alimentary tract A high transfer rate was obtained when appropriate antibiotics were also given	Large numbers of donor organisms were required for colonization; transfer, when it occurred, was at a very low rate
Experimentally, in animals, R. factors can be transmitted at a high rate from E. coli to Salmonella typhimurium	Good donors, good recipients, and good colonizers have to be used
A higher incidence of R + E. coli has been found in animal attendants and kitchen personnel than in the general human population	A higher incidence of R + E. coli has been found in babies and in vegetarians than in adult meat eaters; in general, human beings have more contact with each other than with animals or their raw products

Data adapted with permission, from Smith (from WHO Working Group, 1973, Bremen).

ed additives in animal feedstuffs generally increases its use in veterinary therapy. Figure 1, from a review we published in 1975[10], gives a schematic illustration of interactions between therapeutic agents, on one hand, and bacteria of multiple origin, on the other. These interactions are one of the factors of a constantly developing microbial ecology to which not enough thought is being accorded.

All these questions concerning contaminants and additives in feedingstuffs could hardly be answered without detailed experiments. All these facts will dictate a philosophy rather than a line of conduct, although man's philosophy is generally a determinant of his line of conduct. Finally, of all the additives used in anumal nu-

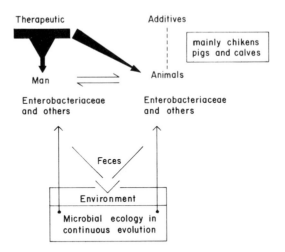

Fig. 1 - Graph illustrating the numerous exchanges of bacteria between the various animal species and man indicates the complexity of the problem to be solved. Who is infecting whom? Where is the danger? Should we accuse therapy, or the additives, or both?

trition, only the antibiotics give rise to objections which, even though they may not always be scientifically well-grounded observations, venture too far as to their possible long-term implications which are, in addition, not easily demonstrable. It is in no way surprising, therefore, that such hypotheses should give rise to irrational and unconscious reactions likely to have political aftereffects within the framework of the current anti-pollution problem. It is also intellectually inadmissible to have a few individuals meet to discuss, in an abstract manner, problems that have no connection with one another and need careful interpretation, and then return final judgements on the questions. Such an attitude borders on inquisition, and once again one has failed to view the problem as a global phenomenon. It is also strange to note that DES has been recommended and authorized in the countries of the Swan Report and Task Force Report.

If we compare feed additives advantages and public health hazards we have on one side of the balance sheet, hard facts now well known; on the other side are hypothetic projections, which are admittedly based on findings that are not only difficult to verify but, above all are impossible to connect with the influence of feed additives in that they are linked to the general ecology. The withdrawal of some antibiotic inducing drug resistance has decreased, if not eliminated, the main threat arisen by additives used in animal feeding. The increasing use of such drugs in therapy increases therefore this threat.

REFERENCES

[1] Anderson, E.S. Aspects of infective drug resistance. Proc. Symp. Royal Soc. Med., London, 1970, p. 53 (PMP Service Ltd., London 1970).
[2] Anderson, E.S. Ecological effects of antibacterial drugs. Resistance to infectious diseases. Proc. Int. Symp., 1970, pp. 157-172 (PMP Service, Ltd., London 1970).
[3] BNF. "Why Additives? The Safety of Foods". The British Nutrition Foundation. Forbes Public, London 1977; 75 pp.
[4] Boon, W.R. Cf. ref. BNF (1977).
[5] Chabbert, Y.A., J. C. Baudens and D.H. Bouanchaud. "Medical aspects of transferable drug resistance. Bacterial episomes and plasmids". Ciba Found. (Churchill, London 1969).
[6] Earl Johnson, A. Changes in Calves and Rats consuming Milk from Caws Fed Chronic Lethal doses of Senecio Jacobaea (Transy Ragwort); Am. J. Vet. Res.; 37: 107-110, 1976.
[7] Ferrando, R. Le microbisme de porcherie. Rev. Méd. Vét.; 101: 93-112, 1952.
[8] Ferrando, R., and R. Truhaut. La toxicité de relais. Nouvelle approche méthodologique d'évaluation toxicologique des additifs

aux aliments des animaux d'élevage. C. R. Acad. Sc. Paris; 27 série D; 279-283, 1972.
[9] Ferrando, R., A. Parodi, Nicole Henry and A.L. Ndiaye. Influence d'un prétraitement du caneton par le DDT sur sa réaction ultérieure vis-à-vis d'un tourteau d'arachide pollué par Aspergillus flavus. C. R. Acad. Sc. Paris; 279 série D; 1131-1134, 1974.
[10] Ferrando, R. Future of Additives in Animal Feeding. World Review of Nutr. and Diet. 22: 183-235, 1975.
[11] Hult, K., Anna Teiling and S. Gatenbeck. Degradation of ochratoxin A by a ruminant. Appl. Environ. Microbiol.; 32: 443-444, 1976.
[12] Malten, K.E. Allergy to antibiotics in minute amounts in food. In: "Antibiotics in Agriculture". Bibl. Nutr. Diet. n° 10. 1968 Karger - Basel - New York.
[13] Masquelier, J. The Bactericidal Action of Certain Phenolics of Grapes and Wine. In: "The Pharmacology of Plant Phenolics". Fairbairn, J.W. (ed.). Academic Press, London - New York; 1959, p. 123.
[14] Neutra, M., J.H. Mener, and L.G. Mayoral. Actions de la malnutrition protéinocalorique sur la muqueuse jéjunale des porcs traités à la tétracycline. Amer. J. Clin. Nutr.; 27: 287-295, 1974.
[15] Prokowskii, A.A., M.F. Nesterin, G.P. Vavilina, E.E. Ryazantseva, and L.Y.A. Solovieva. Medico-biological investigation in the estrogen application for growth stimulation in animal husbandry. Vestn. Akad. Med. Nauk. SSSR.; 27: 3-15, 1972.
[16] Report of FDA task force on the use of antibiotics in animal feed (Veterinary Medicine Office, FDA 1972).
[17] Riedel, G. Influence de l'auréomycine et de la flavomycine sur les poulets axéniques exposés au stress. Germfree Research: Biol. effect gnotobiotic Environment "Proceed. Intern. Symp. 4th, New York, 1972". Academic Press, London - New York, 1973, p. 239.
[18] Sapeika, N. Pharmacodynamic Action of Natural Food. Wld. Rev. Nutr. and Diet.; 29: 115-123, 1978.
[19] Swann, Report of joint Committee on the sue of antibiotics in animal husbandry and veterinary medicine (Her Majesty's Stationary Office, London 1969).
[20] Truhaut, R., and R. Ferrando. La Toxicité de Relais. Toxicology; 3: 361-368, 1975.
[21] Truhaut, R., and R. Ferrando. Le toxicologue et le nutritionniste face aux problèmes d'évaluation toxicologique. Wld. Rev. Nutr. Diet.; 29: 4-41, 1978.
[22] Watanabe, T. Transferable drug resistance: the nature of the problem; in "Bacterial episomes and plasmids", Wolstenholme and O'Connor. CIBA Found. Symp., p. 81-101 (Churchill, London 1969).
[23] Wolstenholme, G.E.W. and M. O'Connor. "Bacterial episomes and plasmids". CIBA Found. Symp. (Churchill, London 1969).
[24] Zakaryan, L.M. Mercuric chloride resistance of staphylococci isolated from sick and healthy persons. Zh. Mikrobiol. Epidemiol. Immunobiol.; 11: 46-50, 1972.

CONTAMINATION OF THE ARGENTINIAN FAMILY BASKET

BY CHLORINATED DEFENSIVES RESIDUES - BIOLOGICAL REPERCUSSION

Emilio Astolfi

ANDEF - R. Gen. Mena Barreto, 663

Jardim Paulista - 01433 - S. Paulo - BRASIL

The contamination of the infant and general population with defensives belonging to the group of organic chlorides is the result of the daily ingestion of residues contained in the usual diet and in other sources, such as the pollution of the environment (soil, air, water, etc), and by the incorrect usage of these substances. In order to establish the magnitude of the human being's exposure, it is possible to evaluate the impregnation level of the organism, or to calculate the amount of residues which entered, daily, through the food.

The presence of defensives in the human fat is not merely a deposit, but rather an equilibrium state between absorption, metabolism, storage and excretion. The amount accumulated in the fat tissues does not increase indefinitely as a consequence of continuous exposure, but rather the curve stabilizes upon reaching a specified limit. This equilibrium level is related to the dose and time of exposure.

Because of the consumption of chlorides in Argentina it was considered to be of interest to carry out this investigation.

The chlorinated residues in the blood of people professionally exposed and not exposed were investigated, as well as those in the fat and blood of the infant and adult populations, in the maternal blood and in the umbilical cord of newborn children (Tables 1-2-3-4-5). In accordance with these findings, the organochlorinated residue levels of the inhabitants of the Republic of Argentina do not differ greatly from those obtained by different authors from other countries in their respective areas of investigation. The average equilibrium level would be around 6 ppm (Table 6).

TABLE 1

CONCENTRATION OF DEFENSIVES IN THE BLOOD OF PEOPLE NOT
PROFESSIONALLY EXPOSED IN THE REPUBLIC OF ARGENTINA (IN ppb)

INDIVIDUALS	Nº OF INDIV.	BETA (BHC) HCH	GAMMA (BHC) HCH	DIELDRIN	PP'-DDE	PP'-DDT
Adults	20	23.01 ± 10.14	0.98 ± 1.16	1.43 ± 1.21	14.53 ± 6.76	3.13 ± 1.70
Children (5-10 years)	18	6.61 ± 3.93	0.43 ± 0.34	0.94 ± 0.92	8.13 ± 4.06	4.21 ± 1.41
Children (1-5 years)	19	6.72 ± 6.57	0.32 ± 0.24	0.54 ± 0.29	5.56 ± 4.83	2.49 ± 2.26

TABLE 2

CONCENTRATION OF DEFENSIVES IN THE BLOOD OF PEOPLE PROFESSIONALLY EXPOSED AND NOT EXPOSED IN ARGENTINA (in ppb)

INDIVIDUALS	CASES	GAMMA HCH (BHC)	BETA HCH (BHC)	DIELDRIN	PP'DDE	PP'-DDT
Prov. of Catamarca Administr. Personnel	10	1.71 ± 1.26	19.35 ± 7.88	2.09 ± 1.25	15.78 ± 8.94	3.88 ± 1.64
People that worked with defensives (a)	10	70.38 ± 67.73	237.70 ± 139.92	1.09 ± 0.60	14.49 ± 6.43	3.86 ± 0.90
People that had worked with defensives before (b)	10	±	50.53 ± 27.51	2.81 ± 3.59	86.11 ± 44.32	11.28 ± 5.41
Prov. of Salta Administr. Personnel	10	0.25 ± 0.25	26.66 ± 11.19	0.77 ± 0.74	13.30 ± 3.64	2.47 ± 1.70
People that worked with defensives (a)	10	3.47 ± 2.99	182.33 ± 107.21	1.16 ± 0.72	17.01 ± 8.88	2.31 ± 1.93
People that had worked with defensives before (c)	9	0.51 ± 0.65	13.65 ± 20.06	0	363.00 ± 367.78	346.36 ± 348.39

(a) Workers who had been using HCH (BHC) for 5 years or more
(b) Malaria control personnel who had used DDT previously but who had not worked with it for 5 years or more
(c) People that worked with DDT for 5 years or more (Malaria Control Program)

TABLE 3

CONCENTRATION OF DEFENSIVES IN THE BLOOD OF
PEOPLE NOT PROFESSIONALLY EXPOSED IN ARGENTINA
(in ppb)

INDIVIDUALS	Nº OF INDIV.	BETA HCH (BHC)	GAMMA HCH (BHC)	DIELDRIN	PP'-DDE	PP'-DDT
Adults	20	23.01 ± 10.14	0.98 ± 1.16	1.43 ± 1.21	14.53 ± 6.76	3.13 ± 1.70
Children (5-10 years)	18	6.61 ± 3.93	0.43 ± 0.34	0.94 ± 0.92	8.13 ± 4.06	4.21 ± 1.41
Children (1-5 years)	19	6.72 ± 6.57	0.32 ± 0.24	0.54 ± 0.29	5.56 ± 4.83	2.49 ± 2.26

TABLE 4
CONCENTRATION OF DEFENSIVES IN THE BLOOD OF NORMAL CHILDREN IN ARGENTINA (in ppb)

Nº OF CASES	BETA HCH (BHC)	GAMMA HCH (BHC)	HEPTA-CHLOR	DIELDRIN	pp' DDE	op'DDT	pp'DDT
37	4.63 ± 1.93	0.39 ± 0.20	0.34 ± 0.11	0.69 ± 0.21	5.62 ± 2.29	0.57 ± 0.20	3.59 ± 1.55

CHLORINATED DEFENSIVES IN THE HUMAN INFANT FAT IN ARGENTINA (in ppm)

	ALPHA HCH (BHC)	BETA HCH (BHC)	GAMMA HCH (BHC)	DIELDRIN	pp' DDE	op. DDT	pp'DDT
52	0.21 ± 0.10	0.49 ± 0.44	0.09 ± 0.05	0.07 ± 0.03	2.36 ± 1.12	0.39 ± 0.18	1.53 ± 0.78

TABLE 5

CHLORINATED DEFENSIVES IN THE BLOOD OF PARTURIENTS AND IN THE UMBILICAL CORD OF NEWBORNS - Avegares (ppm)

INDI-VIDUALS	N° OF CASES	BETA HCH (BHC)	GAMMA HCH (BHC)	DIELDRIN	HCB	pp'DDE	pp'DDD	pp'DDT
Mothers	28	0.0088	0.00202	0.00127	0.0066	0.0090	0.0012	0.0037
Newborns		0.0050	0.00085	0.00092	0.0029	0.0059	0.0005	0.0020

TABLE 6

DDT + DDE IN HUMAN FAT IN SEVERAL COUNTRIES

AUTHOR	COUNTRY	DDT + DDE (ppm)
Hunter & Col.	England	2.2
Maier & Col.	Germany	2.2
Hanes & Col.	France	5.2
G. Fernandez & Col.	Argentina	6.6
Danes	Hungary	12.4
Hayes & Col.	USA	11.0
Wasserman & Col.	Israel	19.2

The present work has the purpose of demonstrating how to calculate the amount of chlorinated defensives residues contained in a family basket in the city of Buenos Aires, and its relationship to the acceptable daily ingestion indicated by FAO/WHO.

The authors gathered and computed the data on defensives residues of the distinct components of the daily diet of a sample family, composed of father, mother and two children of approximately 9 and 14 years of age, in foods which had not undergone any elaboration or cooking process, with the exception of dairy products or derivatives (Table 7).

The data presented here are only valid when related and conjugated: biological contamination, residues in foods and average diet. Then it becomes possible to calculate the amount of residues ingested daily.

In general all the data gathered on organochlorinated defensives residues are within the practical limits (non intentional) or tolerances indicated by FAO/WHO, except for maternal milk, which exceeds them (Tables 8 and 9).

The partial sum provided values much lower than the acceptable daily ingestion for DDT, Dieldrin and Lindano. It is surprising to see the amounts of non-desirable pollutants, impurities of products such as alpha, beta and delta of HCH (BHC), sometimes in proportions superior to the gamma isomer of the product (Tables 10, 11 and 12).

TABLE 7

FAMILY BASKET FOR A SAMPLE FAMILY IN THE REPUBLIC OF ARGENTINA

QUANTITY CONSUMED FAMILY/DAY	CATEGORY OF FOODS	PROPORTION OF THE DIETS'S TOTAL
2000 gm	Milk and dairy products	22%
2850 gms	Vegetables	31%
1000 gms	Meats	11%
1000 gms	Cereals and derivatives	11%
2000 gms	Fruits	22%
250 gms	Sweets and sugar	3%

TABLE 8

CHLORINATED DEFENSIVES FOUND IN COW'S MILK, CHEESE AND BUTTER IN THE REPUBLIC OF ARGENTINA (Gomes Artero & Col.)

COW'S MILK	ALPHA HCH (BHC)	BETA HCH (BHC)	GAMMA HCH (BHC)	DIELDRIN	TOTAL DDT
ppm in fat	0.16	0.16	0.04	0.04	0.20
ppm whole milk	0.0048	0.0048	0.0012	0.0012	0.0054
WHO practical limit	—	—	0.1 ppm (in fat)	0.15 ppm (in fat)	0.05 ppm (whole milk)
Cheese (ppm in fat)	0.08	0.06	0.05	0.05	0.11
Butter (ppm in fat)	0.05	0.06	0.05	0.03	0.14
Practical limit (ppm in fat)	—	—	0.1	0.15	1.25

TABLE 9

CHLORINATED RESIDUES FOUND IN BOVINE MEAT
(Roveda & Col.) (in ppm)

	RESID.	WHO
HCH (BHC)	0.28	2 (tol.)
DIELDRIN	0.049	0.2 (l. pr)

CHLORINATED RESIDUES FOUND IN CEREALS
(Nat. of Grains Board) (in ppm)

	WHEAT	CORN	WHO
HCH (BHC)	0.61	0.01	0.50

CHLORINATED RESIDUES FOUND IN FRUITS AND VEGETABLES
(INTA and Min. Agric. Mendoza) (in ppm)

	LETTUCE	ARTICHOKES	TOMATOES	PEPPER	GRAPES	PEACHES	WHO
DDT	0	0	0.026	0.41	0.01	0.01	7 (tol.)

TABLE 10

ESTIMATE OF THE AMOUNT INGESTED OF DDT

	MICROGRAMS? DAY/FAMILY	MICROGRAMS/ DAY/PERSON	MICROGRAMS DAY/KG WEIGHT
			Avge. Weight = 60 kg
Milk	8.9		
Dairy products	37.0		
Vegetables	31.05		
Fruits	20.0		
TOTAI	96.95 →	24.23 →	0.40

ARGENTINA = daily ingestion: 0.0004 mg/kg/day
WHO = acceptable daily ingestion: 0.005 mg/kg/day

TABLE 11

ESTIMATE OF THE AMOUNT INGESTED OF DIELDRIN RESIDUES

FOODS	MICROGRAMS/ DAY/FAMILY	MICROGRAMS/ DAY/PERSON	MICROGRAMS/ DAY/KG WEIGHT
Milk	1.8		
Dairy products	8.75		
Meat	6.0		
TOTAL	16.55 →	4.13 →	0.068

ARGENTINA = daily ingestion: 0.00006 mg/kg
FAO/WHO = acceptable daily ingestion: 0.0001 mg/kg

TABLE 12

ESTIMATE OF THE AMOUNT OF LINDANO RESIDUES
(Gamma HCH (BHC))

FOODS	MICROGRAMS/ DAY/FAMILY	MICROGRAMS/ DAY/PERSON	MICROGRAMS/ DAY/KG WEIGHT
Milk	1.8		
Tea	24.4		
Dairy products	14.0		
Meat	33.0		
Cereals	459.5		
TOTAL	532.7 →	133.05 →	2.21

ARGENTINA = daily ingestion: 0.0022 mg/kg
FAO/WHO = acceptable daily ingestion: 0.0125 mg/kg

A point worth emphasizing refers to the feeding of the infant, from birth until one year of age. It is well known that the main, and in certain occasions the only food is milk. This is exclusive with reference to the first three months of life, according to the modality of the case; however, in some regions of Argentina, due to the low income level of the population, prolonged maternal lactation is induced, past the first year of life, in order to provide, for as long as possible, the proteic supplement provided by the milk. For this reason, and similarly to other nations, the presence of chlorinated defensives residues in maternal milk was investigated (Table 13) and its comparative study with cow's milk showed greater contamination in the human milk, which coincided with findings in Poland, Israel and Guatemala (Table 13).

If we take into account the amount of milk ingested by the infants, according to their ages during the first year of life, we will notice differences in the ingestion of residues of defensives of the organochlorinated group, according to whether it comes from human milk or cow's milk (Table 14).

The daily amount of milk ingested was calculated based on the data obtained from classical pediatrics textbooks and verified in practice. The milk feeding of the child from birth until its sixth month of life oscillates between 110 and 150 mililiters of milk per kg of body weight/day, with an average of 130 ml/kg.

TABLE 13

CHLORINATED RESIDUES FOUND IN HUMAN MILK IN ARGENTINA
(Garcia Fernandez)

RESIDUES	PPM FAT	PPM WHOLE MILK	RELAT. BETWEEN CONC. HUMAN & COW'S MILK
Alpha HCH (BHC)	0.84	0.012	5.5
Beta HCH (BHC)	2.72	0.042	19.5
Gamma HCH (BHC)	0.35	0.006	8.75
TOTAL DDT	9.13	0.140	45.6
			(Calculated over fat)

TABLE 14

AMOUNT OF CHLORINATED RESIDUES INGESTED
ACCORDING TO LACTATION WITH MATERNAL MILK OR COW'S MILK

MONTHS	WEIGHT kg	QUANT. ML/DAY	HCH (BHC) TOTAL		TOTAL DDT	
			HUMAN MILK	COW'S MILK	HUMAN MILK	COW'S MILK
0-1	3-4	420	0.025	0.004	0.0058	0.002
1-2	4-5	720	0.043	0.007	0.100	0.004
2-3	5-6	900	0.053	0.009	0.126	0.005
3-6	6-8	1050	0.063	0.010	0.147	0.006
6-12	8-10	960	0.057	0.009	0.134	0.005

TABLE 15

RELATIONSHIP BETWEEN THE ACCEPTABLE DAILY INGESTION OF DDT RESIDUES ESTABLISHED BY WHO AND THOSE FOUND IN MATERNAL MILK AND IN COW'S MILK

MILK	AMOUNT CONSUMED LITERS/KG	DAILY INGESTION OF DDT MICROGRAMS/KG	WHO LIMIT MICROGRAMS/KG
Maternal	0.130	18.20	5
Cow	0.130	0.78	5

If the DDT concentration in the mother's milk is of 140 mg/l, the Acceptable Daily Ingestion (ADI) or the Acceptable daily Dose (ADD), for an average daily consumption of 130 ml/kg, will be of 18.2 mg/l.

The conclusion of this work is really encouraging with respect to the good agricultural practices in Argentina as far as the use of chlorinated defensives, which are translated into a daily ingestion within the ADI limits recommended by WHO, although it alerts us to the contamination of the infant (Table 15). Although the chlorinated defensives residues are superior in maternal milk than in cow's milk, and exceed the limits fixed by WHO, this does not imply a risk of acute or chronic intoxication for the lactating infant, for the safety limit is quite wide.

Notwithstanding, it is an indicator of exagerated and improper use at the domestic level. There occurs the paradox that while there has been orientation and education of the farmer, with respect to correct field application, in most cases there is still an indiscriminate usage of such chemical products at the domestic and urban level. None of this must, however, lead to the discouragement of maternal breastfeeding, a beneficial priority.

PROTECTION OF THE CONSUMER

FROM AFLATOXINS IN FOOD

H. Schulze

Univ. of Munich; Veterinaerstrasse 13

D-8000 Munchen 22 - WEST GERMANY

A preventive and comprehensive consumer protection and environmental control primarily concerns the protection from hazards to human health caused by chemical compounds and microorganisms in food. To assess the health-related risk and the total exposure of man to chemical substances, apart from environmental chemicals and additives, we must pay attention to undesirable metabolic products formed by mould fungi. The aflatoxins seem to be the most thoroughly investigated group of mycotoxins. Aflatoxins are mainly formed by aspergillus flavus (which gave them the name) and aspergillus parasiticus. In several countries maximum level regulations for the content of aflatoxins in animal feeds and food have been established. In practice it is difficult to observe these tolerances. Apart from that, these regulations do not seem to be sufficient with respect to the regulated aflatoxin components and the groups of food, which have to be controlled.

Aflatoxins are highly hazardous because of their carcinogenic effect. This was the main reason for the establishment of maximum levels. For carcinogenic substances, maximum levels are generally problematic with respect to the theories of carcinogenesis. Considering an apparently unavoidable natural contamination in this case, there is no real alternative for tolerances if several foods and animal feeds are to remain on the market.

The FAO and WHO are striving towards world-wide harmonization of standards in food law and environmental law. These organizations have jointly set up the Codex Alimentarius Commission, which takes into account both hygiene precautions and the needs of food technology, when agreeing on which food standards to adopt. Such efforts run into difficulty, mainly when food regulations in importing countries do not correspond to environmental and food law regulations in producer-countries. On grounds of health protection, harmful substances demand uniform decisions on what constitutes acceptable daily or weekly intakes. Precise permissible levels for individual foods must take into account the different patterns of consumption. In the example of aflatoxin standards, recommendations are sometimes brought in which directly conflict with these patterns. Protein supplies of Third World countries are endangered by the permissible levels being set too restrictively low.

For this reason maximum level regulations for aflatoxins are primarily legal standards which are based only partly on toxicological threshold values. Therefore, it is not surprising that maximum levels in different countries are varying in a wide range. In 1967 WHO/FAO recommended a limit of 30 μg per kilogram for aflatoxin B_1, B_2, G_1 and G_2. This level seems high, regarding the toxicological data and the patterns of consumption especially in developing countries. It is evident anyhow, that in this recommendation the insufficient protein supply in these countries has been taken into consideration.

This is the situation in the Federal Republic of Germany: the German food and feed laws contain regulations on tolerances for aflatoxins. These regulations are concerned with the protection of animal health and the prevention of damage to the consumer's health by aflatoxins in food. One of the main purposes of the aflatoxin-tolerances in all feed regulations is to prevent aflatoxins from getting into foods of animal origin via food chains. For certain foods, the decree of 1976 sets a limit of 10 ppb for the sum of $B_1 + B_2 + G_1 + G_2$ or 5 ppb for B_1 alone. The maximum level for aflatoxins in feed concerns only aflatoxin B_1 and reaches from 10 ppb (chicken) to 50 ppb (cows). We wonder if the regulations in feedstuffs and in food of animal origin do harmonize. Under this aspects of contamination from aflatoxin M (a metabolic product of aflatoxin B, which has a comparable toxicity). There is a biological interdependence of this kind, in that less than 1% of aflatoxin B is transformed to aflatoxin M. Because of the facts that babyfood on the basis of milk powder represents an "exclusive cost", we must pay special attention to aflatoxins in food. Aflatoxins in eggs and meat are, compared with milk, less important. In Germany, there are concrete considerations of a maximum level for milk. Reliable analysis in series, however, are still a great problem.

A general problem is sampling: the current methods of analysis must be complemented by authorized methods of sampling. An international harmonization would be very useful.

I was able to touch only some problems with aflatoxins. The consequences seem to us most important:

- Despite certain insufficiencies with respect to the control of aflatoxins in feeds and food the German aflatoxin-decree has shown that the maximum levels can be observed. These regulations have led to more responsibility with respect to trade and production.

- With respect to food chains, milk and dairy products must be included in regulations of aflatoxin maximum levels. Aflatoxin M tolerances are of special importance for baby foods and the base of milk powder.

- In order to avoid competitive distortion, we need an international harmonization in the field of sampling and analysis. The record stage could be the harmonization of maximum level regulations.

- Aflatoxin problems cannot be solved by administrative measures alone. We need better information from the farmers concerning infections from toxin-producing fungi during storage. On the other hand we should not forget the possibility of aflatoxin formation by a lack of hygienic conditions in private households. Therefore we need an intensive consumer education in this field.

NUTRITIONAL LOSSES IN STORAGE

AND PROCESSING OF LEGUMES

Kyoko Saio

National Food Research Institute
Min. of Agricluture, Forestry and Fisheries
12-4-1 Shiohama, Koto-ku - Tokyo, 135 - JAPAN

1 - INTRODUCTION

FAO estimate of the world production of grain legume in 1976 was about 51.5 million metric tons, of which 26.1 million tons were produced in developing countries. Yet it is said that at least 20-30% of the production suffer losses caused by post-harvest handlings and storage systems, especially by insects and rodent infestation. According to the review of the literature published during 1964-1977 concerning losses in stored cereals and pulses (Adams, 1977), investigations on pulses were mostly those which dealt primarily with major losses by insects, vertebrates and fungi and secondly with losses seen during processing, packaging, distribution and utilization. And those investigations were carried out in India, Bangladesh, Pakistan and Africa where pulses are their staple food. This convinces us that substantial improvement of storage systems are of urgent issue for the solution of food problems. We must keep in mind that the problem is not so much a qualitative one but a quantitative one.

The Expert Consultation on Grain Legumes Conservation and Processing which was organized by FAO and held at Central Food Technological Research Institute, Mysore, India, last November, put the emphasis on the following aspects: 1) developing resistance to insects and other infestation by improved breeding, storage and handling techniques; 2) better utilization of improved milling technologies for splitting and dehulling; 3) re-evaluation of traditional practices on legume processing in view of cost-benefit and nutritional aspects, and 4) prevention against hard-to-cook beans.

When the basic problems of post-harvest conservation have been solved quantitatively on national or worldwide levels, what would come next is the question of "nutritional" losses in storage and processing. Therefore this paper is based not on grain legumes, in a general review, but rather, I want to focus on nutritional losses in storage and processing of soybean on the basis of our experimental data, as I believe the report on soy would be quite suggestive and could contribute to research on other legumes.

2 - OUTLINE OF LEGUMES

Legumes or pulses, edible seeds of leguminosae, have been eaten by human beings as a food since ancient time. When they are consumed together with cereals, protein quality shows far more improvement than it is eaten alone (Bressani, 1973). Dietary habit of taking legumes with cereals widespread in different parts of the world is quite reasonable.

The principal species belong to the genera Phaseolus, Pisum, Cicer and Vicia, and they have relatively high protein content (20 30%) in comparison with cereals, a low fat (2-5%) and high carbohydrate (55-60%) content. The oilseed pulses, such as soybean (Glycine max), peanut (Arachis hypogea), winged bean (Tetragonolobus purpureus) and a part of lupine (Lupinus spp) are distinct in their group, because of their higher oil and/or protein content.

Amino acid contents and nutritional quality of several legumes are given in Table 1 (Burr, 1975). The table suggests that the fortification with limiting amino acids improves their protein quality and a considerable portion of pulse protein is not hydrolyzed into amino acids in the digestion.

Raw pulses contain a wide variety of toxic substances: trypsin inhibitors, photohemagglutinins, goiterogens, cyanogenetic glycosides, antivitamin factors, metal-binding constituents, estrogenic factors etc. Fortunately, these toxicants can be destroyed during home cooking to an insignificant consistency. The PER of many raw pulses rises from zero to maximum value, slowly declines and reaches zero again during period of cooking (Kakade et al., 1965; Bresanni et al., 1963). This is presumably due to the destruction of heat-labile toxic factors by, firtly, a short period of cooking, and then due to a lowering of protein quality and digestibility by successive excess-heating denaturation. The storage at high temperature and/or high moisture content results in undesirable quality changes such as darkening in color, formation of off-flavour, destruction of amino acids, lowering of protein quality and digestibility and an increase in required cooking time. The problem of hard-to-cook beans is important for many pulses from practical point of view. The effect of storage on cooking time is markedly dependent on the moisture contend during storage (Burr et al., 1968).

TABLE 1

PROTEIN QUALITY OF LEGUMES (Burr, 1975)

	AMINO ACID CONTENT[a]								E/T^d	CHEMICAL SCORE	BIOLOGICAL VALUE	DIGESTIBILITY[e]	NET PROTEIN UTILIZATION
	Isoleucine	Leucine	Lysine	S-containing[b]	Aromatic[c]	Threonine	Tryptophan	Valine					
Common Bean	105	108	132	54°	127	99	105	93	2.39	54	58	73	42
Broad Bean	100	101	119	44°	124	84°°	90	89	2.20	44	55	87	48
Chick Pea	111	106	126	63°	142	94	90°°	92	2.43	63	68	86	58
Cowpea	96	100	126	64°	128	90°°	113	91	2.31	64	57	79	45
Lentil	108	108	132	49°	140	99°°	100	101	2.46	49	45	85	38
Lima Bean	124	116	137	64°	153	104°°	105	104°°	2.65	64	66	78	51
Lupine	110	102	97	61°	119	91	105	81°°	2.18				
Pea	107	97	138	58°	121	102	93°°	95	2.35	58	64	88	56
Pigeon Pea	78	90	141	42°	169	73	58°°	73	2.25	42	57	78	44
Vetch	91	98	106	59°	101	84	—	80°°	1.99				
Peanut	84	90	65°	68	146	65°	108	84	2.03	65	54	87	47
Soybean	114	110	118	74°	133	96°°	133	97	2.46	74	73	90	60

a – Percentage of provisional amino acid scoring pattern (FAO/UN, 1970; FAO/WHO, 1973)
b – Methionine + Cystine
c – Phenylalanine + Tyrosine
d – Grams of total essential amino acid per gram of N. The E/T ratio for some other protein are: casein, 3.25; egg, 3.22; beef muscle, 2.85; rice, 2.61; oats, 2.30; wheat gluten, 1.99.
e – Digestibility for some other protein are: wheat, 91; rice, 98; cow's milk, 97; beef and veal, 99.

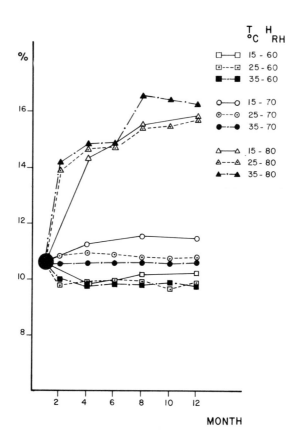

Fig. 1 - Changes in Moisture Content of Grains in Storage. Moisture content; roughly grinded, dried at 130° C for 3 hr.

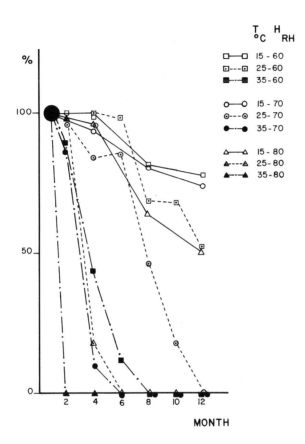

Fig. 2 - Changes in Germination Capacity in Storage. Germination capacity; 50 grains were kept in vessel air-tighted and conditioned RH with water at 30º C for 72 hr.

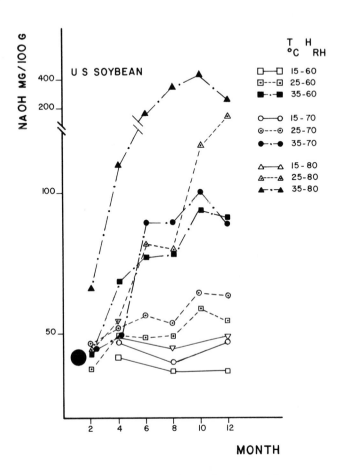

Fig. 3 - Changes in Acid Value in Storage. Acid Value; Soxlet-extracted with ether and titrated with 1/10N ethanol-KOH

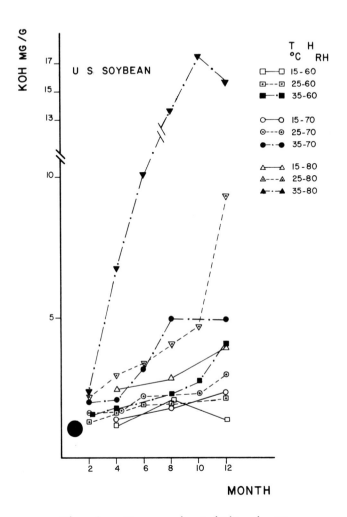

Fig. 4 - Changes in Acidity in Storage. Acidity; extracted with 80% of ethanol and titrated with 1/20N NaOH

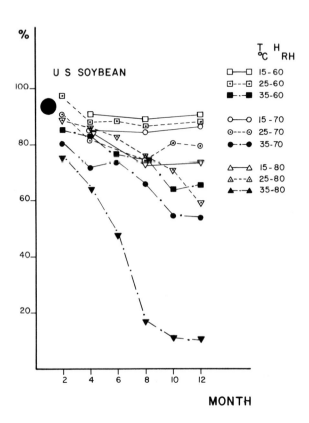

Fig. 5 - Changes in NSI.
NSI; A.O.C.S. method

3 - RESEARCH RESULTS ON SOYBEANS

Agedness of Soybeans in Storage

Soybean has been so important as a food protein resource in Japan that the Ministry of Agriculture, Forestry and Fisheries is now planning to reserve a certain amount of soybeans in its stock for the sake of provision against eventualities. In response to this governmental project, our Institute made the investigation of agedness of soybean quality during storage in various environments (Ohta et al., 1978; Saio et al., 1978). The experimental design is shown in Table 2.

Moisture Content of Whole Beans - As shown in Figure 1, equilibrium moisture contents at 60, 70 and 80% of Relative Humidity (RH) are 9.8-10.2, 10.6-11.5 and 15.7-16.3%, respectively, depending on the temperature. Relatively long period: 8 months passed to reach an equilibrium from whole beans containing 10.6% moisture content.

Germination Capacity - As shown in Figure 2, germination capacity is markedly affected by high temperature and/or high humidity. Soybeans stored at $35°C$ and RH 80% reached zero in germination capacity after 2 month-storage.

Acid Value and Acidity - As shown in Figure 3, acid value of extracted crude oil increased remarkably depending on temperature and/or humidity. The acid value stored at $35°C$, 80%, for 10 months counted 17, which meant that 20% of total lipid were hydrolyzed into fatty acids. Acidity titrated after extraction with 80% of ethanol, increased markedly too, as shown in Figure 4. Increases in these values suggest deterioration of oil in yield and quality.

Extractability from Tissues - As shown in Figure 5, Nitrogen Solubility Index (NSI) decreased remarkably. Extractability of nitrogen and solid into soybean milk decreased as well as NIS, as shown in Figures 6 and 7. Soybean milk was prepared by immersing whole beans in water overnight, grinding the swollen beans with water and filtration. Decrease in extractability from tissue seems to be caused from changing the pH of soybean milk to acidic range and progressing protein aggregation with protein-protein, protein-carbohydrate and protein-lipid interactions. Such reaction took place dominantly in 11S than in 7S component as shown in Figure 8 (soybean 7S and 11S components are main soybean storage protein; glycinin). The pH of soybean milk changed from 6.7 to less than 6.0 and darkening in color of milk was remarkable. Lower extractability is found to influence seriously on yield and quality of either conventional or new soybean foods which have extraction process in its making.

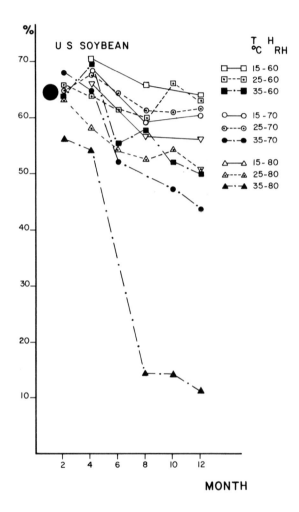

Fig. 6 - Changes in Solid Concentration in Soymilk. Solid; 10 ml of soymilk were dried at 130° C for 3 hr.

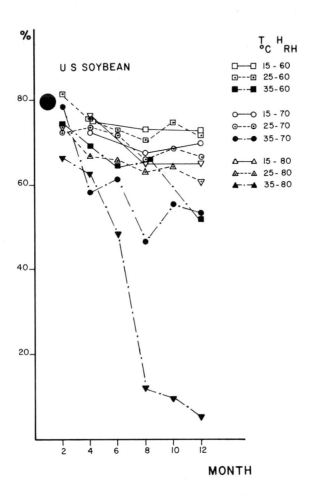

Fig. 7 - Changes in Nitrogen in Soymilk. Nitrogen; semi-micro Kjeldahl method.

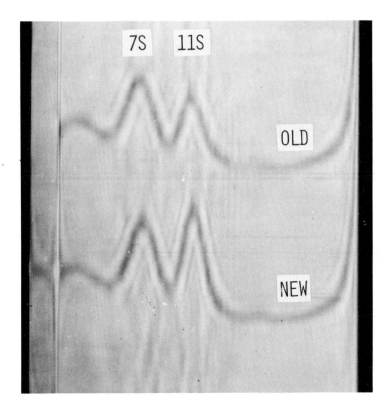

Fig. 8 - Ultracentrifugal Pattern of Soybean at the Start of Storage (NEW) and one Stored at 35º C, 75% of one month (OLD). Sample solution was dialyzed overnight in phosphate buffer, the picture was taken after 55 min in centrifugation at 51,200 r.p.m.

(UCA-1A)

TABLE 2

EXPERIMENTAL DESIGN OF SOYBEAN STORAGE

VARIETY	PERCENT	TEMPERATURE ($^\circ$C)	RELATIVE HUMIDITY (%)	ANALYSIS
U.S. beans (IOM)	moisture, 10.61; protein in dry base, 41.83; crude oil in dry base, 21.51	15	60	every 4 months
		25	70	
Chinese beans	moisture, 10.92; protein in dry base, 41.82; crude oil in dry base, 20.45	35	80	every 2 months

Note: Soybeans were stored in vessels air-tighted and conditioned their relative humidity by the use of H_2SO_4 or $(NH_4)_2SO_4$.

The beans thin-layed in vessels were used for an analysis but not for successive analysis after opening it.

Losses in Water Immersion - In preparation of many traditional soybean foods as well as the other legumes, whole beans are immersed in water so as to absorb water and swell out. Loss of solid in water immersion was notable, as shown in Figure 9. The solid consists of about 50% of carbohydrate, 20% of ash, 10% of nitrogen substances (as protein) and others. Figure 10 shows increase of sugar and Figure 11 increases in ash and nitrogen. The changes in loss between the soybean at the start of storage and one stored at 35°C, RH 80%, for 6 months, were 5 to 30% in total ash, 3 to 16% in calcium, 3 to 30% in magnesium, 1 to 25% in phosphorus, 20 to 30% in iron and 4 to 32% in potassium. The losses mean not only a lowering of effective use but also conspicuous shortage of nutrients such as sugar and minerals for the preparation of fermented foods.

Effects on Food Processing - Tofu, soybean curd, was prepared from soybean milk of stored beans. The hardness of prepared gel decreased markedly and reached at non-coagulative state with the one which was stored at 35°C, RH 80% and for more than 10 months, as shown in Figure 12.

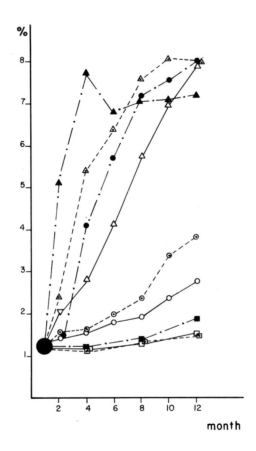

Fig. 9 – Changes in Solid Loss in Water immersion. Solid; 25 ml of extract were dried at 130° C for 3 hr.

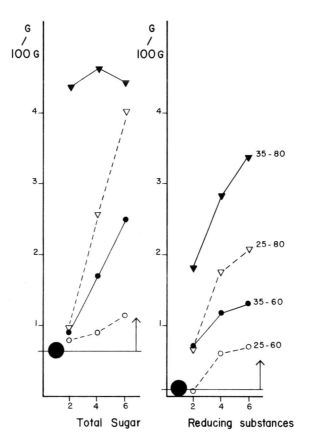

Fig. 10 - Changes in Sugar Loss in water immersed. Total sugar; Phenol-H_2SO_4 method. Reducing substances; Somogi--Nelson method.

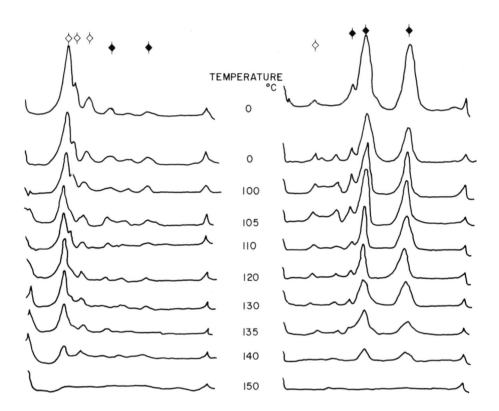

Fig. 11 - Changes in Total Ash and Nitrogen Losses in Water immersed. Total ash; dried and made to ash at 550°C overnight. Nitrogen; Semi-micro Kjeldahl method.

STORAGE AND PROCESSING OF LEGUMES

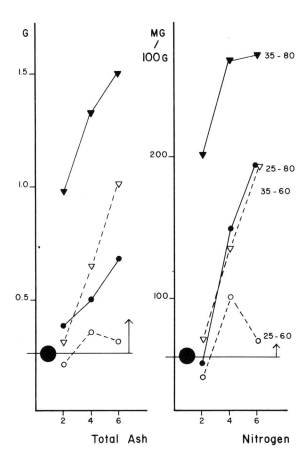

Fig. 12 - Changes in Hardness of Tofu-like Gels. Tofu was prepared by heating soybean milk at 100°C for 10 min, adding 0.5% glucono-lactone and incubating at 70°C for hr.

The hardness of cooked beans by autoclaving at 121°C for 30 minutes, increased as shown in Figure 13.

Such deterioration depends seriously on temperature and moisture content in storage but moisture content seems to affect even more seriously than temperature does. Consequently, lowering moisture content of grains in storage is important.

Figures indicate the results of U.S. beans (IOM) but beans produced in China also show almost the same trends.

Protein Crosslinking Reactions in Storage and Processing

When protein are severely heated or kept too long in improper storage, a fall in nutritional value appears. Recent investigations in this area made it clear the crosslink formation in food protein not only lower the nutritional quality and digestibility of food protein but sometimes introduce toxicity. Such crosslinking reactions are classified as follows:

(1) Maillard reaction can occur between Σ-amino groups of protein and sugar aldehyde groups. With increasing severity of damage, Maillard reaction progresses passing the deoxyketosyl stage and forming brown pigments or melanoidin (Figure 14).

(2) Formation of new isopeptide bonds by reaction of Σ-amino group of lysine with either carboxyl groups of aspartic or glutamic acids (Figure 15).

(3) Base or heat-catalyzed elimination reactions of serine and cystine in proteins, yielding dehydroalanine which condense with cystine into lanthionine, or with Σ-amino group of lysine into lysinoalanine, etc. (Figure 16).

(4) Racemization of amino acids – Amino acids were chemically modified and rendered unavailable (this reaction is not crosslinking but one of the reactions forming non-metabolized substances).

(5) Interaction of peroxidized lipids with proteins may lead to undesirable changes in nutritional quality. It is elucidated to involve reactions of protein with either free radicals, hydroperoxides or breakdown products of hydroperoxides such as malonaldehyde.

The investigation on such crosslink reactions were carried out with food proteins derived from animal sources such as egg, milk and muscles which effects are more serious than vegetable sources. Only a few were done with soybean products (Degroot et al., 1969; Saio et al., 1975; Sternberg et al., 1976; Finley et al., 1977).

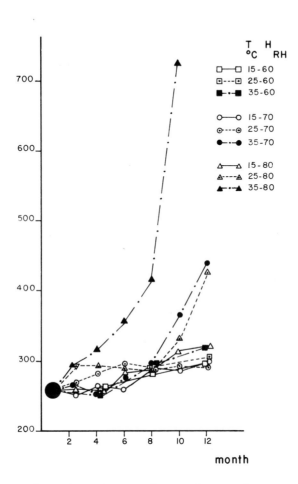

Fig. 13 - Changes in Hardness of Cooked Beans. Beans were autoclaved at 121°C for 30 min.

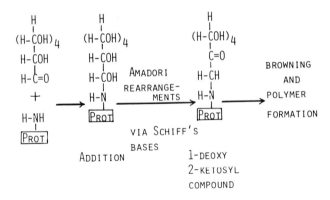

Fig. 14 - Simplified schem of the Maillard reactions
(Hurell et al., in Prot. Crosslinking)

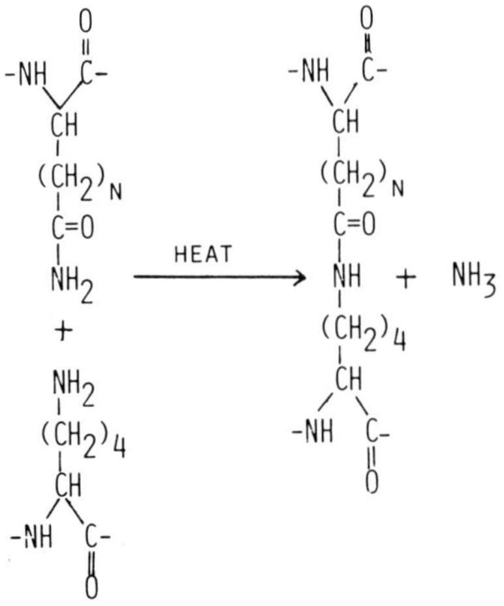

Fig. 15 - Formation of isopeptide bond,
asparanine (N=1), Glutamine (N=2)
(Hurrell et al. in Protein Crosslinking)

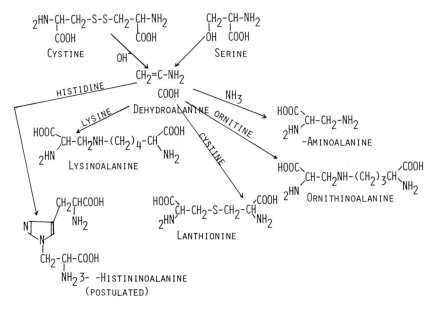

Fig. 16 - New Amino Acids formed through the Reaction of a Dehydroalanine (Whitaker et al. and Finley et al. in Protein Crosslinking)

I would like to introduce here our results on the formation of lysinoalanine (LAL) on alkaline treatment of soybean protein. Basic amino acids and cystine (as cysteic acid) were analyzed with an amino acid analyzer after different treatments. The results show:

(1) The higher amount of LAL was recognized as the time treated protein solution in 0.2N-NaOH at 40°C progressed. Lysine and cystine decreased and arginine and hystidine also decreased slightly (Fig. 17).

(2) The higher amount of LAL was recognized as the pH of protein solution elevated over 11. When treated at 40°C, pH 13 and for 24 hrs., 9% in 7S and 30% in Cold Insoluble Fraction (CIF: fraction having 80-90% of 11S) per total lysine in each protein fraction, changed into LAL (Figure 18).

(3) The higher amount of LAL was recognized as the treating temperature elevated. Lysine and arginine decreased markedly and ornitine (or other amino acids which were eluted near lysine) seemed to be formed at higher temperature-heating (Figure 19).

(4) The amount of LAL formed was connected with cystine content in proteins. In CIF or 11S fractions having higher content of cystine than 7S, the higher amount of LAL was formed even in weaker treatments.

(5) Lysinoalanine was slightly recognized in commercial soybean products when a large amount of sample was applied on a column (Figure 20).

The other results of ours on high-temperature denaturation of soybean protein revealed that the gross-structure of protein subunits degradated to form lower molecular weight substances by heating at over 150°C, resulting in decrease in Amido Black 10B bound to protein and deamidation. Figure 21 shows the SDS-polyacry amide gel electrophoresis patterns of soybean protein heated at different temperature over 100°C (Saio et al., 1976). Most of textured soybean protein products or retort pouched products do not have clear electrophoretical patterns.

4 - CONCLUSION

Legumes are the second largest source of calories and protein after cereals for a great majority of people, especially in developing countries. In legume processing technologies, close attention should be given to the dietary habits of each country or region with different cultures, when new technologies are being introduced and developed. From the standpoint of nutritional quality of legumes, the elimination of anti-nutritional and toxic factors is the primary

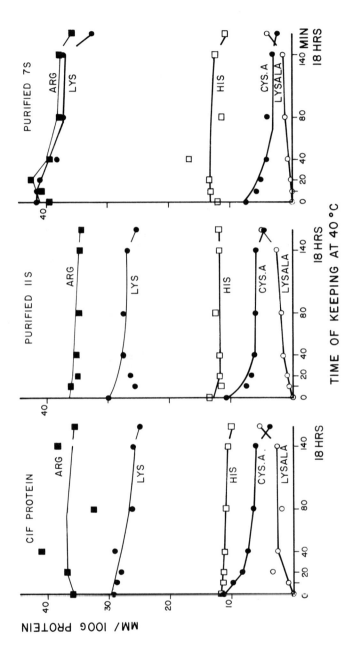

Fig. 17 - Changes of Lysinoalanine, Cysteic Acid and the Other Basic Amino Acids.

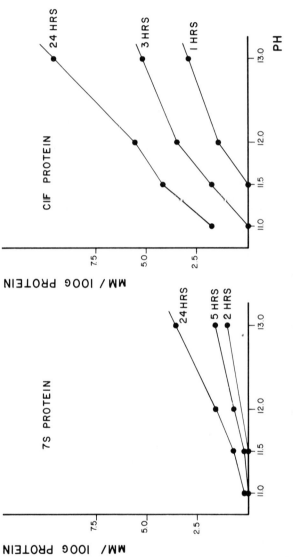

Fig. 18 - Lysinoalanine formed on Alkalin Treatment at 40°C

STORAGE AND PROCESSING OF LEGUMES

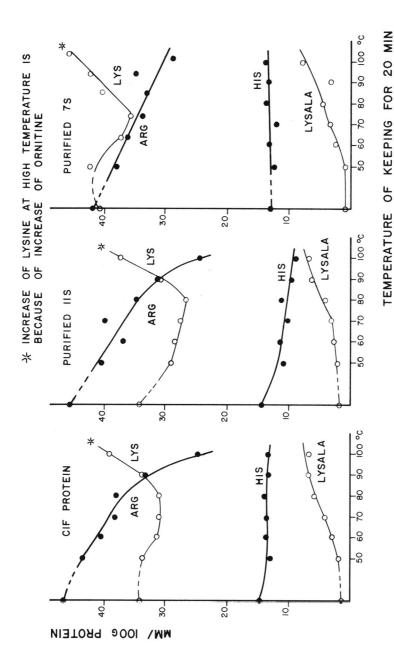

Fig. 19 - Changes of Lysionoalanine and the other Amino Acid on Alkaline Treatments.

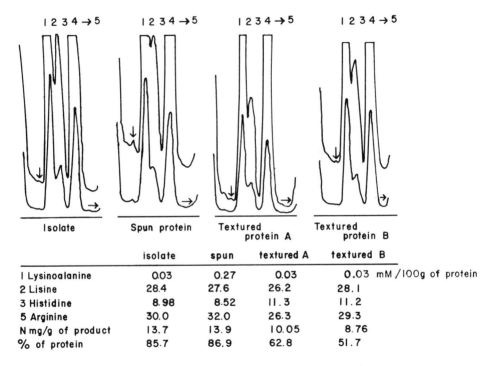

Fig. 20 – Examination of Lysinoalanine in soybean food products

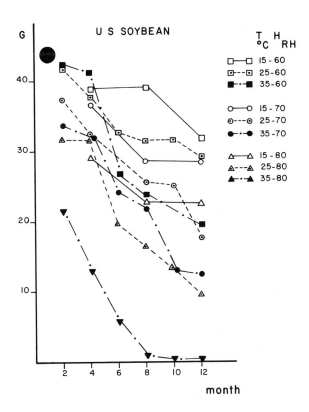

Fig. 21 - SDS-Polyacryl Amide Gel Electrophoretical Patterns of Crude 7S and CIF heated at High Temperatures.

concern and secondly the prevention of nutritional losses in processing and storage. Increased interest should be taken in the lowering of protein quality and digestibility by the interactions of protein with protein, carbohydrate and lipid. In my opinion, excessive severe treatment in processing and improper storage should be avoided. Since such reactions as formation of LAL were so often recognized in familiar foods or ingredients (Sterberg et al., 1976), it remains to be further investigated and discussed in the future whether the current interest in such reactions in almost too sensitive a manner has scientifically good reasons or not.

BIBLIOGRAPHY

Introduction and Outline of Legumes

Adames, J.M., 1977, A review of the literature concerning losses in stored cereals and pulses, published since 1964. Trop. Sci. 19, 1.
Bressani, R., Elias, L.G. and Valiente, A.T., 1963, Effect of cooking and of amino acid supplementation on the nutritive value of black beans. Brit. J. Nutr., 17, 69.
Bressani, R. and Elias, L.G., 1974, "Legume Foods. New Protein Foods", 1A, 230, Altscul, A.M. (ed.).
Burr, H.K., Kon, S. and Morris, H.J., 1968, Cooking rates of dry beans as influenced by moisture content and temperature and time of storage. Food Tech., 22, 336.
Burr, H.K., 1975, "Pulse proteins. Protein Nutritional Quality of Foods and Feeds", Friedman, M. (ed.), 1, 119.
FAO. Report on the FAO Expert Consultation on Grain Legume Processing, held in Mysore, India, 14-18 Nov. 1977.
Kakade, M.L. and Evans, R.J., 1965. Nutritive value of navy beans. Brit. J. Nutr., 19, 269.

Agedness of Soybeans in Storage

Ohta, T. and Takano, K., 1978, Deterioration of soybean quality in storage, concerning with the preparation of fermented foods as Natto. Presented at the Meeting of Nippon Shokuhin Kogyo Gakkai.
Saio, K., Nakagawa, I. and Hashizume, K., 1978, Deterioration of soybean quality in storage, concerning with the preparation of non-fermented foods as Tofu and Aburage. Presented at the Meeting of Nippon Shokuhin Kogyo Gakkai.
Saio, K. and Arisaka, M., 1978, Deterioration of soybean quality during storage at high temperature and humidity. Nippon Shokuhin Kogyo Gakkaishi, in press.

Protein Crosslinking Reactions
in Storage and Processing

Arnaud, M.J., Bracco, I., Magnenat, E., and Finot, P.A., 1977, Distribution of lysinoalanine by whole body autoradiography and cellular localization in the kidney. Acta Pharm. Toxic. 41, 138.

De Groot, A.P. and Slump, P., 1969, Effects of severe alkali treatment of proteins on amino acid composition and nutritive value. J. Nutr., 98, 45.

Hayase, F., Kato, H. and Fujimaki, M., 1973, Racemization of amino acid residues in proteins during roasting, Agr. Biol. Chem. 37, 191.

Karel, M., 1973, Protein interactions in biosystems, Protein-lipid interactions, J. Food. Sci. 38, 756.

Saio, K. and Murase, M., 1975, Formation of lysinoalanine on alkaline treatment of soybean protein. Nippon Shokuhin Kogyo Gakkaishi 22, 30.

Saio, K., Terashima, M. and Watanabe, T., 1975, Food use of soybean 7S and 11S proteins, Heat denaturation of soybean proteins at high temperature. J. Food. Sci., 40, 537.

Saio, K., Terashima, M. and Watanabe, T., 1975, Changes in basic groups of soybean protein by high temperature heating, ibid, 40, 537.

Sternberg, M., Kim, C.Y., and Schwende, 1975, Lisinoalanine: Presence in foods and food ingredients. Science, 190, 992.

Struthers, B.J., Dahlgren, R.R., and Hopkins, D.T., 1976, Biological effects of feeding graded levels of alkali treated soybean protein containing lisinoalanine in Sprague-Dawley and Wistar rats. J. Nutr., 106, 1190.

Van Beek, L., Feron, V.J., and De Groot, A.P., 1974, Nutritional effects of alkali-treated soyprotein in rats. J. Nutr., 104, 1630.

Woodard, J.C. and Short, D.D., 1973, Toxicity of alkali-treated soyprotein in rats. J. Nutr., 103, 569.

Protein Crosslinking, Nutritional and Medical Consequences, 1977, Friedman, M. (ed.), Part B, 1-729.

ROLE OF POST-HARVEST CONSERVATION OF FOODS

IN ACHIEVING NUTRITIONAL GOALS

Ramesh V. Bhat

National Institute of Nutrition

Osmania (P.O.), Hyderabad - 500 007, A.P., INDIA

During the last decade, to increase food availability, the main emphasis was on improving yields of cereals and millets through breeding high yielding varieties. It is now recognized that this alone may not be enough and that it is essential to conserve what is produced. In the field of post-harvest conservation of foods the major thirst during the last decade was on quantitative food losses and assessing the means by which these foods were lost. As a result of these data, it should now be possible to initiate some concrete action programmes to minimize the qualitative and quantitative losses that occur at various stages of food handling such as the farm, during storage, trade and household levels.

There have been several important developments both at the international and at the national levels during the last five years. The seventh UN General Assembly has set a target of 50% reduction in food losses to be achieved within a decade. If successful, this would mean a saving of 40 million tonnes of cereals and 5 million tonnes of pulses. The FAO in 1977 established a $ 20 million fund to combat post-harvest food losses. In addition to these, a Group for Assistance on System relating to Grain after Harvest (GASGA) has been formed. This group includes representatives of the Food and Agricultural Organization, Rome; International Development Research Centre, Ottawa; International Institute of Tropical Agriculture, Nigeria; Food and Feed Grain Research Institute, Kansas, and Tropical Products Institute, London. The major program of this group include the correct assessment of post-harvest losses, adoption of methodologies to reduce them, the creation of awareness among people and training of manpower. The UN University, Tokyo, has identified post-harvest technology as a priority item in its international programs. All these efforts should lead to a proper blend of applied research, training methodology, extension set up and appropriate intervention programmes.

It is perhaps appropriate here to mention the developments that have taken place in India in recent years which relate to post-harvest conservation of foods. An All India Coordinated Project on Post-Harvest Technology was set up in 1972 by the Indian Council of Agricultural Research (ICAR) to develop technology for reducing post-harvest losses. The scheme was initiated mainly at agricultural universities to develop appropriate technologies for handling, threshing, storage, milling and conserving food, as also better utilization of by-products.

A joint committee of the ICAR and the Indian Council of Medical Research (ICMR) has been set up to coordinate research efforts to deal with problems of food contamination leading to disease outbreaks in man. The Central Food Technological Research Institute, Mysore, has been continuing its research on applied aspects on minimizing food losses specially through measures to protect grains against insects damage and through appropriate milling technologies for paddy, coarse grains and pulses. These are in addition to training personnel at the international level.

The Save Grain Campaign has been launched by the Department of Food, Ministry of Agriculture and in this program, practical training in scientific methods of storage and pest control is given to farmers, traders and personnel of cooperatives. Demonstration and publicity in selected areas to popularize correct methods of storage and pest control also form part of the activities.

A recent aspect of considerable importance has been a study to determine possible relationships between improper post-harvest technology on the one hand and disease outbreaks in man on the other. Several disease outbreaks in man have been attributed by the National Institute of Nutrition, Hyderabad, to this factor. Important among these are mycotoxins, weed seed contamination and pesticide residues. An outbreak of human aflatoxicosis resulting in the death of over 100 tribals in parts of Western India in 1974-75 was found to be due to the consumption of aflatoxin contaminated corn stored improperly. Ergot contamination of pearl millet specially of the high yielding varieties is assuming serious proportions. Consumption of such grain by man leads to nausea vomiting, giddiness and prolonged somnolence. Venocclusive disease — a liver disorder — was the result of eating stapel millet Panicum miliare contaminated with weed seeds of Crotalaria spp. In each of these instances appropriate and simple post-harvest technology like proper shelling and storage in case of corn, removal of contaminated ergoty pearl millet by flotation technique and suitable seiving for removal of weed seeds would have prevented the outbreak of the diseases.

Pesticides are often mixed directly with grain to ward off insects. Consumption of such contaminated grains without proper washing has been known to lead to paralysis as observed recently in parts of Central India.

Several significant contributions have been made during the last five years in different centers under the post-harvest technology scheme. The Paddy Processing Research Centre, Tiruvarur in India, has observed that spraying of 10% solution of sodium chloride on standing crop of ripening paddy can reduce the moisture content in the grain from around 25% to about 16% within 48 hrs. Pest incidence in such treated paddy was less even after an year's storage.

Harvesting at the proper time also ensures better returns. The Central Rice Research Institute, Cuttack, has shown that head rice and milling recovery would be greater if early and medium duration varieties are harvested between days of 27 - 33 after 50% flowering, and if late maturing varieties are harvested between days 33 and 39 after 50% flowering.

Sun drying of agricultural produce is a common practice in tropical countries. According to the Rice Processing Engineering Centre, Kharagpur, head rice recovery by this method was about 80% while in mechanical drying it increased to around 90%. The outturn of parboiled rice in conventional mill was only 65% as compared with 68% in modernized mills. Parboiling of rice which has been traditionally practised in India has several advantages including increased nutritive value. Recent studies have indicated that the period of parboiling influences the extent of loss of various mineral nutrients during milling. Also the bran of parboiled rice has more oil (20-25%) than that of raw rice (15-20%). In addition, deterioration of oil during storage was more in raw rice bran oil because of its higher free fatty acid content.

Regarding storage structure, it has been shown that polythene lined bamboo bins are more efficient than indigenous bins. The extent of insect infestation of IR 20 paddy variety in the improved polythene bin was 2.5% compared to 4.5% seen in unlined ones. Better storage can also be achieved by storing paddy in insecticide impregnated gunny bags. Pest incidence was only 0.5% in Phoxton — malathion treated gunny bags as against 2.0% in control bags stored over a nine month period.

An undesirable outcome of the green revolution, from the nutrition stand point, has been the increase in cereal production at the expense of legume production. The decrease in production of pulses is due to more land being taken up by cereals/millets because of factors such as unremunerative prices. Lack of high yielding varieties, high susceptibility to insect attack of legumes and need for increased irrigational facilities. It is not easy to reverse this trend. However, by improving the milling technology, better storage practices and minimizing insect attack, greater amount of pulses could be conserved and made available to meet protein requirements of mankind. Also by using detoxification methods, toxic legumes such as Lathyrus could be better utilized.

Regarding milling technology, it has been shown by Central Food Technological Research Institute, Mysore, that by improved methods of milling yields can be increased by 10 to 20% in cases of chick pea, green pea, pigeon pea and black gram. Coarse millets such as sorghum, pearl millet and ragi (Eleusine coracana) are staples in India for the poorer segments of the population. The processing of these grains like pearling and removal of the outer husk improves their appearance and consumer acceptability.

It is essential at this stage to attempt at a cost benefit analysis ratio. A study recently concluded by the Administrative Staff College of India at Hyderabad has indicated that the actual loss of food grains in India per year is 5.25 million tons. They have also calculated the cost of grain loss per ton of grain in conventional and modern storage structures and found taht this loss can be easily reduced by 50% by using modern storage structures. Storage requirements fot the next ten years have also been calculated. The capacity to store 8.1 million ton of food grans would have to be developed by 1985-86. This would mean an additional investment of rs. 284 crores.

However, the cost should not be counted only by economic parameters in terms of profit calculation. The accompanying social benefits to the community like the generation of employment and improvement in community health that arise as a result of all these factors would not be insignificant. Economists have some simple formulae to calculate the cost benefit ratio. The output value can be calculated by these formulae.

$$\text{Net present value} = \sum_{t=1}^{T} \frac{B_t - C_t}{(1-i)} \quad 0 \quad \ldots\ldots\ldots (1)$$

Where B_t = Benefits at time t
C_t = Cost at time t
i = Interest rate
t = Time

In the next five years such a calculation should be made for the investments that have already been done and this should serve as an indicator for any future mid-course corrections.

ROLE OF POST — HARVEST CONSERVATION OF FOODS

NEED FOR THE HOUR

The need of the hour appears to be the conservation of food at several stages. Efforts for post harvest conservation should start right at the pre-harvest stage by proper selection and breeding of the most suitable varieties and adjusting the time of sowing so that flowering and harvest time coincide with minimal rain fall. Proper natural drying facilities like sunshine, mechanical drying and solar driers would help recovery of more grains. Modern methods of milling and pulses and cereals would minimize loss of food grains as also their nutrient content.

Better storage methods by improving the existing storage structures by appropriate rural technology using locally available cheap resources, minimizing losses by rodents, mocrobes and insects would save considerable amount of food grains. Utilization of by-products through agro-industries not only generates more employment thereby increasing purchasing power but also helps in better conservation. Recovery of rice bran oil is a good example.

Appropriate methods of extraction of oil seeds such as solvent extraction would yield at least 10% more of oil. Although this poses the problem of solvent residues if food grade solvents are not used, it would certainly increase the availability of this valuable commodity. Another sector which needs top priority is the conservation of perishable commodities such as milk, vegetable and fish.

For obtaining best results in the field of post-harvest technology, a system approach that involves coordination between the various disciplines is essential. At the moment, the field of post-harvest conservation appears to be a no man's land, left mostly to the farmer, trader and the common man. Interaction with agronomists, plant breeders, mycologists, entomologists, cereal-chemists, agricultural engineer, food technologists, oil technologists, economists, clinicians, pathologists and epidemiologists would possibly strengthen the efforts.

Finally, we have fanciful slogans — green revolution, blue revolution, operation flood, save grain campaign etc. It would be appropriate now to also have a campaign like the grow more food campaign which we had in India years ago to attain self-sufficiency in food. This can be appropriate named as <u>Conserve More Food Campaign</u>, and I hope this symposium will give <u>a lead in this direction</u>. Man has succeded in producing more foods and he should also succeed in conserving it. If not, this would be the triumph and tragedy of mankind.

SIGNIFICANCE OF NUTRITIONAL LOSSES

IN THE PROCESSING OF GRAINS

> David A.V. Dendy
>
> Tropical Products Institute
>
> London - ENGLAND

May I begin by reminding you of the Principle of Le Chatelier: "When a constraint is put upon a system, then the system will adjust itself to remove that constraint". Two connected examples come to mind in the field of nutrition. When a human body is deprived of its normal intake of nutrient, immediate morbidity rarely occurs: indeed, it is found that the metabolism adjusts itself so as to require rather less of that nutrient. When the UN agencies recommended certain minimum levels for protein and calorie requirements, and bodies such as the PAG were formed to advise on implementation of decisions made, under the constraint of greater scientific knowledge, these minima were gradually lowered[1,2].

Of all people, scientists should avoid sweeping conclusions based upon inadequate knowledge, and I hope to show in the following brief survey, that a pragmatic approach to individual situations is a better use of scarce resources: that these resources are best channelled into the provision of more food by stimulating income among the poor and by reduction of post-harvest losses.

Cereals are man's most important food: a principal source of calories and, for many, of protein. It is therefore surprising that only recently has the systematic study of all the losses occurring in cereal processing and storage been undertaken. In the less-developed countries, cereals are usually grown by subsistence farmers and processed in the farm or in small mills nearby. Half the world's people obtain more than half their calories and protein from cereals. With over 400 million tons of cereals being consumed in less developed countries and gross losses of perhaps 10-20% occurring, even a 10% reduction in these losses would amount to a saving of several million tons of human food.

Estimates made of post-harvest losses have frequently even less factual basis than the estimates of production. However, there has been a growing international awareness of the need for reducing post-harvest losses, e.g., in September 1975, a resolution of the Seventh Special Session of the United Nations General Assembly stated that "the further reduction of post-harvest food losses in developing countries should be undertaken as a matter of priority, with a view to reaching at least a 50% reduction by 1985". From what loss levels this is to occur is not stated. As Kenton Harris has stated in the preface to the draft manual <u>Post-harvest grain loss assessment methods</u>: "'Guestimates' by knowledgeable people have had a useful role in the past; will continue to be used in the future; and are especially useful when timely opinions are needed as to where the more serious losses occur. These guestimates have served a useful purpose. But, with increasingly limited resources requiring benefit-related priorities, there is a need to know more accurately what the post-harvest losses are".

Post-harvest losses of grains are those occurring between completion of harvesting and human absorption. The processes considered are primary; threshing, transportation, storage, milling, grinding, distribution; and secondary: baking, cooking. The influence of these processes on digestion and absorption is also important. Agro-industries such as corn processing and industrial milling and baking, which are of great importance in the metropolitan countries, are of little significance in subsistence economies where most of the world's malnourished live.

In discussing losses one can distinguish between gross loss, nutritional loss, and economic loss. The first two are losses of nutrition, the last need not be. A distinction must always be made between a national loss and a personal loss. A loss may be said to occur when food material is removed from the processing and storage system by spillage, consumption by non-humans (rodents, insects), or conversion to inedible or indigestible material[3]. In the strict definition followed by FAO and others, theft is not considered a loss, but a transfer of ownership! Similarly one might say that shattering in the field or spillage from carts is only a loss if gleaners do not collect the spilled grains. It will become apparent as one discusses loss occurrence that each case must be considered in its own environment and on its own merits. One must take the pragmatic view of measuring losses and reducing them for each situation and not endeavour to provide a rigid framework to fit all cases.

Losses occur whether the grain system is handled by one person — such as the peasant farmer — or each stage separately — as often occurs with small-scale commercial operations in ldc's. Particularly in the case of rice, which is eaten as whole grain, an interaction occurs between the processing stages; thus it is well-known that breakage of the kernel during husking and polishing is very often due

NUTRITIONAL LOSSES IN PROCESSING GRAINS

to maloperation of the drying procedures used for the paddy.

The total disappearance of grain from the system is a gross loss and this occurs in several ways, for example:

- Shattering in the field at harvest. Gleaners may gather the spillage, which can be 5% of the harvest[4].

- Threshing loss by grain being left on the stalk (12% for paddy threshed by beating as measured by TPI[4]). The stalk may be burned, of fed to cattle, when the loss is not total, as the food value reappears in part as milk or meat.

- Consumption of grain by rodents and insects during storage. Rodents can foul the grain to such an extent that the unconsumed grain is no longer fit for human consumption. Even the lower economic value of this grain as animal feed may not be obtained. Storage losses have been better studied than other losses, and the recent bibliographies of post-harvest losses issued by TPI reflect this, thus the bibliography on storage had 265 references whereas that on all other aspects of post-harvest grain losses had only 206[3].

- Losses may be caused by improperly controlled drying. This is particularly true of paddy, where considerable cracking of the kernel may occur in sun-drying or in artificial dryers. The subject has been studied frequently and correct drying conditions to minimize loss are known, but, at the level of the peasant farmer, difficult to implement. Needless to say, inadequately dried grain will deteriorate in storage. Removal of grain from drying floors by birds is also a significant loss.

Losses occurring during rice milling may be gross, nutritional, and invariably economic. In the case of paddy, misuse of machinery may lead, as does improper drying, to high breakage. Let is be remembered, however, that broken rice is not necessarily a bad thing. The commercial miller may obtain a lower price, but the poor man can afford to pay. The peasant, whose wife hand-pounds his paddy, will eat his rice whether it is broken or not. The parboiling process leads to slight losses of material[5] but improved vitamin retention during milling and cooking. However, parboiled rice is not acceptable to some communities.

Losses of food substance which occur during grinding are likely to be caused by spillage or by contamination of a bran fraction with edible flour. The peasant farmer will guard against these losses more assiduously than contract or commercial millers.

The final stage in which gross loss occurs is during washing and cooking: in Korea researchers[6] studying rice measured losses of solids of up to 2.6% during washing. In cooking, if rice is steamed or cooked in water till dry, gross losses will not occur but if cooking water is discarded then there will be a gross loss which TPI have noted may be as high as 10% depending on the age of the rice — and much higher for broken grains.

Nutritional losses have two principal causes and these must be clearly distinguished. The first cause is adventitious and avoidable, and may be typified by the loss of vitamins and protein in weevil-infested grain — weevils are selective feeders and know what is good for them — and in the conversion of food grain to insect detritus of no food value. Adam & Francis at TPI[7] measured a loss of 38% in food values in wheat infested with sitophillys oryzae and stored for only 14 weeks at 27°C. Liscombe found that in heavily infested wheat not only was there a 1.1% loss in weight after 11 weeks storage but a yield loss of over 5% of flour at 70% extraction.

The other very important loss is frequently unavoidable: that caused by food habit. Most of us prefer to eat white bread and polished rice, even when we know that wholewheat bread and brown rice contain valuable B vitamins. We prefer crusty bread, though the lysine content in lowered by Meillard browning, which, with the caramelization of starch to colloidal carbon gives bread crust its delightful colour and delicious taste. We eat toast, though Dr. Cho Tsen has told us that the nutritive value of bread is significantly reduced by toasting[8].

Included in food habit losses are those which are inherent in the food material. Such losses may be measured as the difference between the analyzed content of a particular nutrient and that actually digested. Furthermore, if a cereal or grain legume contains anti-nutritional factors these may not only suppress the digestion of a particular portion of the grain itself, but once in the alimentary tract, similarly interfere with the absorption of other ingested foods.

The most obvious example that comes to mind is the effect of phytic acid on the absorption of minerals. Whilst some would have us eat wholewheat bread in order to gain both the fibre and the valuable vitamins present in the bran, they omit to say that high extraction flours contain phytic acid, 90% of which is in the aleurone layer. Phytic acid immobilizes the calcium, iron and zinc, not only in the bread, but also in the rest of the diet. In Metropolitan countries this rarely presents problems, as the diet is mixed and flour is often enriched by addition of calcium carbonate and also, let it be said, thiamine, niacin and riboflavin. However, where a peasant population relies on wheat as its principal food and grinds and bakes in the village, the full force of the phytic acid acts on the diet, with rickets and zinc dwarfism as two possible sequelae. With no prospect of calcium addition at the mill, and an unfortunate replacement of milk by effervescent pleasure drinks in villages near to main communications, as has been noted by the Institute of Development Studies[9], these unfortunate conditions are likely to increase. One might also mention the problem of rickets among the children of Asian immigrants in the U.K., which, according to Dr. Reinhold[10] is due to eating wholemeal chapatties.

It is not intended here to discuss methods of loss measurement; the draft produced by Dr. Kenton Harris for the AACC under contract from USAID, is an important attempt to set down a systematic methodology. Suffice it to say that whereas gross losses are easily measured, losses of specific nutrients such as vitamins, amino acids and minerals are more difficult, though of more interest to the academic scientist.

Having measured the loss one must calculate the cost of loss reduction. One must be critical of the UN desire to reduce losses by 50% in ten years. With some losses such as those caused by inadequate drying, poor storage structures, or antiquated mills, it may be relatively easy to calculate farm by farm, nation by nation, the quid pro quo of loss reduction and decide whether losses are significant enough for action, and make the necessary investment decisions regarding dryers, stores, fumigants, mills, training and so forth. With nutritional losses the problems are very complex. In the first place because we are dealing with many vitamins, amino acids, and minerals, and the significance of the losses ensuing varies greatly. Secondly, a deficit in one part of a mixed diet is frequently balanced by a surplus in another. To the well-fed of this world nutrition losses are of even less importance than gross losses which may, at least, be of economic significance. Those who do not eat a well-balanced diet tend to be the poor and the vulnerable: especially in less-developed countries.

Let us now consider specific examples:

In Iran the diet of remote villages is based on wholemeal bread, milled and baked in the village. Should one merely tell the villagers that rickets and zinc absorption are caused by their diet and leave them to make their own changes? No. Clearly one must train nutritionist extension workers who will gradually help to improve the diet. In this particular case it will not be practiceable in the short term or necessarily desirable, to replace village milling by commercial milling (with addition of calcium). There is, in any case, no transport system for the wheat and flour. Other methods of increasing calcium intake must be considered for both acceptability to villagers and cost. Obvious actions such as increasing milk and cheese production may be possible if dairy herds can be improved and if sufficient grazing is available. In the case of Asian immigrants in Britain the remedy is readily to hand via the many arms of the Ministry of Health and Social Security.

TABLE 1

LOSSES IN WASHING AND COOKING OF RICE "AKIBARE"

	50% POLISHED	90% POLISHED	
	Washing	Washing	Cooking
Gross (solids)	2.11	2.20	0
N-free extract (carbohydrate)	1.1	1.2	-
Protein	5.4	5.6	-
Lysine	-	6.2	2.2
Tryptophan	-	-	1.9
Calcium	-	20	-
Niacin	42	42	22
Thiamine	21	40	19
Riboflavin	15	24	14

Source: Cheigh, Ruy, Jo, and Kwon, "Nutritional Losses during Washing and Cooking of Rice", Korean Inst. of Sci. and Technology, March 1978.

The examples of losses in rice washing already discussed also demonstrated that around 40% of vitamins are removed (Table 1).

For niacin this is particularly important as even polished rice can supply most of one's daily needs if no losses occurred.

A simple remedy is to discourage housewives from washing and encourage millers to provide cleaner rice. Fortification of rice cannot help subsistence farmers, but the efforts of nutritionists working in the villages can improve the dietaries.

While on the subject of B vitamins in rice, let us emphasize that, though parboiling may be desirable from a nutritionist's point of view, consumers in Far Eastern countries will not eat it. Similarly unpolished brown rice is rarely acceptable to rice eating communities. Where the sequelae of vitamin B defficiencies occur, simples and less costly remedies must be sought rather than a probably futile effort greatly to change peasant diets. As Professor Aylward has remarked "Food is not nutritious until eaten" — a remark which should also be borne in mind by those who would have the poor eat protein concentrates from fish, leaves, fungi etc.

The place for investment in rice loss reduction is, surely, at drying and milling, where at least a 5% increase in mill out-turn is frequently possible by modest investment in improved machinery and in training. This 5% could represent some 20 million tons of food worldwide. By improving the village "toll" or "custom" mill the food intake of the peasant farmers can be improved. The centrifugal huskers being introduced as an alternative to the steel huller will provide a simple and relatively cheap method of lowering gross rice losses in villages.

We may say that nutritional loss is also significant on those rare occasions when a community's diet contains an inhibiting factor such as phytic acid. Nutritional loss is significant when the population concerned receives only the minimum gross calory intake. It is then that a disproportionate loss of a vitamin or an amino acid becomes significant. Generally speaking, such losses though very common, are a symptom of poverty. At the 1976 ICC meeting in Vienna an important paper was presented which demonstrated that there was not really a protein problem, but a food problem. Given an income, a family will buy food. Given a demand, a farmer may grow a surplus for sale. Protein deficiency has frequently been confused with starvation and it is now a matter of nutrition history that much of the high protein concentrates such as soya and milk powder which were put into areas of starvation were metabolized as calories — cereals would have been as useful.

If resources are to be used in loss reduction programmes aimed at improving nutrition, then these resources would be best used to increase the actual quantity of food available. We must not be misled by the scientific interest of loss studies: frequently it is the unimportant losses, especially of nutritional value, which make the most interesting academic study: it is the gross losses, easy to assess and of no great scientific interest which are in need of reduction if the world's poor are to be fed.

REFERENCES

[1] FAO, 1950, Nutritional Study No. 5, Washington.
[2] FAO, 1954, Nutritional Study No. 15, Rome.
   FAO, 1957, Nutritional Study No. 16, Rome.
   FAO, 1965, Nutritional Study No. 37, Rome
   FAO, 1973, Nutritional Study No. 52, Rome.
[3] TPI, 1977, Report G110, "Bibliography on post-harvest losses in cereals and pulses..." (Storage), London.
[4] TPI, 1976, Report to Government of Malaysia, "Study of technical and economic problems of off-season paddy in the Muda area", London.

5. C.S. Shivanna, 1976, "Leaching losses during commercial parboiling of paddy...", J. Fd. Sci. Tech., 94.
6. Cheigh, Ruy, Jo, and Kwon, "Nutritional Losses during Washing and Cooking of Rice", Korean Inst. of Sci. and Technology, March 1978.
7. B.J. Francis & J.M. Adams, (TPI), "A note on the loss of dry matter and nutritive value in experimentally infested wheat" (in press).
8. Cheich, Ryo, Jo, & Kwon, 1978, "Nutritional Losses during Washing and Cooking of Rice, Korean Inst. Sci. Tech.
9. Sue Schofield, 1975, "Village Nutrition Studies", Institute of Development Studies.
10. J.G. Reinhold, 1976, "Rickets in Asian Immigrants", The Lancet.

III. Research in Food Nutrition

REPORT AND RECOMMENDATIONS OF THE WORKSHOP

ON PRIORITIES IN NUTRITION RESEARCH - PANEL DISCUSSION

G.H. Beaton - Chairman

Dept. of Nutrition and Food Science - Univ. of Toronto

Toronto, Ontario - CANADA MSS IA8

In the discussions that followed the formal presentation of papers, a number of issues came into focus. Although no formal recommendations were put forward and no votes were taken, a general consensus emerged on some of these issues. The following report deals with a few of these matters.

A major initial problem in the discussions, and in many similar discussions in other settings, was a divergence in the definition of "nutrition", of "research" and of "nutrition research". Nutrition may be defined very broadly so as to encompass all aspects of biology and society impinging upon food and upon nutritional health. Conversely, it may be defined in a much more restrictive sense as in the relationship between the ingestion, digestion, absorption and utilization of food and its impact upon health. Obviously there are gradations between these extremes. Those members of the panel who were closely associated with field operations and community problems tended to acept a very broad definition while some others felt that a stronger focus and narrower definition was required in any discussion of priorities and of plans for the development of research resources.

An analogous problem arose with regard to the conceptual definition of "research". This could be seen as ranging from closely controlled experimental studies, through more open field or clinical studies and into observational studies of communities or evaluation studies of programs; to some members of the panel, the term might also encompass the individual involved in operational programs who had, and applied, an analytical mind in making day by day operational decisions. Again the proximity of the panelist to operational programs tended to affect the definition accepted.

The problem of definition was exacerbated in considering whether of nor "nutrition research" and "research into nutritional problems" had the same meaning. For example behavioural research, fundamental and applied, is a very important part of research into nutritional problems; it might or might not be described as nutrition research. The distinction was deemed to be important. Not all research relevant to nutritional problems necessarily should be conducted by nutritionists or be conducted within nutrition institutes.

These questions of definition of terminology, not resolved in the present discussion, were seen as fundamental to any delineation of priorities. Conversely, the discussion illustrated that in considering research priorities starting from an examination of nutritional problems of the community, the research required to fully support operational programs might range into many fields of endeavour, whether or not these be called nutrition.

There was a consensus that relevancy, as well as feasibility, must be an important consideration in establishing priorities. It was also recognized that relevance was importantly conditioned by the nature of the needs of particular communities whether these be in developing or developed countries, by the institutional role (an educational institution properly might have different research goals and priorities than would be expected in a service institute associated with the national government), and by the extent of the research investment made by the particular government. Thus, for example, countries with limited resources and major nutritional problems may elect to set priorities that relate very closely to application, drawing upon the findings of more fundamental research conducted in other settings. As resources develop, the same government might choose to reinforce its operational interests with a somewhat more fundamental type of research capable of addressing questions arising from the operational area. Thus, one should envisage a spectrum of research and of research priorities ranging from the very applied to the quite fundamental, although not all of this spectrum would necessarily be found in one country or be sponsored by one agency.

From these deliberations, there was a clear consensus that research priorities, other than in the most general terms, cannot be established internationally. Conversely, international agencies and institutions properly may establish their research priorities in relation to their own goals and resources. There would be an obvious advantage to a harmonization of institutional priorities within a country, among countries and among international bodies with the objective of assuring a continuity, without avoidable redundancy, of the development of knowledge and understanding ultimately relating to the preservation and improvement of human health, the goal upon which all members agreed.

The conduct of research, operational or fundamental, requires the development and maintenance of an effective human resource base. Many of the most pressing nutritional problems present research questions that cannot be resolved in the short term. For both of these reasons there was agreement that when research policies and priorities are established they must be seen as persisting over sufficient periods of time to provide continuity of activity. This does not preclude the gradual evolution of priorities and activities as knowledge increases and as problems evolve. The abrupt and major alteration of research policy and priorities must be avoided.

There was a clear consensus that research priorities should be established as a part of a continuing dialogue between those who conduct research and those who use the findings of research and that the priorities so decreased should be responsive to needs, recognizing the continuum of levels of information generation and application. All agreed that a major problem at present was a very imperfect system of communication that both incites the curiosity of investigators about problems faced in operational areas and conveys research findings to potential users. Clearly, rectification of this problem should be a goal in both research and operations planning.

APPROACHES TO SETTING PRIORITIES

G.H. Beaton

Dept. of Nutrition and Food Sci., Fac. of Medicine

University of Toronto - Toronto - CANADA

In a 1977 report addressed to world food and nutrition problems, a U.S. National Research Council Subcommittee[1] offered a definition of the role of research that has interest for our discussion.

> "The role of research is to broaden the range of choices availble to all who influence world food supply and nutrition."

Clearly this statement is directed toward a particular area of international concern. Equivalent statements could be offered if one were addressing much more specific areas such as the nutritional treatment of clinical disease, the care of the low birth weight infant or the problems of the increasing geriatric population in North America.

This statement, if accepted, has some very important implications in the selection of priorities. In the same report cited above, the group phrased three questions to be addressed in selecting priorities:

> "1. What advances in knowledge will specific areas of research produce, and what is the scientific and technologic significance of these areas?
> 2. If the research produces results, what effect would these likely have on reducing global hunger and malnutrition over the next several decades?
> 3. What supportive action will be required to conduct research for the accelerated activity recommended (e.g. more resources, policy change, organization, etc.)?"

I suggest that one important limitation to this type of approach, and indeed to most approaches to research planning, is that it presupposes a knowledge of the outcome of research. That is, to properly answer the first question one must assume that research in a certain area will provide a predicted type of information which will then have predictable impact in operational programs. A danger of such an approach is that it might seriously underrate research in areas where we have identified serious gaps in information — the real unknowns — as compared to areas in which we see the need to refine information. Nevertheless, questions of this type are a useful <u>part</u> of the approach to setting priorities when achievement of a defined operational target is the goal.

At the final meeting of the Protein-Calorie Advisory Group in the Spring of 1977, a somewhat different set of questions was posed as an aid to the identification of areas warranting augmented research activity[2]. These may be paraphrased as:

1. What is the relevance for nutrition improvement?
2. What is the status of existing knowledge?
3. What is the current extent of scientific and public attention?
4. What is the operational feasibility of implementation?

One may see these questions as having implications similar to those phrased by the NRC subcommittee. However, I think they feature two additional dimensions: extent of current work and feasibility of conducting the research (although both of these may be inherent in the third question posed by the NRC subcommittee).

I am sure that all of us at this meeting can pose pressing questions that seem to need answers. For some of these questions, major progress toward answers may be underway at present; <u>future</u> research priorities may be reduced. For other questions, the existing state-of-the-art, whether phrased as understanding of the problem or availability of appropriate methodology, is such that we are not yet capable of addressing the question. In considering priorities we may have to downgrade this area or place emphasis upon a different level of research, e.g. development of the necessary methodology.

Let me offer you a practical example. Since its publication there has been continuing discussion of the FAO/WHO Report on Protein and Energy Requirements[3]. Two subsequent advisory committees, the most recent being last fall, have considered the report and offered recommendations to the agencies. Clearly, in due course, another committee will be convened to address the question of human requirements for protein and for energy. What are the future research priorities that relate to the specific objective of refining and improving these estimates? Much attention, and much work, has focussed upon the definition of protein requirements for normal humans, and more recently for subjects "adapted" to conditions prevailing in Third World countries. These studies, by and large, use nitrogen balance as the index of adequacy. I suggest that such investigations may have a reduced future priority simply because I expect that there will be considerable new information arising from work already underway.

Contrast this with the situation of estimates of energy requirement. To say the least, the precision of our knowledge of human energy needs leaves a great deal to be desired. It is very rudimentary. Yet current scientific thinking suggests that in nutrition planning, the first objective must be to meet energy needs. Should research on human energy requirements have a high priority? There are probably fewer than ten groups in the world that are currently attempting to conduct major research in this area. The constraint is not lack of interest or failure to recognize the importance of the area. I suggest that a major constraint in this area is lack of appropriate methodology and research design to address the issues. If a priority is assigned it probably should be toward development of new methods and new approaches to determination of human energy needs. I suggest also that this is where the priority may lie in protein requirements. The shortcomings of nitrogen balance as an indicator of protein adequacy are well known but, to date we have no acceptable substitute.

Let me address the matter of priorities from another perspective. What are the prerequisites of good research?

1. A good question.
2. A sound approach to addressing the question.
3. An opportunity to implement the identified approach.
4. An individual with the scientific curiosity, background knowledge, technical competence and personal drive to formulate, conduct and interpret the research.
5. A situation in which these coincide in time and place.

I suggest that in any approach to setting research priorities there must be major concern about the development of human resources and the establishment and maintenance of situations and atmospheres in which scientific curiosity can flourish. I fear that these dimensions of research policy are sometimes set aside in our current drive for planning, for systems approaches and for the relief of constraints to predetermined pathways of "progress".

I suggest that in the future we may achieve greatest benefit from research taht is now in early developmental stages — research that has been initiated by the curiosity of individual investigators, albeit by investigators who have a perspective of the broader nutritional issues and the imagination to see the directions in which their research may contribute. However, I caution that because they are still in an early development stage, a clear assessment of their future impact cannot be made. They do not fit well into a system designed to set priorities at a more operational level.

To cite but one aspect of such research may I suggest some of the newer developments in behavioural research. New approaches to the examination and measurement of determinants of food selection are emerging. While we have long recognized the scope of environmental, economic and social factors that affect food selection, we have lacked approaches that could describe these in quantitative terms and predict their relative contributions. Such approaches are now emerging.

While a great deal of attention has been focussed on the long-term behavioural effects of severe malnutrition, particularly of protein-energy malnutrition, it is only recently that attention has been focussed upon short-term diet mediated behaviour. Current studies are examining the effect of diet on brain neurotransmitters and their effect on behaviour. At present, these studies may be seen as being quite fundamental. However, they may have important future implications in many areas of applied nutrition.

Therefore, I add two criteria to be considered in establishing research policy and considering research priorities.

1. Will the policy favour the establishment and maintenance of an atmosphere and setting in which individuals develop and scientific curiosity flourishes?

2. Will the policy foster a climate in which individuals in research and in operational fields can effectively intercommunicate?

I suggest that while the specific listing of research priorities will vary with the long-term goals of the body setting these priorities, the above criteria should be considered as a part of any policy. We must encourage competent individuals with scientific curiosity. We must facilitate situations in which these individuals are encouraged to direct their curiosity in directions that are likely to be beneficial in the long run. Good research is often opportunistic involving the right individuals in the right setting at the right time. We must avoid any notion that all research can be directed and governed by narrowly defined goals or rigid priorities.

Notwithstanding these remarks, I think we must recognize the right, and the obligation, of individual granting bodies, agencies, or governments to identify general areas in which they wish to invest their research funds. In return, it is our responsibility as scientists to encourage and advise these bodies not only on the priority areas they may select but also on the implication of their policies in terms of the development and maintenance of necessary human resources for both the conduct of research and the adaptation of research findings to operational programs.

REFERENCES

[1] National Academy of Sciences, World Food and Nutrition Study: The Potential Contributions of Research. Steering Committee, NRC Study on World Food and Nutrition, Washington, 1977.
[2] Protein-Calorie Advisory Group of the United Nations. The UN System and the World Nutrition Problem: An Agenda to Test the New Institutional Arrangements. PAG Bulletin 7, No. 3-4, p. 22, 1977.
[3] FAO/WHO, Energy and Protein Requirements. Report of a Joint FAO/WHO Ad Hoc Expert Committee. Who Tech. Rept. Ser. No. 522, 1973.

PRIORITIES FOR NUTRITION RESEARCH

J.M. Gurney - Director

Caribbean Food and Nutrition Institute

P.O. Box 140 - Kingston 7 - JAMAICA

It would seem evident that any increase in basic knowledge can, at some time, be applied to change the human condition. If this is so, there is no 'basic' research that cannot also be called 'applied' or, at the least, 'applicable'. However the converse is not necessarily true. Application of knowledge of nutrition does not only involve chemistry and physiology but also takes place within socio-economic and political contexts. Since human societies and the ecological backgrounds in which they are set vary widely, inevitably the application of nutrition knowledge will, in some of its aspects, differ between societies.

Biochemical, metabolic, and clinical research in nutrition can be carried out wherever in the world there is 'clinical material' and suitable laboratories. However research into the application of nutritional knowledge and into the socio-economic-political aspects of nutrition must be carried out where the problems are. There is therefore clearly a need for centres of excellence performing this latter kind of research within the Third World.

Two problems in particular face nutrition centres in the developing world. The first is the soft-ended nature of much research into socio-economic political nutrition problems and the second is the need to combine research with service. We have other problems of course concerning appropriate staffing, optimal size, funding, facilities, control and so on.

THE SOFT-END OF RESEARCH

I think we must accept that research at the applied end of nutrition cannot always be so 'scientific", in the way the word is often used, as laboratory based nutrition. Some of our work can be subjected to randomized controlled trials (Cochrane, 1972) but much is not subject to objective validation — at least not with the tools we now have. It is perhaps irritating that real life is so complex and apparently disorganized. However our problems are with real life. In my view our nutrition research priorities in the developing world should be with application, however difficult it is to devise indices and controls and to prevent extraneous factors (such as price fluctuations) from influencing our results.

A joint committee of the Agricultural and Medical Research Councils of the United Kingdom recently produced a report entitled "Food and Nutrition Research" (ARC/MRC, 1974). It is intended to provide a basis for discussion in the consideration of policy on food and nutrition research in the United Kingdom. It is an excellent and very useful report eminently suited to its purpose. It would, however, I think, be unwise to base research priorities in the Third World or for International Agencies mainly on this report. The concept of nutrition used is too limited — perhaps too 'pure' — for our needs. Take for example the first sentence of the introduction of the report which reads as follows:

> "The science of human nutrition is mainly concerned with defining the optimum amounts of the constituents of food necessary to achieve or maintain health throughout the full span of life in all groups of the population."

Developing countries will never control malnutrition with such a limited concept of nutrition. Our main concern is not to define how much people need but to ensure that they get it using already existing knowledge as our guide. We should work mainly at the applied end of nutrition and if at present such application is not scientific we should devise means to make it so or accept that it is not.

RESEARCH IN SERVICE

The second major problem I would like to raise is the need to combine research with service. I am speaking here in the context of the Caribbean Food and Nutrition Institute which exists to serve its sixteen member governments, and in my own context as a member of an international organization (the World Health Organization/Pan American Health Organization). While this may put us in a somewhat unusual position I think that the problem I am raising is common to many nutrition institutes. So our problem in terms of research is two fold:

1. Should we carry out research as a separate enterprise at all?

2. How can we ensure that the self-discipline of always questioning our work and testing its effectiveness remains with us in our search for solutions to the nutritional problems that surround us?

The Caribbean Food and Nutrition Institute has five purposes. These are: service, training, information dissemination, coordination and research. It would seem obvious — although in the pressure of work and excitement of action often forgotten — that research should be a part of these other purposes. Our ongoing service, training, information dissemination and coordination activities should be subject to testing, monitoring and evaluation. And we should be seeking, through research, to find and develop new and appropriate types of service, training, information dissemination and coordination.

Thus, research should be an attitude of mind that permeates all our programmes. It should not be a segregated part of our activities.

In the context of the Caribbean Food and Nutrition Institute I would go further. I do not think we should carry out any research unless it directly relates to our other activities. We should not accept research funds, research students or research fellows unless they contribute directly to our service function (using the word service to include training, information dissemination and coordination). We should welcome them if they do so contribute. But we should be careful, as every research activity takes staff time as does every student and fellow. Therefore we must ensure that we do not reduce the service component of our work through taking on research.

The Caribbean Food and Nutrition Institute has eight objectives encompassing: diagnosis of the food and nutrition situation, food and nutrition policy development, strengthening technical nutrition units and programmes, education and training, diagnosis, prevention and treatment, institutional food services, educational materials and research. The research objective is "the conduct of practical, operational research appropriate to objectives one through seven" (CFNI, 1978).

The research I am talking about here is how to get the knowledge we already have, and any new knowledge we may acquire, into the scheme of things — into primary health care at the grass-roots level, into development planning and into nutrition education programmes, for example. This implies forging useful links between the scientists and on the one hand the administrators and technocrats and, on the other the educators and communicators. I think we at the Caribbean Food and Nutrition Institute have succeeded quite well in this process of liaison but I do not think we have been able to develop

enough feedback and testing mechanisms (Williams, 1965) to ensure
that the methods we employ have useful results at the cheapest cost.
This would be, in my estimation, one of our prime needs in nutrition
research: research into how best to get knowledge about nutrition
incorporated into programme development.

Clearly the research objective of the Caribbean Food and Nutrition Institute gives a social commitment to our research and raises
the question whether science, and in particular nutritional research,
can be socially committed and still retain its scientific validity —
and its respectability.

The socio-economic-political aspects of much nutrition research
in the developing world and the need to combine research with service
have implications for postgraduate education. The education of
future staff of nutrition institutes should fit them to be able to
function appropriately. It should not programme them to seek refuge
in a laboratory, where the problems are clear and controllable.
Rather it should encourage them to work at getting things done in
the community and in the administrative structure using basic nutrition when needed. At least part of the postgraduate course should
take place where the problems are.

C.P. Snow, scientist, administrator, novelist and politician,
in his Godkin lectures on science and government (Snow, 1962) describes four types of links between science and the political process.
He describes "open politics" as occurring when ideas are submitted
to the larger assembly, be it a group of opinion (for example the
scientific community) or the electorate. Snow then describes three
types of what he terms "closed politics". These are "committee
politics" in which committee members use their personalities and
arguments to lead to a consensus, "hierarchical politics" which is
typical of bureaucracies with a definite chain of command, and
lastly "court politics" which involves influence on the person who
has the concentration of power.

If we in the Caribbean wish to make the findings of nutritional
research of benefit to the malnourished we have to work through all
four of Snow's types of politics. We must define our operational
research needs bearing these four types of influence in mind — open
politics, committee politics, hierarchical politics and court politics.

MOBILIZING RESOURCES

Lastly I would like to emphasize the importance of mobilizing
individuals and institutions that are not normally associated with
nutrition research. Mission-oriented nutrition research, to use the
phrase used in a report to the Pan American Health Organization

(PAHO, 1977), will need differing expertise in various projects. It would seem appropriate for a centre that carries out such research to develop a network of expertise as far as possible from within the country or countries served. I do not think a nutrition institute should ever be big enough to carry out all the nutrition research needed. Rather I think it should in many instances coordinate the work of others. I would like to give three examples from our own experience in the Caribbean:

1. To bring together a group of specialists to evaluate available knowledge and draw up a Technical manual on a subject. (An example might be a local diet manual for diabetics) or to devise a strategy to solve some problem (e.g. to combat diarrhoeal diseases and energy-protein malnutrition in young children.)

2. To recruit temporary advisors or consultants to work on a specific problem — for example to study the functioning of food and nutrition councils.

3. To develop a food and nutrition surveillance or evaluation system using mainly the existing government networks of services and reporting — rather than developing a new system.

These approaches that involve others in nutritional research do more than simply provide information. They also help the people involved to develop an interest in and knowledge of nutrition in all its complexity so that over time the human resource base needed to reduce the prevalence and severity of malnutrition is strengthened. To adapt a cliché: nutrition research is far too important a subject to be left to nutritionists alone.

BIBLIOGRAPHY

ARC/MRC, 1974. Food and Nutrition Research: Report of the ARC/MRC Committee. London, HMSO.
CFNI, 1978. Caribbean Food and Nutrition Institute and PAHO Area I Nutrition Staff; Programme of Activities 1978 as approved by the Eleventh Meeting of the Advisory Committee on Policy to CFNI. Kingston: CFNI (CFNI-J-3-78).
Cochrane, A.L., 1972. "Effectiveness and Efficiency: Random Reflections on Health Services". Nuffield Provincial Hospital Trust.
PAHO, 1977. Report to the Director: Pan American Health Organization Advisory Council on Medical Research Sixteenth Meeting. Washington, PAHO (HRR 16/1).
SNOW, C.P., 1962. "Science in Government". The Godkin Lectures at Harvard University 1960 with a new Appendix. Mentor, New York.

Williams, L. Pearce, 1965. What is Science? Agric. Sc. Rev.
   3, 3, 18-23 (reprinted in Baroda J. Nutr. 1977, 4, 1).

PRIORITY FOR APPLIED NUTRITION RESEARCH

IN DEVELOPING COUNTRIES

R. Orraca-Tetteh

Nutrition and Food Science Dept.

University of Ghana, Lebon, GHANA

Nutrition is a subject whose definition can be narrow or as broad as one would wish. This is because nutrition is basic to human life, and thus is affected by socio-cultural and economic factors. Nutritionists have traditionally worked within the narrow definition but over the past few decades there has been a recognition of the need to work within the broader definition of the subject in order to solve the urgent problems of nutrition.

The priorities in nutrition research are many, depending on the problems of nutrition in the developed and developing countries. But in most countries, nutritionists have concentrated on researches which appeal to them, depending on their particular area of nutrition work. Though nutrition has several divisions these have not been clearly delineated for specialization as in some other subjects such as medicine or agriculture. This has tended to affect the priorities for research. Some of the areas which must be given special attention in training of nutritionists are Nutritional Biochemistry, Clinical Nutrition, Nutritional Anthropology and Applied Nutrition. These various fields as areas of specialisation can themselves be subdivided.

For various nutritionists in different countries the priorities may be in these different fields of nutrition, but these are necessary for a concerted attack on nutritional problems. For instance at present, the nutritional problems of developed countries are increasing in the area of diseases of affluence and metabolism. Research into such nutritional and metabolic diseases as diabetes, atherosclerosis and obesity are indeed priority areas for nutrition research in developed countries, which in the long run can benefit developing countries.

PROTEIN ENERGY MALNUTRITION

In most developing countries in the tropics, protein-energy malnutrition because of its high prevalance has for a long time had attention focussed on it. The extreme forms of protein-energy malnutrition, Kwashiorkor and marasmus have held sway in nutrition research for obvious reasons. It is now urgent that attention be focussed on the problems of the large number of children who are chronically but moderately malnourished. The effects of this malnutrition on their development in all spheres must be elucidated so that attention of governments can be drawn to the great wastage of potential for the development of their countries.

FOOD PRODUCTION

Food is the concern of nutritionists since there can be no nutrition without food. It is therefore necessary that nutritionists be concerned with the availability of this most important material. In various surveys carried out in many developing countries, it has been shown that most of the populations do not satisfy their requirements for various nutrients. This is particularly true for children, pregnant and lactating women. The obvious recommendations have always been for these groups to increase their intake of these nutrients. Obviously the foods of these areas will provide these nutrients. It is a fact that food production in most developing countries has not been increasing as fast as population growth. Therefore the amounts of food available to the people are far below that required for good nutrition of all sections of the communities. The quality of the food in terms of the good quality protein-rich foods is also low. Thus some of the causes of the inadequacy of the diets consumed are known.

Nutritionsits must therefore bring such facts to the attention of governments, agriculturists and others concerned with providing the necessary inputs and impetus for food production. The basic facts must be collected, analysed and presented to governments. For governments in developing countries to take action nutritionists working in cooperation with others such as economists, agriculturists, health workers etc. must bring their research results to the attention of governments. For initiation of good food and nutrition policies some relevant data must be made available. There is therefore need to investigate the food production and consumption problems in communities.

FOOD CONSUMPTION

Food consumption by people is influenced by various factors which can affect the satisfaction of their nutritional needs. Some of these are prices of foods, the incomes available to the consumers, social and cultural factors. In most developing countries at present, the world inflationary trends have had very adverse effects on the people. Food prices are soaring high because the costs of inputs for food production are increasing constantly. The incomes of the people are not rising commensurate with the rising prices of consumer goods, there is therefore the dilemma of families in obtaining a decent diet. These problems demand attention of both nutritionists and others, if malnutrition and under-nutrition in these communities have to be reduced.

Some of the reasons for malnutrition and undernutrition in the vulnerable groups such as children, pregnant and lactating women can be found in the distribution of food within families. In most of the developing countries fathers usually take food first before other members of the family. The best parts of the meal consisting of the protein-rich foods are therefore consumed by the father. Children and mothers have to be satisfied with the left overs. Obviously the nutrition of these family members cannot be optimum. Investigation into such intrafamily food distribution practices can provide the necessary information and background reasons to enable the necessary nutritional education action to be taken.

EFFECTS OF GOVERNMENT POLICIES

Nutritionists are increasingly being drawn into the investigation of problems created by various governmental policies which have direct or indirect bearing on nutrition of communities. The great increase in urbanisation in developing countries has followed the past trend in developed countries. The movements of population into the urban centres are creating problems for the proper feeding of these urban populations. Governments continue to increase the development projects in these areas to the neglect of the rural areas. The influx of people into the peri-urban and slum areas leads to poor living conditions which affect particularly children who are very susceptible to all manner of infections. The effects of infection and infestations on the nutritional status have been shown to be detrimental to the host. There is need in nutrition surveys for the investigation into such conditions on the nutrition of the people.

In Ghana, a governmental policy which had affected the costs of goods and services and therefore the nutritional status, was the raising of the minimum wage. The government decided that the minimum wages of the labourer or manual worker should be doubled. This meant that a chain events was set in motion which ultimately has affected the price of food purchased by all. Thus the nutrition of these workers has been adversely affected by a governmental policy which has been intended to help these same people.

The export of food commodities in most developing countries reduces the food available to the people. In many developing countries the staple foods which supply the energy of the diet are the starchy roots and tubers such as yam and cassava. These foods are now in great demand in many developed countries, for use in animal feeding. Thus cassava is exported from Ghana as dried chips to continental Europe to be used in providing the energy base of animal feedstuffs. There is therefore a reduction of the supply of cassava available for local consumption, and this has led to high prices for cassava. Thus the policy to try and earn much foreign exchange for the country has adversely affected the nutrition of the population.

NUTRITION EDUCATION

The researches of nutritionists can provide relevant information for solving the food and nutritional problems of the worls. But solutions cannot be effected until the necessary changes in behaviour can be accomplished. Therefore nutrition education is of prime importance in the fight against malnutrition and undernutrition. Unfortunately this is one area of nutrition which has not received the attention it deserves.

It has long been assumed that as long as the necessary information is available, people will use them for their benefit. This has not been proved right and there is the urgent need to train nutrition educators to investigate the reasons for diverse food behaviour and food habits of people, and the relevant methods to get good nutrition education to the people in need of it.

Nutrition education by itself will not achieve the goal of good nutrition for the peoples of developing countries. Evidently the problems of food and nutrition in these areas are part of a bigger problem of development. Thus the need to institute programmes which take into account the whole environment of communities is very urgent. Applied Nutrition Programmes which provide the basic needs of communities have better chances of improving the nutrition and general conditions of people. Nutrition education as part of such programmes where food production and preservation, basic needs such as water, sanitary facilities and health protection are provided for, can help in eradication of malnutrition and undernutrition.

BIBLIOGRAPHY

Berg, Alan, 1973, "The Nutrition Factor: Its Role in National Development", The Brookings Institution, Washington, D.C.
Orraca-Tetteh, R., 1972, The need for a nutrition policy in National Development. Universitas, Vol. 1, No. 3 (New Series), p. 47. University of Ghana, Legon, Ghana.

Orraca-Tetteh, R., 1974, Food and Nutrition as components of the
    Scientific Pre-Requisites for Development in Africa. Journal
    of African Studies, (3): 300. Univ. of California Press.

PRIORITIES IN NUTRITION RESEARCH: REMARKS FROM THE STANDPOINT
OF THE ITALIAN EXPERIENCE

A. Mariani

Istituto Nazionale della Nutrizione

Roma, ITALIA

It is known that between genotype and environment, health promotive factors and external agents of morbidity interreact. Among the formers, already during intrauterine life, nutrition adequacy plays a predominant role in determining physical and mental development, resistance to diseases and life expectation.

Furthermore, in both developing and developed areas, as a consequence of deficiencies of production, processing and distribution of food, on one side, and of errors or excesses of consumption, on the other, food and nutrition and the social, economic and cultural situation of the population are interdepending. Moreover, in critical situations, of the type that can be found in countries undergoing stages of deep economic transformation or transition — similar to what happened and is still happening in some areas of Italy — undernutrition or marginal nutrition may persist, while cultural and technological evolution, modifying traditional models of consumption, may alter food habits hence introducing new factors of risk.

Therefore, it is not surprising that a practically worldwide agreement has been found on targets of general interest for nutrition research. These are

1) functional significance of nutritional status and nutrient requirements for health;
2) relationships between nutrition and epidemiology;
3) evaluation of the nutritional quality of food; and
4) studies on food toxicology or safety for use.

The achievement of these goals appears to be essential, in fact, for the formulation of adequate nutrition policies.

However, difficulties become evident at the decision taking level, when operative choices need to be made, identifying key questions like research topics and methodological approaches.

Actually, in several countries, Italy among them, general directions seem to press so that nutrition research — at least as concerns national institutions — does not involve basic issues. It is common opinion in fact that the aims of research should rather be directed to providing information to be realized through short- — and medium- — term interventions for improving the relationships between nutrition, public health and socio-economic status of the population.

The research programs of the Istituto Nazionale della Nutrizione are largely informed with this type of directions. As regards novel protein sources and their nutritional evaluation, the Finalized Programs of the Italian National Research Council reflect criteria of feasibility and prospects of practical applications.

However, with regard to trends and information that can be derived from nutrition research, we must recognize that in many areas we need yet to wait for the results of studies on often sophisticated experimental models.

Research on the metabolic adaptation of laboratory animals to unbalances of the various nutrients, proves in fact to be of considerable value for the understanding of the problems in defining nutritional requirements for man in the most diverse conditions, extreme or marginal. On the other hand, experiments on new protein sources, subordinately introducing natural toxicants, require basic research. This is the case of the identification of the factors responsible of favism — a problem particularly relevant in Italy — and of the fine mechanisms of their toxic effect, if it is due to molecules naturally present in fababeans or to their metabolites. It is easy to recognize then that in the field of nutrition there isn't much reason in opposing basic and applied research, particularly since it is difficult to cut border lines between the two.

Other problems of priority rise from the utilization of epidemiologic observations. This observation provided already useful indications for the formulation of dietary goals aimed, for instance, to the prevention of cardiovascular diseases. At any rate, it is well known that in industrialized countries changes in dietary patterns, associated with increasing affluence, mainly contribute to the extension of obesity. Research on better methods for estimating fatness is necessary for the progress of the knowledge of all the aspects of obesity. At the same time, research should be focused on the epidemiology of children obesity.

In countries like Italy, where in these last years food has become plentiful and where, consequently, cases of obesity since early childhood have reached a high frequency, this type of research appears of particular interest, also in view of setting up comparative studies on a retrospective basis.

However, little information is available in the literature on whether or not obesity is a single disease. In the context of epidemiological studies on obesity, particular attention should be paid to studies on the familiarity of obesity. The acquisition of such information, together with a more precise definition of obesity, represents the necessary premise for the identification of eventual genetic implications on the pathogenesis of some types of obesity.

At any rate, in the context of studies of the causative factors of obesity (whether genetic or not) it is important to develop experimental research on the variation in energy requirements between individuals and on the biochemical mechanisms regulating energy utilization.

Still, in relation to epidemiological studies, research on the influence of the diet on a variety of chronic diseases appears to be of primary interest. In this connection, it is important also to establish if the diet is simply a causative agent or rather a powerful modulator and a likely source of many converging etiologic factors.

There is evidence that respecting prudent dietary recommendations since childhood and keeping these prudent habits all life long, the risk of diseases, including cancer, can be minimized. Unfortunately it has to be observed that the F.P. of the Italian NRC on preventive medicine do not include the study of these relationships.

In connection to the relationships between nutrition and cancer, however, it appears evident the need of linking epidemiologic observation in men with experimental studies on animals and *in vitro* cells preparations.

As concerns nutrition and disease, the problem of choosing priorities becomes difficult when decisions need to be made in countries where — as demonstrated by the Italian experience — there are groups within the population who may be "at risk" from nutritional deficiencies and where, at the same time, nutritional unbalances, over-consumption and problems of food toxicology, proper of industrialized areas, emerge. Therefore, it is necessary to evaluate first the weight of the various situations.

Monitoring of the nutritional status and study of food and nutritional requirements for health deserve therefore priority in

such situations. Particularly taking into account the variety of
the problems, it is felt the need of developing research and of
modulating methodological approaches. Furthermore, the existence
of marginal conditions and at the same time, difficulties in defining them, pose particular problems.

Population studies, for instance, have revealed in various geographic areas in Italy socio-economic gradients in the somatic development and that a better nourished child has a growth significantly faster than his under-privileged, less well-nourished coetaneous.

This fact wouldn't be so important in itself were not for the
observation that better-nourished and more developed children are
on the psychological condition of being and feeling more appraised
even in terms of their intellectual performance. On the other hand,
less adequately nourished children show a better physical fitness.
Which is, therefore, the functional meaning of these conflicting
indications?

Naturally, the methodological approach has to be refined and
improved mainly to identify — in a general picture of nutritional
monitoring — the cut-off and the trigger points. Although at present the anthropometric methods seem to provide a better confidence
for the evaluation of marginal food conditions, it is clear that
greater efforts have to be made for the standardization of methods
apt to reveal marginal situations of unbalance of individual nutrients, in relation to reciprocal interrelations and to the total
value of the diet.

As regards food quality and safety, in Italy also is felt the
need of establishing the characteristics of new methodologies for
monitoring new products and new processes with regard to

a) assessing the true nutritive value for man of conventional
and novel foods, with reference, in particular, to new protein sources and their food formulation;
b) protecting human health from the risk of chemicals (additives
and contaminants) and establishing tolerance levels to be
accepted on an European scale.

As concerns the first point, evaluation of protein quality, for
example, the ideal methodology has not yet been elaborated. On the
mean time, the proposed utilization of isolates and concentrates
from plant proteins or of novel protein sources urges the industry
and the control laboratories to ask for <u>in vitro</u> tests, characterized
for being inexpensive, fast, reliable and acceptable at an official
level.

Extensive experimental studies, as planned in Italy (P.F. CNR) to verify the correlation between the methods for evaluating biological quality of protein should be encouraged.

Although evaluation of safety appears to be difficult, if not impossible, there is an increasing general sensititivy to the problem. This issue has become more pressing since the enforcement of rather strict and limitative bills, as the American "Toxic Substances Control Act" or the French "Sur le Controle des Produits Chimiques". However, often these bills are not supported by national or international agreements on the criteria and procedures to be followed in evaluating the actual hazard to human health. That is rendered even more complicated by the difficulties of defining food toxicology criteria and methodologies.

On the whole, without deterring new research ideas in areas regarded at present of lower interest, in establishing priorities for research in nutrition, the leading criterion has to be flexibly selected to contribute to the definition of food policies for the country. In this regard, it could be objected that at present in the world — and this is our case too — we have enough information to plan food consumption for the population according to criteria of greater rationality. Orientations of this type have recently been expressed in Italy as Government guidelines to be applied to the production, processing and distribution of food. However, a careful critical analysis reveals that all the available information in the field of nutrition is incomplete, particularly with regard to the knowledge on the existence of particular situations and/or the role played in the various geographical areas by differences in socio-economic conditions, sex, age, type of activity and so on.

Paraphrasing opinions on the cultural situation in our country, it can be affirmed that the need of priorities in nutrition research is supported by the acknowledgement of the existence of a certain degree of "scientific malnutrition" and/or of a "low density" of scientific knwoledge in the field.

In order to elaborate, develop and/or modify nutritional policies, it has to be observed that a priority has to be directed to research on the impact of the new trends and interventions adopted. As concerns the social and economic aspects of dietary planning, there is evidence for instance that in our country, due to the changing nature of food quality and their choices, that are likely to be available, as a consequence also of our belonging to the EEC and of its directives, it is likely or it would be desirable to modify the patterns of food consumption by individuals. Little is known about how to influence food habits, except by economic means. In order to improve health and the social situation, high priority is, therefore, attached to research of the factors that condition the selection of foods and their relative importance and of ways of modifying undesirable nutritional patterns. Not to make worthless all the discussion on nutrition priorities, first of all it is necessary to create the conditions for the development of research in the field. It has already been observed in the UK that since 1945 the rapid advance of knowledge, that has come from research in cellular and molecular biology, has had the effect of discouraging the interest of research workers in the field of nutrition, where the problems are by nature long-term; thus, the situation has been compounded by the scarcity of career opportunities. An even more difficult situation is that established in Italy, where for budget limitations the Istituto Nazionale della Nutrizione has not been able not only to increase but even to maintain its own research staff and where research in nutrition, carried out by Universities and other technical scientific institutions, has only a limited and episodic significance. Larger contributions to this area are instead desirable. It is well recognized, in fact, that being nutrition an eminently interdisciplinary field, the contribution to research in nutrition should involve backgrounds and institutions not traditionally utilized up to now.

At any rate, where applicable it is essential to strengthen already existing national organizations for research in nutrition, or create suitable councils or consortia when, and this is the case of scientifically more advanced countries, several efficient research centres operating in the field exist.

It is essential too that medium- and long-term programs of research in nutrition be complementary.

To this purpose, national organisms deputed to plan nutrition research should enlarge their intervention power, independently from peculiar aims as it is presently, in consequence of operating now on the public health field and now on the agricultural one, involving all the aspects relevant to human nutrition. To reach that goal, to avoid scattering and overlapping and to realize a unitary policy, in every country and jointly in the world, the creation of an international consortium of national institutions for nutrition research would be particularly useful. That consortium should have mainly the role of coordinating and promoting research and of sponsoring projects of recognized interest, that cannot find support at a national level.

PRIORITIES FOR NUTRITION RESEARCH

IN A DEVELOPING COUNTRY — TURKEY

Orhan Köksal

Institute of Nutrition and Food Sciences

Hacettepe University - Ankara - TURKEY

Since more than a century, basic research on Nutrition and Food Sciences has been carried out with increasing interest and attention. During this period energy and nutrient needs, their properties, physiological effects and metabolism have been determined. In addition to these, various investigations have been completed in order to find out energy and nutrient content of food and diets and factors affecting nutritive values. It has been said that there are few topics left to be studied in the field of basic research on nutrition and food sciences.

Meanwhile, applied research studies have started by utilizing the knowledge accumulated through basic research on the subjects related to nutrition and food sciences. The importance of research directed towards determining the criteria of the nutritional status of individuals and or the community have been emphasized. So, activities on applied research in various aspects of nutrition have gained importance.

Despite the vast amount of research studies, the number of people who suffer from malnutrition is still increasing in many countries. Therefore attention and efforts ahve been directed towards "the role of research in combatting malnutrition and its effects in a country".

Research topics which can be used for planning and operations to combat malnutrition are being placed in the following four profiles:

1. Researches covering the functional significance of nutritional status of individuals and community.

2. Researches related to determination of quality, safety, adequacy and acceptability of foods and diets.

3. Researches related to the evaluation of intervention programs implemented to improve the nutritional status.

4. Researches determining the effects of government policies on the feeding of community, especially on nutritional status and food consumption of low income groups.

Research topics related to these areas have been carefully mentioned by various reports.

Research topics that would be a necessity in the planning and implication of programs to struggle with the problems of malnutrition are covered by the report prepared by a Study Team and called "World Food and Nutrition Study"[1], and other two technical reports published by WHO, "Food and Nutrition Strategies in National Development"[2] and "Methodology of Nutritional Surveillance"[3].

Nutrition research that will be given the priority among many subjects included in each of four profiles listed above could be established in each country according to their conditions, needs and resources. In this respect, the following points have utmost importance:

(a) Criteria to be used in establishing priorities for nutrition research.

(b) Provision of resources for planning and realization of nutrition research.

(c) Securing the most proper use of the data obtained from the research.

I would like to point out our own experiences and to talk about some studies and their effects carried out in Turkey.

CRITERIA TO BE USED IN ESTABLISHING
PRIORITIES FOR NUTRITION RESEARCH

People or organization who are going to determine the criteria, to be used in establishing priorities, is more important than the criteria itself. Even when the criteria is scientific and conforms with reality, the possibility of giving priority to the subject to be studied and the profitable use of the results derived from it, would be very small if the decisionmakers are not interested. Therefore it is necessary that the criteria to be used for establishing priorities has to be acceptable by the decisionmakers, planners and economists.

We are living in a century that every effort is being evaluated by its economical aspect and profit. So, cost-benefit consideration should be kept in mind and the main criteria will be determined in this respect.

In many developing countries, great efforts are being made for industrialization and increasing production. Emphasis is given to establish modern factories and power supply plants. Moreover, by providing proper nutrition to meet energy and nutrient requirements of people for promotion of manpower are neglected. Expenditures for studies concerned with industrialization have been tremendous. Yet, the need for studies to extend our understanding of human nutrition has been under-estimated by administrators, politicians and planners. They do not believe in research in this field and express their view by saying "Why brother with complicated research and why not pay more attention on providing some food to people who suffer from severe malnutrition".

Indeed, to provide and distribute food to malnourished people, it is necessary to know, which people suffer from malnutrition? How much and what foods needed, and how self-sustaining nutritional improvement in such populations can be generated? Without a more rational understanding of the etiology and epidemiology of malnutrition, it is not possible to identify reliable and efficient solutions and alternative interventions.

During the last decade, and even now in some places, generally the food and nutrition problems were considered simply as problems of food production regarding quantity and quality. Nowadays the emphasis has shifted towards the consumer rather than the producer. Thus, especially in developing countries it is inevitable to plan studies to provide answers for the following questions which are fundamental.

1. What kind of nutritional problems affect what segment of the population with what severity and prevalence?

2. What are the significance each of these problems for the people affected? and what are the consequences of failure in reducing or eliminating these problems?

3. What are the factors involved (social-cultural-educational-economical-agricultural, etc.)?

4. How and to what extend would direct or indirect food and non-food interventions solve each or some of the problems and at what cost?

Generally planners and policy-makers do not allocate any money from the national budget for the intervention programs without having reliable evidence about the problem existing. But it is also hard to persuade government officials on organizing nation-wide nutrition surveys and securing the necessary funds. Therefore, the efforts of researcher should be directed towards convincing and sensitizing planners and decisionmakers about the usefullness and the necessity of nutritional studies. In the meantime, government and research institutions in developing countries should be encouraged by international organizations, (by U.N. organizations — WHO, FAO, UNICEF) to carry out nation-wide nutrition studies and support them by providing training for personnel and contribute financially. In Turkey UNICEF and CARE were granted for encouraging and supporting a national nutrition and food consumption survey.

After obtaining the data concerned with nutrition status and food consumption, second priority should be placed on researches that will be helpful in the implementation of various intervening programs.

PRIORITIES FOR INTERVENTION PROGRAMS IN DEVELOPING COUNTRIES

In developing countries where industrialization and increased production are accepted as the most important elements for development, planners, politicians and economists are showing resistance to intervention programs which are necessary in preventing and combatting malnutrition problems prevailing especially among vulnerable groups. They believe that if the production increase and if laborer and the poor can earn more, then malnutrition would be eliminated in consequence. In countries where liberal and mixed economy systems are in effect, governments expect that the private sector or the people themselves establish the intervention programs and find solutions against malnutrition. They do not even consider preparing legislation to set principles that would enable intervention programs or would either force or support private sector for preparing such programs. Therefore researchers who are investigating the value of intervention programs should not be disappointed and discouraged that their findings and suggestions have not been considered and used by the government officials for the national plans. Sooner or later the time will come when the outcome of these studies will be appreciated and utilized in planning applied programs.

In Turkey, the attitude of the governments towards some of these studies related to intervention programs, were as follows:

1. Direct Feeding Programs and Free Food Distribution

The proposal concerning the continuation of the school feeding program and free baby food distribution in MCH centers by the government has not been implemented seriously yet. Previously these programs had been supported by UNICEF and CARE, in elementary schools and MCH centers between 1956-1975. Before that proposal, applied research projects regarding to school feeding and MCH nutrition programs have been carried out using food mixtures available in the country. The necessity and the advantages of the projects were obvious[4,5]. But, allocation of funds in the national budget required for these programs has not received acceptance over the past three years. On the other hand, results of detailed nutrition surveys[6,7,8,9] indicate that 20-40% of preschool children and 15-25% of elementary school children show growth retardation and signs of PEM in Turkey.

2. Nutrient Supplementation and Fortification Programs

(a) According to the studies, in Turkey with a population over 40 millions, approximately 3 million people show various degrees of simple goitre signs. Goitrogenic areas have been determine and the relationships between the ecological factors and simple goitre have been investigated extensively. In order to prevent simple goitre, technique and standards to be followed in the preparation of iodized salt and legislation has been established in 1968. Several proposals concerning the production and distribution of iodized salt along with education programs needed for the instruction of this intervention program has been prepared by several scientists who have emphasized the need for government coordination and cooperation. Unfortunately, the responsible government bodies failed to see the importance of the problem and no action has been taken regarding the goitre problem which is increasing gradually.

Iodized salt produced by a private firm did not bring any solution to the problem, since it was too expensive and was not available in the goitrogenic areas.

(b) Iron deficiency anemia is widespread among preschool children, child bearing age women and pregnant mothers in Turkey. For preventing anemia, research proposal related to enrichment of staple foods, such as bread flour, with iron has been sent to ministries and institutions concerned[10]. Also the positive effects of using iron pills by pregnant mothers have been tested and reported to Ministry of Health[11]. No reply has been obtained from the government to the results reported and projects proposed regarding this very serious problem.

(c) Wheat bread is the staple food of Turkey. People meet 50% of their energy needs through bread[9]. Soya beans are also produced and there is a good production potential. The result of an applied study regarding the addition of 5-6% soya flour (protein content 50%) to wheat flour indicated that fortified bread has a positive effect on growth and development as well as on resistance to infections. It has been also shown that the bread fortified with soya flour can be kept for longer periods without getting stale and its acceptability is excellent[12]. Proposal for adding 5% soya flour to bread flour has not been implemented by the government yet.

3. Nutrition Education

(a) During the application of school feeding programs in Turkey between 1956-1975, the government had intentions to provide nutrition education to elementary school children. Attempts were made to train the elementary school teachers in nutrition to supervise the programs. However, evaluations have shown that the education directed to children did not have any influence on nutritional status and feeding patterns of their parents.

(b) A new nutrition education program directed towards families, especially females who are responsible for feeding the family, has been formulated by Ministry of Health with the assistance of Hacettepe University Food and Nutrition Institute and UNICEF. Audio-visual teaching materials and methodology of this education project have been tested and the effectiveness of the educator, who is midwife is being evaluated as an operational research study.

OBJECTIVES OF THE NUTRITIONAL RESEARCH
AND APPLICATION OF THE FINDINGS

The main objective of nutrition research is to provide information to reduce or eliminate malnutrition in developing countries. Nutrition researchers will be happy and satisfied if their findings are used for these purposes. However, it has been well known that findings of many valuable studies are hidden in files.

How and with what way planners and politicians can be convinced and sensitized about nutrition research and its results? It seems to me that planners and politicians are not interested in this subject unless severe malnutrition, hunger and famine problem are present. Malnutrition in the form of undernutrition prevalent among infants, children and pregnant mothers of poor social classes has no danger for their political life and themselves. Researches related to obesity, coronary heart diseases, hypertension, diabetes, gouttes and other degenerative and metabolic diseases are more interesting and important for the decisionmakers. They themselves are often under the risk of these diseases. Therefore the best approach towards obtaining the resources for what we are aiming at, would be establishing a research center for the investigation of diseases prevalent among the decisionmakers and make use of the facilities to study other problems as well. As we see that in developed countries to get financial support, institutions seem to be involved in the diseases of the well to do but meanwhile they also work on malnutrition and its consequent effects.

REFERENCES

Report of Study Team 9 (Nutrition), World Food and Nutrition Study, U.S. Government Printing Office, Washington, D.C., 1977.
World Health Organization, Food and Nutrition Strategies in National Development. Technical Report Series, 584, Geneva, 1976.
World Health Organization, Methodology of Nutritional Surveillance. Technical Report Series, 593. Geneva, 1976.
Köksal, O. (1971), Nutritional problems in Turkey during the weaning period and some solutions. Turkish J. Pediatrics. 13: 59.
Köksal, O. (1976), Effects of nutrition on infant and maternal mortality and other vulnerable groups in special developing areas. Turkish J. Pediatrics, 18: 72.
Köksal, O. (1972), Türk Halkının Beslenme Durumu, Sorunları ve Nedenleri (Nutritional Problems and Causes of Turkish People), Türkiye Tıp Akademisi Mecmuası - Archives of the Turkish Academy of Medicine Rapor: III-2.
Gürson, C. and Neyzi, O. (1967), İstanbul Bölgesinde Çocukluk Yaşlarında Beslenne Durumu. Besin Simpozyumu, TUBİTAK, 1969.
Uzel, A., Yücecan, S., Ekinciler, T., and Özbayer, V. (1972). Edirne İlinde Beslenme Araştırması, Beslenme ve Diyet Dergisi 1: 155.
Köksal, O. et al. (1977), Nutrition in Turkey. Report of Turkish National Nutrition and Food Consumption Survey - 1974. Hacettepe University Institute of Nutrition and Food Sciences, Ankara.
Institute of Nutrition and Food Sciences. Proposal Letters to Ministry of Health and Ministry of Agriculture (Toprak Mahsülleri Ofisi). Fortification of Bread with Iron, 1975.
Pekcan, H., (1974), Kazan Sağlik Ocağı Bölgesinde Demir Yetersizliği Anemisi Görülme Sıklığı, Belirtileri ve Tadavi ile Olan İlişkisi (Thesis). Hacettepe University - Department of Community Medicine (in mimeograph).
Köksal, O., (1972), Türkiye'de Beslenme Sorunu ve Yetersizliklerinin Çözümlenmesinde Tedbir ve Tavsiyeler. Türkiye Tıp Akademisi Mecmuası Rapor III-2: 1-9.

NUTRITION RESEARCH PRIORITIES

FOR NORTH AFRICA

>Zoheir Kallal - Director

>National Inst. of Nutrition and Food Technology

>Tunis, TUNISIA

The countries of North Africa share many common features. The 1975 FAO Food Balances indicate that mean intakes of calories and protein are adequate with the staple of the diet being cereals in each. They also share a common religion which has strict food taboos and influences food habits. As a result of these commonalities, the nutrition problems of this area are similar. Wealth from oil and rapid development has enabled some of the countries to be in a position to now solve some of these problems.

Despite the improved economic situation of the region these countries differ markedly from the developed countries where the nutrition intake standards have been established. Their higher incidence of disease and environmental stress may necessitate higher intakes of certain nutrients. At the same time, it is necessary to know what local dietary characteristics have facilitated the adaptation to this environment and to prevent complete adoption of the western diet and its atendant high incidence of nutritionally associated diseases such as hypertension, atherosclerosis and cancer. For nutrition planning purposes and predictions of foods needs, food production, and importation goals, it is essential to have a clear understanding of these differences in nutrient needs for the region.

Even though there has been marked economic improvement in the region, malnutrition is a frequent problem in the area particularly among pregnant and lactating women, infants and pre-school children. Growth retardation and anemia, the most common problems, can be due to inadequate diet but may also be caused by disease. Which nutrients are in short supply and whether the state of health or the state of nutrition is most important are questions which must be answered for each locality. On the basis of the answers to these questions, it will be possible to devise the best type of intervention program for combating malnutrition. If the problem is one of inappropriate food selection, what are the food habits which should be changed or encouraged; what is the best way to produce these changes and to insure that the change has actually taken place? Perhaps there are some foods which can be modified or supplemented to improve their nutrient content, or special foods can be developed for certain population groups.

But malnutrition does not exist in isolation nor can all problems be solved by one institute or ministry. Nutrition should be an integral part of preventive medicine and family planning services where it has proven to be supportive; nutrition considerations should also be a part of decisions made in other sectors of the government. The problem arises in how to best introduce nutrition into on-going health programs and to convince ministries for education, agriculture, and the economy of the impact of their programs on nutrition and vice versa. What kind of cooperative projects or programs can be developed to encourage the cross-fertilization of ideas to produce mutually supportive work.

The state of nutrition of a country is constantly changing and is influenced by multiple factors. Hence, in order to have a flexible nutrition program responsive to a country's needs, it is essential to have an on-going system to monitor some of these factors at the local level. What are the best indicators and which of these can be used in North Africa? How does one translate these indicators into impact statements which will influence nutrition policy for the country? It is a major advance to proceed from a nutrition survey to an on-going system for monitoring and then correcting nutrition inadequacies. This is a pressing subject for applied research and development throughout the world if the food shortages predicted for the near future are to be averted.

NEW ZEALAND NATIONAL DIET SURVEY:

PRELUDE TO A NATIONAL NUTRITION POLICY

J.A. Birkbeck

Dept. of Human Nutrition, Otago University

P.O. Box 56, Dunedin, New Zealand.

The recent completion of a dietary and anthropometric survey of adult new Zealanders has paved the way for discussion of a possible National Nutrition Policy document. While seeking to benefit from the experience of other nations, who have already formulated or proposed such a policy, it is hoped to develop a proposal more soundly based and less controversial than has been the case elsewhere. This paper reviews some results of our survey and the difficulties in the way of a small developed nation in establishing a National Nutrition Policy and Nutritional Surveillance.

THE SURVEY

The sample was drawn from the 1976 census, to represent in age, sex and socioeconomic status adults between 20 and 64 years of age, excluding pregnant and lactating women and anyone not in normal health. We aimed for a 0.1% sample, with a final survey of 1938 individuals, rather more than required. Demographic characteristics of the completed sample adequately represented the population as a whole.

Subjects were studied in their own homes, providing demographic data; 24 hour, activity associated dietary recall; and anthropometric measurement of stature, sitting height, weight, triceps and subscapular skinfolds, and mid-upper-arm circumference.

Time does not permit an extensive review of the results. In body dimensions, New Zealanders resemble American or British Caucasians; obesity is a common occurrence, and there is evidence of a secular increase in adult height. Median body weight and skinfold measurements rise sharply between those in their 20s to those in their 40s, with a small drop after age 50 years. Although the definition of overweight or obesity is arbitrary, all measures point to a high prevalence especially in women.

Compared to recommended intakes, the New Zealand diet is very generous in energy and protein. A significant proportion of subjects fell below recommended intakes of thiamine, niacin, pyridoxine, and especially of folacin and cobalamins, and also of calcium and zinc (especially in women). Many had fairly low ascorbate intakes, although this was probably partly seasonal.

Cholesterol intakes were high (median in males varied from 550-680 mg/day) and the P:S ratio was very low (median 0.18 to 0.20).

Fat was the largest single source of energy in all age groups, ranging from 39-45% of total energy intake. Alcohol played an important role, up to 10% of energy intake in men. Protein calculated as energy yielded 13-16% of total intake.

Usual sources of nutrients are being studied in some detail. Intake of some nutrients, often those apparently marginal in the population, may be dependent in large measure on a limited range of food items. 40-50% of all calcium intake comes from milk and cheese, and 30% of folate comes from vegetables and a further quarter from bread, both items which appear to be diminishing in popularity, although the potato (yielding 6% of the total) holds its own.

There was a negative correlation (sometimes significant) between energy intake and obesity.

COMMENTARY

The interpretation of such results raises many problems not directly related to survey methodology per se. Central to interpertation is the use of food composition tables. Even the most comprehensive of these suffer major shortcomings. Tables always give single values for nutrient composition of a given food, despite the probability of substantial differences in individual, regional and national samples of the food, even for unprocessed foods. In future the magnitude of such differences must be evaluated and the tables give information about the probable degree of variability to be anticipated. Processed foods are almost impossible to characterise unless legal requirements force divulgence of the composition, at least to appropriate central agencies. In the case of fortified foods, tables must reflect the unfortified status, the current level of fortification being recorded separately.

Overall, there is a desperate need for a central databank and clearing house for food composition information on an international level. The establishment and operation of such a service could be a valuable activity of IUNS.

Given these inadequacies, interpretation of apparent nutrient intakes becomes difficult. Only a broad view of dietary adequacy can be formed, perhaps in terms of quartiles only. Yet in our survey more than 25% of the population appeared to have inadequate intake of say calcium or folate as judged by recommended intake values.

The current definitions of Recommended Dietary Allowances fail to recognise the great dearth of information regarding the absolute requirements for an individual nutrient. They also signally fail to recognise such complexities as control of absorption by dietary composition or gut regulatory systems, or the interaction of nutrients within the body. Clearly on this basis recommendations for nutrients such as iron or calcium must be severely qualified in future.

Our survey has not thus far usilized any biochemical estimations. These should ideally reflect adequacy of both the most critical or "rate-limiting" function of the nutrient, about which in most cases we know virtually nothihg; and the adequacy of body stores for the majority of nutrients which are stored. The problem here is to define satisfactory nutritional status". This is easy to define on paper:

1. All functions dependant on the nutrient in question are functioning at optimum levels.

2. If the nutrient is stored in the body, the stores are sufficient to permit satisfactory function during temporary periods of inadequate intake or periods of temporarily increased requirement or both, such as pregnancy, lactation or intercurrent infection.

3. The level of intake or storage is not such as to engender deleterious consequences.

Trying however to convert these points into practical quantitative definitions is another matter; there is far too little information available on all these aspects. Furthermore we must recognise that there are probably individual, likely genetically determined variations in both nutrient requirements and nutrient handling. There is a desperate need for the study of what might be termed "nutritional genetics". It does not seem justified to promote levels of nutrient intakes which may be difficult to achieve in a population just because a subset of that population has a genetic variation which increases their requirement or reduces their ability to handle common levels of intake (see for example the cholesterol controversy).

Probably the best we can hope for is the specification of a range of intake of each nutrient, perhaps with qualifications about the sources of the nutrient, with the proviso that it must form part of a mixed diet which simultaneously fulfils all other requirements. If the diet does not meet these criteria, then the individual may be able to adapt but this would require special consideration (vide FAO categories of protein sources).

Where does all this leave us in regard to a National Nutrition Policy?

The logic of a National Policy in countries plagued with undernutrition needs no justification, but what of more fortunate nations? There has been a great deal of concern among developed nations that overnutrition, selective undernutrition, or nutrient imbalances may be in part responsible for much morbidity and premature mortality among their populations. Although this argument has centred principally around the relationship of diet to hyperlipidaemia, degenerative angiopathy and related matters, it extends widely to include adiposity, dental caries, gastro intestinal malignancy and many others. If it is accepted that even some of these relationships are valid, then the formulation and implementation of a National Nutrition Policy would be an essential facet of preventative health care. It has recently been suggested that a National Policy is doomed to failure, and that a piecemeal, pragmatic approach would be more effective. This is illogical. Unless a "master-plan" exists, a fragmentary approach is very likely to result in mutually conflicting actions, and the omission of important goals.

Before such a policy can be formulated, much background information is needed. A preliminary diet-anthropometric survey of healthy adults, and of healthy children, will indicate levels of nutrient intake and thus potential danger areas, and also clarify the important food sources of those nutrients to show how changing food habits may influence intake. The current falling milk consumption in our country has clearly important repercussions on calcium intake.

Nutrients whose intake levels appear likely to be unsatisfactory can then be further evaluated by biochemical measurement of nutritional status.

The accumulated information can now be integrated and compared with evidence from other countries, and local morbidity and mortality statistics, to determine what action may ameliorate the situation.

At this stage, a National Nutrition Policy can be formulated. It must not be so conservative as to be ineffective, but neither must it embrace inadequately founded, currently fashionable theories, nor lay excessive emphasis on only a few areas such as degenerative vascular disease.

Promulgation of a policy document, and action to implement its recommendations, is however insufficient. Although very few programmes have given much attention to this aspect in the past, it is obligatory to establish a monitoring system to evaluate the effectiveness, with feedback to modify the policy where necessary.

Monitoring requires two kinds of surveillance: nutritional surveillance and (ill-) health statistics surveillance. The former need not be expensive, but should rather provide a small team to undertake regular, systematic population sampling of anthropometric measurements, dietary habits and biochemical parameters. The Agency can also be responsible for collecting statistics on such matters as child growth and health, and assessing health statistics and related material.

New Zealand has then taken the first steps towards the formulation of a National Nutrition Policy, and a government committee is actively considering the principles of such a Policy.

CRITICAL CONSIDERATIONS FOR

NATIONAL NUTRITION SURVEYS IN AFRICA

Alfred Zerfas

School of Public Health - Dept. of Nutrition - UCLA

Los Angeles, Ca. USA

Since 1975, the School of Public Health, UCLA, under contract to the US Agency for International Development, has assisted four African countries perform cross-sectional national nutrition surveys.

These surveys include anthropometry, hemoglobin estimation and selected clinical signs in children under five years of age and their mothers, to provide estimates of undernutrition point prevalence. We used a statistically valid multistage cluster sample for each country and for selected areas within each country. We also examined demographic, socio-economic, diet and health indicators to provide clues for possible variables related to undernutrition, defined by anthropometry.

Some critical aspects of these surveys are reviewed.

1. Sampling for these surveys may be facilitated by using a valid sample frame from another recent or ongoing national survey. Registration lists and maps of selected sites must be carefully scrutinized.

2. Adequate measurement techniques are of critical importance. Formal standardization tests to review performance are highly recommended.

3. Anthropometry provides valid 'hard' data. For example, based on the sample, the government can be informed how many children under five years actually have undernutrition. The weight-for-height index specifies acute and height-for-age index specifies long-standing undernutrition, based on the National Academy of Science, USA reference data. A refer-

ence, well-nourished group is also measured in the surveyed country. Even then, country nationals often find it difficult to interpret results.

4. For logistic reasons, surveys are usually done during the dry season, most often a time of relative food sufficiency. Hence, the level of acute undernutrition may describe the optimal situation. Usually, there have been no firm plans for follow-up studies.

5. Weight, height and age are the basic anthropometric measures for children. Adequate age estimation is often difficult, due to the absence of reliable records. Arm circumference should be included for baseline information in possible future surveillance. Fatfold measures should be done where obesity may be suspected.

   For technical and logistic reasons, we were unable to weigh mothers. However, we found the arm circumference was a useful tool in assessing undernutrition in this group. Maternal height is strongly recommended, particularly in countries with many different tribal groups.

6. The point prevalence of breast feeding, sources of other milk, special (weaning) foods and shared (family) foods for the child, according to age, can be readily obtained.

   Qualitative diet recall methods ascertain the types of foods eaten by the child and family. These indicate the foods "available" to the family and whether the child receives them or not, an important consideration for nutrition education.

7. Objective health indicators, such as the child's temperature, may be included where feasible, to emphasise the importance of acute infections with nutritional status.

8. The validity of retrospective mortality information is always open to question. We have nevertheless used mortality indices for in-country comparisons, rather than a national description.

9. We fully realise that valid causal information can only be obtained from a longitudinal assessment. Due to the lack of planning on the project for this type of assessment and in order to satisfy at least in part, government requests for nutrition-related information, we examined these on a cross-sectional survey. In most instances, results confirmed what was already suspected with regards to the correlates of undernutrition in young children. These results

could then reinforce program priorities to reduce undernutrition in the country.

THE NEED FOR PERIODICAL REVISION

OF THE NATIONAL NUTRITIONAL STATUS

F.A. Gonçalves Ferreira

Inst. Nac. de Saúde - Centro de Estudos de Nutrição

Av. Pe. Cruz - Lisboa - PORTUGAL

Governments need to be alerted, with regularity, about the nutritional status of their countries' population, so that they may orient the national food policy in its three concrete aspects: food, nutrition and economics, which cannot be dissociated.

Nutrition is one of the basis for the general health of individuals and greatly affects, in either a positive or a negative sense, the prevention and reduction of the severity and diffusion of diseases in communities. The study of the relationships between food consumption and health and nutritional status, or health level, as well as its evolution throughout the life span of people, is essential in an organized society which is pledged to the improvement of the well-being of its members.

The big food and nutrition world problems present themselves under two aspects or models, both very critical:

- the culturally-behind populations of the undeveloped countries lack the essential foods, due to technical incapacity of production, economic acquisition difficulties, or even due to lack of knowledge about the selection and utilization of what is available. It is in these populations — which grow faster than their production does — that there is great prevalence of situations of hunger or generalized deficiencies and classical diseases of malnutrition.

- the populations of the developed, rich and industrialized countries have an excessive consumption of food (calories) and of foods which cause degenerative and metabolic diseases (sugar, saturated fats, distilled alcoholic beverages), and are already suffering the effects of malnutrition by excess, under the form of new generalized diseases of alarming severity.

The countries placed in a position intermediary to these two models are rapidly advancing towards the second, from cities to rural areas. It is urgent that we impede this advance, and help the population move towards a new model — that of rational feeding, with the characteristics of being sufficient, balanced and economical.

The food and nutrition problems to be clarified immediately, in these countries, include:

- the assessment, in terms of food availability and consumption, of how the population is feeding itself;
- the research, at national, regional and social group levels, of existing deficiencies and inbalances, and of the consequences these may have on the health and well-being of the people;
- the definition of a rational food policy, to establish justified measures for internal food production, imports of essential foods and distribution for regular comsumption, in controlled quantities, quality and price, according to the needs;
- the establishment of food education measures, at the national level, utilizing the experience and teachings already gathered by developed countries;
- qualitative and quantitative food research so that safety food "stocks" may be formed, and also to adapt production policy in the field of agriculture, fishing and livestock with the objective of reaching a balanced self-sufficiency or, at least, reducing the most severe deficiencies.

There is no one measure or precise indicator on which to base the assessment of nutritional status, and thus, it will be necessary to collect data from several sources. This data gathering should be continuous because the effects of nutrition, in terms of national health, can be of short or long range.

Many countries do not have concrete data available on individual and family consumption of the foods which, according to the official food statitiscs, would be available to the population.

Experience shows that the practical and correct way of surpassing this backward situation is to immediately carry out a national level survey, planned with precise objectives as to the collection of informative data on food conditions and nutritional status, in an adequate period of time, and taking into account the diversity of the population's characteristics. In reality, it should be a national priority to know how the population is feeding itself and the level of its nutritional status, by means of direct studies about the general and specific food conditions of the population as a whole and in differentiated groups, with the consequent evaluation of how these conditions reflect on their health.

Parting from this knowledge it is indicated to proceed with the periodical revision of the nutritional status, utilizing a few adequate indicators.

The indicators for the continuous assessment of nutritional status are already collected from routine statistical data in many countries, and are related to food supply as well as to different health aspects of the population:

a) information on food supply is provided by statistics of availability and consumption, in general, and by specific food surveys and family expenditure surveys which include foods;

b) health statistics provide elements relative to birth (perinatal mortality, born dead, birth weights), mortality, taking into account certain causes of death, and morbidity, collected from the health services (health centers, hospitals) and, whenever possible, information about the people and groups themselves.

But some of these indicators may need to be completed by means of special studies in all countries.

The priority which should be given to nutrition, as the basis for the population's health, implies that the level and changes in nutritional status should be known throughout time. For this, governments should promote the means to obtain the necessary information elements, which include:

a) the so-called running-indices, related to statistics about food supply and consumption and the health of the population and which provide information about the existing foods, ways in which they are consumed, and severe effects in the health of individuals, beginning from intra-uterine life;

b) periodical nutritional surveys, at the national level and, with shorter regularity, in populational sectors or groups, to include anthropometric, biochemical, hematological and socio-economic information;

c) special studies for the clarification of infant feeding problems (hypernatremy and dehydration, breast feeding), specific diseases (rickets, lack or excess of protein, insufficient fiber, obesity), cardio-vascular diseases and dietetic factors in the changes to foods which are rich in fats, sugar and calories and low in bread.

Simultaneously, on the other hand, the population should be well informed about the rational food and nutrition problems which directly affects them, so that they may accompany what is being done to improve their nutritional status and even, more important, may contribute in a conscious and voluntary way for this to happen without delay.

ASSESSMENT AND MONITORING OF NUTRITIONAL HEALTH

Arnold E. Schaefer, Ph.D. - Director

Swanson Center for Nutrition, Inc. - Suite 101

8401 West Dodge Road, Omaha, Nebraska 68114 - USA

Some of my friends in Government continually refer to the ICNND surveys as those very detailed expensive studies. At no time in the hisroty of U.S. assistance in nutrition has a program reached so many countries (35) at such a small investment (less than eight million dollars). The country nutrition surveys conducted during the tenure of ICNND 1955-67 by the very nature of the limited time involved, usually two to three months of actual field time, furnished data on the nutritional status and problems occurring at that time. To obtain more accurate assessment of the overall nutrition problems it was obvious that additional stydies and evaluation programs needed to be implemented during various seasons of the year. The salient feature of the surveys was that laboratory facilities and supplies were left in the country and key specialists were trained in nutrition assessment procedures. Recomendations were developed by the country specialists in concert with the Ministries of Agriculture, Health, Defense and Planning and directed toward maximizing use of local resources.

I repeatedly hear — "We need a cheap, easy, quick method of assessing nutritional status". Wouldn't it be great if a simple immunological type test could be performed to say "yes" or "no malnutrition"? My good friends in U.S. AID have been consistent in referring to the ICNND surveys as those "expensive time-consuming surveys". This can be excused on the basis of ignorance. I can now observe how little they get in terms of data, or training or follow-up action when one hires non-country consultive firms to do a speedy survey (merely a rehash of other data) which now gets classified as a "Systems Analysis Approach".

My good friend Dr. Horowitz, former Director of PAHO, continually reminded me that we needed an "inexpensive method". Obviously we all agree on the merits of economy. My sincere comment was "Would you be willing to support a study of the living children at the same rate of financing that PAHO spent on studying the dead (mortality index of children in Latin America)?" The funding for nutrition appraisal in 1970-72 was less than $20,000 per year compared to the cost of the mortality study which was fifty times this amount. The answer is still obvious — look at this year's PAHO budget for nutrition assessment.

As nutritionists we have failed to convince first the Health Directors on the social, economic and humanistic cost of malnutrition. How then can we expect to have legislators and Government bureaucrats respond with sympathy and action?

Likewise, FAO as well as our own USDA and AID support large food consumption surveys. The objectives have merit, however, invariably they are over-extended, i.e., "costly" on the basis of the soft data they provide and analysis of the data is usually delayed two to four years (in the U.S. this lag time is usually five years or longer).

What alternatives exist? COuld one obtain the necessary food consumption, cost and market data from a smaller sample frame and establish sample market basket surveys on a continuing monitoring basis? If so, one could then utilize this technique to estimate food and nutrient intake data as available at local market areas. Could one use the same "sample frame", merely adding on other nutrition assessment methods such as physical, anthropometric and biochemical?

CENTRALIZED-MONITORING SYSTEM

There is no single simple approach for assessment of nutritional health status. Since nutritional status of a population group is not static, there is need for continued surveillance of the effects of the changing food supply, economic and social status, on healtn of the populations. Fundamental to any monitoring system is the need for the following:

1. A data collection and analysis center with liaison to the Departments of Health, Statistics, Agriculture, Census, Finance and Planning.

2. Health, nutrition and agriculture personnel assigned the task of monitoring and evaluating.

3. Short term specialists — utilization of extra Departmental and University personnel — for conducting scheduled surveys,

special monotoring studies and for analysis of data and conduct research when needed.

The organization must provide for close and continuing contact with planners, data sources and research people. The primary priority for such a system is to make meaningful recommendations for current national planning and then to evaluate the consequences.

The first requirement for planning a nutrition assessment or a monitoring program is to define clearly and precisely the objectives since this is essential to identify the sample frame, choice of methods to be employed, interpretation and utilization of findings. Let us assume that the key objectives are:

1. to provide sound scientific data to define the prevalence and risk or malnutrition, its location, causes;

2. to define current and potential resources (food, manpower, education, funds) for practical solutions;

3. to establish a baseline of nutrition and health data from which future monitoring and evaluation can be made in reference to the success or failure of intervention programs.

METHODS OF COLLECTION

The pathogenesis of nutritional deficiency disease provides the challenge for us to employ methods of assessment that will predict potential hazard and will identify those individuals or groups that are "risk candidates" for developing health problems. The pathogenesis of malnutrition emerges from the following series of events:

1. Inadequate nutrient intake or impaired absorption and utilization;
2. Depletion of tissue reserves;
3. Interference with normal biochemical, metabolic reactions;
4. Functional changes;
5. Clinical-physical signs and symptoms.

SIMPLISTIC ASSESSMENT AND MONITORING TECHNIQUES

The methods employed obviously are dependent upon the resources available. In many cases this may well be limited to one assessment method such as anthropometric measurements in combination with a general food availability survey.

In most situations resources are limited, thus one needs to utilize all existing data collection programs, i.e., mortality, morbidity, and other health and population records, agriculture reporting systems, food export-import data, production estimates, and census data.

Monitoring of nutritional health to be meaningful should be based on a sample population which enables one to make general inferences to a larger population segment. Stratifications of the whole population can be based on prior information suggestive of subpopulations more nearly homogenous than the total population. Again, simplicity of survey reference methodologies to be employed and cost in terms of money, personnel, and time are directly related to sample size.

The Pan American Health Organization in 1973 developed a working document on the general guidelines for establishment of a data system for the assessment of nutrition and health status. I was actively involved in the development along with D. M. Guzman of INCAP and the nutrition staff of PAHO. The report contains guidelines for sample size. For example, let us assume that a full scale nutrition survey is not possible and it is desirable to screen small communities in a given region in order to select locations for action programs on a reasonable priority basis. In this case, the children under five years of age can be used as an indicator of community nutritional problems. Assessment could consist of measures of height and weight for age in combination with an abbreviated food consumption and food pattern questionnaire. If resources permit, then the next priority item would be a hemoglobin and/or hematocrit determination. If a problem of vitamin A deficiency is suspected, then a cursory review of the mother's observation on whether the child suffers from night blindness could be included. If the blood data indicates anemia (moderate to severe risk), then a detailed follow-up study should be made of a selected group to establish the etiology of the anemia and to determine corrective treatment and preventive measures. One can often obtain this type of data on a detailed hematological study of ten to twenty children.

Example of sample size:

Communities with less than 800 inhabitants - all preschool children
Communities with 1,200-1,500 inhabitants - 50% of preschool children
Communities with 3,000-5,000 inhabitants - 20% of preschool children

Priority of Methods: Simple to more complicated

1. Anthropometric Measurements
   1) height and weight for age
   1) arm circumference

2) skinfold thickness
2) head circumference

2. Dietary Studies
    1) frequency food pattern data
    1) food availability-cost (agricultural census data)
    2) twenty-four hour recall on subsample

3. Socio-Economic Data
    1) locale description and individual family data

4. Biochemical Studies
    1) hemoglobin/hematocrit
    2) subsample for added tests if resources are available
       — serum iron, ferritin, transferrin saturation index
       — serum vitamin A
       — serum albumin
       — serum vitamin C
       — RBC folate

5. Physical Examination
    1) abbreviated examine recording "nutrition indicator lesions" of malnutrition
    2) disease survey (parasitology, ect.)

REQUIREMENTS TO INSURE VALIDITY AND RELIABILITY

1. Sample - it is obvious that definition of sample structure is imperative. It should consider random cluster stratification of vulnerable groups such as 0-5 year olds, pregnant mothers and, if resources permit, entire family groups. Likewise, within the clusters random probability sampling procedures must be followed.

2. Standardization of survey techniques — even the simplest of methods such as height-weight measurements require very rigid specifications, continuous standardization of equipment, and trained personnel whom likewise must be standardized in their measurement techniques. Laboratory tests require "inter" and "intra" controls and standards.

3. Uniformity of recording data — data collection formats must include coding and editing and provide establishment of a standard data processing, storage and retrieval system.

REQUIREMENTS FOR FOLLOW-UP ACTION

The most urgent necessity is to provide a quick flow of data to yield a meaningful report on findings, causes and potential action needs. This data must be furnished all agencies having a role in insuring the health and improving the economic and education status of the population.

It should define areas of priority action, research, and the need for in-depth studies, and provide guides for evaluating effectiveness of action (intervention) programs.

SUMMARY

Why? What for? What Will Be the Follow-up Action?

These questions should be asked before ever launching a national monitoring service. Merely collecting data and writing status reports invariably accomplishes little if anything as regards improving the nutritional health of the population.

Governments must be responsive to the findings. This requires a commitment by those agencies or departments that have operational responsibility (Agriculture, Health, Education, Finance). There are enough examples of monitoring which result in zero action.

NEW APPROACHES TO THE ASSESSMENT OF NUTRITIONAL STATUS — SELECTION AND UTILIZATION OF NUTRITION INDICATORS — THE JAMAICAN EXPERIENCE

V.S. Campbell

Scientific Research Council

Kingston, Jamaica, West Indies

INTRODUCTION

The nutritional status of segments of the Jamaican population has been assessed directly by clinical, biochemical and anthropometry and indirectly through the collection of vital statistics, e.g. infant mortality rate, food consumption and inter-related factors.

It has been found that birthweights are frequently low; death rate/1000 live births in 1976 was in the region of 20.3. Nutritional deficiency has been indicated as the cause of death of 200 children 1-4 years in 1976. About 7% of the children birth to 3 years suffer from medically detectable malnutrition (Gomes Grade II), some 30% are on the borderline (Gomez Grade I) and approximately 1% is severely malnourished. (See Appendix I for Population Structure). Little is known of the nutritional status of the School child other than the fact that children from lower socio-economic households do not grow as well as their more privileged counterparts.

In the adult population, 45% of pregnant and lactating women are anaemic. These problems have been associated with poverty and factors pertaining to food availability.

In selecting nutrition indicators, emphasis has been placed on the food supply system and health programmes. It is pertinent to give a brief resumé of Jamaica so that the measures adopted can be viewed in perspective.

Jamaica, a tropical country, has a population of approximately 2 million and covers an area of 4,411 square miles. Forty-six (46) percent of the population (830,000 persons) are under 14 years and

41% of the population lives in towns. One-half of the population in the 14-24 year age group is employed and in general some 11% of the total population is unemployed. There are 1.2 million acres (486,000 hectares) of arable land. Only 6% of the land in farms is devoted to domestic crops, most holdings comprising <5 acres, while the large farms of 500 or more acres concentrate on export crops. Thirty-one percent (31%) of the labour force (890,000 persons) is employed in agriculture, forestry and fishing. There is a per capita national disposable income of J$1,223.00 (US$765).

THE FOOD SUPPLY SYSTEM

There is a high import dependence for basic foods, animal feeds and fertilizer. In 1975 of a total import bill of $212.6 million, $118.1 million dollars were spent for food. With astringent foreign exchange budgeting the amount of food imports fell to $70 million in 1976. There is a policy to import cheap foods which might have a negative effect on the development of local supplies.

There is a high wastage of farm products (20-30%), and loss of cereals due to rodent and insect destruction is approximately 15%.

Cereals (rice, wheat flour, cornmeal) provide approximately one-third of the nation's energy and protein intake and 95% of this food category is imported.

There is a market economy: a complex food distribution system and attendant fluctuations in prices and frequent shortage of foods. Fluctuations in prices of staple foods have, to a certain extent, been rationalized through controls on basic commodities, e.g. cereals, but there has been no attempt to stabilize or control prices of locally grown agricultural products.

FOOD COSTS

Food balance sheets however show that there is ample food available but as these are based on averages it is well appreciated that there are those who receive far below minimum requirements.

In mid-1977 it was estimated that $23 per week would be necessary to satisfactorily feed a family of 4 (2 adults, 1 school child, 1 toddler) using primarily locally produced foods. This was above the then minimum wage of $20 and certainly several orders of magnitude greater than the public assistance dole of $4 per person per week.

The ratio of the cost of food to family income in 1975 shows that some 60-70% of the income of low income earners was spent on food, but this food did not necessarily meet the nutrient requirements of the individuals.

The Scientific Research Council has since 1969 kept an annual, and in some cases half-yearly, record of the cost of a day's meals comprised of cereals, legumes, salted meat/fish, sugar, fruit and vegetable for a family of 5 having the structure as stated above with the addition of an adolescent and found that in December 1969 the cost of those meals was $1.70 while in June 1978 it was $7.90, certainly due to the related independent variable inflation.

DIETARY INFORMATION

There is a taste (preference) for imported foods — tradition and pricing being major factors. The latter is no longer entirely true as the much loved saltfish (cod) has now soared to the princely price of $3/lb. Low income groups' (< $2000 per year) intake of foods fall short in energy by 27% and protein 14%. This grouping constitutes 70% of the population.

MEASURES ADOPTED TO IMPROVE FOOD SUPPLY SYSTEM

1. Increased level of state involvement in food production and distribution.
2. Production drive — formation of cooperatives and provision of more lands for domestic food production (land lease).
3. Bulk buying of grain and other nutrition-related raw materials.
4. More staple foods made available to low income areas at subsidized prices.
5. Monitoring of food supplies and prices.
6. Formulation and adoption of a food and nutrition policy.

HEALTH INDICATORS

Growth and Body Dimensions

In Jamaica, reports on birth weight have indicated that they are frequently low with a median of 3.1 kg. and in the obstetric ward of one hospital, 1 in 8 of all newborn babies weighed less than 2.7 kg (80% of standard).

Weight for age is the most commonly used indicator. Combinations of height/weight, arm circumference for height or weight for head circumference are used more infrequently. Weight for age data is collated from maternal and child health centres, hospitals and the wider community through the use of health aides of which there are some 1,169. The data are collated on the World Health Organization (WHO) Growth Chart and classified by the Gomez system of nutritional status.

The 1967 report of the Ministry of Health's Surveillance programme for 10 of the 13 parishes is given in Table 1.

Nutritional status of children under 3 years is monitored by anthropometry at the maternal and child health centres and in their homes. Some children of necessity will have both types of coverage while others will not. For all parishes less than 42% of the under 3 year old population attend clinic and of these only approximately 50% receive optimum coverage. In general, less than 20% of the under 3 year population receives adequate health coverage. The target is to increase coverage of the under 3 year old age group to 80% by 1983. The health care system for early detection, prevention and promotion begins at the community level (primary health care) so that there will, hopefully, be less at the secondary (hospital) and tertiary (special treatment) levels.

Preliminary data from a survey (1977-78) of 3,000 children 0-3 years show the profile detailed in Table 2 under the Gomez system of nutritional status classification.

The researchers, Gurney et al., indicate that these data show marked improvement over 1970 data. It is to be noted that socio-economic and dietary data were not obtained.

The availability of appropriate weighing machines remains a problem and with changing hair styles, how practical are height data for the older child?

For the school child there is sparse data. Mial et al. in their continued evaluation of urban school children from different socio-economic strata show that children from the more privileged households are taller and heavier than those in less privileged circumstances. The results of a recent survey of the effectiveness of the school feeding programme should expand information in this area.

Biochemical

Haemoglobin levels in adults primarily have been determined although researches into the trace mineral content of blood and implications in malnutrition are on the increase.

A quick copper sulphate method for screening ante-natal patients for anaemia has been devised and is now in use by health auxiliaries.

TABLE 1

PROPORTION OF CHILDREN AGED 0-35 MONTHS[a] IN EACH PARISH BY NUTRITIONAL STATUS DURING 1976[b]

PARISH	NUTRITIONAL STATUS (GOMEZ) (Percentage)					
	Above Normal %	Normal %	Grade I %	Grade II %	Grade III %	(N)
*K.S.A.C.	10.7	67.0	19.4	2.5	0.5	10,399
*St. Thomas	7.4	60.7	27.2	4.1	0.7	2,307
*St. Mary	7.3	63.4	24.9	4.0	0.5	2,815
*St. Ann	8.1	64.8	23.1	3.6	0.5	2,339
*St. Elizabeth	6.8	61.1	26.4	4.9	0.8	4,193
*Claredon	7.7	64.0	22.6	4.7	1.1	5,149
St. James[1]	1.6	62.4	32.2	3.4	0.3	5,326
Hanover[2]	0.3	71.2	24.8	3.4	0.3	4,544
Westmoreland[3]	0.7	71.7	21.9	5.2	0.6	2,674
*St. Catherine	9.0	63.1	21.7	5.0	1.1	3.670

[a] Except Hanover, 0-47 months age group.
[b] Except K.S.A.C. & St. Catherine, 1975 data used.
* Clinic attenders only.
[1] Whole population, Mondego Bay excluded i.e. only rural St. James considered.
[2] Whole population.
[3] Whole population; Lambs River and Grange Hill excluded.

N = Number of children.

TABLE 2

NUTRITIONAL PROFILE OF SURVEYED CHILDREN

AGE YEARS	NORMAL %	GI %	GII %	GIII %
< 6/12	78.6	14.9	6.0	0.5
> 6/12	65.6	26.2	6.8	1.4
1-2	56.1	34.8	7.9	1.1
2-3	53.7	37.0	8.4	0.9

Immediate treatment through the distribution of an iron complex is effected and the patient is informed without delay. Appropriate technology for routine screening for anaemia is thus in vogue. A haemoglobin level of 10g/100 ml has been taken as the norm for the population as the suggested international levels of 11-13 g/100 ml have been found too high for the levels normally found.

SUMMARY

The target groups of the 2 million population under nutrition surveillance, in descending order of priority, are children 0-4 years, mothers and school children. Moderate protein-energy malnutrition in children and anaemia in adults have been identified.

- Practical socio-economic, health and dietary indicators applied include food cost: family income, food supplies, prices of staple foods, growth of young children, haemoglobin levels in adults, vital statistics and primary health care coverage.

- To regulate the food supply in a market economy conditioned by inflation, scarce foreign exchange, low food production, high unemployment and a per capita disposable income of $1,223 (J) ($765 US), there has been increased levels of state involvement in employment, food importation, production and pricing.

- The weight for age index of child growth has been most practical and general health coverage has been expanded through the establishement of maternal and child health centres (15 in 1975-76) and the use of Community Health Aides, affording 76% coverage in some areas.

- A quick copper sulphate method of screening ante-natal patients for anaemia is in use.

- The death rate in the 1 year age group was 15.16/1000 in 1977 compared with 20.3/1000 in 1976.

- Levels of nutrition in pre-school children have improved as shown by 1978/1970 data.

ACKNOWLEDGMENT

The cooperation of the Ministry of Health and Environmental Control primarily the Nutrition Department, the Caribbean Food and Nutrition Institute, the Department of Statistics, and the Tropical Metabolism Research Unit, UWI, in providing the data and colleagues in the Scientific Research Council for collating and typing is gratefully acknowledged.

APPENDIX I

POPULATION STRUCTURE OF JAMAICA

AGE GROUP	NUMBER (in '000)	PERCENT
0 - 1	54.6	2.7
1 - 3	232.6	11.5
4 - 9	300.9	14.9
10 - 15	244.1	12.1
16 - 19	165.9	8.2
20 - 39	561.0	27.8
40 - 49	160.7	8.0
50 - 59	143.1	7.1
60 - 69	91.7	4.6
70 + years	61.8	3.1
TOTAL*	2,016.4	100.0

* 1975 estimate.

BIBLIOGRAPHY

A Food and Nutrition Policy for Jamaica, 1975.
WHO. Pocketbook of Statistics, Jamaica, 1977.
Commonwealth Caribbean Medical Research COuncil, Report of Twenty-second Scientific Meeting, Belize, 1977.
Commonwealth Caribbean Research Council, Report of Twenty-third Scientific Meeting, Barbados, 1978.
Gomez, F., Galvan, R.R. et al., J. Trop. Pedia. 2, 77, 1956.
Gurney, J.M., Fox, H. and Neill, J., A Rapid Survey to Assess the Nutrition of Jamaican Infants and Young Children in 1970. Trans. of the Royal Society of Trop. Med. & Hyg. Vol. 66, No. 4, pp. 653-663, 1972.
Jamaica Department of Statistics, Annual Report, 1976.
Jamaica Department of Statistics, Expenditure Patterns of Working-class households (unpublished), 1975.
WHO. Methodology of Nutritional Surveillance. A Report of a Joint FAO/UNICEF/WHO Expert Committee. Technical Report Series 593, WHO, Geneva 1976.

REPORT AND RECOMMENDATIONS OF THE WORKSHOP ON "NEW APPROACHES
TO THE ASSESSMENT OF NUTRITIONAL STATUS. SELECTION AND
UTILIZATION OF NUTRITION INDICATORS"

A. E. Schaefer

Director - Swanson Center for Nutrition, Inc.

8401 West Dodge Road Suite 101; Omaha, Nebraska 68114 USA

Nutrition assessment procedures for population groups in the developing countries must of necessity to kept simple and yet provide sound scientific data to define the prevalence and risk of malnutrition (over and under nutrition) its location and causes. The surveys should define the current and potential resources (manpower, food, money, education) for practical solutions. Likewise it should establish a baseline of nutritional health data from which future monitoring and evaluation can be made in reference to either the success or failure of intervention programs.

Currently there is no one simple procedure to evaluate nutritional status. The discussion included a few newer techniques which have merit whenever resources are available — such as (1) determination of prealbumin as an early indicator of protein malnutrition; (2) serum ferritin as an indicator of iron status and (3) ultrasonic measurements of lean body mass for obesity assessment. The approaches currently used included programs involving every local community (Philippines) in which every child is given a height, weight, age measurement. This included to data over 4.5 million children with the data base being used at the local level to record the problem in individual homes and to evaluate and monitor progress. This is an example of the data findings moving upward to the central government instead of the usual procedure of having the data findings "sent-down" from the central government. The "HANES" health and nutrition examination survey in the United States is a multi-procedual survey conducted on a random sample of the entire population. The procedures are similar to those developed by the ICNND International studies including anthropometric, biochemical, clinical, and dietary assessment along with other health, social and economic data.

Special emphasis was given by many of the participants to the necessity of including in the very limited anthropometric data assessment information on food availability, habits and customs including breast feeding. Herein the priority groups are the 0-5 yr. old child and the mothers. If resources are available the priority activities should include assessment of anemia, then protein and Vitamin A determinations (if necessary on a sub-sample). If Vitamin A deficiency is suspected special query of the mother should include as to whether the child suffers from night blindness. Survey personnel should be aware of key nutritional deficiency lesions such as enlarged thyroids, "Bitot Spots" basic, skin lesions, edema, etc.

The developed countries must be concerned with assessment of over nutrition and its relationship and consequences to other chronic diseases (diabetes, cardiovascular and renal disease and cancer). Thus it is prudent for them to include special methodology for assessing the degree of fatness or lean body mass, as well as serum lipid analysis and measures of physical fitness.

CONCLUSIONS

Sampling in the developing countries can often be facilitated by utilizing other survey bases, i.e. agriculture or household economic surveys. Methods and personnel must be standardized at the beginning and continuously throughout the study. This must be rigid and precise for even such simple methods employed for anthropometry. Basic anthropometric measurements should include height, weight for age, area circumference and when possible skin fold measurements. Interpretation of the data requires a "United Approach" — this is still lacking, and these countries often employ various interpretative guides. Fatfold measurement should be included to assess leanness, as well as obesity. Dietary assessment should at a minimum include qualitative recall of available foods and distribution within the family. This should include breast and infant feeding practices. Other health data should be noted such as mortality, morbidity, and infectious disease rates.

RECOMMENDATIONS

1. Food nutrient composition data banks need to be expanded and updated — especially on processed foods.

2. There is need to refine dietary standards to more adequately reflect requirements instead of "safe dietary allowances". Iron, Vitamin A, Calcium, trace minerals require special investigation as regards imbalance and on the influence of other foods on their availability (absorption and utilization).

3. Standard growth curves for special ethnic-cultural groups need to be developed.

4. IUNS should assist in developing a "standard language" in defining what is meant by malnutrition, its types and degrees.

5. Emphasis should be placed on potential early indicators of fetal risk. This to include sophisticated enzyme analysis as well as standards on mothers' height, previous pregnancy outcome, pregnancy weight gain, etc.

6. Continous efforts to develop improved simple biochemical screening tests to indicate potential body nutrient depletion in the early stages of development.

THE FOOD AND NUTRITION SITUATION

AND THE PERSPECTIVES FOR FOOD DEMAND IN MOROCCO

M'Bareck Essatara

Chief of Human Nutrition and Food Economics Dept.
Institut Agronomique et Vétérinaire
Hassan II - B.P. 704 - Rabat, Agdal - MOROCCO

The state of food and nutrition on a countrywide level can be understood by several methods:

- the balance sheets of food availability consist of measures utilizing an average year which bring to light various uses of food production that can be translated into terms of mean energy and nutrient contribution;

- nutrition surveys including consumption surveys and clinical surveys.

I. THE FOOD AND NUTRITION SITUATION

The household consumption budget survey (Sec. d'Etat au Plan, 1973) carried out on 6,309 households, of which 47% were urban and 53% rural, showed that the total private consumption expenditures was 663 Dh/person/year in rural areas, 1,378 Dh/person/year in urban areas and 900 Dh/person/year for the country as a whole.

Cereals

Cereals constitute the basic foodstuff of the Moroccan population. One quarter of the food budget goes to these products (25.3%).

Expressed in quantities of cereals the mean consumption is 216 kg/p/yr for the whole country, 245 kg in rural areas and 159 kg in urban areas, the level of subsistence being 50% and 2% respectively.

Meats

Meats occupy the second rank after cereals in terms of food expenditures (22.1%); the consumption in urban areas is nearly double that of rural people. The level of subsistence is rather low in general at 8.3%, except for the rural portion, where it can reach 50% for fowl and rabbits.

Fish

Although Morocco is a large producer of fish, the level of consumption is rather low, 1.8 kg/p/yr in rural areas, 7.1 kg in urban areas and 3.1 for the entire country. Specific share of total food costs is 1.3% of the budget.

Eggs

Egg consumption is low at 36 units per person annually in the urban areas, 14 for the rural and 21 overall.

Milk and Cheese

Consumption is not high and represents only 4.6% of food costs. Expressed in equivalents of milk by the litre, consumption is 29.8kg per person per year. The urban population consumes mainly fresh milk (20.1 litres). However, the rural population uses whey for the most part.

Fats and Oils

Specific expenditures for edible fats and oils represent 8.4% of the average food budget. The quantities consumed are from 11.8 kg per person per year for rural areas to 15.94 for urban areas, and 13.11 kg for the population as a whole.

Sugar and Sweetened Products

Maroccans are heavy consumers of sugar. They pay 11.9% of the food budget for this substance, urban sugar consumption is 31.2 kg/p/yr, 26.5 kg for rural people and 29.7 kg for the whole country.

Vegetables

Specific expenditures for vegetables represent 8% of food costs;

root tubercules and bulbs come at the head of the list.

Fruits

8.8% of an average food budget is reserved for purchases of these products. The quantities are 45.8 kg/p/yr for the rural, 47.7 kg for the urban, and 46.5 kg for the country.

Tea, Coffee and Aromatic Plants

Rural consumers reserve 6.6% of their food budget for acquisition of these products, their urban counterparts use 5.3% and the whole country average is 6.1%.

Total consumption is 5.5 kg/p/yr which includes 1.49 kg of tea in rural areas 8.9 of which 1.0 for tea in urban and 1.6 of which 1.3 is for tea in the whole country.

Spices and Condiments

Average consumption is 7.3 kg/p/yr for urban areas, 10.7 for rural people and 9.6 for the whole country. This represents 1.8%, 1.9% and 1.9% respectively, compared to total food costs.

Table 1 shows the profile of the food ration according to location and for the country as a whole.

Energy and nutrient resources expressed per person per day for the rural, urban and country are shown in Table II; the nutritional balance of the average food ration is given in Table III.

Next, one must compare the energy and nutrient contribution to the recommended contribution or recommended allowances or needs in order to understand the dietary status of the population. The latest FAO/WHO (1973, 1974) norms have been used in calculating the recommended intake of energy and nutriments for the Moroccan population (Laure, Essatara, Jaouadi, 1977; Sec. d'Etat au Plan, 1971). These recommendations are shown in table IV.

A conparison of Table I and II on the one hand and Table IV on the other show that the energy needs are covered a priori for the whole country and the rural areas, but not for the urban people.

At first glance the need in nutriments seems to be covered for the country as a whole, rural and urban except for calcium where there is deficiency, iron which is low in urban areas, vitamin B_2 (riboflavin) which is deficient at the country level and Vit. PP

TABLE 1

FOOD RATION FOR 1970 – 1971 IN kg/PERSON/YEAR

FOOD SOURCES	RURAL AREAS	URBAN AREAS	TOTAL
Cereals	245.0	159.0	216.4
Meats	14.8	24.2	17.9
Fish	1.8	7.1	3.6
Dairy products	32.3	24.6	29.6
Eggs (units)	14.0	36.0	21.0
Fats	11.8	15.9	13.1
Sugars	31.2	26.5	29.7
Sweetened products	0.2	1.0	0.5
Vegetables	71.2	123.6	88.7
Fruits	45.8	47.7	46.5
Non-Alcoholic beverages	0.2	4.4	1.6
Alcoholic beverages	–	2.2	0.8
Tea, Coffee and aromatic plants	5.5	8.9	6.6
Spices and condiments	10.7	7.3	9.6

TABLE 2

DAILY ENERGY AND NUTRIMENTS INTAKE PER PERSON

	RURAL AREAS	URBAN AREAS	TOTAL
Energy (Kcal)	2,600	2,202	2,466
Proteins (g)	75.0	62.7	71
Lipids (g)	45.1	58.7	50.2
Calcium (mg)	298	295	296
Iron (mg)	15.9	11.4	14.3
Vitamin A (I.V)	2,696	4,425	3,281
Thiamin B_1 (mg)	1.8	1.0	1.0
Riboflavin B_2 (mg)	0.6	0.4	0.5
Niacin PP (mg)	18.1	10.6	15.6
Ascorbic acid (mg)	38.3	76.2	50.6

TABLE 3

NUTRITIONAL BALANCE OF FOOD RATION

		RURAL AREAS	URBAN AREAS	TOTAL
Energy (%)	Carbohydrate	72.8	64.8	70.2
	Cereal	66.3	52.3	62.3
	Protein	11.5	11.4	11.5
	Lipid	15.7	24.0	18.3
Proteins (%)	Plant proteins	86.6	73.2	82.7
	Cereal proteins	78.4	59.2	72.7
	Animal proteins	13.4	26.8	17.3
	$\frac{\text{Animal}}{\text{Total}}$	13.5	26.8	17.3
	$\frac{\text{Animal}}{\text{Plant}}$	15.6	36.6	21.0

TABLE 4

RECOMMENDED DAILY INTAKE
BY PERSON FOR MAROCCAN POPULATION

Energy (Kcal)	2,240
Proteins (g)	52.85
Calcium (mg)	460
Iron (mg)	15.1
Vitamin A (IV)	1,970
Thiamin B_1 (mg)	0.90
Riboflavin B_2 (mg)	1.30
Niacin PP (mg)	14.8
Ascorbic acid (mg)	26

(Niacin) lacking in the urban sector.

Analysis of Table III shows that the food ration is more unbalanced than it is energetically insufficient, at least for the averages across the country, rural and urban. Yet we know that these means, as usual, mask the reality. The diet is characteristically relatively excessive in carbohydrates and cereal calories especially in rural areas, and has a relative deficiency of calories of protein origin no matter what the sector. The share of proteins of good biological quality in total protein intake remains low. The share of calories of lipid origin is also relatively low.

However, the level of consumption related to income explains more fully the dietary situation. As shown in Table V, the rate of national caloric coverage of 110% seems at first glance to provide a satisfactory dietary situation, but this merely hides a different reality. In effect, the mean of caloric availability just reaches or does not quite touch the estimated needs of the poorest income class, representing half the population. Among these classes the lowest one-third of the population (33%) have mean levels of availability below 80% of caloric needs.

On the other hand, for another third of the population 33% the mean in caloric availability exceeds 130% of need, reaching as high as double the recommended amount for 8% of Moroccans. As for the share of total proteins, although the national mean is 134% of need, a third of the people do not receive the recommended amount in proteins, 8% barely makes it, while a fifth (21%) of the people use more than double the recommended level.

The protein-calorie deficit is well confirmed by the national nutrition survey of 1970-71 at least for children up to four years of age (Bull. de la Santé Publique, 1973). It appears that at the national level 41.58% of children exhibit moderate protein-calorie malnutrition (weight deficit of 20-40%) and deserve nutritional rehabilitation. These rates are 44.78% in rural and 33.34% in urban areas.

Nearly 5% of children (4.66% overall, with 5.36% rural and 2.36% urban) displaying severe protein-calorie malnutrition (weight deficit greater than 40%) are in mortal danger and should receive hospital treatment. In all, 46% of the children (50% rural and 36% urban) are affected by protein-calorie malnutrition.

Concerning nutritional needs in calcium and the problem of rickets, rickets is frequent in children at least among those less than five years of age, due to an insufficient intake of calcium acompanied by a deficiency in vitamin D, resulting essentially from the clothing habits and the way of life.

TABLE 5

PERCENTAGE RATE OF RECOVERY OF DAILY REQUIREMENT PER PERSON ACCORDING TO INCOME CLASS

TOTAL INCOME CLASS DH/PERSON/YEAR	POPULATION %	ENERGY Kcal %	PROTEIN g %	LIPIDS g %	CALCIUM mg %	IRON g %	A I.V %	B_1 mg %	B_2 mg %	PP mg %	C mg %
moins de 214	8.3	646 / 29	17.8 / 32	11.0	81 / 18	3.9 / 26	777 / 39	0.3 / 33	0.1 / 8	4.0 / 27	13 / 49
214 - 310	8.3	1,267 / 57	34.1 / 56	20.9	135 / 29	7.5 / 50	1418 / 72	0.8 / 89	0.2 / 15	8.3 / 56	19 / 74
310 - 384	8.3	1,652 / 74	45.6 / 76	27.7	185 / 40	10.1 / 67	1422 / 72	1.1 / 122	0.3 / 23	10.9 / 74	24 / 91
384 - 466	8.3	1,784 / 79.7	50.8 / 90	30.7	211 / 46	10.6 / 70	2336 / 119	1.1 / 122	0.3 / 23	12.2 / 82	28 / 108
466 - 542	8.3	2,033 / 91	56.6 / 99	36.2	247 / 54	12.3 / 81	2323 / 118	1.3 / 144	0.4 / 31	13.9 / 94	35 / 133
542 - 627	8.3	2,236 / 99.8	65.4 / 113	37.3	254 / 55	13.2 / 87	2247 / 124	1.4 / 156	0.4 / 31	14.5 / 98	34 / 132
627 - 732	8.3	2,582 / 115	69.8 / 125	49.2	318 / 69	14.8 / 98	3047 / 165	1.5 / 167	0.6 / 46	15.8 / 107	46 / 178
732 - 864	8.3	2,790 / 125	79.2 / 146	53.9	319 / 69	16.2 / 107	3750 / 190	1.9 / 211	0.6 / 46	17.3 / 118	54 / 208
864 - 943	4.2	2,952 / 132	84.4 / 156	55.9	333 / 72	17.1 / 113	3799 / 183	1.8 / 200	0.6 / 46	17.7 / 120	58 / 222
943 - 1042	4.2	3,024 / 135	87.2 / 165	59.7	381 / 83	17.9 / 119	3971 / 202	2.0 / 222	0.7 / 54	20.3 / 137	64 / 246
1042 - 1171	4.2	3,076 / 137	90.4 / 177	62.9	344 / 75	17.9 / 119	4505 / 229	1.9 / 211	0.6 / 46	17.8 / 120	65 / 250
1171 - 1358	4.2	3,961 / 177	115.7 / 215	76.4	461 / 100	23.4 / 155	4462 / 227	2.7 / 300	0.9 / 69	24.8 / 168	73 / 282
1358 - 1641	4.2	3,395 / 152	101.8 / 207	79.5	391 / 85	18.3 / 121	5446 / 276	1.9 / 211	0.6 / 46	19.1 / 129	88 / 325
1641 - 1977	4.2	3,024 / 143	122,0 / 220	86.6	495 / 108	23.1 / 153	6213 / 315	2.6 / 289	0.8 / 62	22.9 / 155	99 / 382
1977 - 2682	4.2	4,442 / 198	130.8 / 278	107.1	557 / 121	24.8 / 164	6502 / 330	2.8 / 311	1.1 / 85	24.4 / 165	116 / 448
2682 et plus	4.2	4,460 / 119	131.5 / 326	132.0	643 / 140	23.0 / 152	8619 / 438	2.2 / 244	1.1 / 85	21.8 / 147	150 / 579
TOTAL	100.0	2,466 / 110	71.0 / 134	50.2 / —	296 / 64	14.4 / 95	3281 / 167	1.6 / 178	0.5 / 38	15.6 / 105	51 / 195
—	—	2,240 / 100	(1) / 100	—	460 / 100	15.1 / 100	1970 / 100	0.9 / 100	1.3 / 100	14.8 / 100	26 / 100

(1) It was calculated for each income class according to operational N.P.V. of the food ration of each income group.

Obs.: The recommended protein intake depends on the composition of food intake.

4% of the children have definite clinical rickets. On the national level 23% of the children are thought to be rachitic.

This has been a brief overview of the dietary and nutritional state of Morocco at least through the year 1970-1971. How can we evaluate what has happened since then? In point of fact we have very little data with which to study the subsequent developments.

II. PERSPECTIVES ON DIETARY DEMAND

In order to study and analyze the probable evolution of food demand in the more or less distant term, and draw the resulting conclusions we must consider the incidence of future demand on the various concerned sectors. In the following pages, we are limiting ourselves to an examination of food consumption through the 1982 horizon, which coincides with the final year of the next economic and social development plan, with 1977 taken as the base year (Jaouadi & Essatara, 1977a, Jaouadi & Essatara, 1977b).

Without getting into methodological details, it can be clearly stated that the demand for food in 1982 is based on the rates of probable growth in the economy and the population of the country, given that these two elements constitute the two major factors affecting dietary consumption.

Table VI reproduces the food demand for 1982 expressed by person and for the rural and urban areas and the country. One can more clearly show the development of food demand between 1977 and 1982 measuring it against an indicator vis-a-vis the index of 100 based on consumption levels in 1977 (Table VII). We have also found it useful to measure the rural and urban share of potential food consumption in 1982 (Table VIII).

These results call for the following remarks:

The projections of demand fit consumer behavior for a given income level according to the breakdown of purchases if the consumer can freely obtain goods on the market as he desires; the demand side does not therefore prejudge the supply side. This means, therefore, that the projections represent potential demand rather than effective demand or consumption. These two concepts coincide only if supply grows at the same rate as demand and if the hypotheses on the increase in population and income levels are achieved.

If this comes to pass, we can say that consumption per person will undergo a more or less heavy increase according to the products, and that this increase will reach the rural population as well as the urban population.

TABLE 6

PROJECTED FOOD CONSUMPTION DEMAND IN 1982

FOOD SOURCES	RURAL AREAS		URBAN AREAS		TOTAL
	kg/per capita/ year	Total in metric tons*	kg/per capita/ year	Total in metric tons*	kg/per capita/ year
Cereals	252.8	3,034.1	160.5	1,493	4,527
Sugar	32.6	391	26.3	245	636
Legumes	6.2	74	6.8	63	137
Vegetables	79.2	950	129.5	1,204	2,154
Fruits	54.8	658	57.8	538	1,196
Meats	15.2	182	24.7	230	412
Dairy products	43.5	522	32.7	304	826
Fish	2.4	29	8.9	83	112
Fats	9.4	113	12.5	116	229

* metric ton = 1000 kg.

TABLE 7

NATIONAL PRESENT AND PROJECTED TOTAL FOOD DEMAND

FOOD SOURCES	CONSUMPTION IN 1977	PROJECTED DEMAND OF CONSUMPTION IN 1982
Cereals	100	119
Sugar	100	123
Legumes	100	124
Vegetables	100	125
Fruits	100	129
Meats	100	136
Dairy products	100	124
Fish	100	128
Fats	100	129

TABLE VIII

PERCENTAGE DISTRIBUTION OF FOOD DEMAND BETWEEN RURAL AND URBAN AREAS

FOOD SOURCES	RURAL 56.3% OF THE TOTAL POPULATION	URBAN 43.7% OF THE TOTAL POPULATION
Cereals	67%	33%
Sugar	62%	38%
Legumes	54%	46%
Vegetables	44%	56%
Fruits	55%	45%
Meats	44%	56%
Dairy Products	63%	37%
Fish	26%	74%
Fats	49%	51%

If as we first suggested, the annual increase in income is to be the same for both rural and urban, then consumption increases per head will be higher in rural areas where the degree of saturation of needs in the base year was still low compared to urban areas.

For the country as a whole, food demand will be generally strong and will vary according to product from +19% to +36% between the base year and the horizon year. These facts lead one to be concerned that imports, especially for certain products (cereals, sugar, oils, etc.) will show a rapid jump in quantity during the next five-year plan in so far as national production does not follow the rate of increase in food demand. This heavy increase in global consumption is linked to the rate of demographic growth which is very high.

Potential demand in 1982 is treated separately for rural and urban areas. The rural milieu will consume more than proportional to its population in cereals, sugar and milk, i.e. essentially the energy foods which are relatively cheap. The products will be consumed less than proportionate to population by the urban dwellers, who instead will be large-scale consumers of practically all the other foodstuffs.

Note however that this heavy increase in dietary demand might have another interesting aspect. It should stimulate national production and permit an increase in the incomes of rural people, if measures are taken to support small-scale farmers so they can produce more, sell more, and thus increase their cash income.

The increase in consumption per person leads to the idea that the dietary and nutritional situation in the near term will be characterized by a quantitative and qualitative increase, at least at the national level given the hypotheses than supply satisfies demand. But this does not mean to say that the disparities will be less strong than in the past. Theoretically, with a growth rate of the urban population around 5%, one can see that a significant fraction of the total population is shifting towards a better diet each year.

Nevertheless, it is likely that disparities will exacerbate within the urban population which is tending to become less and less homogeneous and of which a goodly number of people will have lost the better features of a rural diet without being able to attain the quality of the typically urban diet.

Witnout wishing to anticipate what will be the rate of satisfaction of food demand for the various products, it is possible to consider the hypothesis that supply will be relatively insufficient; if so:

- we will see a modification in the price relations of different foods;
- we will then be in for multiple substitutions notably of the secondary type of products involved which show a price difference easily perceived by the consumer. In this case, the phenomenon of substitution will be such as to stimulate production of certain foods (white meats replacing red meats, for example). Analysis of the perspectives of supply would give us the information needed to know what substitutions might take place.

It is understood that level of potential demand for each product is not to be viewed as a production objective because one must also take into account other uses of foods (consumption needs of tourists, non-food uses, new types of use, etc.).

The study of food consumption can be of great value to us for several reasons. Exploitation of this field for the purposes of nutritional analyses does not evidently constitute the unique objective of the approach. In effect, there are indicators other than nutritional on which one can draw and which have impacts on agriculture, on commerce food industries and agro-industries etc.

The search for a characterization of the food and nutrition situation present and future, by means of a simple interpretation of levels of average consumption, where means are expressed as calories or nutriments, cannot give more than a rough indication of a vital and complex problem.

However, we must recognize that we have thus far succeeded in showing that the objectives assigned to sectors having a relationship to production, transformation and food legislation as well as those charged with protection of consumer health or with rationalizing food behavior in fact these objectives look first of all toward satisfying the nutritional needs of the Moroccan population.

It would be valuable to know the dietary and nutritional state of the Moroccan population for predicting the following:

- for the medium-term a survey with multiple objectives for the whole country of which the results could be made available for planning purposes;
- to be efficient in the medium-term with a transacional indicator of nutritional change we should measure the variation of consumption levels among the portions of the population that are clearly beneath their physical caloric needs;
- but in the short term there should be put into place a sample nutritional surveillance survey on a permanent basis, to trace the changes in consumption, in particular for that half of the population which does not receive its minimum energy needs.

BIBLIOGRAPHY

1. Bulletin de la Santé Publique, 1973. n.s. 54, Rabat.
2. Division des Affaires Economiques du Ministère de l'Agriculture et de la Réforme Agraire, 1977. Projection de la demande des produits alimentaires 1982-2000. Rabat.
3. FAO/OMS, 1973. Besoins énergétiques et besoins en proténies. Rome.
4. FAO/OMS, 1974. Manuels sur les besoins nutritionnels de l'homme, Rome.
5. Jaouadi, M.T. & Essatara, M'Bareck, 1977b. Bilans de Disponibilités pour la consommation Humaine au Maroc (1971-1975). Le Maroc Agricole, 8ème Année, Octobre 1977 No. 99.
6. Jaouadi, M.T. & Essatara, M'Bareck, 1977a. La consommation alimentaire au Maroc situation actuelle et perspectives pour 1982. Le Maroc Agricole, 8ème Année, 1977, No. 98.
7. Laure J., Essatara M'Bareck & Jaouadi, M.T., 1978. Besoins et apports en nutriments au Maroc. Bull Econo. et Social, (in press).
8. Ministère de l'Economie Nationale, 1961. La consommation de l'enquête 1959-1960. Rabat.
9. Préfecture de Rabat-Salé, 1977. Archives des prix des produits,

consultes, en 1977, Rabat.
10. Secrétariat d'Etat au Plan et au Développement Régional, 1973. La consommation et les dépenses des ménages au Maroc. Avril 1970, Avril 1971. Vol. Alimentation et Nutrition, Rabat.
11. Secrétariat d'Etat au Plan et au Développement Régional, 1977. Recensement général de la population et de l'habitat. La population active, résultat de sondage du 1/10è, Rabat.

CONTRIBUTION OF ANTHROPOLOGY TO NUTRITION:

INTRODUCTORY REMARKS

>Norge W. Jerome
>
>Dept. of Community Health
>University of Kansas Medical Center
>Kansas City, Kansas, USA

INTRODUCTION

Traditionally, symposia and other sessions on nutrition with an emphasis on anthropology have dealt with the "strange" or "exotic" beliefs, value systems and behaviors of exotic peoples in the "emerging" nations", or of ethnic and/or underpriviledged groups in modern, industrialized societies. These sessions often attempted to describe and explain the lifeways and behaviors of these peoples in relation to food use and nutrition. Unfortunately, undue emphasis on the exotica and on behaviors stripped of their sociocultural and environmental contexts have inadvertently reinforced the biases of Western-trained scientists and professionals towards non-Western peoples. As a result, problem-solving exercises, discussion sessions and intervention programs based on these symposia have been largely misdirected.

This symposium departs from the traditional. Our goal is to describe, discuss and demonstrate anthropological concepts, methods and procedures in the evaluation of nutrition and health programs. In some instances, we will demonstrate through case studies how anthropological concepts lead to analysis and interpretations of program design and implementation that escape the nutritional scientist and practitioner. We will further demonstrate that anthropological concepts and (specifically participant observation methods) nutrition program planning, implementation and evaluation often highlight unintended consequences of program utilization.

In recent years nutritionists and public health officials have become increasingly concerned about the role of social, cultural and economic factors not only in the etiology of malnutrition and related health programs, but also in nutrition program design. In this symposium we suggest that these factors should also be included as important features in nutrition programs either as a post hoc review, or as an on-going "process evaluation.

The first paper (Pelto and Jerome) discusses some major issues - theoretical and methodological - in program evaluation from an anthropological perspective; specifically from the perspective of fieldwork research. Our aim is to provide nutritionists with information about methods and theories that can be incorporated into evaluation studies.

Marchione's evaluation study of Jamaica's Community Health Aid program represents an excellent example of the use of social anthropology in the evaluation of an applied nutrition program. He describes a dual approach to evaluation — micro-level or internal, and macro-level or external — wich appraises the values of program designers and the institutional framework of the country operating the nutrition program.

The third paper describes a nutrition program in Iran and points out some pitfalls that can arise when a community-oriented program excludes the community in its design. Steady and Eide's paper focuses on women in rural African settings. The authors state that work done by women as producers and reproducers, and the flow of energy from such work, may represent an organizing frame of reference for nutrition/anthropological activities. They present two conceptual models as examples of energy flow for women as producers and reproducers in two types of economic systems.

Together, the papers show how anthropological theory and methods may enhance nutrition program design and implementation. These papers further demonstrate that anthropological concepts and methods can be particularly useful in evaluating nutrition programs.

AN ANTHROPOLOGICAL PERSPECTIVE

ON NUTRITION PROGRAM EVALUATION

Gretel H. Pelto, Ph.D.* and Norge W. Jerome, Ph.D.**

*Dept. Nutr. Sci. - Univ. of Connecticut - Storrs,
 Connecticut - U.S.A
**Dept. Community Health, Univ. of Kansas Med. Center
 Kansas City, Kansas - U.S.A.

1. INTRODUCTION

Although most administrators of nutrition service programs now agree that they should undergo periodic evaluation, there are many disagreements about the specifics of evaluation strategies, such as mode, timing, sequence and frequency of the evaluation process. Also, there is a lack of consensus on the program units to be evalluated, on the stage at which program evaluation yields maximum benefit, and particularly on the functions and roles of the program evaluator. Traditionally, a program's success is measured by a predetermined set of outcome criteria which often masks the real achievements of individuals enrolled in the program. Although some accomplishments of individual participants are obvious and overwhelming, they are often dismissed as "interesting developments" associated with the program's utilization.

The failure to account for, and measure the varying levels of achievement among program participants indicates an inability to conceptualize, capture and measure the micro-social systems that exist in every community and at all levels of social organization. Since every individual enters a nutrition program with his or her own personal needs, skills, abilities and goals, it is expected that program gains will vary with the individual's objectives, priorities, knowledge and other resource base. These individual baseline variations and associated changes that follow program input must be identified, described and measured in nutrition program evaluation. If the individual participant within his/her household context and community structure is specified as the focal unit of analysis, it should be possible to develop and implement evaluation strategies that produce meaningful results to program administrators, planners, evaluators and participants.

II. THE STRUCTURE OF EVALUATION RESEARCH

Evaluation of nutrition programs have varied across a wide spectrum of methodological style-from the subjective impression of program implementors to carefully controlled studies that use an experimental or at least quasi-experimental design. Regardless of the degree of rigor, the ostensible purpose of evaluation is to asess the extent to which change has occurred in the conditions to which the program is directed. Thus, a central feature of evaluation is the selection of key indicators of nutritional, dietary or health status. These indicators are then assessed to determine the degree of change.

The choices of appropriate indicators for evaluating nutrition service programs (e.g. anthropometric measures, clinical indices, dietary quality, morbidity rates, biochemical assays, or measures of cognitive functioning) are themselves controversial issues (Klein et al., 1979; Austin, 1977; Muskin, 1979). Measures that are appropriate in one context may be completely inappropriate in another because of differences in the time frame of the study, of the nutritional status of the population, and other factors. There may be debate, therefore, about the usefulness of a particular indicator in any given context.

However, it is generally recognized that a change in an indicator in the direction of improved status signifies program success. For example, the demonstration by Edozien (1976) that maternal participants in the WIC program (a nutrition supplementation program in the United States) gave birth to heavier babies has been regarded as positive evidence of the success of that program.

When the evaluation follows the basic principles of scientific research design, the measurements of change in key indicators appears to be straightforward. Baseline data, collected before an intervention program is undertaken, serves as the standard against which the degree of change can be measured. The incidence of low birth weight or the percent of obese adults, for example, provides the base rates to measure success of programs aimed at reducing the incidence of low birth weight or of obesity in a given population.

It is important to note that success is measured in terms of a statistically significant change in the mean value for a population. Therefore, the failure to achieve a degree of change that meets the criterion of statistical significance is often interpreted as a sign of program failure.

The method of evaluating success through measurement of change in the population value for key indicators can be very useful under certain restricted conditions. It is particularly appropriate in the situation in which the intervention is planned as an experimental study of the relative efficacy of two alternative modes of intervention. In that situation, a comparison of mean change in the indicators under Intervention A versus Intervention B enables one to decide which form of intervention is more effective, although the statistics alone do not reveal the mechanism of effectiveness.

Another condition in which this mode of assessment is useful is in monitoring the continuing progress of a program that has already been established as effective. This functions in much the same way as the monitoring of incidence rates of infectious diseases provides feedback about the current state of innoculation programs.

The foregoing examples are limited to highly specific and relatively unusual circumstances. Along with a number of others, we suggest that this model of evaluation may be inappropriate, inefficient and even destructive in many contexts of nutrition program evaluation.

In the first place, this style of evaluation tends to ignore the political nature of evaluation. There is increasing recognition that both planning and evaluation are, by their very nature, political acts (Ugalde and Emory, 1979). The concept of the "rational plan" or the "rational evaluation", carried out in an apolitical context, is largely a myth, regardless of the type of political system in which it occurs. Failure to recognize this reality is partly responsible for the hostility that program implementors often express toward the evaluation process.

Since the political nature of evaluation research has been the subject of considerable discussion, we will not elaborate these points here. Rather, we highlight another feature of the mode of evaluation outlined above - measurement of change in mean values of key indicators for a population - feature that had been less full described.

When emphasis is placed on the mean values of key indicators, the range of variation that the mean summarizes is nearly always obscured. Significant change in a population value can occur as a result of several different patterns of change: a small increment "across the board" in all segments of the population or large increments in various subgroups. Alternatively, a significant change in a sub-group may be completely obscured if the majority of cases remain unchanged (Marchione, 1977).

We suspect that in some previous evaluations the failure to demonstrate effectiveness may stem, at least in part, from a failure to separate, analytically, the degree of change in various subgroups. The implication of this simple statistical problem is not merely a call for more refined analysis of the standard deviation of the distribution of values, although we would certainly support such analysis. A further implication is the need for a broader theoretical and methodological framework for interpreting and analyzing the effects of nutrition intervention programs. We are not referring only to the importance of the economic context of nutrition planning, although this is certainly a primary aspect of such a framework.

In the next section we will present an outline of some theoretical perspectives, Section III is valuable not only for understanding etiological processes but also for planning and evaluating intervention programs.

III. AN ECOLOGICAL PERSPECTIVE

As in many fields of enquiry, nutritionists have become increasingly interested in using an ecological approach to the analysis of nutrition issues. The founding of a journal, in the early 1970's, titled The Ecology of Food and Nutrition, preceded by the multi-volume regional analysis of malnutrition entitled Ecology of Malnutrition by Jacques May are but two examples of this interest by nutritionists. Within anthropology, ecological studies and ecologically-oriented theory have assumed a role of considerable prominence in the last decade or so.

There are many different conceptions or interpretations of the meaning of an "ecological approach", but a common theme is the emphasis on viewing human nutritional or cultural behaviors in relation to environmental features. Built on the basic foundation of systems theory, human ecologists study the interaction of cultural and social organization with the physical environments in which these social groups function, with the goal of understanding the adaptive nature of human cultural responses and their significance for cultural evolution.

Often biocultural in approach, ecologically-oriented anthropoligists study problems such as the biological consequences of human subsistence systems. For example, the study of Livingstone (1958) on the relationship of the spread of agriculture to the incidence of malaria and the maintenance of sickle cell trait in West Africa has become a classic example of biocultural interaction. Rappaport's study (1968) of the interrelationships among religious ritual, warfare and pig production of the Tsembaga Maring in New Guinea is an example of ecological-cultural interactions that have direct nutritional consequences. When the nutritional requirements of an expanding pig population threatened human food resources, ritual processes came into play which dramatically reduced the number of pigs through ceremonial feasting, and which had the additional function of bringing a temporary halt to inter-tribal warfare.

The Ecological Frame-of-Reference

The ecological approach or "frame of reference" can be described in various ways. Cone and Pelto (1967) have diagrammed it as in Fig. 1.

The diagram is intended to convey the idea that each of the components is in interaction with other components and forms a system such that a change in one component will bring about compensatory changes in other components.

Applied to questions of food and nutrition the framework reinforces the widely-accepted idea that the nutritional status of any given population is the result of a complex set of social, environmental and biological forces (Pelto et al., 1980). This framework can be useful for organizing the data for a comparative analysis of nutritional status of a hunting-gathering subsistence group as compared to that of a horticultural society.

In moving from a comparative analysis of populations to the analysis of nutrition problems of a specific group (i.e. a contemporary village or city) the same basic concepts apply. We can ask: "How do features of social organization and technology affect the nutrition of people in this community?" "What is the relationship of features of the physical environment to nutrition?" "In what ways does the social environment, the wider political-economic world outside the community affect local nutrition?" "How do people's ideas about food affect nutritional status?" and so on.

It cannot be assumed that the components for each of these questions are uniform within the community. For example, it is unlikely that everyone within the community has access to the same technological inventory, or have the same ideas about or interest in food. Thus, we would emphasize the concept of intracultural diversity as a basic part to the ecological analysis (Pelto and Pelto, 1975); Pelto and Jerome, 1977). The question, then, becomes "What are the systematic inter-relationships of different values in each of the main components as they affect and are affected by nutritional status?"

Resource Utilization and Household Structure

In most societies, the basic mechanism for meeting biological, psychological and some social needs is the household or family. There are some exceptions (but not many) to the generalization that in most societies the household is the basic adaptive unit (Pelto and Pelto, 1976; Scrimshaw and Pelto, 1979). Through the household, many activities are organized — child care, food procurement and preparation, provisioning of clothing and shelter, (more variably) religious, social and other economic activities. A main point is that households <u>always</u> perform multiple functions.

Within a community, even in relatively less complex societies, there are household differences in styles or "strategies" of resource utilization. Differences in family size and composition, in skills of individual and household members, in psychological characteristics, physical capabilities and health history all affect the ways in which households allocate and utilize scarce resources. In complex, stratified societies these differences are often very marked and are frequently the subject of sociological and anthropological study.

Economic and community development projects are often conceptualized by planners as efforts to expand the resource base of a community or region. However, what is less often considered is the likelihood that the new resources will be differentially utilized by community members because of differences in the adaptive strategies of households. This micro-differentiation can lead to consequences that are unintended or unanticipated by program planners (Poggie and Lynch, 1974; Mueller-Wille et al., 1978).

Examples of this process are numerous, and among the most striking are recent analyses of the so-called Green Revolution. It has been documented that the introduction of new seeds and other aspects of new agricultural technology in some Third World contexts has brought about increased social stratification and actual loss of resources for many small farmers (Franke, 1973; Griffin, 1974). While features of the larger political-economic system (e.g. in loan policies of banks) often play an important role in bringing about these outcomes, it also appears that relatively small differences within peasant communities can have significant effects on program utilization (DeWalt, 1979).

This latter point is strikingly illustrated in a study of the effects of an agricultural development program in a Mexican ejido. Through careful, quantified analysis, as well as with ethnographic data De Walt (1979) has demonstrated that the small farmers within a superficially homogeneous community responded in different ways to the new resources introduced by the development plan. Four distinct adaptive strategies can be identified, ranging from a full-scale commitment to commercial production of animal feed to the minimal utilization of one or two features fo the developer's program. De Walt suggests that differences in program utilization can be systematically related to differences in household characteristics, including differences in "cosmopoliteness" and in socio-economic resources.

Nutrition intervention programs, whether they take the form of direct food assistance, nutrition education, or other activities, can also be regarded as the introduction of a new resource into the community. Just as is true of other types of changes in the resource base, a nutrition program will be differentially utilized because of previously existing conditions within households. "It may be fully utilized by some, ignored by others and few ... may select only those elements that seem applicable to their specific needs at a specific time" (Jerome, 1978). A major task of program planners/evaluators is to elucidate and interpret the process of differential utilization and impact.

Manifest and Latent Functions in Intervention and Development Programs

Sociologists and anthropologists of an earlier period (Linton, 1936; Parsons, 1951) developed the concept of "latent vs. manifest function" to characterize an interesting feature of many social phenomena — the contrast between the stated or ostensible purpose of some behavior or cultural feature (its "manifest function") and those aspects of its functions that are not stated or perceived by the group in question ("latent functions"). For example in the Tsembaga situation (above), the "manifest function" of the ritual pig feast "was to serve the gods" but its latent functions, as outlined by Rappaport, were to return the ratio of pigs to food resources to a manageable level and to facilitate social communication.

Latent functions of planned interventions have rarely been systematically considered by development planners and evaluators, although they have been often recognized on an informal basis. In some situations, however, the latent functions of an intervention may have considerably more impact than its ostensible purpose. An interesting example of this phenomenon is the case of a planned, new city in Central Mexico (Poggie and Miller, 1968). The purposes for building the city were to attract industry and workers out of the congested area aroung Mexico City and to establish a pleasant alternative for urban migrants. Although some industrial development did take place in Ciudad Industrial, the city never grew to its projected size and appeared to fail to develop a range of social and recreational characteristics that would make it an attractive place. The planners did not consider the experiment to be a success.

However, the analysis of the changes in the surrounding communities (a study that was made possible by the serendipitous availability of baseline data) showed that the intervention had a powerful effect on the region, leading to greatly expanded economic opportunities. A diversified network of goods and services developed in response to the presence of the new population center and as small communities within the region, established economic, social and political ties with the center. In terms of economic and community development, the latent functions of the intervention were more important and more successful than its manifest original purpose.

Often, the latent function of planned intervention are not always so positive, and the anthropological literature contains many examples of situations in which a well-intended intervention led to negative consequences for local people (Paul, 1955). In a carefully developed experiment, Chavez and colleagues have demonstrated that improving the nutritional status of undernourished mothers and their infants in a Mexican community led to a reduction in the interval between births — hardly a benign outcome in a situation in which population growth is a serious problem (Chavez and Martinez, 1973). Similarly, although not well documented, the relationship between use of infant formulas, reduced suckling and early return of ovulation is another example of a contrast between the manifest and latent functions of introducing new infant feeding techniques into communities with poorly developed birth control programs.

In the United States, the Expanded Food and Nutrition Program of the Department of Agriculture provides an example of nutrition intervention program in which the latent functions affected non-nutritional aspects of people's lives. A key feature of the program is the recruitment and training of women from "underpriviledged" communities to serve as community nutrition aides in providing nutrition education and other services to members of their own community. Whatever the impact of the aides on nutritional status in the community (a subject of debate), it is clear that the program has had important effects for the aides themselves. The position provides them and their families with income; in that respect the program can be regarded as an income maintenance program. It also provides them with job mobility, and enhances their prestige in their community; and finally, it should be noted that the aides serve as role models, especially for younger women and girls (Jerome, 1971).

The latent functions of programs of planned change derive from the fact that communities, like individuals, are systems composed of interdependent and interrelated parts. The implication of our earlier statement about the multiple organizational tasks of households is that they are also small systems. Thus, nutrition interventions, whether driected at families or targeted to types of individuals (e.g., pregnant women; young children) will have multiple (latent) effects on family functioning (Scrimshay and Pelto, 1979).

The relationship of maternal nutritional status to child spacing as described above, is one example of the effect of nutritional intervention programs on family composition. In a short time frame, the effect is likely to have a negative feedback on nutrition by increasing the number of people who must share a limited food supply. However, in some circumstances, the long-term nutritional effect may be positive is surviving children become productive in family food acquisition.

The introduction of new types of food-related technology can have both direct and indirect nutritional consequences on a population. For example, it appears that the introduction of the family food freezer may have the direct nutritional consequences of reducing intakes of salt and sugar by populations which had used salting, pickling, and sweetening in traditional methods of food preservation. At the same time, the freezer could preserve traditions of home baking, which were likely to disappear under the time and energy pressures of working women. In West Finland, for example, an occasional week-end baking day, with the porducts placed in the freezer, may reduce family consumption of commercially-produced baked goods (Pelto, 1978). Although the nutritional consequences of home-produced vs commercially-produced baked goods are not particularly clear in this instance, the contrast has certain implications for nutrition education strategies.

In earlier periods, in both North America and parts of Europe, the introduction of mechanized milking equipment affected male and female roles in dairy production; often, with mechanization, milking was redefined as a male, rather than a female task. We may hypothesize that the expansion of electrified and complex kitchen equipment (such as the "food processor") may bring about expanded participation of males in food preparation in some cultural contexts in which men have been noticeably absent from the kitchen. Similarly, the definition of the outdoor barbecue grill as a male food-preparation domain (a common occurrence in North American European families) can have consequences for food consumption patterns and buying habits that may have nutritional consequences. In general, the indirect consequences mainly affecting family functioning and roles and social interaction patterns.

Social Organization and Cultural Institutions

Some food behaviors may be "deeply embedded" in complex social institutions, as in the case of the pigs among the Tsembaga. As examples from industrialized societies, the North American "hamburger-and-milk shake" pattern and, in parts of Europe, the new, grilled sausage style of home entertainment, may be very strongly affected by special patterns that have nothing to do with food (Fishwick, 1978).

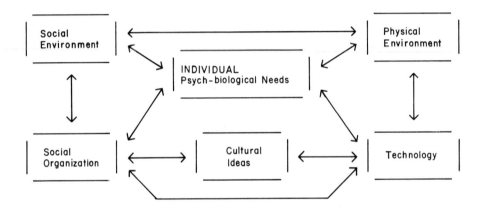

Fig. 1 - Ecological Frame-of-Reference.

Biological Characteristics and Nutrition Intervention

The problem of lactose-intolerance is well known but deserves continued attention as an example of a nutrition intervention strategy that failed to consider human variations in biological characteristics. The fact that many adults in Africa, Asia and elsewhere have been given milk powder and other lactose containing foods in ignorance of the physiological problem it can cause, has perpetuated the assumption about "peasant ignorance", "traditionalism" and other ethnocentric thinking among nutrition planners and project personnel. Individual and group differences in substance tolerance, nutritional requirements and other biological aspects of food utilization must be regarded as potentially very important — a point of view that is made more salient in the ecological frame-of-reference (Margen and Ogar, 1977).

The Systematic Nature of Change

The foregoing discussion conveys the idea that all interventions (both planned and unplanned) are likely to have multiple effects - intended and unintended. The systematic nature of household and community functioning means that changes in food use and nutrition-related activities are likely to have miltiple effects, through their effects on household roles, social, economic and even recreational activities, as well as through the effects of changed physiological functioning of individuals. These effects, in turn, may feedback to nutritional status of household members in either positive or negative ways.

While it is not possible for program planners to anticipate all of the latent functions of their interventions, it is very important that a-equate information on social and cultural realities be acquired prior to program implementation inorder to make intellignet estimates of probable effects. Moreover, it is incumbent on planners/evaluators to monitor programs for assessing both manifest and latent effects as they occur.

The theoretical issues we have raised in this section imply new models or strategies for nutrition intervention programming and evaluation. In the following section we will discuss the methodological issues that are involved in such a model.

IV. CONTRIBUTION OF ANTHROPOLOGICAL METHODOLOGY TO EVALUATION RESEARCH

From the theoretical perspectives outlined above, we propose that health and nutrition intervention programs should be designed and evaluated from a holistic perspective. Narrowly-focussed programs, planned and executed without intimate knowledge of the cultural-ecological setting, often fail to achieve their objectives. Furthermore, the traditional relationship of program execution to evaluation, as well as the restricted focus of evaluation research further reduces a program's capacity to acquire helful information. We are not alone in recognizing these problems and many others have articulated the need for new models of health and nutrition intervention programs and evaluations (Klein, 1979; Marshall and Polgar, 1976; Foster and Anderson, 1978; Dubos, 1959; Rossi, 1973).

A broad data base is required for more effective programming and evaluation. This demands a style of research that is the hallmark of anthropological fieldwork. Ethnographic research, utilizing the participant-observation method, requires that the researcher participate actively in community life, and make systematic observations of on-going behavior, and conduct formal, structured, interviews on daily life. The qualitative data obtained from participant observation provides the context in which quantitative data are assessed and interpreted (Pelto and Pelto, 1978).

Participatory research is the key that unlocks the mysteries that surround the subtle and unintended changes associated with program implementation. It is also the strategy that facilitates appropriate program development, design and delivery. And the key to effective program evaluation.

Ideally, participatory research should precede program design. This type of research involves a partnership between the researcher and members of the local community and includes community participation in the activities of the researcher. By specifying community needs and priorities and communicating them to the researcher for incorporation into a program's design and implementation, the community increases its opportunity to achieve its goals. Similarly, a program that is derived from community goals and priorities stands a better chance of succeeding than one that is developed without input from the target community. These principles also hold for program evaluation. Every program subjected to evaluation must be assessed on the basis of community needs in general, and the specific needs of subgroups and individuals within that community.

Participatory research or "community penetration" involves the gathering of systematic information through intensive field research in the target community. It provides first-hand information on the nature, structure and organization of households in the community, on household and community needs and priorities, on variations in household resources, in the use by households of community resources, and on the various coping and adaptive strategies of households and families. In addition, this type of research yields important information on community economic structure and systems, the resource base of communities, leadership structure and role models, on social networks within a given community. This research strategy also furnishes data on mode and styles of communication within the community; sources of information; and the social, religious, ceremonial and ritual life of community members. Research of this type also yields information on recreational and leisure patterns; on areas of conflict and cooperation, and on economic and sex role differentiation within the community. This body of ethnographic data describing the heart and soul of the community helps planners in the design of programs that mesh with ongoing social life. Moreover, the data can assist planner/evaluators in predicting and assessing the multiple effects-manifest and latent - of program implementation.

The qualitative ethnographic data described above differs from socio-demographic data that are often obtained to determine distributions of age, income, gender, formal education achievement levels, and occupational and employment patterns. The rich ethnographic information also differs from the more quantitative data on mean values of key indicators, which have been used traditionally to determine program success or failure.

Good program planning and evaluation research requires the use of the three types of data referred to above. However, conventional evaluation research has been limited to the gathering of data of the latter types with the exclusion of the rich qualitative ethnographic material described above. The challenge, of course, is to integrate the three approaches so that the positive elements of the three approaches can be utilized to full advantage in the nutrition program planning/evaluation process.

The apparent problems in integrating the three approaches are readily resolved. First, the perceived differences between the descriptive data obtained from ethnographic research using the participant observation method and the more quantitative data obtained from broad community surveys are more apparent than real. The two types of data are similar in that (1) both are obtained from the residents of the community being studied and (2) both are based on a series of verbal responses or other types of measures of the community residents. However, the two types of data differ in the following important respects: (1) the qualitative data are also derived from a series of <u>repeated observations</u> of verbal and non-verbal events in the focal community and (2) responses by community residents furnishing qualitative data are derived from questions formulated from current on-going phenomena personally experienced by the researcher. Too often, sociodemographic data are obtained from a pre-determined set of questions that are based on theoretical models which may have little resemblance to the realities of community life.

Secondly, the planner/evaluator always interprets sociodemographic statistics within a sociocultural and political context although such interpretations are seldom made explicit. However, these sociocultural interpretations are often processed on the basis of limited contacts with residents, non-representative sampling, administrator biases, and budgetary and political concerns. The ethnographic approach thus provides the planner/evaluator with a more explicit, objective, systematic and scientific approach to data processing, interpretation and decision-making.

How valid is this approach? The validity of participatory research lies in the repetitiveness of the observations, the explicitness of the strategy and findings, and like all other research, the conscientiousness and objectivity of the researcher. The investigator has an obligation to fully describe the basis on which conclusions are derived so that everyone may weigh and assess the reliability and validity of the conclusions.

In addition to producing reliable information, this approach also has the advantage of producing simple, cost-effective, practical solutions to problems that are often considered enormous to policy makers and program administrators. For example, the recommendation that Commodity Foods (a now defunct USDA Food Donation Program) be distributed more frequently than once per month to the residents of an urban county in Kansas, grew out of the field observations that Kansas program participants could not possibly store the bulk packaged foods in their traditional storage space (Jerome, 1972). Consequently, the recipients stored food under beds and in clothes closets in the adults' bedroom to keep the items out of sight of young children who had no concept of eating "just enough for today". This method of storage, hardly desirable, helped "to stretch the food for the month". Another recommendation, that powdered Kool Aid be enriched with ascorbic acid, also grew out of the observations in a community study that that particular product (with no added nutrients at the time) was the beverage of choice in a low-income urban community; at that time, ascorbic acid was the most limiting nutrient in the people's diet (Jerome, 1967).

Although these recommendations may appear to be the simplest form of intervention, the information on which they are based emerged from the researcher's presence in the community on an ongoing basis, sharing in the lives of the respondents, personally coping with the same problems as the respondents, and attempting to find simple solutions to problems as they arise. These recommendations could not have emerged from a theoretical model developed in an office or from securing sociodemographic data from a questionnaire or form. However, in reference to the first example, the data could elucidate why, for example, some recipients were largely dissatisfied with the program and why children whose mothers received an adequate amount of "commodity foods" (absence of a father was a precondition for receiving food under this program) continued to be malnourished. The data could further explain why the more dependent and less active children in a household could show a higher degree of malnutrition than those who were less dependent (and therefore more able to find the food that was stored in bedroom closets).

A recent report by Masisi (1978) confirms the value of participatory research and the integration of multiple approaches in nutrition intervention programs. As part of the Chiwanda Nutrition Education Project in Tanzania, village needs were identified and communicated through ongoing interactions between villagers and project workers prior to program implementation. A plan to address village economic, food and nutrition needs was then developed cooperatively between project administrators and villagers, endorsed by Chiwanda leaders, political party and government functionairies, and subsequently followed by an intervention strategy, i.e., (a) fruit growing; (b) vegetable growing; (c) poultry raising and (d) training seminars in farming and nutrition. The project was implemented in five stages and evaluation of the preceding stage took place automatically as a new stage was launched. As a result, although the project was originally labelled "Nutrition Education" "instruction on nutrition came late by design... (the intervention program was designed so that instruction in nutrition was undertaken when working materials were available") (idib., p. 10). When working materials (eggs) were available, a system of egg distribution and utilization was instittuted and nutritional instruction focusing on "the functions of food in the human body; food nutrients and their sources; _eggs_ as an important source of protein; why _eggs_ should be eaten; the consequences of not eating _eggs_ or other protein-rich foods; the need for pregnant mothers and children to eat _eggs_ regularly; provision of different _egg_ recipes" was put into operation.

Masisi's report illustrates one instance where instruction on protein as a nutrient was integrally related to community need, community readiness, and availability of the "teaching material". Nutrition education did not take place in a vacuum. The intervention program exemplifies how the three types of data discussed above can be integrated in program planning, implementation and evaluation.

Similar reports showing the value of participatory research in combination with other data gathering approaches have come from Ghana (Oracca-Tetteh and Watson, 1978) and from Ethiopia (Clark, 1977). Referred to as the "action-oriented approach" to "nutrition education" by Oracca-Tetteh and Watson and as the "problem solving program design" by Clark, the participant observation method has proven to be extremely useful in gathering baseline information on community needs and in designing intervention programs to address those needs. Generally, the participant observation method has been shown to be more than a research method for anthropologists; it is also a sound strategy for organizing and sensitizing communities around their felt needs and assisting them in identifying problems. It is a method that the planner/evaluator can use for planning, designing, implementing and evaluating community-based nutrition programs.

The data collection plan we have laid out in this paper bears a strong resemblance to the type of research carried out by anthropologists and other social scientists in studies of social change. In some respects, the evaluation of planned change can be simply regarded as a special type of this general domain of research. There are, to be sure, some important differences between targeted, planned change and change that is brought about as a result of social forces. In the former case there is generally greater control over the input, the time frame is often shorter, and it is purposefully directed to bringing about a more positive situation, while undirected change has no such evaluative component.

It may be useful, however, to conceptualize evaluation as a special type of social change research specifically because a social change perspective would encourage the examination of a wide range of phenomena in the evaluation process. The likelihood of discovering latent, as well as manifest functions is increased when different aspects of family and community life are assessed within a broad framework. In addition, the approach advocated increases the likelihood that all of the important, relevant variables affecting program utilization and outcome will be systematically assessed. Finally, this approach increases the prospects for intelligently estimating the longterm effects of intervention.

This latter point is particularly important for the matter of policy guidelines. Political consideration often requires evaluators to present results within a time perspective that is inadequate to demonstrate definitive results. The rich description utilizing qualitative and quantitative data would provide a far more satisfactory basis for continous policy-making than does the narrowly-focused data base of the assessment of change in key indicators.

BIBLIOGRAPHY

Austin, J.E, 1977, The Perilous Journey of Nutrition Evaluation. Position paper for the National Academy of Sciences Workshop on Effective Interventions to Reduce Infections in Malnourished Populations, June, 1977.

Chavez, A. and Martinez, C., 1973, Nutrition and Development of Infants from Poor Rural Areas. III: Maternal Nutrition and its Consequences on Fertility. Nutr. Rep. Intl.

Clark, N.M., 1977, An Analysis of an Integrated Approach to Education for Community Development. In: "Teaching Nutrition in Developing Countries". Shack (ed.), Meals for Million Found.

Cone, C.A. and Pelto, P.J., 1967, Guide to Cultural Anthropology. Scott, Foresman Company, Glenview, Illinois.

DeWalt, B.R., 1979, "Modernization in a Mexican Ejido: A Study in Economic Adaptation". Cambridge University Press, New York.

Dubos, R., 1959, "Mirage of Health: Utopias, Progress and Biological Change. Harper and Row, New York.

Edozien, J.C. et al., 1976, Medical Evaluation of the Special Supplemental Food Program for Women, Infants and Children (WIC). Chapel Hill, North Carolina: Department of Nutrition, School of Public Health.

Fishwick, M., Ed., 1978 "The World of Ronald McDonald". Popular Press of Bowling Green State University, Bowling Green, Ohio.

Foster, G.M. and Anderson, B.G., 1978, "Medical Anthropology". John Wiley and Sons, New York.

Franke, R.W., 1973, The Green Revolution in a Javanese Village. (Unpublished Ph.D. Dissertation 1973, Harvard University - Cambridge).

Griffin, K., 1974, "The Political Economy of Agrarian Change: An Essay on the Green Revolution". Harvard University Press, Cambridge, Massachusetts.

Jerome, N.W., 1967, Food Habits and Acculturation: Dietary Practices of Families Headed by Southern Born Negroes Residing in a Northern Metropolis. (Unpublished Ph.D. Dissertation 1967, University of Wisconsin - Madison).

Jerome, N.W., 1971, Consumerism and Professionalism in Contemporary Health and Nutrition Programs. Summer Institute, University of Kansas Medical Center, Kansas City, Kansas.

Jerome, N.W., 1972, Adaptive Responses to Poverty in Relation to Food Use and Consumption Patterns. Proceedings of the Symposium on Breaking the Cycle of Poverty and Malnutrition. HEW Publication Number (HSM) 72-8110: 2-5.

Jerome, N.W., 1978, Nutrition Education: An Anthropological Viewpoint. Thresholds in Education 4: 6-7.

Klein, R.E., et al., Eds., 1979, "Evaluating the Impact of Nutrition and Health Programs". Plenum Publishing Corporation, New York.

Linton, R., 1936, "The Study of Man". D. Appleton-Century Company, New York.

Livingston, F.B., 1958, Anthropological Implications of Sickle Cell Gene Distribution in West Africa. American Anthropologist 60: 533-562.

Marchione, T.J., 1977, Food and Nutrition in Self-Reliant National Development: The Impact on Child Nutrition of Jamaican Government Policy. Medical Anthropology 1: 57-80.

Margen, S. and Ogar, R., Eds., 1976, "Progress in Human Nutrition". Vol. II, AVI Publishing Company. Westport, Connecticut.

Marshall, J.F. and Polgar, S., Eds., 1976, Culture, Natality and Family Planning. Carolina Population Center, University of North Carolina. Chapel Hill, North Carolina.

Masisi, Y.K.C., 1978, The Chiwanda Nutrition Education Project: An

Approach to Peasant Animation. Department of Nutrition and Food Science, University of Ghana, Legon. Paper presented at the Workshop on Nutrition Education: Rethinking Food and Nutrition Education Under Changing Socioeconomic Conditions, University of Dar-Es-Salam, Tanzania, June, 1978.

Müller-Wille, L. et al., Eds., 1978, Consequences of Economic Change in Circumpolar Regions. Occasional Publication No. 14. The University of Alberta Boreal Institute for Northern Studies, Edmonton, Alberta.

Muskin, S., 1979, "Educational Outcomes of Nutrition and Health Programs. Evaluating the Impact of Nutrition and Health Programs". Klein, et al., Eds. Plenum Publishing Corp., New York.

Oracca-Tetteh, R. and Watson, T.D., 1978, A Reassessment of the Nutritional Status of the Population of Baati (August, 1976) after the Initiation of a Modest Nutrition Program. Department of Nutrition and Food Sciences, University of Ghana, Legon. Paper presented at the Workshop on Nutrition Education: Rethinking Food and Nutrition Education Under Changing Socioeconomic Conditions, University of Dar-Es-Salam, Tanzania, June, 1978.

Parsons, T., 1951, "The Social System". The Free Press of Glencoe, New York.

Paul, B.D., Ed., 1955, "Health, Culture and Community". Russell Sage Foundation, New York.

Pelto, G.H., 1978, Dietary Modernization in West Finland. Paper presented at the annual meetings of the American Anthropological Association, Los Angeles, California, 1978.

Pelto, G.H. and Jerome, N.W., 1977, "Intracultural Diversity in Nutritional Anthropology. Health and the Human Condition: Perspectives in Medical Anthropology". Hunt and Logan, Eds. Duxbury Press, Boston.

Pelto, G.H. and Pelto, P.J., 1976, "The Human Adventure: An Introduction to Anthropology". Macmillan Publishing Company, New York.

Pelto, G.H. et al., 1980, "Methodological Issues in Nutritional Anthropology." In Nutritional Anthropology, Jerome et al., Eds. Redgrave Publishing Company, Pleasantville, New York.

Pelto, P.J. and Pelto, G.H., 1975, Intracultural Diversity: Some Theoretical Issues. American Ethnologist 2: 1-18.

Pelto, P.J. and Pelto, G.H., 1978, "Anthropological Research: The Structure of Inquiry". Cambridge University Press, New York.

Poggie, J. and Lynch, R., Eds., 1974, "Rethinking Modernization: Anthropological Perspectives". Greenwood Press, Westport, Connecticut.

Poggie, J. and Miller, F.C., 1968, Social and Cultural Aspects of Modernization in Mexico. University of Minnesota, Minneapolis.

Rappaport, R.A., 1968, "Pigs for the Ancestors: Ritual in the Ecology of a New Guinea People". Yale University Press, New Haven.

Rossi, P.H. and Williams, W., Eds., 1972, "Evaluating Social Programs". Seminar Press, New York.

Scrimshaw, S.C.M. and Pelto, G.H., 1979, "Family Composition and Structure in Relation to Nutrition and Health Programs. Evaluating the Impact of Nutrition and Health Programs". Klein et al., Eds. Plenum Publishing Corp., New York.

Ugalde, A. and Emory, R., 1979, "Political and Practical Issues in Assessing Health and Nutrition Interventions. Evaluating the Impact of Nutrition and Health Programs". Klein et al., Eds. Plenum Publishing Corp., New York.

ANTHROPOLOGICAL PERSPECTIVES IN NUTRITION AND HEALTH PROGRAMMES.

The Khombole Project, Senegal, 1957-78.

>Igor de Garine
>
>Centre National de la Recherche Scientifique
>
>Paris, FRANCE

INTRODUCTION

The theme of this symposium is definitely centred on applied research. As requested, the question to be answered is: "How has anthropology been contributing to nutrition and public health programmes, and how could it contribute to them?" The answer is not purely a technical one.

The case study I shall be dealing with is the Health and Nutrition Education Programme in Khombole, Senegal. Launched in 1958, it is still running under the name of Centre for the Study and Application of Public Health and Child Protection Methods for Rural Areas. Having initially at its disposal ordinary and modest means, the Centre received from 1959 to 1968 funds from bilateral and international technical assistance. Since 1970, it has survived strictly on national funds. We are not dealing here with a star project enjoying many technical improvements and drawing benefit from the latest intellectual fashions, but an ordinary project already 20 years old and representative of the attempts made in French-speaking Africa during the last decades. Its evolution demonstrates equally well its success, drawbacks, and the normal drop-offs following a return to national financing.

OBJECTIVE

It was not a strictly geographical project attempting to eradicate malnutrition in a definite area. Drawing its inspiration from an experiment already carried out in France at Soissons, the purpose was to create a pilot centre dealing with teaching, applications and research on programmes about mother and child care in rural areas, embracing, as the mother and child cannot be separated from their environment, the whole family (Le Centre Rural de Protection Familiale de de Khombole, 1957, p. 71).

Four secondary objectives were set up in 1958:

1. Seasoning young medical doctors and social and medical African personnel to bush conditions.
2. Optimizing the action of local medical social services.
3. Training the population in the fields of nutrition and public health, dealing first with pregnant women, family mothers and, with the collaboration of social and education services, reaching the total population.
4. Carrying out research and public health and nutrition problems in tropical environment.

Progressively the medical program of Khombole led it to become a pilot zone (Raybaud, 1968, p. 8) where various public health and nutrition programmes were studied, and where education methods which could be extended to the whole of Senegal and other West African states were sought.

ORGANIZATION

There are two parts closely collaborating:

a) The Centre of Demonstration and Operational Research on Public Health Programmes in Rural Areas, related to the Institute of Social Pediatrics of Dakar University and, up to 1970, the International Children's Centre.

Since 1962, the Khombole Centre benefits from the creation of a School for Environmental Sanitation Agents, financed by bilateral and international aid.

b) The medical activities of Khombole-Thienaba is under the responsibility of the Ministry of Public Health and Social Affairs of Senegal, and it has the usual funding and means: at Khombole — an in-patients ward, a maternity service; in the bush — three health posts.

Collaboration between the two sections is established on the following basis: field research is carried out in the medical circumscription and benefits form its help, personnel and field knowledge.

The various extension methods set up at the Khombole Centre are expanded to the villages solely through the ordinary means at the disposal of the medical circumscription.

CHOICE OF THE AREA

It was made in 1957 by the two promoters of the project according to three criteria: (i) to be representative of Senegalese rural areas, or at least of the groundnut-growing region; (ii) to be easily reached from Dakar by professional researchers, students, and, I would add, visitors; (iii) to be able to rely on local backing. This has been achieved since 1956, when the local medical officer dedicated himself to the project and kept it going until now. I believe his name should be mentioned here as a token of respect — the african doctor, F. Coly.

PHYSICAL ENVIRONMENT

The Khombole-Thienabe area is located in the Thiès region, a little over a hundred kilometers east of Dakar, to which it is linked both by road and railrord. The whole area enjoys a tropical Sahelo-Sudanese climate: one dry season from November and May; one rainy season from June to October. Normal rainfall is about 600 mm spread over 35 days. Since 1968 the area suffers from drought and the rainfall is limited to 450 mm, with important ecological consequences. Soils are sandy with argillaceous patches. The main food crop is finger millet (Pennisetum). For the last hundred years, groundnuts have been grown intensively as a cash crop in the region. Cultivation areas are today impoverished by continuous use without fertilizers or manure. Human density reaches 85 inhabitants per square kilometer. The natural growth has practically vanished.

THE HUMAN ENVIRONMENT

The Thienaba area is crossed in its middle by the concrete road from Thiès to Diourbel, on which are located the administrative centre of Thienaba and the small town of Khombole, with which we shall be dealing. It numbered about 5000 inhabitants in 1967, today about 10,000. The rest of the population (about 50,000) dwells in rural villages, sparsely located. Two ethnic groups are equally represented: the Wolof and the Serer, occupying two hundred villages, formerly grouped into two cantons, today divided in smaller rural communities. The two ethnic groups mentioned are quite different, they constitute independent territorial entities and are not equally adapted to modern ways of life and market economy.

THE NUTRITION AND HEALTH ENVIRONMENT

The previous experience acquired by the Institute of Social Pediatry of Dakar University, the Organization of Food and Nutrition Research in Black Africa (O.R.A.N.A.), localized in Dakar, and local medical services. allowed a poor health situation in the area to be forecast — child mortality reached about 50%, supposed to be due mainly to malnutrition stemming from irrational traditional behaviour and attitudes.

ANTHROPOLOGY IN THE PROJECT

On October 28th, 1958, the corner-stone of the Health Centre was laid at Khombole. One year later the project hired a social anthropologist for two consultant field missions.

i) From November 1959 to May 1960, he was asked to study the conditions of domestic life, focusing on the sociocultural background of food habits, and to formulate recommendations about possible education programmes adapted to the area. He was also asked to justify retrospectively the choice of Khombole as a pilot zone.

ii) From July to October 1961, in relation to the introduction of protein-rich foods to meet poor family needs, he was asked to study the socio-economic acceptability of two protein-rich foods: dried fish, from shark flesh (Centrophorus), named Poisson 45' which unfortunately did not mask the fact that it is was derived from about the only species of fish not eaten traditionally (Dupin, 1961); and Flour No. 21, millet flour supplemented with peanut oilcakes (unfortunately aflatoxin infested). He was also asked to carry out, during the rainy season, the various studies on food he had previously undertaken in the dry season.

The main commitment was to gather baseline data on socio-economic and cultural aspects of food and nutrition in order to determine focal points for education programmes, mainly the content of such programmes and the shape they should take to be acceptable.

MEANS AND TECHNIQUES

During the first mission, four investigators were made available (two Wolof and two Serer); during the second, two more investigators were in charge of the food consumption survey. Data about food and food behaviour were considered from etic and emic viewpoints — What is food behaviour? What is the traditional viewpoint about it?

Research work came under two main headings:

1) The study of phenomena directly connected to food behaviour — mainly in the economic field, production, technology, consumption.

2) A general study of traditional systems — sieving through the various cultural fields in order to determine those factors which might have a direct bearing on the subject under consideration.

From the cultural and economic viewpoint, it appeared to be necessary to distinguish three homogeneous areas: the small town of Khombole; the Serer and Wolof villages, among which two of each were selected. Most of the studies were carried out on a core of eight-eight domestic units selected on a non-random basis, according to the working possibilities they offered. They were, however, chosen in order to represent the various traditional units and socio-cultural classes operating in the society. The investigations came under two main headings:

a) Study by direct observation and questionnaire on socio-economic aspects — detailed census and genealogy, description of habitat and means of production, inventory of material belongings, tools, clothes, cattle, crops, income, expenditure, etc. in each compound.

A qualitative survey of food consumption was carried out in each family group over a period of one to ten weeks, recordding the various food and dishes consumed and the money spent in this respect.

b) Study by direct observation and interview on production, preparation and food consumption techniques — socio-economic structure, magical and religious systems, events in the life cycle of individuals, and feasts having a bearing on food consumption, etc.

During the second mission, the investigations were continued on the same compounds and completed by a quantitative food consumption survey. In order to gather the reaction of potential consumers to the two proposed protein-rich foods, a study of commercialization and distribution of food at local level was carried out.

THE RESULTS

It would be a euphemism to say that they were fully taken advantage of. One should remember that, at that time, the evaluation fashion was very recent. S.P. Hayes Junior's book, "Measuring the Results of Development Projects", was published by UNESCO only in 1959.

1) On a general basis, it was not possible for the anthropologist to justify the choice of Khombole as representative of bush areas in Senegal (road and rail-road connections make quite a difference), but it was possible to show the complexity of the area: opposition between Wolof and Serer styles, between urban Khombole and rural villages in the zone, and how these reciprocal relations were typical of the groundnut-growing region. This contributed to putting the emphasis on rural Wolof and Serer villages, as well as on Khombole, in selecting the "sample" on which work has been carried out from 1960 to 1978, and to the implementation of an itinerant system of action, which had already been decided.

2) On a specific basis, the analysis of food facts from these various angles provided information about the etiology of malnutrition and substantiated some of the hypotheses about:

a) consumption - adverse food practices during weaning and treatment of various sickness (fever and measles), and an unbalanced distribution of food during meals.

It also demonstrated a number of efficent traditional recipes from the nutritional viewpoint. These observations were taken into consideration in educational programmes and various publications (N'Doye et al., 1973).

b) production - storage and commercialization.

Analyzing the economic life at village level showed the influence of production and commercialization of foodstuff on nutritional level and attracted attention to a number of imbalances: precarious ecological equilibrium, disproportion between food crops and cash crops (groundnuts), scarcity of monetary income and its uneven distribution throughout the year, resulting in very little money being available for food purchases. (This is one of the reasons why the protein-rich programme was rapidly dropped).

It was also possible to record low productivity of food crops and primitive post-harvest technology, random commercialization of foodstuff on a purely speculative basis, all of which generated a low-protein diet and serious seasonal food shortage. The study also revealed a system of values based on conspicuous consumption and prestige activities unfavourable to capital accumulation and nutritional forecast. It also stressed the outstanding importance of a usurious credit system deeply detrimental to the farmers' interest.

These various aspects had a heavy bearing on food and nutrition at family level which forecast, at planning level, the need to act on production, storage and commercialization of food products; at evaluation level, the necessity was to determine secondary objectives and indicators for various aspects of the programme in the area of material culture as well as that of information, attitude and behaviour. Nothing was undertaken along these lines.

SOCIAL STRUCTURE, STATUS AND SCALES OF VALUES

Analyzing social structure, authority and decision-making processes among various groups of the population, contributed to applying the educational programme by easing a too feministic approach, helping to include men in the project, promoting decentralization of action towards rural villages, strenghthening the inclusion in the programme's activities of the main traditional female leaders — the village midwives — who became the focal point of action in the bush (Villod, 1965). The latter aspect had been decided since 1957.

However, no notice was taken of oppositions and complementarities between rural and urban areas, farmers and white-collar workers, subsistence and wage earning populations, progressive Wolof groups as against traditional Serer. These dynamics between the two groups could have been used as levers by the programme at the level of the beneficiaries as well as the project staff.

PROGRAMME ACHIEVEMENTS

Studies and surveys. This was a fertile area and provided good results. Most of the research was centred on six pilot villages, three from each ethnic group, which were compared to about ten test villages from 1965 onwards. Many topics were covered in relation to the health and nutritional situation of mother and child in African tropical environment. Most belonged to the medical field, ranging from the study of the evolution of total protidemia and electrophoretic fractions (Blatt et al., 1967) among children, to the observation of measles epidemics (Debroise, Sy, Satgé, 1967). Outside the medical field, six food consumption studies were carried out (Raba, 1957; Garine, 1961, 1962; Gros, Dupin et al., 1962); Gros and Toury, 1964; Favier, 1966), and one psychological study of the relationship between mother and child at weaning period (Zempleni, 1967).

Demography appeared as one of the most successful areas and the studies undertaken brought a contribution to knowledge about child mortality according to season, age, breast feeding, fertility (Cantrelle et al., 1967, 1969; Debroise et al., 1967).

ACTION TECHNIQUES

Respecting the priorities which had been set, most of the action was taken in the field of education — nutritional and health education, diet counselling. To a lesser extent, sanitation (latrines), antipaludian measures (Nivaquine), various medical services — mainly injections against measles (in four villages) — were also included (Cantrelle, 1969, p. 44).

The above actions were carried out: (i) in Khombole through its dispensary and maternity service; (ii) in the three rural dispensaries; (iii) in the bush villages. Two mobile one-day consultations allowed eight villages to be visited per week, supposed to result in the total coverage of the area in due course.

In practice, most of the action was undertaken in six pilot villages, two of which stopped collaborating in 1966, leaving only one Wolof village in the "sample". In another area, the training of traditional midwives and, later on, of "animateurs ruraux" (rural extension counsellors), allowed the programme to spread, first to 40%, then to all the villages of the circunscription.

The impact of the various aspects of the programme on malnutrition and health level is positive but could never be evaluated precisely for lack of an adequate system.

BASELINE AND EVALUATION

The family groups upon which the initial surveys bore could have served as a basis for comparison and be included in the project either in the pilot, or in the test, villages: Once the action programme was decided, it should have been possible to determine intermediary objectives and set appropriate indicators allowing the measurement of change at various levels, information, attitude, behaviour, material culture and technology. This was not undertaken systematically. The fact that a few of the homesteads belonging to the original survey were integrated in the project was rather a coincidence. One of the initial Serer villages was included largely because the main Serer investigator in charge of the demography files was born there. Consequently, no benefit was derived from the exploitation of the information and the contacts stemming from the initial sample. This oversight was not justified on methodological grounds since neither the villages nor the compounds in the pilot zone were selected at random. No attempt was made to gather systematic data on economic, social and religious life on this new collection of family groups, as had been done during the initial study.

The "sample", a purely operational one, consisted of the group of children from 0 to 14 years: 2000 from Khombole, 2000 from the pilot villages, 7000 from the test villages. Two types of file were compiled: one dealing with each compound having a child in that age bracket, including information about demography, mortality rate, pregnancies, etc.; one for each of the children in the sample (including anthropometric and clinical information) (Raybaud, 1970, p. 25 and following; Cantrelle, 1969, p. 46).

INDICATORS

The section and evaluation programme developed around the factors most easily accessible to medical and paramedical specialists and around easily quantifiable data. Two types of indicators were sought out: rate of death mainly between 0 to 4 years; periodic assement of growth — weight/height ratio every two months.

These data on 11,000 children, studied over a period of more than ten years, created a rather expectional corpus and allowed a number of comparisons: horizontal comparisons between different groups of the sample according to various factors, sex, age, ethnic group, birth data within the year; longitudinal comparisons between years and various seasons.

SHORTCOMINGS

As expressed by the researchers involved in the project (Cantrelle and Diagne, 1969, p. 55), it proved very difficult to isolate the specific results of the various actions undertaken for lack of intermediate objectives and appropriate indicators, those chosen being too general.

The refinement of medical anthropometrical and demographical analysis is in sharp contrast with the empiristic way in which the socio-economic and cultural aspects were dealt with, although they were highly pertinent. After the initial period, only two systematic attempts were carried out in the field of social sciences: 1) in 1964 a questionnaire survey about living conditions of children in rural areas of Senegal was carried out among 500 villagers, bearing mainly on the priority given by the members of the family to child feeding, health and primary education (Crapuchet and Paul-Pont, 1967); 2) in 1970 an attempt was made to evaluate the efficiency of diet counselling upon a sample of 30 families (Satgé et al., 1970).

The choice of growth in relation to age as a privileged indicator of the improvement of living conditions and hygiene proved insufficient to evaluate the impact of the different factors at work. It is in a rather sporadic way that attempts were made to correlate it with understanding by the mother of diet counselling and infant feeding. However, this was the proper track which should have been followed as soon, and as thoroughly, as possible.

As early as 1967, researchers taking part in the project insisted on the importance of socio-economic factors, and especially on the connection between food production and mortality level in the area. After emphasizing the importance of ecological factors as compared to the results of various parts of the project, Cantrelle wrote: "improvement of the economic conditions would have a higher indirect action on lowering mortality level than direct medical and health actions" (Cantrelle and Diagne, 1969, p. 60).

This type of statement should have brought about reorientation of the project according to the findings gathered, which demonstrated how nutritional and health levels were linked to food production, monetary income, budgetary systems, storage techniques, commercialization.

As regards evaluation, it was easy to sort out indicators at the level of material culture, artifacts, behaviour, attitudes, level of knowledge. For instance, it was possible to seek correlations between young children's nutritional status, income, existence of specific funds handled by women for buying food, actual food consumption, existence of a separate dish for children, etc. It would have been easy to carry out periodically food consumption and family surveys and various inventories, as had been done during the data-gathering period. It is rather surprising that nothing was attempted along these lines, probably for the simple reason that nobody with professional experience was available in the project to deal with socio-economic and cultural factors. It is rather unfortunate that the Khombole project did not attempt to take advantage of the large number of social scientists developing socio-economic and anthropological work in Senegal from 1966 to 1976 (Bensaid, 1966; Pelissier, 1967; Gastellu and Delpech, 1964; Lericollais, 1969; Roch et al., 1971, '75, '76).

It should not be concluded that the Khombole project did not meet with success; the fact that socio-cultural aspects and evaluation were considered in a purely empirical way does not signify that it failed. On the contrary, it reached most of its goals but it remains impossible to prove it scientifically.

ASSESSMENT

Successes

A positive one, not only in the area of research, as we have shown, but also practically.

Infant mortality decreased more in the pilot villages than in the test villages — 185% as against 247% (Cantrelle, 1969, p. 55). The decrease is mainly noticeable during the first months of life and seems to correlate with the disappearance of deaths from umbilical tetanus, which may be traced to the action of newly-trained village midwives.

The various education attempts carried out steadily in Khombole have created a climate of confidence and aroused rural dwellers' consciousness of the importance of child food and health problems.

The level of nutrition information has increased among family mothers (which does not necessarily materialize in actual behaviour being adapted) (Satgé et al., 1970).

It might even be said that the Khombole experience contributed to bring about an awareness at national level. The demands with regard to food sufficiency and modern medical care are today the highest priorities in village communities, to the extent that national means are quite unable to satisfy them.

Maternal Care

The system of traditional midwife training has spread to many public health posts throughout Senegal. In 1978, relying on purely national financing, the Khombole medical program still counts with 200 active midwives, regularly controlled, in most of the rural villages of the area.

Lack of Reproducibility

Here lies one of the main difficulties in reproducing the project. The financial means at hand in Khombole, which is still a demonstration area dependant on the Chair of Social Pediatrics of Dakar University, allow the maintenance of a liaison vehicle. It also maintains an above-average motivation among the personnel involved, who are frequently visited and controlled, and keeps alive the feeling of taking part in a show programme (as regards the auxiliary personnel), and in partaking in scientific work liable to be published (for professional staff). Outside this environment, the motivation to participate in development programmes in rural areas appears to be inversely proportionate to the amount of school and professional training received. Nothing could be undertaken to counteract the tendency of the elite to desert the rural areas for the city lights. In spite of governmental efforts towards developing the countryside Rural Expansion Centres (M. Gueye, 1975), today it is hardly possible to come across a female rural extension agent with good training who is willing to travel to the bush, and have first-hand contacts with villagers. Along the same lines, training courses organized in Khombole failed to awaken among young medical doctors a vocation for rural medicine.

Unsolved Problems

The Khombole project did not give the well-needed opportunity to fight the most important factors bearing on malnutrition. These do not lie within the realm of information and knowledge but mostly in the crude economic situation. The imbalance between food and cash crops continues to be favoured by economic policy. The decay of ecological conditions for ten years makes even more difficult and frustrating the situation of villagers who are now well informed about what they are lacking with regard to nutrition and public health, and are longing to put into practice what they have learned. As a matter of fact, the stress due to drought on the living conditions in the area offers an unexpected opportunity to evaluate retrospectively the effects of the educational programme. If a clinical, nutritional and food consumption survey on samples belonging to the pilot and to the test villages were carried out, it is likely that the differences between various children's status would be significant. As P. Cantrelle wrote: "If food resources are scanty, it needs a high level of education among families to allow them to distribute food in order to keep a proper ration for the children" (Cantrelle, 1967, p. 147).

It should be noted that the factors which joepardized most seriously the results of the programme were those outside its initial scope. Some of these factors were difficult to control, such as drought, but others could have been taken into consideration by the project, such as production, storage, commercialization of food products, and, in another area, motivation of the elite.

A Work of Caution

This lack of versatility and desire to remain self-contained on public health and nutrition education grounds, is quite typical. It translates the desire of the managing team to stay clear of thorny fields such as economy and sociology, which may have political implications. It also stems from the widespread with among specialists belonging to exact sciences, to stay away from those involved in social sciences, among which it is difficult to isolate various parameters and to reach precise quantifications.

POSSIBLE CONTRIBUTION OF ANTHROPOLOGY

Today this position cannot be held easily and it is commonly admitted that the development of food and nutrition programmes implies the analysis of local and regional situations in their total complexity. Satisfying the primary need of nutrition involves most of the cultural machinery. As Marcel Mauss might have stated, it means dealing with a "total social phenomenon". It requires an indutive approach to determine the many factors at work, ecological as well as technical and socio-cultural. An etic viewpoint is easily adopted here.

Furthermore, it is necessary to assert the way beneficiaries and the involved personnel, who are individuals integrated in given cultures and sub-cultures, perceive these relations and also their own, to the programme. This viewpoint (emic) is not easily accepted by specialists outside social sciences. The anthropologist who studies social and cultural life from both angles in communities considered as integrated units, is in a good position to unveil the factors involved and make proposals for action both at microscopic and macroscopic levels (Marchione in Pelto and Pelto, p. 237). The former is easily admitted and the fole of the anthropologist as a psychological action agent is best understood by other professionals and authorities: he is supposed to convince people to become involved in the programme and is in charge of smoothing out conflicts. He is not expected to question the validity of the programme or suggest reorientations for various reasons. Unfortunately, this is exactly where he could help. To do so, he should not be brought in at a late stage and remain confined to a closely restricted area, both geographically and thematically. He should go beyond the role, which he has too often been asked to play, of mere alibi or affidavit for the programme.

NEED FOR TEAMWORK

Bringing an anthropologist or any other social worker into a development programme is a serious matter and will not allow for comfort. As Margaret Mead might have put it: "introduting into the culture of people to improve their lives is also a serious dealing". Therefore, the anthropologist should be involved at a very earliest stage of the programme, from the planning stage throughout the whole life of the project. He should also be brought in, at later stages, for periodical (or post-mortem) evaluation. Of course, he should also accept to cooperate within an interdisciplinary team and take the trouble to make himself understood by other professionals and administrative officers. He should not revert to his most sophisticated techniques in order to improve the rest of the team. This is not a simple matter but can be more easily achieved if the anthropologist's status is given full recognition and he is granted a policy-making responsilility in the project. It is, of course, possible to give a general framework of his possible contritution and I have attempted to do so at length elsewhere (Garine, 1978, p. 55...).

TENTATIVE FRAMEWORK

- First, at the preliminary stage: testing the original hypothesis about the proposed project, gathering baseline data about socio-cultural and economic factors connected with food and nutrition. Analyzing them and drawing action-oriented conclusions. Helping the insertion into the field of other types of surveys and studies, taking part in interpreting their results according to local and general problems.

- At the executive stage: participating actively in the planning, contributing to the insertion of the programme at local, regional and national level, including the political administrative aspects. Determining target groups, primary and secondary objectives and the appropriate and simple indicators for the various parts of the programme. Helping in the recruitment and training of the executive personnel and in the motivation of leaders and authorities connected with the project at various levels. Helping in establishing an evaluation system and seeing that it remains a dynamic creative process.

- At later stages: taking part in the surveillance of the programme by asserting its objective results and the outlook held by beneficiaries, involved personnel and responsible authorities.

BROADENING THE SCOPE

Dealing with aspects which are usually foreign to the economic planner or the nutrition specialist, one should be aware that the anthropologist-in-the-project may point out unexpected interactions which might jeopardize the aims of the programme. In this respect, he might even question broad governmental options contrary to the problem the project is meant to solve. This is important and should be considered to be positive.

There seems to be a general consensus in developing countries to break away from nutritional and public health projects which succeed merely in reducing political and social tensions, or placating specific groups rather than really gettint down to the true causes underlying malnutrition. In the case study to which I have referred, the process was quite unconscious but attempting nutrition improvement in the groundnut growing area in Senegal is closely tied to the ecology, which is difficult to modify, and to economic options, which can be changed. Bettering the nutritional situation implies improvement of food production at the expense of the old-fashioned groundnut cash crop, from which the government receives of its foreign exchange (which is not necessarily being used to improve the lot of peanut planters) (Mersadier, 1968).

Improving the income situation and the budgetting abilities of local people implies doing away with the credit system which is contrary to the farmers' interests (Roch, 1976), but which involves many local powers. Advocating happy work in the bush through cooperative and rural expansion centres necessitates something being done to motivate and tightly control the personnel involved.

CONCLUSION

In this framework, dealing with malnutrition solely at educational level and tracking down "irrational food habits and traditional taboos", is just giving well-meaning specialists a nut to crack and avoiding going to the heart of the matter, which is questioning the establishment of priorities and privileges as well as thinking habits and administrative routine. Breaching the gap between urban and rural areas and improving nutritional status does not solely rest at the level of material culture, devising efficient technical devices; it implies dealing with the reference system and ideologies. It means shifting economic priorities, and even attempting to induce change in the established scales of values and social ranking. Going further than academic speculation in these matters implies a strong motivation and a continuous involvement on behalf of the highest national authorities. The IXth FAO/WHO report on Food and Nutrition Strategies in National Developments (1976) is quite enlightening in this respect. It states, p. 39, "even in the absence of acceptable measures to fight the fundamental causes of malnutrition, it is however possible to improve the immediate situation of the malnourished through public health strategies".

This could apply rather well to described case. Leaving aside humanitarian considerations, this might also be what should be discontinued except in cases of high emergencies. We have reached the stage where there is no doubt that development policies are a matter of national sovereignty and should lie in the hands of national authorities. One also wonders whether development agencies of all brands should not be a trifle more particular about the programmes they accept to promote and finance, but this is a difficult issue.

The same joint report goes on suggesting that "governments should promote those policies allowing the bulk of the population, especially the poorer strata, to be given access to goods and know-how, as well as to benefit from new technological and capital inputs which will both reduce poverty and consolidade economic growth" (p. 26).

If this is the case, no doubt anthropologists will be able to make a contribution to development without betraying their deontological code and the people for whom they are speaking. If the situation remains as it was, it will become increasingly difficult to enlist good-quality professionals and, if this is the case, they will undoubtedly ask awkward questions. It needs an exceptionally clear and impervious conscience to be able to advocate protein-rich food for children and track down food taboos in a poverty-stricken population, where the officially established minimum salary barely allows the purchase of the starchy staple.

BIBLIOGRAPHY

Aretas, Col. (1959), Sur la Supplémentation alimentaire africaine traditionnelle avec la Farine de Tourteau d'Arachide, Dakar, 8 p.

Bastide, R. (1971), Anthropoligie Appliquée, Payot, Paris, 247 p.

Baylet, R., Dauchy, S., Debroise, A. and Lenoc, P. (1967), Situation médico-sociale de l'arrendissement de Khombole: III-IV Environnement infectieux, in: "Conditions de vie de l'enfant en milieu rural en Afrique". Colloque Dakar, fév. 1967, C.I.E., Paris - Réunions et conférences 14, 46-50.

Bensaid, A. et al. (1966), Besoins nutritionnels et politique économique, Réflexion à partir d'une enquête détaillée dans trois villages sénégalais, Inst. des Sciences Economiques Appliquées, Dakar.

Blanc, J., Coly, F. et al., Situation médico-sociale de l'Arrendissement de Khombole, Sénégal, in: "Conditions de Vie de l'Enfant in Milieu rural en Afrique", Dakar, fév. 1967, Centre International de l'Enfance, Réunions et Conférences XIV, p. 41-55.

Blatt, M.J.; Debroise, A. and Satge, P., Étude de la protidémie totale et des fractions électrophorétiques en milieu rural, in: "L'enfant en milieu rural", p. 174-176.

Cantrelle, R., Diagne, M., Raybaud, N. and Vignac, B. (1969), Mortalité de l'enfant dans la Région de Khombole-Thiénaba (Sénégal), Cah. ORSTOM, Ser. Sc. Hum., Vol. VI, No. 4, pp. 43-74

Cantrelle, P., Diagne, M., Raybaud, N., Villod, M-Th. et al. (1967), Mortalité de l'enfant en Zone rurale au Sénégal, in: "Conditions de Vie de l'Enfant en milieu rural en Afrique", Dakar, fév. 1967, Centre International de l'Enfance, Réunions et Conférences XIV.

Cantrelle, P., Raybaud, N., Villod, M-Th. and Diagne, M.A. (1967) Procédé pour obtenir des taux de mortalité dans les pays ne disposant pas d'un système complet détat civil, in: "L'Enfant en Milieu Tropical", No. 37, p. 19-27.

Couty, P., (1972), Employ du Temps et Organisation du Travail agricole dans un village Wolof mouride - Darou Rhamane II, in: "Maintenance sociales et Changement économique au Sénégal, I. Doctrine économique et pratique du travail chez les Mourides", Travaux et documents de l'ORSTOM, No. 15, pp. 85-132.

Crapuchet, S. and Paul-Pont, I., (1967), Enquête sur les Conditions de Vie de l'Enfant en Milieu rural au Sénégal et en Gambie, L'Enfant en Milieu tropical, No. 39, pp. 3-24.

Cros, J., Dupin, H., Toury, J., Richir, C., Quenum, C. and N'Doye,

Th., (1962), "Enquête sur la consommation des lipides à Khombole (Sénégal)", Orana, Dakar, 17 p.
Cros, J., Toury, J. and Giorgi, R., Enquêtes alimentaires à Khombole (Sénégal), in: Bull. de l'Inst. Nat. d'Hygiène, Tome 19, 1964, No. 4, pp. 629-680.
Debroise, A., Dan, V:; Cros, J., Coly, F., Raybaud, N. and Villod, M.Th., Croissance staturo-pondérale de l'enfant de 0 à 7 ans en zone rurale au Sénégal, in: "L'Enfant en milieu rural", pp. 109-116.
Debroise, A., Raybaud, N., Villod, M.Th., Coly, F., Dan, V. and satgé, P., Mortalité dans six villages surveillés par l'Institut de Pédiatrie Sociale, in: "L'Enfant en milieu rural", pp. 130-133.
Debroise, A., Sy, I., and Satge, P.; La Rougeole en zone rurale, in: "L'Enfant en milieu rural", pp. 149-156 and "L'enfant en milieu tropical", 1967, 38: 20-36.
Dupin, H. (1961), "Étude des Conditions d'Utilisation de la Farine d'Arachide dans l'Alimentation familiale ou l'Alimentation des Collectivités au Sénégal", Orana, Dakar, Avril 1961.
Dupin, H. (1965), "Expériences d'Education sanitaire et nutritionnelle en Afrique", Tiers Monde Études, IEDES, PUF, Paris, 1965, 118 p.
Dupin, H. (1968), L'Alimentation, l'État de Nutrition et les Tendances actuelles de Consommation alimentaire en Afrique Intertropicale, in: Développement et Civilisations, IRFED, Imp. Allain, Elbeuf, Paris, No. 35, sept. 1968, pp. 21-30.
Favier, J.C. (1966) "Enquête de Consommation alimentaire dans deux villages du Sénégal", Orana, Dakar.
Garine, I. de (1961), "Budgets familiaux et Alimentation dans la Région de Khombole (Sénégal)", Orana, Dakar, 41 p.
Garine, I. de (1962), Usages alimentaires dans la Région de Khombole (Sénégal), Cah. d'Études africaines, 10, pp. 218-265.
Garine, I. de (1964), Planificacion y evaluacion des los programas de Nutricion aplicada en America Latina (Brazil, Chile, Paraguay), Doc. FAO/60130-64-HR, Rome; 66 pp.
Gueye, M. (1976), Passage d'une structure d'encadrement de production à une structure de promotion participative des popolations rurales du Sénégal, Doc. du Séminaire FAO/FNUAP sur la planification agricole et la population, Tanger 3-14 nov. 1975, FAO/UN/TF/INT 142 (UPA).
Garine, I. de (1978), Population, Production and Culture in the Plains Societies of Northern Cameroon and Chad: The Anthropologist, in: Development Projects Current Anthropology, Vol. 19, No. 1, March 1970.
Hayes, S.P. Jr., (1959), Measuring the Results of Development Projects - A manual for the use of field workers, Monographs in the Applied Social Sciences, Unesco, Paris.
Latham, M.C. (1972), "Planification et Evaluation des Programmes de Nutrition Appliquée", Études de Nutrition de la FAO, No. 26, FAO, Rome, 129 p.

Larivière, M. et al., Étude de l'excrétion entérique des parasites chez les enfants de deux villages en zone sahélo-saharienne au Sénégal, in: "L'Enfant en milieu rural", pp. 50-54.

Le Centre rural de Protection familiale de Khombole, in: Bull. Méd. de L'A.O.F., Journées médico-sociales de Dakar, 26-39, Oct. 1957, pp. 5-50

Lericollais, A., (1969), "S.O.B. en Pays Sérère (Arrondissement de Niakhar)", ORSTOM, Dakar.

Mersadier, Y., (1968), Les Céréales senégalaises, leur Consommation et leur Transformation industrialle, Rome, Rapport FAO, NU:SF/SENS, WS/81575, 38 p.

N'Doye, T., N'Doye, A.M.B., Camara, D.F. and Doulibaly, M.A., "Guide de Régime du Sevrage au Sénégal". Min. de la Santé Publique et des Affaires Sociales, Dakar, 113 pp.

Pelissier, P., "Les Paysans du Sénégal, les Civilisations agraires du Cayro à la Casamance". Impr. Fabrègue, Saint-Yrieux 5HV) 1967, 939 p.; bibl.

Pelto, P. and Pelto, G.H. (1978), "Anthropological research - The Structure of Inquiry". Cambridge University Press, Cambridge, London/N.Y./Melbourne, 333 p.

Raba, Capitaine A., (1957), "Enquête clinique nutritionnelle dans la Région de Khombole (Sénégal)", Orana, Dakar, sept. 1957, 20 p.

Raybaud, N., (1971), L'Observation de la croissance des enfants africains dans une zone de Santé publique en milieu rural, Khombole 1963-70 - Approche méthodologique à partir d'une étude critique, Mémoire de la Section paramécicale, École Nationale de la Santé Publique, Rennes, 104 p., bib.

Roch, J. and Rocheteau, C., (1971), Economie et Populatoin, le Cas du Sénégal, Cah. ORSTOM, Sc. Hum. VII, No. 1, p. 63-73.

Roch, J. (1972), Eléments d'Analyse du Système Agricole en Milieu Wolof Mouride: l'exemple de Darou Rahmane II in Maintenances sociales et Changement économique au Sénégal, I. Doctrine économique et pratique du travail chez les Mourides, Travaux et documents de l'ORSTOM, No. 15, p. 35-66

Roch, J., (1975), Les Migrations économiques de saison sèche en bassin arachidier sénégalais; Cah. ORSTOM, Serv. Sc. Hum., Vol. XXI, No. 1, p. 55-80.

Roch, J., (1976), La Richesse paysanne en Bassin arachidier sénégalais: Inventaire et Essai d'évaluation des biens familiaux, Cah. ORSTOM, Sc. Huml, Vol. XIII, No. 4, p. 383-408.

Satgé, P., Mattei, J.F. and Vuylsteke, J. (1970), Les Conseils de Régime: Critique des Résultats, L'enfant en milieu Tropical, No. 70, p. 16-23.

Satgé, P., Coly, F., Raybaud, N., Villod, M.T. and Touré, M., (1964), Une Centre d'Étude et d'application des méthodes de santé publique et de protection de l'Enfance en Brousse, Khombole, Sénégal, L'Enfant en Milieu Tropical, No. 19, p. 5-14.

Senecal, J., Le Centre de Santé de Khombole, Journées Africaines de Pédiatrie, Dakar, 12-18 avril 1960, Inst. de Péd. Soc. de l'Univ. de Dakar et Centre Int. de l'Enfance, Paris. pp. 129-79.

Senecal, J., Place des Diarhées en Pathologie infantile, Journées Africaines de Pédiatrie, Dakar, 12-16 avril, 1960, Inst. de Péd. Soc. de l'Univ. de Dakar et Centre Int. de l'Enfance, Paris, pp. 129-137.

Senecal, J., Larivière, M:; Dupin, H., Faladé, S., Hocquet, P. and Coly, F., Résultats d'une enquête sur l'état de santé des enfants dans l'arrondissement de Khombole. I. État général et nutritionnel; II: Étude parasitaire, Bull. Soc. Méd. Afr. Noire, Langue Franç., 1961, 6, pp. 195-223.

Stratégies de l'Alimentation et de la Nutrition dans le Développement national, IX? Rapport du Comité mixt FAO/OMS d'Experts de la Nutrition, FAO/OMS, Rome 11-20 décembre, 1974.

Touré, M., (1962), L'Expérience sénégalaise d'Education de la Mère en matière d'Alimentation de l'Enfant âgé de 0 à 6 ans, in: Séminaire sur l'Éducation en matière de Santé et de Nutrition en Afrique au Sud du Sahara, Pointe-Noire, 5-12 juin 1962, C.I.E., Paris, pp. 62-66.

Villod, M.Th. and Raybaud, N., L'Education des Matrones au Sénégal, in: L'Enfant en milieu tropical, 1965, 27, 7-12.

Zempleni-Rabain, J., Le Sebrage chez l'enfant Ouolof - Antécédents et implications in Enfant en milieu rural, 212-217.

EVALUATION IN A CARIBBEAN CONTEXT: SOCIOCULTURAL FACTORS

AFFECTING A COMMUNITY HEALTH AIDE PROGRAM

Thomas J. Marchione, Ph. D.

Dpt. of Anthropology - Case Western Reserve University

Cleveland, Ohio - USA

INTRODUCTION

The uses of social anthropology in the evaluation of applied nutrition programs is not totally new. A small but significant amount of work done in conjunction with international public health programs in Latin America and Africa has been reported upon over the last thirty years (Foster, 1962; Adams, 1955; Cassel, 1955).

This paper is a report of a study which follows in the patterns established by the above researchers in that (1) the research was focused upon a health care program to improve nutritional status; (2) intensive research was done over a relatively long period; (3) participant observation methods were employed; (4) a holistic view of the program context was taken. The study departs from and, in my view, advances the tradition by (1) using more statistical and extensive sampling and analytical methods; (2) including the innovation bureaucracy and governmental structure not merely as context but as part of the analysis; (3) incorporating the evaluation concepts developed by other disciplines; and (4) demonstrating an instance where an anthropologist was consulted in the early stages of program implementation and was not brought in to "troubleshoot" in post hoc fashion after failures became obvious; (5) dealing with the nutritional-biological (physical anthropological) aspects of the program. Furthermore, the results of the study suggest some insight into the increasingly popular use of nutrition aides and village level health workers in programs for nutrition improvement around the world.

The case considered is an evaluation of the Jamaican Government's Community Health Aide Programme in the parish of St. James (one of Jamaica's 13 parishes) during the years 1973 to 1975. In 1973, I was engaged in the program* as a technical officer (PAHO) working for the Caribbean Food and Nutrition Institute. In the follow-up study in 1975, I returned as an anthropological researcher.

THE RESEARCH CONTEXT

A cardinal principle of anthropological approach is that research should begin with a grounded and holistic understanding of the particular socio-cultural and physical environmental context. Without such contextual grounding, data cannot be transformed into information. This is especially true for those reading with an eye to the comparison and application of programs in other contexts.

The geographical focus of this research is the parish of St. James in the western end of the semi-tropical island of Jamaica. Its population was estimated at 115,000 in 1975, representing 6% of the nation's population of 2.01 million people. The parish capital, Montego Bay, is the main population center with 40% of the parish population and a major international tourist attraction. The history and social structure of the parish parallels in many ways that of the island as a whole. Since the time of European contact in the fifteenth century, the region has been trasnformed by island-wide and international changes; nevertheless, cultural and economic dependence upon Europe and North America has persisted. The "black" cultural segment of the population has to date remained overwhelmingly in the majority and overwhelmingly more poor than "brown" and "white" segments (Smith, 1965: 168-175). The status hierarchy places a few capitalists highest and the mass of manual agricultural laborers lowest. Since 1972 the "democratic socialist" national politics of the Peoples National Party (PNP) has dominated the political and governmental aspects of the parish and has represented a challenge to this historical pattern. This later development was a departure from the previous ten years, following independence, when the Jamaican Labor Party (JLP) vigorously used the traditional framework to pursue outside foreign investments in tourism in Montego Bay (Kuper, 1976).

The JLP contribution to St. James was the construction of the 400 bed Cornwall Regional Hospital in Montego Bay. Rural and preventive services such as public health nutrition remained virtually unimproved. Malnutrition especially in young children was approximately 10% in Gomez II and III and hospital malnutrition wards were overcrowded (Marchione, 1973, 1977a).

---

*The British spelling of "program" is used only in the official title.

HEALTH AIDES AND NUTRITION NEEDS

The Community Health Aide Programme was implemented in late 1972 immediately following the victory of the PNP party. St. James and adjoining Hanover Parish had over 300 health aides by January of 1973. Since then over 1300 aides have been trained and deployed island-wide, and plans for additional aides are being implemented. The program has been presented as a concerted attack upon primary health care delivery problems, especially young child nutriton problems. In the early 1970's community survey data suggested that for children under two, malnutrition was an associated or underlying cause of death in over 50% of all deaths in the Kingston area; and if one speculated about the synergism of nutrition and infection, the importance of the nutritional problems were of even greater importance (Puffer and Serrano, 1973: 183). Although Jamaica's infant mortality rate ranks fourth out of twenty-four nations in Latin America and the Caribbean (Sivard, 1974) community surveys showed that lowering infant mortality rates had not been matched by reduced young child malnutrition (Fox, Campbell and Harris, 1968; Gurney, Fox and Neill, 1972). In 1968-9 eight per cent of hospital beds were filled by malnourished children (Cook, 1971). In 1974 it was reported that nutritional illness caused the longest hospital stay (National Planning Unit 1974: 66). Besides the financial costs, new knowledge about mental and behavioral effects have an added urgency to the problem (Birch and Richardson, 1972: 64-71; Richardson, Birch, Grabie and Yoder, 1972).

The basic program plan was to provide a force of new health care workers at the local level. Women with relatively little education were to be recruited, trained for two months, and returned to their home communities to work in first aid and preventive health work. This included nutrition education such as encouraging breast feeding, earlier introduction of solid food to infants and more use of kitchen gardening. The Community Health Aide (CHA) would work under the supervision of the existing network of Public Health Nurses. However, the CHA was also given health clinic responsibilities in conjunction with Clinic Nurses and District Medical Officers, thus attempting to link curative and preventive health activities at the periphery of the health delivery system. They were to be paid about one thousand dollars* a year, a considerable savings over professional health workers. An added bonus expected was that aides would be less likely to leave community service for more prestigious roles than would more highly trained health personnel.

The concept of village health worker or health aide of this type was not basically new; aides such as these have been employed

---

*The Jamaican dollar was equivalent to $1.11 U.S. in 1973, and since then has fallen to an official rate of $.65 U.S.

in other countries for some years. The type of role developed could be broadly classified as a health auxiliary worker (Fendal, 1973; WHO, 1968: 6-8). Specifically, the CHA fits the type because she is a paid worker, with less than full professional qualifications in that field, and she assists and is supervised by a professional worker, in this case the Public Health Nurse. WHO also employs the term "Village Health Worker" for persons engaged in the multipurpose, locally-based functions similar to the CHA (WHO, 1974). Recent Village Health Worker/Auxiliary experience in Iran, Venezuela, and Guatemala provides interesting parallels which may have influenced the Jamaican case (Ronaghy and Solter, 1974; PAHO, 1973: 24-39; Djakanovic and Mach, 1974: 62-67). In these countries, auxiliary workers have been employed without simultaneously transforming the basic political structure in the direction of a more concerted authority system directed toward self-reliance such as occurred in China and Tanzania (Sidel and Sidel, 1974; Mellander, 1973; Van Etten and Rikes, 1975; Djukanovic and Mach, 1975: 57-62).

However, the initial concept for the Jamaican CHA appears to have come directly from the U.S. and European health aide experience (Ennever, 1969; Wise, 1968). Specific reference is made by the creators of the first pilot CHA Programme in Jamaica to Dutch "Maternity Aides" and to Montefiore Hospital's Neighborhood Medical Care Demonstration Project in New York City (Ennever, 1969: 199-200).

LEVELS OF EVALUATION

The research design includes four sub-levels of evaluation (Ricci, 1975: 454-55). I have further grouped the sub-levels into two levels each with two sub-levels (see Table 1). The first two sub-levels are basically focused upon matters _internal_ to the program, at the micro-level, i.e., asking to what extent the program is achieving its stated goals (called "goal-attainment evaluation"; see Schulberg and Baker, 1971: 61-62). Sub-level number one, called "effort evaluation" (Suchman, 1967: 61-62), examines the functioning of program personnel according to defined job descriptions. The second sub-level is called "outcome evaluation". This concerns the degree of measurable progress made in reaching desired nutritional and dietary goals in the target population (Suchman, 1976: 62; Donebedian, 1966: 188).

The third and fourth sub-levels of the evaluation are concerned with processes _external_ to the program or macro-level factors impinging on the program. The third sub-level is called "evaluation of theory in action" — this is a concern with the degree of correspondence between program activities and known major etiological factors or causes of malnutrition in the particular social and physical environment. The fourth sub-level, called "priority evaluation", is

TABLE 1

LEVELS OF EVALUATION USED IN THE HEALTH AIDE STUDY

LEVELS OF EVALUATION		CONCEPTUAL FOCI	METHODOLOGIES	MAIN DATA SOURCES
Internal	Effort Evaluation	Training Effectiveness Description of CHA's, CHA Supervision, Distribution, Field Activities of CHA's, Household Satisfaction	Content Analysis Educational Testing Questionnaire Structured Interview Participant Observation Informal Interview	Training Materials CHA Work Records CHA's Themselves Sample of Health Aide Households Public Health Nurses
Internal	Outcome Evaluation	Behavioral Changes Health and Nutritional Status Changes Health Care Seeking Changes	Structured Interview Anthropometry Survey Food Intake Recall Quasi-Experimental Evaluation Design	Random Sampling of Contact and "Non-Contact" Householders in 1973 and 1975
external	Evaluation of Action Theory	Household Ecology Health and Nutrition Behavior Nutritional & Health Status Epidemiology Macro-Economic System	Structured Interview Anthropometric Survey Food Intake Recall Participant Observation Trend Analysis	Random Samples of Young-Child Households Archival Materials Published Research
external	Priotiry Evaluation	Health & Nutritional Planning, Decision-Making, and Administration, Macro-Political System	Informal Interview Participant Observation Content Analysis Trend Analysis	Government Officials Health Professionals Archival Materials Published Research

assessment of the stated goals relative to other purposes, or priorities, served by the program in the wider social-political system.

Applying this framework suggests the following inquiries:

(1) Effort Evaluation. How frequently were health aides visitint househoulds? What types of households were vitited? What was done in the household and how satisfied was the recipient? What CHA characteristics were most important to effort?

(2) Outcome Evaluation. Did the Community Health Aide Programme change household dietary, health seeking, and/or nutritional status of target children in the desired ways?

(3) Evaluation of Action Theory. Did the CHA Programme address the proper objectives given the social epidemiology of malnutrition (i.e. ill-health) among Jamaican children? To what extent were the child health problems due to social and economic processes outside of program control or, indeed, outside of Jamaican national control?

(4) Priority Evaluation. Did political considerations override the health care objectives and health/nutritional goals of the program? Who makes the decisions about health care in Jamaica and how did that affect the CHA Programme?

The methodologies reflect the eclectic nature of the theoritical and practical concerns underlying the research design. Different methods, and combinations of methods, were necessary at each level of evaluation (see Table 1). The result is a mix of quantitative and qualitative methods and data which are the basis not only of good evaluation research, but also of good anthropological and sociological research in general (Ricci, 1975: 453; Pelto, 1970: xiii; Denzin, 1970: 301). Because of the limits of space I will discuss only the effort evaluation sub-level and the priority evaluation sub-level in this report.

EFFORT EVALUATION — METHODS

Effort evaluation (sometimes called process evaluation) has been compared to the measurement of the number of times a bird flaps his wings without any attempt to determine how far the bird has flown (Suchman, 1967: 61). The general idea in this study was to describe how well the Community Health Aides performed their assigned daily tasks, such as home visiting, how well they were received by the community, and how well they were supervised. The key question was, what personal characteristics of recruits and what aspects of training were related to good performance in the field? Together, these constitute a system of independent variables designed to produce a given outcome in the recipient population.

The independent variables are defined according to administrative goals. It was of course assumed that the Community Health Aides would not only be available to, but also seek out needy householders to promote a variety of health and nutritional measures — especially to those households in deep rural areas. It was necessary to develop precise definitions of goals and objectives from a formal job description, and through sets of informal interviews with program organizers: From these, it was possible to define specific target groups to be benefited and to operationalize a set of variables to be measured or activities to be counted. For instance, a key variable was the frequency of CHA visits to households with children under three years of age.

Using the contents of initial interviews and related materials, I formulated a questionnaire and educational testing instrument to gather a variety of information about the personal characteristics of the health aides, their work in the field, and the effectiveness of the training.

With the data from the questionnaire, it was possible to select a random sample of 20 CHA's stratified by age and locality. Census books kept by each CHA were obtained and a random subsample of 10 preschool child-rearing households was selected from each CHA's area. This, a sample of 200 households was selected for systematic interviewing in early April of 1973. A follow-up survey of 300 households was also conducted in 1975.

PRIORITY EVALUATION — METHODS

The evaluation of priorities requires a shift of focus away from the specifically stated goals of the program to the socio-political environment in which the program is set. At this level, one probes the ways in which even an ideal program would of necessity be modified because of its inconsistency with other national goals or systems. Some potential conflict areas investigated in this study were as follows:

(1) Economic interests;
(2) professional medical interests; and
(3) political party interests.

The methodology used for this level of analysis was considerably less formal than the others discussed so far. I relied upon archival sources such as national policy documents, news accounts, and numerous secondary sources. In addition, some statistical evidence was gathered through an analysis of election returns.

My own participation gave me access to much valuable information through interviews with Jamaican officials and non-officials.

One of my key informants was a Jamaican postgraduate medical student at the University of the West Indies, another a middle-level Jamaican official in the Ministry of Health. Other persons contacted included the following:

1. President of the Nursing Association of Jamaica;
2. Project Officer, U.S. Agency of International Development;
3. Member of Parliament for St. James;
4. Parish Councilors in St. James;
5. Private and Government Physicians;
6. Public Health Nurses in St. James;
7. Secretary of Western Medical Association; and
8. Chief Medical Officer.

Numerous other inferences were gathered through my contacts with the local peasants throughout the parish.

EFFORT EVALUATION - RESULTS

The principal variable in determination of the effort of Community Health Aides was the frequency that households at high nutritional risk were visited in the aides assignment area, i.e., "contact frequency". Contact frequency average .38 visits per week or one visit every 20 days. However, this ranged from as much as one visit per week to one visit per 50 days. The activities of the aide in the household did not differ as a function of contact with the household, the activities were primarily advice and instruction, referrals, and first aide treatment. Consequently, it was felt that qualitative differences between aides visits although important were not crucial in assessing aide effort (see Table 2).

Of course the amount of contact an aide had with any single household in her assignment area was simply a function of the size and population of the assignment area. Some areas were heavily covered with health aides while others had a very sparse coverage. In inverse proportion to coverage was the effect of supervision. Areas with a high density of health aides tended to overload the supervisory capacity of the public health nurses in charge. This tended to produce an adverse effect on the contact frequency of the health aides.

For recruitment purposes, however, it was important to discover the effects of different types of persons on the contact frequency. The aide assignment area and the supervisory load variables were controlled statistically and the relationship of aide contact frequency and five other key variables was determined. These were age, locality orientation of aide, training examination score, relative social economic status, and acceptance by the community in aides area. The results of this analysis are given in Table 2.

TABLE 2

VARIABLES IN THE EFFORT EVALUATION ANALYSIS (N = 20)

VARIABLES	MEANS	UNITS	CORRELATIONS WITH CONTACT FREQUENCY	
			Simple r	r controlling for 7 & 8
1. Contact Frequency	.38	weekly visits per household	—	—
2. Age of the CHA	27.25	years	.25	.49*
3. Exam Score	87.25	percent correct	-.75**	-.61**
4. Locality Orientation	2.10	1.00 is least, 3.00 is most	.45*	.57*
5. Relative SES	1.08	1.00 is no difference in SES between aide and community	.06	.27
6. Community Acceptance	82.30	percent of CHA's households	.12	.04
7. Aides/Population	4.89	aides per 2500 persons	.42*	—
8. Supervisory Load	27.00	per public health nurse	-.17	—

* Significant at the .05 level, F-Test
** Significant at the .01 level, F-Test

Age

Older health aides were harder workers in the field than were the younger and more inexperienced women. Roughly, it was women over the median age of 25 who made better workers. Age was related to numerous attributes of the aides. Of course, the older aide was most likely to be a settled community member. One out of the nine aides under the age of 25 was married, while eight out of the eleven aged 25 or over were married. The older women among the health aides also had considerably more child-rearing experience and were, consequently, much more credible health workers in the high risk households.

Locality Orientation

Eighty percent of the health aides listed their service community as their home address. This did not insure, however, that the aide would be oriented to the locality in social behavior and values. Jamaica is a small island with a long history of outside exposure. Typical examples of outside contact were a period in school in Kingston or Montego Bay, a common-law husband or boyfriend living in Montego Bay with a truck or taxi, or parents who lived abroad. Locally oriented contacts included church, marriage to local shopkeeper, farmer or trademan. As anticipated, locality orientation as defined as the number of local versus outside cosmopolitan contacts was a key factor affecting visiting effort. The more locally oriented health aide had a significantly higher contact frequency with the households in her assignment area. Furthermore, locality orientation was higher among older women ($r = .39$, $p < .05$). This suggests a reason why older aides were harder workers.

Training Examination Score

The relationship between the examination score of health aides following their training and latter contact frequency was unanticipated. It appears that the health aides who did the best on the examination following the training period were the least likely to be regular and frequent home visitors to the households in their respective areas. It was precisely the less literate, and those with more traditional concepts of nutrition who scored lower on this examination. The examination, however, was as much a test of the intracultural differences between the recruits as judged by the public health nurse trainees as it was a test of technical knowledge. In a sense the exam can be thought of as a test of the "professional-orientation" of the health aide as judged by public health nurse trainers (Walsh and Elling, 1969). Support for this argument became clear in analysis of data gathered two years after the exam was administered. A follow-up of the sample of twenty aides showed that three had left to go on to other jobs and five had taken job qualifying examinations for more prestigious health-related roles. Women who left and those who planned to leave the job of Community Health Aide had significantly higher training exam scores two years earlier. This effect was further confirmed in an analysis of aide visits to a random sample of 300 households in the 1975 follow-up. Aides with lower training scores were much more likely to visit households frequently than were high scorers (df = 2, chi square test, p < .001).

Both interview and statistical methods showed that age of the aide and their locality orientation were related to their professional orientation. Older and more locality oriented women were less professionally oriented (lower scores on the training exam). A success for the program was the lack of relationship existing between contact frequency and professional orientation on the one hand and relative social economic status of the aides on the other hand. This can be explained by the fact that most aides were very similar to their clients in cultural background and social economic status. It is important to note that the level of program acceptance was very high in the community (82%). However, although professionaly oriented aides were significantly more accepted by community residents, this did not enhance their effort.

PRIORITY EVALUATION - RESULTS

Generally speaking the health aide program was established with a minimum of consultation with the communities to be served or the bureaucrats who were to implement the program at the local level. It was typical in its top down approach to public health intervention programs. It should not be surprising then that the national social and political hierarchies had great impacts upon the operation of the program at the community level. I will discuss just two of these problems which have already been implied in the discussion of the effort evaluation.

Professional Dominance

Medical professionalism had profound effects on the establishment of program priorities, program activities, and mobility aspirations of the health aides. From the time of the first pilot health aide program in Jamaica in 1967 until the final implementation in St. James Parish, the role of the Community Health Aide was considerably diluted, and less and less input from the community to be served was granted. Initially, the pilot project started at the University of the West Indies Department of Social dna Preventive Medicine provided for candidates to be chosen on the recommendation of citizens of the area to whom they were known (Ennever and Standard, 1969: 195). Even at this early and modest stage opposition to the health aide was vigorous from the nursing profession, especially from private nursing services. The authors of this program eventually stressed that the health aide would not conflict with nursing roles and would be under strict and direct supervision of public health nurses. Later recommendations by the Pan American Health Organization which would have given the aide responsibilities for midwifery similar to the Venezuelan Simplified Medicine Program were not implemented (Felszer, 1971: 7). Considerable antagonism occurred among trained nursing staff when the community began to call the health aides the "new nurses" and began to demand more clinical services than the aides were trained or allowed to provide. Similar antagonisms arose regarding the high pay given to the health aides compared to assistant nurses. Predictions of health professionals regarding the general inability of health aides to keep health problems confidential and the potential for medical malpractice by relatively unsupervised aides proved to be unfounded.

In fact, the evidence regarding the effects of the selection and training of health aides suggested that transmitted values of the nursing profession attenuated needed CHA effort at the community level — a behavioral manifestation of the proclivity of Jamaican nurses to migrate to the United States and European countries. There were 1,888 nurses in public service in Jamaica in 1974, many of whom were trained at the University of the West Indies. In parishes such as St. James nursing posts usually have exceeded the number of trained persons willing or available to fill them. Nevertheless, the largest single professional, technical and kindred occupational group of emigrants in that period (Palmer, 1974). This reflects the continuing cultural dependence of the profession on standards of practice and desirable lifestyles established in metropolitan countries.

More evidence of this professional dominance was the input and virtual control of parts of the program by U.S. based medical schools. This often occurred at the expense of the involvement of medical professionals trained in Jamaica.

Political Influences

The political influences over the health aide program reflect that decisions are often made in response to values quite different from the nutritional and dietary requirements of the target population. We have seen how professional dominance diluted community service goals. Similarly, patterns of national political structure can undermine the establishment of a program which is responsive to particular community needs. One of the most difficult problems of this nature in Jamaica was the influence of two party politics. The impact of party politics in the Caribbean, especially Jamaica, can not be understressed. Politics in lower class Jamaica is not only an important social activity, it is lifeblood (Stone, 1974: 80; cf.

Foner, 1973; Kuper, 1976). Loyal clients receive employment, elegibility for government program, promotions, government contracts, and numerous other favors distributed by central government political patrons (Jamaican Governors Commision of Enquiry, 1973: 91). Three types of party influences on the Jamaican health aide program were investigated: (1) the recruitment and consequent deploymen of CHA's based on political party patronage; (2) the influence of politics on CHA-Community relations, and (3) conflict between program personnel associated with different political parties — actual or perceived.

The maldistribution of health aides in fact could be partially traced to the fact that certain areas which were in political favor provided more recruits to the program. Since the program was designed to return individuals to their home communities, maldistribution of aides tended to become a problem. Overall, the average number of aides in government versus opposition districts was not grossly imbalanced in St. James Parish. However, one needy area loyal to the opposition JLP received one aide while one area considered a government party stronghold with similar needs and of similar size had 14 aides. Study of another parish showed a ten-fold difference in aide density in government stronghold versus opposition areas.

The influence of party politics on CHA-Community social relationships was one of the main points which critics said would undermine the program's nutritional goals. The aides, it was claimed, would preach government politics rather than good nutrition in the households of the opposition party. This problem never materialized to any great extent. Although some antagonisms were clearly present between members of the opposition and the health aides in one or two communities, overall program acceptance was very high and aides concentrated on their essential health and nutrition tasks.

Finally, the aides were largely of one party, and their supervirors were largely hired by the previous opposition government. Consequently, the logic of party politics dictated that the public health nurses were opposition party members. Whether this was indeed true or not, it colored the emotional relationship of health aides with their supervisors. Minor supervision problems were easily escalated into party antagonisms. These personnel problems were not totally debilitating ones, but to resolve them satisfactorily at the local level required an understanding of the national political structures.

SUMMARY AND IMPLICATIONS

This evaluation has implications both of a methodological and substantive nature. Methodologically it illustrates some uses of social anthropology in the evaluation of applied nutrition programs, and proposes some innovations in research design and methods of evaluation study. Substantively, this study draws some general conclusions about the uses of community health aides in applied nutritional programs.

With respect to the methodology of evaluation, I found that it was important to recognize that that there were two fundamental types of evaluation — micro-level or internal, and macro-level or external. At the micro-level evaluation was based upon the program's goals and objectives or, in anthropological parlance, the internally derived values of program designers. This type of evaluation which is very familiar to most of you, resulted in information which helped to refine the process of recruitment and administration of health aides by the Jamaican Government. The macro-level or external evaluation is derived from a social scientific appraisal of the effects of the institutional framework of the country in question on the operation of a nutrition program. This dual approach recognizes that the scope of evaluation is often too limited. Attention has tended to be too focused on the bottom of the administrative system. "This serves to lock in the system early, prevent change and ensure control and accountability that is directed from the top downward" (Guttentag, 1975: 8). In other words, grave inadequacies in the national social structure and the implementing bureaucracy are masked; and the responsibility for programming problems is placed at the feet of small inadequately financed and conceived nutritional improvement programs. In the Jamaican case, inadequacies of the Community Health Programme were traced to conflicting priorities and interest groups characteristic of the Jamaican society. Medical professional guild interests, foreign influences over medical values, and medical personnel mobility aspirations had measurable influences on the operation of the program in the communities served. Furthermore, the political structure resulted in patronage patterns and interpersonal difficulties which could have easily been blamed on the recruiters or the personalities of individual aides or public health nurses.

The evaluation of the program in Jamaica suggests that community health aide programs for nutritional improvement will vary in shape and potential depending not only on the physical environmental, but also the social structure of the countries implementing such programs. It reaffirms the fact that village level worker programs can not be mindlessly transferred from society to society without fundamental modification (New and New, 1975; Ronaghy and Solter, 1974). For example, in plural democracies the principle of taking recruits from the community in which they will serve may prove to be inefficient because the patronage patterns will tend to distribute jobs more heavily in strongholds of the party in power; this is not the case in one-party socialist countries such as China where such programs have had great success (Sidel and Sidel, 1973; Mellander, 1973). Furthermore, the roles of non-professional in countries with relatively strong medical professions may be so truncated as to make their potential contribution to nutrition slight relative to the cost.*

In conclusion, there is undoubtedly a need to train non-professional, nutrition workers for the benefit of the world's people, especially the rural or poor. However, implementation and evaluation of such programs can not be separated from evaluation of national social and political systems. Neither nutritional programs nor the nutritional health of people can be separated from the material and social/institutional environments in which people live; and furthermore, these environments are not neutral, they are politically manipulated. If one honestly seeks to satisfy the basic needs of people in communities, those who will benefit should have the freedom to define objectives and the means to manipulate their own environments to achieve a better life.

REFERENCES

Adams, Richard N., 1955, A Nutritional Reserach Program in Guatemala. In: "Health Culture and Community: Case Studies of Public Reactions to Health Programs". Benjamin D. Paul, ed., Russell Sage Foundation, New York, pp. 435-458.

Birch, Herbert G. and S. Richardson, 1972, The Functioning of Jamaican School Children Severely Malnourished During the First Two Years of Life. In: "Nutrition, the Nervous System, and Behavior". Scientific Publication No. 251, PAHO/WHO, Washington, D.C.

Cassel, John, 1955, A Comprehensive Health Program Among South African Zulus. In: "Health, Culture, and Community: Case Studies of Public Reactions to Health Programs". Russel Sage Foundation, New York, pp. 15-41.

Cook, Robert, 1971, The Cost of Malnutrition in Jamaica. Ecol. Food and Nut. 1: 61-66.

Denzin, Norman K., 1970, "The Research Act". Aldine Publishing Company, Chicago.

---

* Some of you may be curious about the impact of the Jamaican Community Health Aides despite the difficulties regarding implementation discussed in this paper. That is, you are wondering about the results of the outcome evaluation. I do not space here to discuss the full extent of this result. I would not have considered presenting this paper to you if the aides were making major contributions to the improvements of the nutritional status of the communities they served. The changes in the Jamaican economy due to its relation to international markets have had greater effects in my estimation (Marchione, 1977a).

Djukanovic, V. and E.P. Mach, Eds., 1975, Alternative Approaches to Meeting Basic Health Needs in Developing Countries. WHO, Geneva.

Donebedian, Avedis, 1966, Evaluating the Quality of Medical Care. The Milbank Memorial Fund Quarterly 46: 166-206.

Ennever, O., M. Marsh and K. Standard, 1969, A community health aide training programme. The W. Ind. Med. J. 18: 193-201.

Felszer, Mario, 1971, Draft project for training programme of community health aides. Unpublished document. PAHO, Kingston, mimeographed.

Fendall, N.R.E., 1972a, "Auxiliaries in Health Care: Programs in Developing Countries". The Johns Hopkins Press, Baltimore.

Foner, Nancy, 1973, "Status and Power in Rural Jamaica: A Study of Educational and Political Change". Teachers College Press, New York.

Foster, George, 1962, "Traditional Cultures and the Impact of Technological Change". Harper and Row, New York; 1968, "Applied Anthropology", Little Brown, Boston.

Fox, Helen, V.S. Campbell and J. Harris, 1968, The Dietary and Nutritional Status of Jamaican Infants and Toddlers. Bull. of the Scient. Res. Counc. 8: 31-51.

Gurney, J.M., Helen Fox and John Neill, 1972, A rapid survey to assess the nutrition of Jamaican infants and young children in 1970. Trans. Royal Soc. Trop. Med. Hyg. 66: 653-662.

Guttentag, Marcia, T. Kiresok et al., 1975, "The Evaluation of Training in Mental Health". Behavioral Publications, New York.

Jamaican Governor's Commission of Enquiry, 1973, "Report of Commission of Enquiry into the Award of Contracts, the Grant of Work Permits and Licenses and Other Matters". Government Printer, Kingston.

Kuper, L., 1976, "Changing Jamaica". Routledge and Kegan Paul, London.

Marchione, Thomas J., 1973, An evaluation of the Nutrition and Family Planning Components of the Community Health Aide Programme in the Parish of St. James, Jamaica. CFNI, Kingston.

Marchione, Thomas J., 1975, Anthropology as Feedback from Community to Health Establishment? Unpublished Paper presented at the American Anthropology Association Meeting. San Francisco.

Marchione, Thomas J., n.d., Factors Associated with Malnutrition in the Children of Western Jamaica. In: "Nutritional Anthropology". N. Jerome, R. Kandel and G. Pelto, eds., Marcel Dekker, New York (in press).

Marchione, Thomas J., 1977a, Food and nutrition in self-reliant national development: the impact on children of Jamaican policy. Med. Anthr. 1: 57-79.

Marchione, Thomas J., 1977b, Health and nutrition and self-reliant national development: an evaluation of the Jamaican community Health Aide Programme. Unpublished Ph. D. Dissertation.

Mellander, Olaf, 1973, Nutrition and health care in preschool chil-

dren: report from China. Envir. Child Health 19: 253-257.
National (Central) Planning Unit, 1971, "Economic Survey. The Government Printer, Kingston.
National (Central) Planning Unit, 1975, "Economic and Social Survey, Jamaica 1974". Government printer, Kingston.
New, Peter K. and Mary L. New, 1975, The links between health and political structure in the New China. Human Organ. 34: 237-251.
PAHO, 1973, Medical Auxiliaries. PAHO, Washington, D.C.
Palmer, R.W., 1974, A Decade of West Indian Migration to the United States, 1962-72: An Economic Analysis. Soc. Econ. Studies 23: (December).
Pelto, Pertti J., 1970, "Anthropological Research: The Structure of Inquiry". Harper and Row, New York.
Puffer, R.R. and C.V. Serrano, 1973, "Patterns of Mortality in Childhood". PAHO, Washington, D.C.
Ricci, Edmund H. and James, E. Nesbitt, 1975, Policy-evaluative research: some methodological and political issues. In: "Topias and Utopias in Health". Stanly Ingman and Anthony Thomas, Eds. Mounton, The Hague, pp. 445-475.
Richardson, S.A., H. Birch, E. Grabie and K. Yoder, 1972, The behavior of children in school who were severely malnourished in the first two years of life. J. Health and Soc. Behav. 13(3): 276-284.
Ronaghy, Hossain and Steven Solter, 1974, Is the Chinese "barefoot doctor" exportable to rural Iran? Lancet 1: 1331-1333.
Schulberg, Herbert C., Alan Sheldon and Frank Baker, Eds., 1969, "Program Evaluation in the Health Fields". Behavioral Publication Inc, New York.
Sidel, Victor W. and Rugh Sidel, 1973 "Serve the People: Observations on Medicine in the People's Republic of China". Josiah Macy Jr. Foundation, New York.
Sivard, Ruth Leger, 1974, "World Military and Social Expenditures 1974". Institute for World Order, New York.
Smith, M.G., 1965, "The Plural Society in the British West Indies" The University of California Press, Berkeley.
Stone, Carl, 1974, "Electoral Behaviour and Public Opinion in Jamaica". Institute of Social and Economic Research, University of the West Indies, Kingston.
Suchman, Edward A., 1967, "Evaluation Research: Principles and Practice in Public Service and Social Action Programs", Russell Sage Foundation, New York.
Van Etten, Geert M. and Alanagh M. Raikes, 1975, Training for Rural Health in Tanzania. Soc. Sci. and Med. 9: 89-92.
Walsh, James Leo, and Ray H. Elling, 1968, Professionalism and the poor-structural effects and professional behavior. J. Health and Soc. Behav. 9: 16-28.
Wise, Harold G., 1968, The family health worker. The Amer. J. of Publ. Health, 58: 1828-1838.
World Health Organization, 1968, Training of Medical Assistants and

Similar Personnel. Technical Report Series No. 385. WHO, Geneva.
World Health Organization, 1974, Training and Utilization of Village Health Workers. Working Document. WHO, Geneva.

EVALUATION IN A MIDDLE EASTERN CONTEXT:

A CASE STUDY

>   Catherine Geissler
>
>   Dept. of Nutrition - Queen Elizabeth College
>
>   University of London - ENGLAND

INTRODUCTION

   This paper points out some pitfalls associated with evaluating nutrition programs that were developed by decree. Evaluation is hampered when baseline data on program participants are lacking and program objectives and design do not coincide with the needs of society. Ethnographic data derived from well conducted anthropological studies might be used to translate consumer aspirations to political leaders and heads of states. In many instances, the felt needs of a community or society may correspond with the views and aspirations of a national leader. Iran has been selected to illustrate these points.

SOCIODEMOGRAPHIC INFORMATION ON IRAN

   The Middle Eastern country considered in this paper is Iran. It is a semi-arid country roughly the size of Europe in which the 98% Muslim population of 34 million (1978) is situated in villages and towns on the periphery of the Central desert and on the Caspian littoral*. Most villages, of which there are more than 65,000, are small with less than 250 inhabitants. Two-thirds of all villages lack commercial roads and can be reached only by jeeps or by animal and are therefore difficult of access. An estimated 1/4 million are semi-nomadic. Iran is an oil-rich monarchy which has experienced a

---

* The political, social and economic conditions in Iran have undergone significant changes since this paper was written and submitted for publication.[1]

very rapid growth in GNP and a rapid rate of urbanization in recent years. Fifty-four percent of the population was urban in 1976, compared with 38% in 1966 when the total population was 26 million.

The capital, Teheran, has grown from 3 to 4 1/2 million between 1966 and 1976. Until recently, this urban migration was accompanied by large-scale unemployment and underemployment. However, since the boom following OPEC oil price rises in 1974, a spate of construction and development activities has resulted in labor shortages in some areas of activity and the importation of both unskilled and skilled labor.

Agriculture has declined despite agrarian reforms in 1963 when land was confiscated (with compensation) from large landowners who had not mechanized and distributed to agricultural workers who were expected to pay for it on easy terms over a number of years. The wealth from the large land holdings was transferred to urban industrial and commercial enterprises and banking, thus leaving agriculture underdeveloped. Over 20% of the rural population is landless and 60% have holdings of less than 10 hectares (over 20% < 1 hectare)[2].

Agriculture depends greatly on irrigation which, traditionally, has been supplied mainly by ghanats. These ghanats or underground conduits that tap water from mountain sources and carry it many miles require constant maintenance and labor input. The system has been ineffective since land reform in 1963.

Land reform was part of a series of reforms called the White Revolution, with the stated objectives of improving the condition of the deprived masses and of creating a modern industrial state. Other sections of the revolution introduced the Education Corps, Rural Extension Services, and the Health Corps. Extension Services cover only 14% of villages and tend to concentrate on larger farms. The Literacy Corps has raised the level of literacy to 55% in urban centres and 20% in rural areas. The rural literacy rate has remained constant over the last 5 years due to the low number of schools in rural areas, and the move of the literate population into town.

School enrollment, especially of girls, is much lower in rural than in urban areas. Girls are kept from the schools because of the lack of women teachers while religious and social constraints make it difficult to send women to teach in rural schools.

The infant mortality rate has declined from 216% in 1962 to 100% in 1976, for the country as a whole. However, rural rates have been estimated to be higher — i.e. at 120% in 1976.

The staple food in the Iranian diet is high extraction wheat bread, although rice is popular around the Caspian Sea. The relative importance of meat, rice, dairy products, fruit and vegetables in the diet varies with income and seasonal availability.

FOOD AND NUTRITION ISSUES IN IRAN

Information about nutrition problems in Iran comes from three main sources:

1) National nutrition surveys conducted by the Food and Nutritrition Institute or Iran amongst the rural population and the urban poor during the last 15 years. Data obtained from these surveys include information on diet based on five-day weighed food intakes of selected families, anthropometric measurements, and clinical signs of malnutrition. Very little socioeconomic data have been collected.

2) Statistical Centre Surveys with data on income and household food consumption of families in some provinces in Iran.

3) Specialized studies with information on specific nutritional problems such as goitre, zinc deficiency and marasmus.

Conclusions based on data obtained from the three major sources cited above are as follows:

- energy consumption is less than 80% of standard in about half of both rural and urban families;

- approximately one-fourth of the families in poorer areas such as the southern provinces have average per capita dietary intakes of less than 1,500 Kcals;

- intakes of energy and other nutrients increase with income throughout the eleven income categories defined in the Statistical Centre Surveys. Within these income categories, intakes are higher in rural than in urban populations. An increasing incidence of marasmus has been reported in infants admitted to hospitals in Teheran.

- the other main nutrient deficiencies reported were for riboflavin, vitamin A and vitamin C. Intakes of riboflavin are particularly low in all but one province. Thirty percent of rural, and a greater proportion of urban families consume less than a third of the riboflavin requirements. Intakes of vitamin A are also less than the requirements for vitamin A; however, the poor southern provinces are the most affected. Medical symptoms confirm a vitamin A deficienty. The intake of vitamin C is also seasonal. A third of families receive less than a third of the requirements, but there are no overt signs of deficiency.

- Mineral deficiencies are also prevalent. Zinc deficiency has been reported around Shiraz where the problem has been studied. It is related to the consumption of unleavened, high extraction bread. Goitre is also prevalent in many areas.

The available data show clinical evidence of zinc deficiency in some children in at least one area, i.e., retarded growth and sexual immaturity, and there is evidence of anaemia in a small percentage of children in several parts of the country. Deficiencies of vitamin A, riboflavin and iodine appear to be widespread. Anthropometric measurements also show deficient growth in many geographic areas.

The limited socio-economic data collected in the countrywide nutritional surveys have been used to describe the population rather than to correlate them with food intakes. As in most large-scale surveys, the socio-economic data were not adequate to explain the nutritional findings, nor was the selection procedure adequate for reliable measures of incidence.

The Nutrition Institute of Iran has now embarked on a regional survey around Isfahan. The aim is to eliminate the inadequacies of previous surveys and to provide information useful for nutrition program planning. This is being done by collecting more detailed socio-economic data on defined groups selected to represent major socio-economic strata. These include the various socio-economic strata in urban areas, and in three types of villages — those that are either entirely agricultural, mixed agricultural and industrial, and whose economy is based on commuter labor in urban industry. So far, sociologists but no anthropologists have been involved in the current regional survey.

IRAN's SCHOOL FEEDING PROGRAM

The two major nutrition programs that exist in Iran are those established by decrees of the Shah in 1974 and 1975, when increased oil revenues permitted this social expenditure. The first is a school feeding scheme which has received in-depth evaluation, will be discussed in this paper. The second decree, which will not be discussed in this paper, established a supplementary food program for pregnant and nursing mothers at risk, i.e., low income women.

The 1974 and 75 decrees establishing the school feeding programs granted 500 mls of milk plus a piece of cake as a midday snack to all children enrolled in the first nine grades of both private and government schools. This was done without consulting the Nutrition Institute.

Since the initial decree, the program has undergone several changes in its administration and in the type of food given. Due to an inadequate milk supply and a limited number of milk pasteurization plants, it was clearly not possible to maintain the supply of milk warranted by the decree. In addition, even where pasteurized milk was available, children were not used to drinking milk in the quantities served. For this reason, a decision was made to reduce the amount of milk to 250 mls per children. Currently, an allowance of 10 Rls per student per day is given, and regional decisions are made on what foods to buy so that the daily ration corresponds roughly with the energy and protein values of the revised edict. This amounts to 300-400 kcals and 8-10 gms of protein per child. Within these norms, a choice of foods is made to fit regional criteria of availability, price and shelf-life. Milk is now a minor menu item. This de-emphasis on milk in the modified program is a logical move.

A typical example of a week's menu of supplementary food is shown below, in Table 1.

Apart from supplying food energy, this program does not do much to counter the specific widespread deficiencies such as vitamin A, riboflavin, iodine and zinc. Nor is it targeted to the needy. The cost of this program is over $200 million per year. A large amount of money is wasted in supplementing the diets of those who do not require it. An attempt to counter these discrepancies is being made in an evaluation of the program by the Nutrition Institute. This evaluation program will be discussed later.

TABLE 1

TYPICAL EXAMPLE OF A WEEK'S MENU
OF SUPPLEMENTARY FOOD

DAY	FOOD	
1	Almonds	25 gms
2	Cake	80 gms
3	Cheese	22 gms
	Biscuit	20 gms
4	Biscuit	75 gms
5	Pistachios	30 gms
6	Milk powder	40 gms

LIMITATIONS IN DESIGNING THE SCHOOL FEEDING PROGRAM

The reasons for choosing school feeding as an intervention program — especially one based on milk — are obscure; but it is apparent that nutrition per se was not given prime consideration. The nutrition program was not undertaken on the advice of nutritional bodies. I suspect that like many other programs, this one was established for pragmatic reasons such as political visibility, maintenance of local food distribution patterns, availability of food or funds supplied by international or other agencies, imitating the prevailing intervention fashion, and so on.

It might be argued that these reasons for nutrition program design and implementation serve as substitutes for a sound data gathering and evaluation approach to nutrition programming. But this need not be the case since anthropological methods and strategies can be usefully employed in pre-program data gathering.

The lack of social or anthropological information in the nutritional data collected in Iran over the last 15 years reflects the general approach to applied nutrition research programs followed almost everywhere during that period. More recently the growing realization that nutrition intervention programs are having a very little impact on malnutrition throughout the world has led to different approaches to nutrition studies and programs. Sociocultural concepts and indices are finding currency in nutrition program design but particularly in evaluating nutrition programs. The latter is well illustrated in the Iranian case.

The special value of anthropology in the planning and evaluation of nutrition programs stems from the anthropologist's research methods which involve close and prolonged contact with the people concerned. Among other advantages, the ethnography of a community will assist program planners in differentiating between a community's felt social needs from its biological needs. This ethnographic method is by definition time-consuming. However, it can be fruitfully combined with methods used in large-scale surveys.

EVALUATING THE IRANIAN SCHOOL FEEDING PROGRAM: A MULTIDISCIPLINARY FOCUS WITH EMPHASIS ON ANTHROPOLOGY

The Institute of Nutrition proposed to the government that the school feeding program should be comprehensively studied to iron out problems that had developed. These included: 1) wastage of milk that was not consumed; 2) rejection of foods such as imported cheese that were not liked — even in areas with the least food resources such as Baloutchistan; 3) high level compliance by some teachers with the Imperial decree to feed every child regardless of need;

4) lack of decision-making powers about appropriate foods at the local level.

Probably many of these problems could have been avoided if the program planners had taken into account social customs at the outset. It is doubtful whether anthropological studies would have been needed, since most of the problems with the program required only a minimal knowledge of local traditions.

In 1976, two years after the start of the school feeding program, the Nutrition Institute secured a contract to evaluate its impact on the recipient population. Although the lack of baseline data makes this a complex task, the evaluation included the following components:

1) synthesis of available data pertaining to the health and dietary intake of school children;
2) studies of lactose intolerance;
3) evaluation of milk acceptability in school children by questionnaire in 5 provinces;
4) study of food preferences by questionnaire in all provinces;
5) development of a procedure to determine which children should be included in the program; and
6) estimate of the financial cost of changes in the program.

Other items considered for evaluation:

1) effect of the program on absenteeism;
2) opinions of public, parents, schoolmasters, etc., on the program;
3) study of wastage and cost of importation and subsidies;
4) effect of the program on appetite;
5) long-term nutritional effects of the program.

Practically every component requires methods based on nutritional sciences as well as the social and behavioral sciences.

The decision on what aspects of the program to evaluate depends on who requires the evaluation. Evaluation may be required, as in this case, to solve problems of practical administration and to avoid wastage. This type of evaluation is of particular interest to funding organizations such as government and international agencies. Planners, nutritionists and anthropologists are interested for practical and theoritical reasons in the social and nutritional effectiveness of the program. Some of the social aspects of Iran's school feeding program are its impact on government popularity; its effect on school attendance — especially of girls; the social impact of possibly increasing school attendance by girls; the introduction, acceptance and use of new foods in an educational context; the consequences of depending on donated foods from abroad, and so on.

Such bodies as FAO and USAID require an assessment of cost effectiveness for planning and for the comparison of different programs. Costing is certainly an important aspect of evaluation.

But rational nutritional and economic considerations are only part of the story. How important is detailed nutritional, economic and anthropological data for a program whose main aim may not be to solve nutritional problems, but to advertise a show of concern for the people? Such non-nutritional, social and political aspects of nutrition programs receive very little attention, although they are of paramount importance in their choice, function and effects.

Academic nutrition has for long recognized the need for anthropological information in applied nutrition, although it has been used less than we would wish. Economists have recently come to the fore in nutrition planning. But politics is as much an anathema to academic nutritionists as sex was to Victorians. Its existence is recognized, it may be deemed important, but it cannot be talked about.

The reasons for this are fairly clear. There are dangers of implied political criticism in any such discussions, and scientists generally view themselves as only technical experts — the Pundit to the Prince. However, our understanding of the implications of nutrition programs is greatly impoverished by this taboo. Anthropology should have an important role to play in designing and evaluating nutrition programs; so does political science.

The School Feeding program discussed in this paper is similar to others devised by a central authority; they are basically paternalistic. The main role of the anthropologist in such programs seems to be to translate the customs, beliefs, desires of the people to the central authority. If the goals of national leaders are to assist their constituencies in achieving their felt needs — then anthropology can be extremely functional in that regard. This point has been made by the International Labor Organization[3] and is now being followed by FAO in some of their nutrition planning studies. Perhaps anthropological studies would be less important in communities that had more control over their own lives.

REFERENCES

[1] Iran Almanac and Book of Facts, 1977, 16. ed., Echo of Iran, Teheran, Iran.
[2] Aresvik, Oddvar. The Agricultural Development of Iran, New York Praeger, 1976:
[3] ILO, Employment, Growth and Basic Needs, Geneva, 1976.

EVALUATION IN AN AFRICAN CONTEXT: SPECIAL EMPHASIS ON THE

WOMAN PRODUCER AND REPRODUCER — Some Theoretical Considerations

Wench Barth Eide

Inst. for Nutrition Research

University of Oslo, NORWAY

1. INTRODUCTION: THE NEED FOR INTEGRATING
   NUTRITION AND ANTHROPOLOGY

This paper supports the views expressed in Pelto and Jerome's article and expands on the discussion of the potential contributions of anthropology to the study of nutritional conditions. We also intend to illustrate the reverse — the application of tools and techniques from nutritional science to anthropological studies on food and nutrition. These will be accomplished by citing case materials from rural Africa.

To some extent, both nutrition and anthropology already share a common heritage through the development in Western scientific traditions. Therefore, by virtue of their common historical roots they are already integrating despite the obvious differences in philosophies, subject matter and methods. We recognize earlier attempts by a few nutritionists and anthropologists to join forces in studying malnutrition in Africa (for example, Richards and Widdowson, 1936). However, there is an urgent need to find new ways of integrating the two disciplines for nutrition research and evaluation. Concepts, theories and methodologies emanating from such integration should better reflect the various levels of interaction in the biochemical/physiological, ecological and socio-cultural aspects of nutrition. At the same time, a reorientation of both disciplines is needed in order to free them from the limitations imposed by their Western traditional roots.

We employ a bidisciplinary focus for considering how studies of malnutrition in general, and in the African context in particular, have been conducted over the last 20 years (see PAG report, 1977).

The central points of our critique can be summarized as follows:

1) Typically, nutrition studies have been conducted with natural science "laboratory" approach — even for those on human beings living in real life settings and under natural conditions;

2) nutritionists have tried to reduce the causes of malnutrition to a few specific factors and have been negligent in studying interrelationships among different conditioning factors;

3) nutrition studies have deemphasized qualitative in-depth studies on the causes and effects of malnutrition, while over-emphasizing quantitative studies;

4) nutrition studies have pretended to be value free and thus "neutral" although the values of the researchers are quite obvious to an objective reader or to one with different traditions;

5) nutrition studies have always had a Western scientific bias, i.e., they have been conducted essentially by researchers from Western countries, or at least by those who had been trained in Western universities, with Western philosophies, standards and methods;

6) nutrition studies with a medical focus have been largely conducted by male researchers who often lack access to women's spheres of activities and are often insensitive to women's problems. The more limited dietary surveys without a medical focus have usually been carried out by female researchers who often lacked scientific prestige.

Many nutritionists will probably have difficulty in recognizing the critical character os the six points listed above. Rather, these points may be regarded as strategies for achieving the objectives and scientific "ideals" or their profession. The ideals have guided their concepts and processes of identifying problems; indeed their research designs and their methods of solving problems and evaluating programs. Although much of the field research carried out by nutritionists has been useful, we maintain that it has reflected a limited perspective on the realities in which malnutrition occurs. This assertion will be substantiated in the pages that follow.

We have selected a complex of issues pertaining to women's work and performance for illustrating how food and nutrition systems lend themselves to nutrition-anthropological analyses. We will use "the women's issue" to illustrate how a narrow disciplinary focus can lead to inadequate problem identification and conceptualization, and consequently to improve choice of research. Further, we will demonstrate how these inadequacies continue to prevent an appropriate understanding of the societal processes and mechanisms that lead to malnutritrion.

2. ISSUES AND PROBLEMS IN THE AFRICAN SETTING WITH SPECIAL REFERENCE TO WOMEN

Some generalizations can be made about the role of women in the African setting, although we recognize the disparity between African societies in terms of social organization, specific historical development and ecological variations. The role of women is further complicated by the fact that it has to be understood within the context of development processes and as it has been influenced by both external structural factors and indigenous socio-cultural processes.

The historical development of Africa can be broadly understood through modern theories of underdevelopment. The colonial penetration of Africa extracted surplus that was expropriated to the metropolitan centers (Frank, 1969; Amin, 1976; Rodney, 1974). In transferring surplus, male labor was extracted from agricultural production and channeled into wage earning activities. By necessity, female labor was channeled into filling subsistence needs, thus increasing women's work load in agricultural production. The colonial penetration also resulted in substituting the cultivation of more demanding crops with less demanding ones — regardless of nutrient value — and increased the responsibilities of women within the household. Jette Bukh (1977) provides an example from Ghana concerning changes from yam to cassava cultivation.

Cash cropping further led to unaccustomed land shortages; not only was the best land reserved for this type of cultivation, but women were also deprived of land they had traditionally cultivated. The introduction of private ownership of land further limited women's access to it.

Ecological variations also affect the woman's role in production, and particularly the types of crops under cultivation. Three main geographical zones of cultivation can be identified: the African "green belt" corresponding to the Sudan-Sahelian zone, the African plantation belt in the tropical forest from West to East, and the African "reserve" belt corresponding to the Southern part of the continent. Every factor associated with Africa's historical development and with increased work load among African women has been present in each geographic zone.

Some might argue that the "women's issue" is really part of the class issue, and a minor one in comparison to the more pressing problems of general poverty and hardship. While this may be so, it must be pointed out that women often experience poverty and hardship even more severely than men. The consequences of poverty are more serious for women because they have — contrary to commonly held beliefs — a larger responsibility for ensuring the satisfaction of basic needs in many societies than do men (PAG Report, 1977). This greater responsibility usually continues even when economic processes

alter the patterns of production — and consequently the traditional
pattern of division of labor.

Women's role in the household requires close study in order to
understand the specific processes that place particular children "at
risk" to malnutrition. Nutritionists and social scientists alike
have neglected such studies. This reflects the long established male
bias in all research. This bias has sometimes been given a most un-
selfconscious and un-substantiated rendition concerning women producers in Africa as the following statement from de Garine (1971)
illustrates: "The traditional division of labor may seriously impair
productivity, as when agricultural work is performed mainly by
women".

Because women as producers and procurers of food are so fundamental to African rural economics we challenge such an isolated statement of women's productivity. We must then ask: What types of information are available about women's role in food and nutrition systems
in the current social science and nutrition literature?

Descriptive information about the role of women in food production and other tasks related to food throughout Africa is still rare
although new information is on the increase. Early anthropological
studies failed to fully describe women's various roles in production
activities; instead, the dynamic nature of the interaction between
food and colonial exploitation were often obscured in functionalist
interpretations. More recent studies of the roles of African women
in production have defrosted many of these interpretations; and data
documenting and analyzing the role of women in food production and
other tasks related to food throughout Africa are becoming available
(Boserup, 1970; Bukh, 1977, Pala, 1975; Skjønsberg, 1977; UN/ECA,
1974; see also PAG Report, 1977).

3. THE IMAGE OF WOMEN AS IT HAS BEEN DEVELOPED WITHIN NUTRITION RESEARCH

The focus of nutrition research in Africa, as well as in many
other regions, has contributed to an extremely limited and distorted
image of women and their roles in nutrition. In this context,
women's biology and reproductive functions have been the basic research "object", while their social roles in nutrition have been
limited to their activities as cooks for their families, and to some
extent, nutrition educators of their children. Women are rarely
viewed in broader perspectives in the nutrition literature. For
example, male oriented variables such as "fathers" occupation are
usually used to identify socio-economic determinants of malnutrition,
while ignoring women's extremely important economic (though often
non-monetary) functions in food production systems. Failure to fully
recognize the woman's role in each economic system often leads to

erroneous conclusions on socio-economic determinants of malnutrition.

In sum, it can be said that women have not been considered social actors in nutritional change, but merely as passive recipients of good food for foetal and child development, or as equally passive receivers of a type of nutrition education that has been expected to yield automatic behavioral results. This limited perception of women's roles in food and nutrition systems may be partly responsible for the largely poor results of nutrition intervention programs which are often directed toward women.

Questions on sex roles in relation to nutrition therefore becomes particularly relevant to nutrition planning and evaluation issues. The paradox in planning for better nutrition is that those most involved in nutrition related activities, i.e., women, are usually ignored in the decision-making processes in many countries (see PAG Report, 1977).

In a very real sense, then, women may be viewed as "the true guardians of the family's nutritional needs". The family emerges as an important institution in matters associated with food and nutrition systems since food is often consumed within the family unit. Since woman's role is crucial in this unit, the nutritional status of the family and to a certain extent the society, is likely to be determined by the various elements and forces which act upon and influence the role and behavior of women.

4. A SPECIFIC EXAMPLE: AN INTEGRATED VIEW OF WOMEN'S DAILY TIME AND ENERGY EXPENDITURE

It is possible to identify nutrition related problems that can be better understood through the perspectives of other disciplines, in particular anthropology. We have identified one problem that focuses on the role of women in nutrition which we will use to illustrate how the insight gained from an anthropological perspective could shed new light on an "old" nutritional problem.

The problem selected is familiar to both nutritionists/anthropologists, but with different emphasis: the amount and balance of energy in a given system. Anthropologists have been interested in the energy budgets in societal systems, while nutritionists have been interested in the energy balance of "systems" represented by individuals. We will present a model adjustable to a variety of economic situations, and discuss how an integration of anthropological and nutritional perspectives on energy budgets can help link the physiological with sociological/anthropological aspects of women's work in food and nutrition systems and also elucidate how factors and conditions at the macro and micro-economic levels of the society influence food and nutrition systems and human behavior within these systems.

Our point of departure is that women traditionally perform numerous time- and energy-consuming tasks during the day. In poor societies, this is true for all women but it is particularly manifest in rural areas in Africa where women are heavily involved in food production and other tasks concerned with food (Skjønsberg, 1977). This complex of tasks determine the degree to which women can involve themselves specifically in child care (Sharman, 1970; PAG Report, 1977), and also respond to innovations, educational programs and the like. Consequently, consider time, motivation and physical capacity as primary determinants of women's participation in nutritional programs directed to them, and in their functioning as agricultural workers and food producers.

It has been shown that women in Africa work more hours than men. This bit of data may be sufficient to indicate that exploitation of women exists, but does not say much about how this condition may affect the nutritional levels of family members. Time allocation of women's daily tasks is at issue here.

Nutrition educators are often interested in people's motivations, particularly as these affect attitudes and behavior. They are less interested in the social context from which these motivations arise although the social situation may be more closely associated with behavior than with inappropriate media or messages*.

By contrast, nutrition-physiologists are often interested in studying energy expenditure and time/energy budgets of indivudual. Such studies have been conducted in connection with different work activities, particularly of men. We still lack a substantial data base on energy expenditures of women as they pursue a variety of occupational and social activities. This is unfortunate, because this type of information would have given us an opportunity to consider, at least hypothetically, how specific patterns of energy expenditure might be associated with women's use of time. More particularly, information on women's patterns of energy consumption and expenditure could provide clues on how individual women in specific environments and social and economic situations adjust their time and energy to accomodate their various activities and the priorities they set for themselves.

The following questions are among those that could be asked concerning women's work and performance as they relate to food and

---

* Several papers at a recent workshop held in Tanzania on "Food and nutrition education under changing socio-economic conditions", brought up the issue of structural constraints to changing behaviour as contrasted to the belief in the message and the media as sole determinants for change (Jonsson, 1978, Malentnlema, 1978; IUNS Workshop, Dar-es-Sallam, June, 1978).

nutrition and as they assume a bidisciplinary focus:

- What determines women's <u>time expenditure</u> and <u>allocation of time</u>?
- What determines the <u>physical output</u> (working capacity) of a woman involved in food production/provisioning besides child bearing and rearing?
- What determines their <u>motivation</u> (to innovations, educational programs, visits to health stations, follow-up of rehabilitation center instructions etc.)?

The following hypotheses relating to time expenditure, physical output and motivation could be posited about the present reality of many rural women in Africa today:

High time expenditure may be due to:

- numerous responsibilities in the home and in the field or farm;
- unbalanced division of labor and responsitilities between the sexes;
- application of inappropriate technology;
- excessive walking distance to farm, fuel and water sources.

Low physical output in food production and provisioning may be associated with:

- excessive physical input in basic activities other than food production and provisioning;
- high energy demands for physical input due to use of inappropriate technologies;
- unbalanced distribution of food at the family level with consequent malnutrition among females;
- closely repeated pregnancies with its consequent drain on women's physique.

Low motivation may be due to:

- general tiredness and ill health (e.g. low energy intake: iron deficiency anemia);
- lack of decision-making power over own produce or cash from produce;
- irrationality of specific programmes as perceived by the woman.

We have here a mixture of potential factors, or aggregates of factors for the study of women's performance in food production and nutrition in various contexts, particularly with reference to rural Africa. Some of these potential factors could be fruitfully studied by anthropological/sociological methods, while others could be studied by methods developed by nutritionists. The various factors could certainly be organized into the ecological frame of reference (Cone and Pelto, 1967) referred to in Pelto/Jerome's introductory paper*.

The search here is for a unifying idea that makes it conceptually <u>logical</u> to integrate different methodological approaches and techniques of the two disciplines. We believe that an alternative perception of <u>the value of women's work</u> provides an organizing frame of reference. Traditionally, women's work has been attributed a <u>use value</u> rather than an <u>exchange value</u>. Consequently, women's labor has not been valued within the context of the labor market in an extent economic system. Instead, women have been perceived as reproducing the labor force rather than functioning as active participants within it (Steady, 1977)**.

The <u>use value</u> concept of women's work has also dominated the work of nutritionists whose research seldom extended beyond women's biology and reproductive capabilities. Unfortunately, this research tradition continues today despite arguments that nutritional considerations should be included in economic planning and national development programs. By neglecting the <u>potential exchange value of women's work</u>, nutritionists and nutritional planners exclude a vital force in national development programs.

Let us now examine more closely how the factor <u>energy</u> — in terms of working expenditure and in terms of food value — could provide one

---

* We find this frame of reference both clarifying and potentially useful as long as it is made clear that the (alternative) central box on nutritional status should be subdivided in actual measured status on the one hand and its "precursors": <u>consumption</u> in terms of food intake, and <u>performance</u>, such as energy expenditure, pregnancy patterns, and feeding practices. Such "performance" components are those through which environmental, technological, organizational and cultural variables finally work to produce a certain measurable nutritional status. This will also better clarify where and how the special research skills of social scientists and nutritionists should be employed.

**The normative expectations surrounding women's roles as wives and mothers, which obviously limit their opportunity for developing knowledge and skills through educational and other programs, may have been generated by the colonialists, whose wives did perform mainly in the domain of the kitchen and nursery room. Despite differences in the African situation, the norms set by colonialists could have reinforced the continuing existence of patriarchal ideologies that exist in some African societies (Steady, 1968).

bridge to a better understanding of women's performance as an intervening variable between macro-processes and the nutritional microuniverse. Energy factors will be used as components in a conceptual model to illustrate the interplay between use value and potential exchange value in a consideration of women's work.

Conceptual Model (1)

While it is conceptually possible, though operationally laborious, to study the flow of energy in a relatively "closed" system such as the ideal subsistence economy, it becomes increasingly difficult to do so as the system "opens". Studies of "open" or complex systems (such as the U.S. agriculture system) are almost impossible due to the "hidden" inputs of energy into the technical and organizational machinery that forms the basis of these systems. Nevertheless, we will employ the concept of energy flow (as shown in Figure 1) to systems where women are part of the network of social actors: individual women are not being considered here. We will first consider the conditions in complete subsistence economies (an almost artificial situation) where all the food consumed by the family is obtained by "home production", and nothing is sold. To simplify the task (but not diverting too far from the realities of rural Africa) we will also consider that the food is entirely produced by the woman in the household (thus ignoring polygamic situations) and we further consider the individual woman, rather than the household, as the "management unit" as far as food is concerned.*

Approaches to the Problem

Reduce A? The biological energy cost must be assumed to be fixed during pregnancy and lactation and has been estimated (Thomson & Hitten, 1973). However, the cost of physical activities during pregnancy and lactation is not know, e.g., whether these physiological states alter the utilization and expenditure of energy from that of a "normal" state during given activities. On the basis of energy measurements of pregnant women in New Guinea, Norgan et al. (1974) suggested that these women adapted to their situation by reducing their physical activities (sitting more, walking less). The degree to which women in general would do the same would probably be determined by cultural practices and expectations surrounding pregnancy (PAG Report, 1977).

---

* Although this is oversimplification of most situations, some justification for it exists (e.g. Barth, 1967, in his analysis of economic spheres in Darfur).

Reduce B? Fuel and water are both absolutely essential in the carrying out of basic daily tasks, and we would presume that the time and energy involved is fairly fixed. As fuel becomes more scarce and must be fetched at greater distances from home, changes would probably be registered as a net-increase in energy expenditure.

Reduce C? The woman may go to the field or farm less often than is necessary if these are far away; this results in sub-optimal cultivation and harvesting. Swantz (1975) gives an example of this from Tanzania where fertile land on the slopes of Kilimanjaro is almost entirely used to produce cash crops (coffed and bananas) instead of maize. Land in the more distant lower plains has been put to use for maize production by women. This new development must have a serious effect on poor women without transportation. In such cases, societal constraints — not personal choice — limits the frequenct of crop cultivation.

Reduce D? Time consumed in general home activities are more variable than that consumed in fixed walks to the field or to water a source. More or less time and energy can be put into food preservation and preparation, and more or less attention and care can be given to children. In the latter case, in particular, if a mother cannot visualize immediate results from certain child care activities as compared to fetching water, as one example of immediate productive activity we may assume that she would easily "save" time and energy by reducing time expenditures on less productive activities. Other activities which might fall in the "saving" or "cutting corners" category when time and energy become scarce, include educational classes, clinics, meetings, etc.

Reduce E? A moman might reduce her labor input in the field, but this would result in a reduced amount of food for the family. Since the disadvantages resulting from this type of activity is quite obvious to the woman involved, we postulate that she would try to maintain a high (relatively) input here. Answers to questions on the relative input of energy for each activity require further research. The following question is particularly important: What is the relation between D and E above? i.e. To what extent would keeping up the economic activity of food production affect child care and child nutritional levels?

Discussion of Figure 1, Example I: Subsistence System

We agree that the model is not perfect. For example, through this limited "subsistence model" we have considered women's work and energy expenditure as freely exchangable between different tasks, and assumed that the woman has free choices in deciding what to do (here: to allocate her time and energy resources). Howver, in practice this is not necessarily so, because different tasks may belong

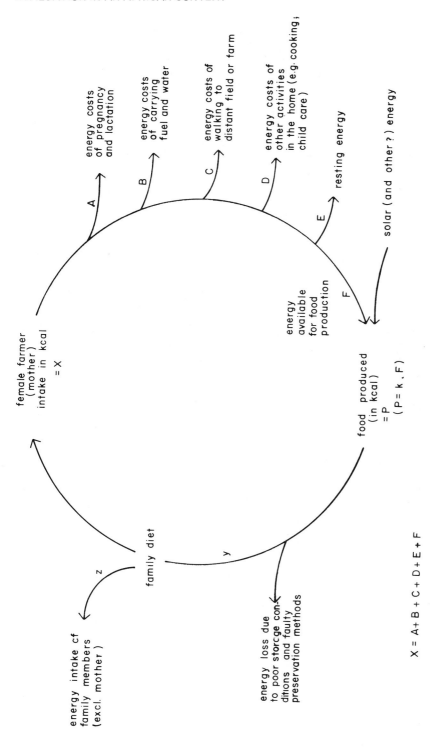

Fig. 1 — Food procured from subsistence farming only.

to different "spheres". Within any society this social compartimentalization presents "barriers" to a free exchange of tasks. In addition, there might be social sanctions and moral reprobations in cases where an individual attempts to cross such barriers (Bohannan and Dalton, 1965; Barth, 1967).

We may therefore hypothesize that there will be certain moral or other sanctions directed to a woman who tries to reallocate her attention, time and energy to an activity that a society or family member considers less important than the traditional one. For example, even if the woman wants to spend more time in taking care of her children, she may be expected to place a high priority on cultivation in the field or farm. Such expectations could impose severe limitations on the amount of time spent in child care. Again, even if a woman knows that the cultivation and use of green leafy vegetables would contribute to her children's health, her time may have to be used in petty trading in order to get enough cash to purchase staple food and food products. In addition, women may have their own strategies for optimizing their situations in very rational ways, although the basis of the rationality may not be understood by male food planners or Western (or Western trained) male and female nutrition educators (Lamphère, 1974). Often, information directed to the "ignorant" mother on what action should be taken to ensure good child nutrition, is ignored when the social and economic context dictating the action are ignored by the nutrition educator.

Numerous social barriers to exchangeability between different spheres of women's work abound. In the case of rural Africa, however, the "domestic" and the "production" shepres and the reproductive/productive spheres may be one and the same. Program evaluation must be viewed within such a context.

Conceptual Model (2)

Let us turn to a second situation where some of the food produced by the woman is sold on the market. The situation is schematically represented in Figure II.

Discussion of Figure II, Example II:
Subsistence and Market Systems Combined

For food sold on the market, there is a relationship between cash value and energy value. Within the total energy flow, part of women's work, or energy allocated to food production attains a cash value. The cash obtained can be exchanged for new calories (i.e., other food items) for consumption by the family. Part of this will be consumed by the woman herself.

EVALUATION IN AN AFRICAN CONTEXT 631

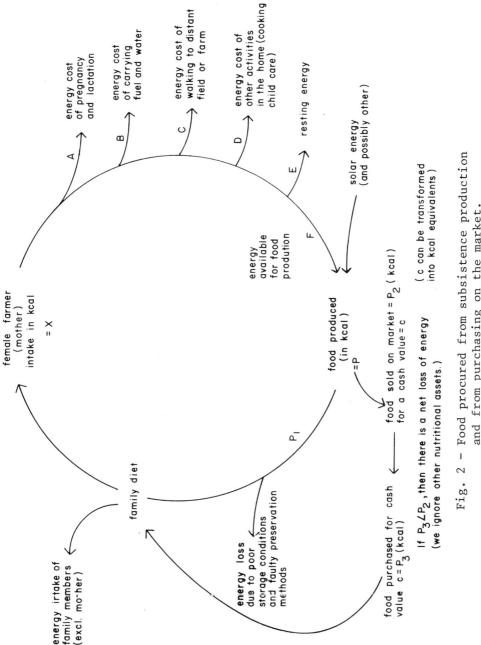

Fig. 2 — Food procured from subsistence production and from purchasing on the market.

This transformation of energy from the woman's labor (energy expenditure) in food production, harvesting and processing (through food energy sold, rebought and partly re-fed into the woman) into energy for further work in food production etc. will result in a larger or smaller net loss. The latter may be due to metabolic losses from low efficiency in energy utilization, from poor food storage conditions, or from food processing wastes. Theoretically, these could be estimated and accounted for. Losses may also occur from purchasing non-food items and food products with minimal energy value. However, these losses are being ignored in this very simplified model. We are concerned here with market exchange in order to determine whether the price obtained for the food energy sold is reasonable enough to recover the energy expended for obtaining the goods exchanged.

If a net loss is incurred, it should be possible to identify the specific factors associated with the loss, i.e., a determination of environmental and market conditions that preclude the achievement of energy equivalents or gains. One may conclude that these same factors are directly responsible for the exploitation of women's labor expressed as exchange value, the nature and origin of these factors could then be systematically evaluated. These factors could be related to a variety of conditions such as local middle-men systems, high profit margins on industrial products, or the high prices of products produced by costly modern inputs in agricultural production systems. These factors have already contributed to putting the woman producer at a disadvantage in the marginally fertile land and with minimal resources.

Following this line of argument, for action purposes one should also examine alternative methods of utilizing women's labor to better advantage or, to retain our terminology, maximizing its potential exchange value. In turn, this would also mean maximizing the local food supply without jeopardizing child care and feeding. Options available for optimizing the role of women in both spheres without exploiting these — as is being done today — are in theory legion.

5. USING THE MODELS FOR LINKING WOMEN'S ROLE IN NUTRITION TO A WIDER DEVELOPMENT PERSPECTIVE

The two models are but two examples of a wider range that could be derived from economic and social situations in which women find themselves, and which have been determined by historical and political processes interacting with ecological and cultural conditions.

For example, direct interventions like food aid or supplementary feeding programs would bring additional dimensions to the models for energy flow. Often, the food provided substitutes or competes with local patterns of food production, marketing, energy expenditure, time allocation and cash flow. An especially intriguing example comes from "foof for work" programs in which men are provided food in exchange for labor. How does this interfere with the cash available to the family? Would this development increase women's work in food production?

Using the idea of energy budgets as one theoretical bridge between nutritional and anthropological methodologies and perspectives, we may ask the following questions:

1) What data do the hypothetical models require explanatory purposes?
2) What political considerations do the models generate?
3) How would the models facilitate nutrition research?

There is an obvious need for data on energy expenditure patterns of women in agricultural work in Third World countries. Some work on energy expenditure by women during various activities at different agricultural cycles and undergoing various physiological stresses such as pregnancy is in progress in Guatemala. However, good data substantiating the models are currently unavailable. At the same time, we recognize that there is a limit to the number and types of energy studies that should be undertaken. Without going into details on this point, we would caution against using this approach to its extremes by spreading respirometers and other devices all over the African subcontinent or other parts of the Third World! Nevertheless, it seems warranted to suggest that a minimum amount of data be provided to clarify and expand the data on the actual work and energy input and expenditure of women in various farm and domestic activities.

What political considerations do the models generate?

We sould like to see a few systematic studies carried out to test our conceptual models, and to assess the likelihood of "uneconomy" in agricultural development policies in women's situations.

Studies of women's situations have demonstrated that although they are the main food producers, they are seldom reached by modern agricultural developments. Women lack training in agriculture and have need for agricultural extension services, appropriate agricultural tools, and agricultural credits and loans. The provision of such services require a fundamental re-orientation of priorities within the existing agricultural extension service, a totally new criteria for obtaining credits and loans, and a much stronger emphasis on the development of a technology appropriate for producing for local consumption and especially suited to women. Such re-orientation would conflict with the objectives of a national economy based on income from export agriculture. An emphasis on food production for local use does not necessarily ensure a large enough surplus to be exported, or to be fed into secondary food production to satisfy the demand from the rich within the country.

At the domestic level, a "rational" way out of the women's dilemma, as seen from a Western perspective, would be to change the patterns of division of labor, so that women could get more help from their husbands in basic food production, child care and other domestic work. The original division of labor designed to optimize the use of human resources appears to have become conditions of blatant inequality in contemporary economic settings; this is often explained and justified as a "continuing traditional values".

How would the models facilitate nutrition research?

The models we have presented offer one opportunity for the concerned nutritionist to begin to understand how human beings, in this case women, are tied to the wider network of economic, social and political processes which have a bearing on the nutritional micro-universe. Much can be learned from the nutrition-anthropological study conducted by Gross and Underwood (1971). They used energy considerations to illustrate the consequences of sisal production on the nutritional situation of children of male workers in the energy intensive technological operations of sisal in Brazil. In fact, the authors — a nutritionist and an anthropologist — have employed the same general reasoning as we have in this paper in linking nutritional status to macro-economic factors in "development".

In order to identify other examples of the same nature, nutritionists must be willing to learn from anthropologists how to look at totalities rather than fragments of realities. The concept of energy budgets in women's productive activities within a social system assumes a holistic view, while conventional energy studies are by nature fragmented and limited in generating social policy. For example, while it may well be proven that feeding workers on the job is useful for productivity and thus for profit, or for some limited goals of a development program, when looking at the program in its totality, adverse effects on the individual, family and community may also be indicated. Payment in food may reduce cash income which might otherwise be used to buy food for the family.

6. USE OF MODELS IN EVALUATION

Diagnosing a problem correctly is an important step towards its solution. Similarly, appropriate evaluation of an action program, or the ability to identify or predict the (or latent)* impact of certain program policies is equally important. The hypotheses posited in this paper and the models proposed are appropriate for designing both nutrition research and evaluation programs.

It is easy, however, to get involved in the analysis of methods of evaluation per se and forget the purpose of the evaluation. A better evaluation methodology must not be thought of as an end in itself.

BIBLIOGRAPHY

Amin, S., 1976, "Neocolonialism in West Africa". Pathfinder Press, New York.
Barth, F., 1967, "Economic Spheres in Darfur. Six Themes in Social Anthropology, A.S.A. Monographs". Travistock Publications, London.
Bohannan, P., and Dalton, G., Eds., 1965, "Markets in Africa". Anchor Books, New York.
Boserup, E., 1970, "Women's Role in Economic Development". St. Martin's Press, New York.
Bukh, J., 1977, Report to the PAG Project on Women in Food Production, Food Handling and Nutrition. United Nations, New York.
de Garine, I., 1972, The sociocultural aspects of nutrition. Ecology of Food and Nutrition 1: 143-164.
Durnin, J.V.G.A., 1976, Sex differences in energy intake and expenditure. Symposium on sex differences in response to nutritional variables. Proc. Nutr. Soc. 35: 145-154.
Frank, A.G., 1969, Latin America: Undervelopment or Revolution. New York: Monthly Review Press.
Gross, D.R. and Underwood, B.A., 1971, Technological change and calorie costs: sisal agriculture in northeastern Brazil. American Anthropologist, 73: 723-740.
Jonsson, U., 1978, The Planning Approach to Nutrition Education: National Potentials and Constraints. Paper presented at the IUNS Workshop, Rethinking Food and Nutrition Education under Changing Socioeconomic Conditions, Dar-es-Salaam, Tanzania.
Knutsson, K.E., 1974, Det Anthropologiska Perspektivet - Reflektioner Omkring et Ennas Identitet. Vetenskapliga Perspektiv.
Lamphere, L., 1974, Strategies, Cooperation and Conflict among Women in Domestic Groups, in: "Women, Culture and Society". Rosaldo and Lamphere, eds., Stanford University Press, Stanford.
Malentnlema, T.N., 1978, Who Is Ignorant? Paper presented to the IUNS Workshop, Rethinking Food and Nutrition Education under Changing Socioeconomic Conditions, Dar-es-Salaam, Tanzania.
Norgan, N., Ferro-Luzzi, A. and Durning, J.V.G.A., 1974, The energy and nutrient intake and the energy expenditure of 204 New Guinean adults. Phil. Trans. R. Soc., London 268: 309-348.

---

* For a discussion of the use of manifest versus latent consequences in nutrition evaluation, see Pelto and Jerome, 1978.

Pala, A.O., 1975, A Preliminary Survey of Avenues and Constraints on Women's Involvement in the Development Process of Kenya. M 75-2449 IDA, University of Nairobi.

Pelto, G. and Jerome, N.W., 1978, Paper presented at the symposium on Contribution of Anthropology to Nutrition, Eleventh International Congress of Nutrition, Rio de Janeiro, Brazil.

Richards, A.I. and Widdowson, E.M., 1936. A dietary study in Northeastern Rodesia. Africa 9: 166-196.

Rodney, W., 1974, How Europe Underveloped Africa. Howard University Press, Washington.

Sharman, A., 1970, Nutrition and social planning, J. of Develop. Studies, 5: 17-19.

Skjønsberg, E., 1977, Women and food and the social sciences. Oslo, mimeographed paper.

Steady, F.C., 1968, The social position of women: selected West African societies. Unpublished B. Litt. Thesis, Oxford University, England.

Swantz, M. Henricsson, U. and Zalle, M., 1975, Socioeconomic causes of malnutrition in Moshi district. Research Paper No. 38, Bureau of Resource Assessment and Land use Planning (BRALUP), Dar-es-Salaam, Tanzania.

Thomson, A.M. and Hytten, F.E., 1973, Nutrition during pregnancy. Wld. Rev. Nutr. and Diet. 16: 22-45.

UN Economic Commission for Africa (UN/ECA), 1974, Africa's Food Producers: The Impact of Change on Rural Women. Paper prepared by the Women's Programme Unit, Human Resource Division, for the American Geographical Society.

UN Protein-Calorie Advisory Group (PAG), 1977, Women in Food Production, Food Handling and Nutrition. United Nations, New York, mimeographed paper.

THE EXPERIENCE OF GREECE

Antonia Polychronopoulou-Trichopoulou

Dept. of Biochemistry and Nutrition

Athens School of Hygiene - Athens 602 - GREECE

Greece is a country of about nine million people of whom one third live in small villages of less than two thousand inhabitants. For Greece as for most other countries there are no satisfactory morbidity data and therefore the health of the population can only be evaluated with mortality statistics. Table 1 shows age specific mortality rates in Greece (per 100,000 population), and their rank among the corresponding rates of the 27 European countries for which data are available (rank 1 indicates the lowest and rank 27 the highest rate). It can be seen that until the age of 20 mortality in Greece is high, whereas for adults mortality rates are among the lowest in the world. The low mortality of adults is mainly due to the low incidence of cardiovascular diseases and certain forms of cancer, mainly breast and colon cancer. It may not be coincidental that these diseases are known or suspected to be associated with nutrition.

A comparison of age specific death rates in the urban and in the rural population of Greece reveals that for the younger age groups mortality is higher in rural areas, while the opposite is true for the older ages. The excess premature mortality in rural areas is attributed to infectious diseases but there is a widespread impression that malnutrition may occasionally be involved. On the other hand the higher mortality of adults in urban areas is accounted for by the correspondingly higher death rates from cardiovascular diseases and malignant neoplasms.

It appears therefore that in Greece undernutrition may exist in some rural areas. A similar problem may also exist among the lower socio-economic groups in urban areas, but there are no sufficient epidemiological data to support or refute this notion.

TABLE 1

AGE SPECIFIC MORTALITY RATES IN GREECE (PER 100.000 POPULATION) AND THEIR RANK AMONG THE CORRESPONDING RATES OF THE 27 EUROPEAN COUNTRIES FOR WHICH DATA ARE AVAILABLE (1976)
(rank 1 indicates the lowest and rank 22 the highest rate)

	0	1-4	5-14	15-24	25-34	35-44	45-54	55-64	65-74	75+	ALL AGES
Mortality	2960.1	98.4	40.1	67.1	86.3	151.0	407.8	1043.1	2840.6	9836.9	841.7
Rank	22	17	13	3	5	1	1	1	1	2	3

Several ad hoc studies indicate that in the rural population nutritional problems do in fact exist and relate to economic factors as well as to inadequate nutritional education.

Kondakis in 1964 found a few cases of severe malnutrition, as well as many more with subclinical forms of the condition, in a remote rural area of Northern Greece, and he substantiated his findings with biochemical examinations. In 1968 Lapatsanis et al. reported a relative high prevalence of rickets in a number of Greek villages. Several authors, in studies undertaken between 1965-1975, have found consistently lower values for height, weight and skin-fold thickness among children of rural areas, when compared with children of similar age and sex in urban areas. Very recently Papanicolaou found severe growth retardation among children of a mountainous village and attributed it to calorie malnutrition since protein intake was found normal.

Findings of the above and other similar studies lead to the establishment, in the early sixty's, of the first national nutritional intervention program, centered in the rural areas. The program provided for a supplementary lunch for all children of school-age. The program was not particularly successful; there have never been clearly specified objectives and the population has responded with limited enthusiasm.

The above program was in operation for only about five years. At the present time only a limited nutrition intervention program exists for the residents of villages of Mount Olympos, where iodized salt is provided for the control of endemic bronchocele.

In urban areas at the present time there seem to be no clearly defined malnutrition problems. Some authors found differences in height among army recruits from different socio-economic groups and there have even been reports of rickets among children of deprived population groups of Athen (due to inadequate education rather than strictly economic reasons). However other workers noted that somatometric indices of children of urban population in Greece already exceed the figures considered as "standards". In the light of this evidence and in view of the fact that large slums do not exist in our cities, major nutritional intervention programs have not been implemented in the urban population of Greece, except in an educational context.

In summary it appears, that among children in Greece occur very few cases of clinical malnutrition. On the other hand subclinical or biochemical forms are more frequent (iron deficiency anemia, hypovitaminoses etc.) and may be due to economic reasons as well as to inadequate nutritional education.

The changing nutritional habits of Greek population may create far greater problems than those which exist, today. The relatively low incidence of coronary heart disease in Greece has been atrributed by Keys et al., Christakis et al., Voridis et al. and others to the "low animal fat — low cholesterol" diet of the Greek population and the relative low incidence of large bowel cancer has been linked by Manousos et al. to the high fiber diet of this population. But the diet of the Greek population, and particularly of the urban segment is now quickly changing in all these respects. The diet is becoming richer in refined carbohydrates, cholesterol and animal fat and poorer in fiber. Indeed the Greek diet is changing in quite the opposite direction from that indicated in the U.S.A. "National Nutrition Policy". These dietary changes may be responsible for the graduate assimilation of the spectrum of the Greek population to that of the more developed Europeau population (Table 2). It appears therefore important that a national nutrition policy be formulated in Greece encompassing multiple objectives directed to different segments of the Greek population.

TABLE 2

DEATH RATES (PER 100,000 POPULATION) FROM 14 IMPORTANT CAUSES, IN GREECE, IN 1961 AND IN 1970, AND THEIR RANK AMONG THE CORRESPONDING RATES OF 26 EUROPEAN COUNTRIES FOR WHICH COMPARABLE DATA ARE AVAILABLE

(rank 1 indicates the lowest and rank 26 the highest rate)

CAUSE OF DEATH	1961		1970	
	Rate	Rank	Rate	Rank
Tuberculosis	18.0	15	9.5	16
Acute respiratory infections	1.2	20	1.2	10
Influenza and pneumonia	41.4	12	48.3	13
Other infective and parasitic diseases	15.1	23	13.1	24
Other diseases of respiratory system	16.5	8	40.2	14
Diabetes melitus	10.3	16	22.0	21
Diseases of nervous system and sense organs	25.7	24	18.3	25
Cirrhosis of liver	12.2	17	15.8	18
Other diseases of digestive system	31.0	13	24.6	12
Complication of pregnancy (per 100.000 live-born)	79.6	20	28.3	15
Malignant neoplasms including lymphomas	105.7	4	131.4	6
Accidents and violence	38.2	3	44.4	2
Cerebrovascular disease	83.6	5	117.3	8
Ischaemic heart and other diseases of circulation	124.8	1	171.5	2

DIFFERENCES IN ENERGY EXPENDITURE IN

SELECTED WORK UNDER CONTROL AND AD LIBITUM

W. Wirths

Lehrstuhl fur Angewandte - Ernahrungsphysiologie

Romerstr. 164 - 5300 Bonn 1 West Germany

Obesity is one of the most important problems in all developed countries and for a growing part of the population in developing countries. Sometimes, individuals have the opinion that the energy metabolism is different from person to person under apparently the same metabolism and work conditions.

Durnin et al. (1973) believe that the energy requirements of man and his balance of intake and expenditure are not unknown. The authors conclude this from results of studies of food intake and energy expenditure which show that in any group of twenty or more subjects, with similar attributes and activities, food intake can vary as much as two-fold; references by Widdowson (1947 - 1962), Rose and Williams (1961), Wynn-Jones et al. (1972), Ashworth (1968).

In those surveys — so Durnin et al. (1973) — where intake and expenditure are measured, there is often good agreement between the two estimates for the average of the group, but usually very large discrepancies between individual intake and individual expenditure. The results of careful studies in a number of countries suggest that some people, perhaps through some mechanism of adaptation, are able to be healthy and active on energy intakes which, by current standards, would be regarded as inadequate.

On the other hand, there are also studies in which subjects have been given large quantities of additional food with little or no increase in body weight, Miller and Mumford (1967), Miller et al. (1967). In contrast, there are the difficulties experienced by the obese in reducing body weight in spite of a drastic reduction of food intake, and the well recognized fact that many fat people eat no more, and sometimes less, than those who are not obese.

These observations underline the extent of the ignorance about the mechanisms by which energy balance is maintained.

There is a possibility that some of the measurements of energy expenditure are erroneous to an appreciable degree. There are certainly many sources of error in any of the present-day techniques of assessing energy expenditure. The errors caused by indirect calorimetry are of different nature and result more from the difficulty of maintaining normal behaviour in the subject and the considerable problems of measuring all the activities of a day. It may be that these difficulties lead to an overestimate of total energy expenditure in many cases.

METHOD OF OWN EXPERIMENTS

For the determination of the energy expenditure in respiration studies by work, we apply the Max-Planck-respirometer. It is an instrument used in the field, consisting of a small dry gas meter, weighing 2.5-3kg, which is carried on the back or on the chest like a haversack. The subject, doing his ordinary work, breathes through the meter. The meter contains a device by which a small fraction of the expired air, about 0.3 to 0.6%, is diverted into a small rubber bag. This sample is subsequently analysed (Wirths, 1974).

Experiments with the Max-Planck-respirometer were made to determine the energy expenditure, either by integral — or partial — ("steady-state") method.

In the technique of the measurements it must be noted, that the energy expenditure by work beginning is not in the same intensity. In light and moderate work there is a balance between oxygen input and oxygen requirement in 3-5 min after beginning of work. Then, energy metabolism reaches a steady state.

In the "steady-state-method", or partial method, the measurement of expiration-air will be given a start after an initial time period, for such a time — but not more than 25 minutes — that enough air is expired in the rubber bag for analysis. Energy consumption must be accounted in addition of parts of the different kinds of work.

In work with higher energy expenditure we have to apply the integral-method. This method has no initial period. The air-collection begins with the experiment. After the end of the work, the oxygen consumption continues in a post period until the oxygen requirement is balanced. After deduction of basal metabolism for the whole time the working calories are only transmitted in the real working time.

Moreover it is requisite to attend the Simonson-effect (Christensen and Högberg, 1950) by beginning of work. This initial overload by start-condition can be in the first 15 seconds more than two times higher as the steady state.

Pulse frequency measurements in order to determine the circulation with pulse frequency meter charge occurring in professional activities were done in several population groups for completing the judgement in circulatory system initially in static work.

We don't use the method of heart rate alone, without the Max-Planck-respirometer. The use of heart rate to predict energy expenditure has been shown to involve errors which may be as high as 10 till more than 30%, depending of the subject, the type of activity, the adaptation of training and the body weight.

The assessment of total energy expenditure of workers does not only depend on energy expenditure during working time but also on that of leisure time. Non-occupational activities can be of substantial importance considering the great number of working persons concerned, the share of occupational energy expenditure and total work calories. We found especially in a lot of employees above all with light and moderate work that these people do much work in leisure time at various seasons, sometimes house — or garden — work, sometimes house mechanics and repairs, sometimes sports. The different kinds of work are not only for their own but also as bootleg labour or illegal employment. It is nearly impossible, to find out exactly the real energy expenditure during leisure time.

It is one of the main reason, that the methods commonly used are not precise enough and cannot be validated. The measurement of the energy expenditure poses many technical, logistic and individual problems. The most important problem is to find out really competent subjects. Just there is an other problem that requires large teams of skilled staff and very expensive investigations. Perhaps this is one of the reasons that hence so far only small populations of individual men and women have been studied. Durnin and coworkers are also of the opinion that it is difficult to finance such work.

Twelve trustworthy male test persons, nearly in the same age (23-27 years), with different body weight, were for this experience series available. The test persons are shift-workers from a steel work and students.

Four kinds of comparable tests were made under standard conditions and ad libitum individual: walking on level, solid ground with light clothing in high lace boots (walking speed 5 km/h); walking with 10 kg resp. 20 kg load on level, solid ground with light clothing and high lace boots (walking speed 4 km/h); running on level, solid ground (running speed 10 km/h).

In the ad libitum series the subjects had no pacemaker, but own watches to control themselves. The stretch of road was an other as in the first experience series and unknown for the subjects.

RESULTS

We compared male volunteers with about 70, 80, 90 and 100 kp body weight. The persons had to do at first their work under control. The results of the four tests in working calories per minute are given in Table 1. Besides the averages the ranges are demonstrated in the paper and the standard deviation too. In relation to reference body weight the 80 kp man is about 115%, 90 kp 150%, 100 kp 195%.

The respiratory tests took place always 15 min after start of work. The test time was between 10-15 minutes. The volunteers had constant companions in form of a pacemaker on a bicycle. After examination a small sample of breath air would be transferred in a glass-receiver with a vacuum for later gas-analysis in laboratory.

Afterwards we compared the same test work about individual ad libitum conditions. The results of the four tests in working calories per minute are given in Table 2. Besides the averages the ranges and the standard deviation are named in the complete manuscript.

In relation to reference body weight of 70 kp (100%) the 80 kp test person is 110%, 90 kp 135%, and 100 kp about 170% in walking on solid, level ground with light clothing in high lace boots (walking speed 5 km/h).

In the test of walking with 10 kp load on level, solid ground with light clothing and high lace boots, walking speed 4 km/h, the relation is ad libitum as follows: 70 kp 100%, 80 kp 109%, 90 kp 129%, 100 kp 154%.

In the test of walking with 20 kp load on level, solid ground with light clothing and high lace boots, walking speed 4km/h ad libitum, the relation to the reference man with 70 kp body weight (100%) is 80 kp 109%, 90 kp 119%, 100 kp 132 %.

In the test of running on level, solid ground, running speed 10 km/h with light clothing and high lace boots ad libitum, the relation of the load, expressed in energy expenditure to the reference person with 70 kp body weight is for test person with 80 kp 105%, 90 km 112%, 100 kp 120%.

TABLE 1

ENERGY EXPENDITURE IN SELECTED WORK (UNDER CONTROL)

KIND OF WORK	BODY WEIGHT (kp)	kcal/min	kJ/min	RANGE kcal/min	Sd kcal/min
Walking on level solid ground with light clothing in high lace boots; walking speed 5 km/h	70	2,20	9	1,9 – 2,5	± 0,200
	80	2,53	11	2,2 – 2,9	± 0,518
	90	3,30	14	3,0 – 3,8	± 0,283
	100	4,30	18	4,0 – 4,6	± 0,200
Walking with 10 kg load on level solid gourd with light clothing and high lace boots; walking speed 4km/h	70	3,42	14	3,2 – 3,7	± 0,417
	80	3,93	16	3,8 – 4,1	± 0,573
	90	5,10	21	4,9 – 5,3	± 0,141
	100	6,60	28	6,2 – 6,9	± 0,261
Walking with 20 kg load on level solid ground with light clothing and high lace boots; walking speed 4 km/h	70	5,20	22	5,0 – 5,4	± 0,141
	80	6,03	25	5,8 – 6,3	± 0,718
	90	7,80	33	7,6 – 8,1	± 0,179
	100	10,13	42	9,8 – 10,6	± 0,938
Running on level solid ground (running speed 10 km/h)	70	9,35	39	9,0 – 9,8	± 0,267
	80	10,82	45	10,5 – 11,3	± 0,717
	90	14,12	59	13,8 – 14,4	± 0,779
	100	18,38	77	17,8 – 19,0	± 0,577

TABLE 2

ENERGY EXPENDITURE IN SELECTED WORK (AD LIBITUM)

KIND OF WORK	BODY WEIGHT (kp)	kcal/min	kJ/min	RANGE kcal/min	Sd kcal/min
Walking on level solid ground with light clothing in high lace boots; walking speed 5 km/h	70	2,10	9	1,8 – 2,4	± 0,200
	80	2,30	10	2,0 – 2,6	± 0,228
	90	2,80	12	2,5 – 3,2	± 0,283
	100	3,50	15	3,2 – 3,8	± 0,200
Walking with 10 kg load on level solid ground with light clothing and high lace boots; walking speed 4 km/h	70	3,22	14	3,0 – 3,4	± 0,138
	80	3,52	15	3,3 – 3,8	± 0,132
	90	4,15	17	3,9 – 4,6	± 0,259
	100	4,97	21	4,4 – 5,4	± 0,302
Walking with 20 kg load on level solid ground with light clothing and high lace boots; walking speed 4 km/h	70	4,80	20	4,4 – 5,2	± 0,268
	80	5,22	22	5,0 – 5,6	± 0,126
	90	5,70	24	5,3 – 6,0	± 0,276
	100	6,33	27	6,0 – 7,1	± 0,466
Running on level solid ground (running speed 10 km/h)	70	8,50	36	8,2 – 8,9	± 0,310
	80	8,90	37	8,1 – 9,3	± 0,415
	90	9,40	39	8,8 – 9,9	± 0,447
	100	10,18	43	9,9 – 11,0	± 0,501

The average in the two test-series of the four different types are demonstrated in the figures 1-4. The energy expenditures in the ad libitum work are smaller; for instance in walking on solid, level ground, walking speed 5 km/h, in the subjects with reference weight (70 kp) 95% of the energy expenditure of the expenditure in standard conditions, 91% for the person of 80 kp, 85% for the person of 90 kp, 81% for the person of 100 kp.

In another comparison we can show the evolution of the subjects with reference body weight of 70 kp is in the first test 95%, in the second 94%, 92% in the third and 90% in the fourth test.

A further result is to compare the performance of the person with the reference body weight.

The percentage or ad libitum work in the average of standard conditions is in the first kind of work in the test series 95%, in the second 92%, 94% in the third and 91% in the last.

Moreover a good statement in performances over long periods is the expenditure in working calories per kp body weight in one hour (Table 3) and to compare the differences between the energy expenditure by work under control and ad libitum. In the first case (walking speed limit 5 km/h) there is only a difference in 6 Kcal/h, by the 70 kp subjects. In the last case (running speed limit 10 km/h) the difference in the 70 kp subject is about 51 kcal/h.

Results about energy expenditure per kp body weight are given in Table 4. Calculations about energy expenditure per lean body mass are not finished.

All experiments took place under the same outward and approximate same physiological conditions of the subjects. The experiments started 60 min after a short breakfast with about 600 kcal with 20 g protein (13% of the energy) 20 g fat (31% of the energy) and 81 g carbohydrates (56% of the energy). Measurements of the basal metabolic rate of the subjects had been made before the test breakfast, as well as measurements of the basal metabolic rate including the specific dynamic action 30 and 60 min after mealtime intake. The sequence of the reported tests was at first the second, than the third, the first and at last the fourth test with the highest working energy expenditure. All subjects were in a good training condition. Their usual daily energy expenditure was between 3300 and 3600 kcal.

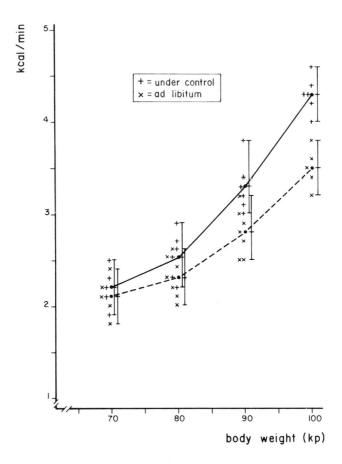

Fig. 1 - Energy Expenditure
m, walking speed 5km/h

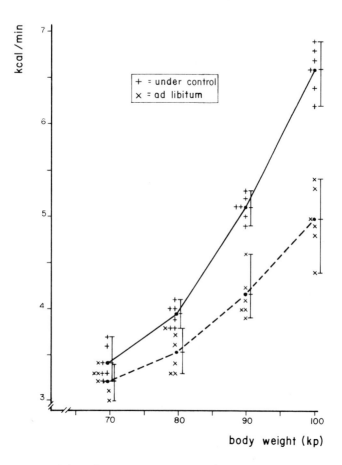

Fig. 2 - Energy expenditure
m, walking speed 4km/h,
10kp load

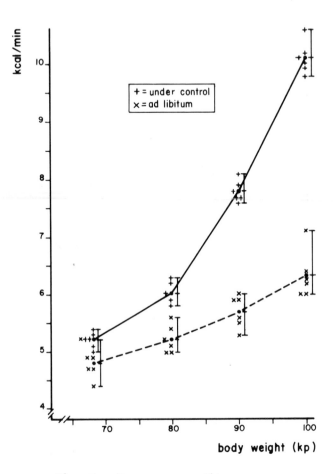

Fig. 3 - Energy expenditure
m, walking speed 4km/h,
20 kp load.

DIFFERENCES IN ENERGY EXPENDITURE IN SELECTED WORK

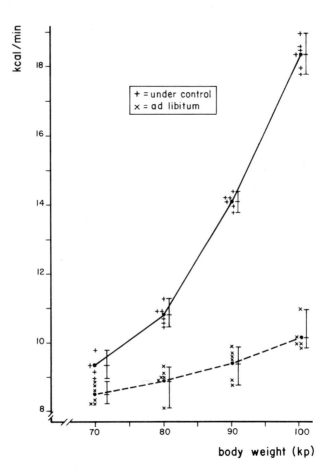

Fig. 4 - Energy expenditure
m, running speed 10km/h

TABLE 3

ENERGY EXPENDITURE IN SELECTED WORK

WORK	BODY WEIGHT (kp)	UNDER CONTROL		AD LIBITUM		DIFFERENCE	
		kcal/h	kJ/h	kcal/h	kJ/h	kcal	kJ
Walking on level solid ground with light clothing in high lace boots; walking speed 5 km/h	70	132	552	126	527	6	25
	80	152	636	138	577	14	59
	90	198	828	168	703	30	125
	100	258	1080	210	879	48	201
Walking with 10 kg load on level solid ground with light clothing and high lace boots; walking speed 4km/h	70	205	858	193	808	12	50
	80	236	987	211	883	25	104
	90	306	1280	249	1042	57	238
	100	396	1657	298	1247	98	410
Walking with 20 kg load on level solid ground with light clothing and high lace boots; walking speed 4 km/h	70	312	1305	288	1205	24	100
	80	362	1515	313	1310	49	205
	90	468	1958	342	1431	126	527
	100	608	2544	380	1590	228	954
Running on level solid ground (running speed 10 km/h)	70	561	2347	510	2134	51	213
	80	649	2715	534	2234	115	481
	90	847	3542	564	2360	283	1184
	100	1103	4615	611	2556	492	2059

TABLE 4

ENERGY EXPENDITURE IN SELECTED WORK PER kp BODY WEIGHT/h

WORK	BODY WEIGHT (kp)	UNDER CONTROL per kp body weight/Kcal	AD LIBITUM
Walking on level solid ground with light clothing in high lace boots; walking speed 5 km/h	70 80 90 100	1,89 1,90 2,20 2,58	1,80 1,73 1,87 2,10
Walking with 10 kg load on level solid ground with light clothing and high lace boots; walking speed 4 km/h	70 80 90 100	2,93 2,95 3,40 3,96	2,76 2,64 2,77 2,98
Walking with 20 kg load on level solid ground with light clothing and high lace boots; walking speed 4km/h	70 80 90 100	4,46 4,52 5,20 6,08	4,11 3,92 3,80 3,80
Running on level solid ground (running speed 10 km/h)	70 80 90 100	8,01 8,12 9,41 11,03	7,29 6,68 6,27 6,11

CONCLUSION

The results of our experiments show that the energy expenditure under standard contitions is much higher than in the working elements ad libitum speed and distance. Performance with highest energy expenditure has the lowest percentage in piece of work or the strongest decline in performance.

Body weight is closely related to the energy expenditure of specific activities involving movements of large parts of the body. But there is an important difference by giving results in energy expenditure of gross body weight or energy/kp body weight (Durnin, 1972). Although the data are very different in gross body weight, however they are very similar when converted to kcal/kp under control and yet more ad libitum.

Who is able to give an exact answer, whether a reference body weight as in our case in man 70 kp, in several countries certainly 55 kp, is the best base for such comparisons. We tested against this other subjects with the so called ideal body weight of the Metropolitan Life Insurance Company (1959) in Max-Planck-Institute in bicycle-ergometer-tests and in treadmill. They were all unable for normal output (4.2 working calories/min) over longer test-times than 4 hours. They had all a smaller oxygen capacity as people in same age with 10-15% above the so called ideal body weight.

If it is allowed, in other work-elements, activities and industrial occupation transmitting such differences in energy expenditure between standard conditions and really practice facts, it could be one of the decisive causes that the energy expenditure from man to man in specious conditions is different. Therefore it gives a solution to the question "how much food does man require".

Many persons overestimate their personal energy expenditure, which is a main reason for a positive energy balance succeeded by overweight.

BIBLIOGRAPHY

Ashworth, A., (1968), Brit. J. Nutr. 22: 341
Christensen, E.H. and Högberg, P., (1950), Physiology of skiing, Arbeitsphysiol. 14: 292.
Durnin, J.V.G.A., (1972), The "real" energy requirements of men and women in different occupations, in: "Alimentation et travail", S. 112-113, Masson et Cie., Paris.
Durnin, J.V.G.A., Edholm, O.G., Miller, D.S., and Waterlow, J.C., (1973), How much food does man require? Nature, 242: 418.

Edholm, O.G., (1977), Energy balance in man, J. Human Nutr., 31: 413-431.
Miller, D.S. and Mumford, P.M., (1967), Amer. J. Clin. Nutr., 20: 1212.
Miller, D.S., Mumford, P.M., and Stock, M.J., (1967), Amer. J. Clin. Nutr. 20: 1223.
Rose, G.A. and Williams, R.T., (1961), Brit. J. Nutr., 15, 1
Woddowson, E.M., (1947), Spec. Rep. Ser. Med. Res. Coun. No. 257.
Woddowson, E.M., (1962), Proc. Nutr. Soc., 21, 121.
Wirths, W., (1974), Evaluation of Energy Expenditure and Nutritional Status in Dietary Surveys, Nutr. Diet., 20: 77-91, Karger Basel.
Wynn-Jones, C., Atkinson, S.J., and Nicolas, P., (1972), Proc. Nutr. Soc. 31: 83A.
Wirths, W., (1977). Energieumsatz bei muskulärer Arbeit, Akt. Ernährungsmedizin 2: 46-50.

A STUDY OF THE ENERGY EXPENDITURE, DIETARY INTAKE AND PATTERN OF

DAILY ACTIVITIES AMONG VARIOUS OCCUPATIONAL GROUPS. IV: WEIGHTLIFTERS

Ma. Patrocinio E. de Guzman

Food & Nutr. Research Inst., Nat. Sci. Develop. Board

Ermita Manila, PHILIPPINES

INTRODUCTION

Weightlifting is a sport which involves muscular effort to raise certain weights on a barbel over the head. Today, it is one of the sports for Olympic competition and is practiced internationally under standard regulations. Competitors are classified into different categories with their corresponding body weights namely: flyweight - up to 52 kg., bantamweight - up to 56 kg., featherweight - up to 60 kg., light heavyweight - up to 110 kg., and super heavyweight over 110 kg.

There are three regular lifts which test the combined strength and athletic ability of the contestant. They are as follows: (a) the press, which calls for lifting the weight from the floor, stopping at the chest and then pushing overhead on signal of the referee, the legs may not move in the press once the weight is at the chest; (b) the snatch, which involves raising the weight all the way from rest to straight arms overhead in one continuous motion; (c) the clean and jerk, wherein the weight must be lifted from the floor to the chest and then overhead under control for two seconds. The highest total weight (kilogram) achieved on the three tests decides the winner.

Aware of the role of nutrition on physical performance and stamina, athletes must be properly fed to be able to compete well and match not only the skill but also the endurance of their competitors. Recognizing this fact, there has been serious concern on the adequacy of the diet of Filipino athletes. It is for this reason that the Philipine Athletic Ameteur Association requested the Food and Nutrition Research Institute to study the dietary intake and daily pattern of activities as well as the energy expenditure of the weightlifters quartered at the Rizal Memorial dormitory, to serve as basis for determining their nutritional needs.

METHODOLOGY

A. General Instructions

A few weeks before the 6th Asian games in Tehran, Iran, seven weightlifters on training for the weightlifting competition were taken as subjectes in this study. Because of the samll size of the population, all weightlifters who were then on training were included. They were designated into different categories accordingly to their body weights. On of them belonged to the middle heavyweight division; one to the light heavyweight, three to the flyweight division and two to the middleweight division.

The evaluation of food intake and energy expenditure was made only during the days of their workouts (Monday, Tuesday, Thursday and Saturady). This deviation from the usual seven weekdays evaluation period done on our previous studies[1,2,3] is due to the time constraints allowed for the study.

B. Measurement of Food Intake

The food intake of the athletes was calculated for four days using the individual food weight method. All foods served to the subjects including snacks were recorded toghther with their weights after they were dished out on the trays for each meal. To determine actual food intake, all plate waste was weighed and then deducted form the initial recorded weights. On days off, the subjects were made to recall all foods eaten and then similar foods were bought and weighed. The food intake was evaluated for calories, protein, fats and carbohydrates using the Philippine Food Composition Table[4].

C. Measurement of Energy Expenditure and Time Spent

The assessment of the total daily energy expenditure was made by the factorial method as described by Durnin and Brockway[5]. The method includes a daily record of the time spent in all activities performed by the subjects for one week and is entered in special activity diary forms. The total metabolic cost of these activities were derived based on the number of minutes spent on a particular activity. The metabolic cost of the activities was either measured directly or by indirect calorimetry using the Max Planck respirometer or from published tables[6].

Weightlifting took only 10 seconds at a time. Samples of their expired air were collected during that time. The determinations for the energy cost of each activity was done by collecting the samples from the subjects while performing major tasks and basic activities. With the use of the Kofranyi-Michaelis respirometer, calibrated before use, a sample of 0,6% of the total expired air was collected in rubber bladders which usually lasted for 8 minutes for occupational and strenuous activities and 10 minutes for the basic activities. Since some of the exercise did not last as long as 8 minutes, the succeeding exercises and activities were included during the time of the collection to fill in the time needed for the collection of samples os the occupational activities. Each activity was measured in duplicate and the mean of the two was taken as the final result. All samples collected were analyzed immediately thereafter using the Beckman E2 analyzer to avoid CO_2 diffusion across the collection bladder. The energy expenditure while sleeping was based on the basal metabolic rate calculated by the method described by Fleisch[7] and Passmore and Durnin[6]. Other values not measured by indirect calorimetry were obtained by using values given by Passmore and Durnin[6] or other published tables.

RESULTS AND DISCUSSIONS

The average age, height and weight of the subjects were 25 years, 164 cm. and 67 kg., respectively. The metabolic cost of the different basic activities of the weightlifters expressed in calorie per kilogram per minute was determined. Sitting, standing at ease and walking entailed 1.44, 1.82, and 4.98 kcal/min. respectively. In terms of kcal/kg/min., these are 0,021, 0.027 and 0,075 respectively. Most of their energy was spent during walking which accounted for about 60 percent of the total energy cost.

In Table 1 it can be seen that the weightlifters have the lowest energy expenditure during sitting and standing compared with shoemakers, farmers, jeepney drivers and clerks and typist. However, they have the highest energy expenditure during walking among these groups, followed closely by farmers and shoemakers.

It should be noted that the seven weightlifters have negligible energy expenditure standard deviations during sitting, standing and walking. This means that these athletes have almost the same energy expenditure, per kilogram of body weights regardless of whether they are bantamweight, heavy weight or super heavyweight.

Mean energy cost of various exercises activities of the weightlifters is shown in Table 2. These are only among the many exercise activities done by the weightlifters and those with the most number of observations are presented.

TABLE 1

COMPARISON OF ENERGY EXPENDITURE
(kcal/kg/min)
OF FIVE OCCUPATIONAL GROUPS

GROUP	ACTIVITIES					
	Sitting		Standing		Walking	
	Mean	S.D.	Mean	S.D.	Mean	S.D.
Shoemakers	0.028	0.003	0.029	0.001	0.071	0.007
Farmers	0.032	0.007	0.032	0.007	0.072	0.013
Jeepney Drivers	0.030	0.004	0.032	0.004	0.055	0.008
Clerks and Typists	0.026	0.004	0.028	0.005	0.056	0.010
Weightlifters	0.021	0.000	0.027	0.000	0.075	0.000

TABLE 2

MEAN ENERGY COST OF VARIOUS EXERCISE
ACTIVITIES OF WEIGHTLIFTERS

ACTIVITY	kcal/kg/min	no. of OBSERVATIONS
Front Squat	0.088	3
Bench Press	0.090	6
Frog Kick	0.091	5
Good Morning	0.091	7
Push Jerk	0.095	3
Form Snatch	0.099	7
Dumb Bell Exercise	0.104	6
Dead Lift	0.106	5
Sitting Press	0.124	4
Leg Press	0.132	4
Power Clean & Push Press	0.103	7
Calisthenics	0.105	7
Back Squat	0.109	4
Squat	0.119	7
Flip Snatch	0.124	6
Overhead Support	0.125	2
Jogging	0.149	7

It can be seen that jogging gave the highest energy expenditure at 0.149 kcal/kg/min. Next, the leg press with 0.132 kcal/kg/min., then the overhead support at 0.125 kcal/kg/min. Both flip snatch and sitting press gave 0.124 kcal/kg/min. Jogging registered the highest energy expenditure of all their activities since it involves shaking of the entire body whereas in leg press, heavier weights are being lifted which involves only the leg muscles. The leg press is intended to increase the stamina of the lifter and is the best exercise for the lower body togehter with squatting. In overhead support, the weights are suspended in an iron bar at eye level and is being lifted with one swift thrust of the arms ovehead. It can be noted that the number of observations made on each activity is not the same because of variation in the type of exercise done.

TABLE 3

MEAN DURATION OF ACTIVITIES AND AMOUNT OF ENERGY EXPENDED IN VARIOUS ACTIVITIES OF THE WEIGHTLIFTERS

	TIME AND ENERGY EXPENDED	
	Min	kcal/kg/min
Sleeping	633	755
Personal Necessities	62	101
Sitting Quietly	79	112
Sitting Activity	379	570
Standing Quietly	52	110
Standing Activity	73	154
Walking	117	596
Cycling	1	99
Weightlifting	26	202

Table 3 shows the mean time and energy expended in the different activities of the weightlifters in one day.

The weightlifters' daily pattern of activity includes 9 to 10 hours of sleep. Their sitting activity entailed an average of 6 to 7 hours in which time they usually hung around the office doing some typing, reading, writing, but mostly watching television, going to movies and playing basketball for recreation. The workouts started at eight in the morning till 12 noon. They started the day doing about 10 minutes of calisthenics followed by lifting increasing loads of weights for only 10 seconds at a time. During workouts some of them just sat around or remained standing while waiting for their turn to lift the weights which accounted for the very short time consumed in occupational activities. They usually end up the daily workouts jogging around the stadium for five to nine minutes a day. The subjects were usually off in the afternoon. They did not do much standing activities except when playing basketball, ironing their clothes, riding a bus and whatever help they can extend at home. The rest of the time they spent resting and doing some of their basic activities. To have the proper strength for the morning exercise, they were not allowed to stay out late at night.

From the figures shown in Table 4, the average intake of this group amounted to 2784 kcal per day. Of the total calorie intake,

TABLE 4

MEAN CALORIE INTAKE AND PERCENTAGE CALORIE
DISTRIBUTION PER DAY

	PRO (kcal)	FAT (kcal)	CHO (kcal)	ENERGY (kcal)
Mean ± SD	346+49	682+129	1756+388	2784+46
% OF Total Calorie	12.4%	24.5%	63.1%	100%

63.1% come from carbohydrate, 24.5% from fats and only 12.4 from proteins. The percentage of carbohydrate in the diet of the weightlifters studied approximated the amount of carbohydrate usually prescribed for a non-athlete (40-60%)[8]. The value of 63.1% for carbohydrate in the study was approximately the same as that obtained by Matawaran et al[9] which was 63% for the filipino athletes that he covered. Fat and protein intakes (24.5% and 12.4%, respectively) where lower than those prescribed for a non-athlete, the normal levels of which are 30-35% for fats and 14-15% for protein[8]. From the different reports reviewed, athletes usually are fed with larger amounts of meat or meals high in protein content, are advised. However, protein requirement does not increase during actual athletic competition nor influence performance capacity[10,11]. In fact, competent physicians recommend reduced protein intake in the meals preceeding any event to avoid problem of urinary excretion since protein is a source of fixed acids which can be eliminated only by urinary excretion.

It is believed that for prolonged exercise, carbohydrate is the best source of fuel. That is the reason why an increase in carbohydrate intake of our weightlifters is recommended during competitions. The inclusion of more tubers and noodles in their diets as well as cakes and pastries for dessert would be most helpful. Rice cakes, bread or biscuits and milk are also recommended.

The daily calorie intake and expenditure of the weightlifters per day averaged over 4 days are given in Table 5. The average intake per weightlifter per day is 2785 calories and the average expenditure is 2732 calories but there is no significance to this difference. However, these values are lower than the results of the previous study on athletes[10] where the mean intake was 3007 kcal and mean energy expenditure, 3181 kcal.

TABLE 5

DAILY MEAN ENERGY INTAKE AND EXPENDITURE OF WEIGHTLIFTERS

SUBJECT	INTAKE (kcal)	EXPENDITURE (kcal)	DIFFERENCE (E-I)
SR	3387	2404	− 983
NT	2280	2448	168
AR	2511	2139	− 372
JS	3135	3037	− 98
AG	2534	3228	694
RM	3288	2435	− 853
CN	2357	3430	1073
MEAN ± SD	2785 ± 468	2732 ± 492	− 53 ± 762

SUMMARY AND CONCLUSION

Measurements of the total energy expenditure and dietary intake were made on seven weightlifters quartered at the Rizal Memorial stadium prior to the 6th Asian games in Techran, Iran. The energy cost of both occupational and non-occupational activities performed by the subjects was determined. Data on the time activity was likewise determined to estimate the total energy cost for the day. Metabolic cost for the basic activities expressed in calories per kilogram per minute gave a value of 0.021, 0,027 and 0,074 respectively. Most of their energy was spent during walking which accounted for 60% of the total energy cost.

Comparison of mean energy expenditure for basic activities (sitting and standing) of farmers, shoemakers, and jeepney drivers showed that the weightlifters have the lowest energy expenditure. However, for walking, the mean energy expenditure of the weightlifters which was 0.075 cal/kg/min., was found to be much higher in comparison with the mean energy expenditure of the farmers and shoemakers which are 0.072 and 0.071 kcal/kg/min., respectively. The seven weightlifters have negligible standard deviations during sitting, standing and walking. This means that the groups of athletes have almost the same energy expenditure per kilogram of body weight, regardless of whether they are bantam-weight, heavyweight or super heavyweight.

Jogging, one of the many activities of the weightlifters, gave the highest energy expenditure at 0.149 kcal/kg/min., followed by the leg press, overhead support and flip snatch, which were 0.132, 0.125 and 0.124 kcal/kg/min., respectively.

An evaluation of the dietary intake of the weightlifters showed that the per cent calorie intake ratio derived from carbohydrate in the diet more or less approximated by recommended normal level for non-athletes while protein and fat intakes were lower than the prescribed range. A high protein intake is recommended only during the pre-training and training periods. Moreover, at the meal preceeding the event, protein intake should be reduced to avoid the problem of urinary excretion. A high carbohydrate pre-game meal is recommended to give better results.

Average energy intake per day over 4 days exceeded the average expenditure, the difference however was not statistically significant. From the results of the study made. weightlifters may be classified as belonging to the active group but less active when compared with other athletes previously studied.

REFERENCES

[1] Guzman, P.E. de, S.R. Dominguez, J.M. Kalaw, R.O. Basconcillo and V.F. Santos. A Study of the Energy Expenditure Dietary Intake and Pattern of Daily Activity Among Various Occupational Groups. I. Laguna Rice Farmers

[2] Guzman, P.E. de, S.R. Dominguez, J.M. Kalaw, M.N. Buñing, R.O. Basconcillo and V.F. Santos. A Study of the Energy Expenditure, Dietary Intake and Pattern of Daily Activity Among Various Occupational Groups. II. Marikina Shoemakers and Housewives.

[3] Guzman, P.E. de, J.M. Kalaw, R.H. Tan, R.C. Recto, R.O. Basconcillo V.T. Ferrer, M.S. Tumbokan, G.P. Yuchingtat and A.L. Gaurano. A Study of the Energy Expenditure, Dietary Intake, and Pattern of Daily Activity Among Various Occupational Groups II. Urban Jeepney Drivers.

[4] FNRC, NSDB. Food Composition Tables Recommended for Use in the Philippines, Handbook I. 4th Revision, June, 1968, Manila, Philippines.

[5] Durnin, J.V.G.A. and J.M. Brockway. Determination of Total Daily Energy Expenditure in Men by Indirect Calorimetry: Assessment of the Accuracy of a Modern Technique. Brit. Jour. Nutr. 13: 41-53, 1959.

[6] Passmore, R and J.V.G.A. Durnin. Human Energy Expenditure Physiol. Rev. 35: 801-840, 1955.

[7] Fleisch, A. Le Metabolism Basal Standard et Sa Determination au Moyen du Metabocalculator. Helvet. Med. Actv. 18:23-44, 1951.

[8] Guzman, P.E. de, S.R. Dominguez, V.F. Santos, V. Lavides and J.M. Kalaw. Filipino Athletes, Their Dietary Intake and Total Energy

Expenditure. PJN, Oct-Dec., 1972.

[9] Matawaran, A.J., de Lara, M.A. and J.T. Caluag. Food Intake of Some Filipino Athletes, PJN, July-Dec., 1969.

[10] Mayer, Jean and B. Bullen. Nutrition and Athletes. Proceeding of the 6th International Congress of Nutrition. Edinburge, Aug., 1963.

[11] Astrand, F. Diet and Athletic Performance, Federation Proceedings, 26: 1772-1777, 1969.

WATER AND ELECTROLYTE BALANCE IN EXERCISE

D. Costill

Human Performance Lab., Ball State University

Muncie, IND, U.S.A.

Over the past ten years producers of athletic drinks have made many claims, based on a minimum of facts, to assit in promoting their products. According to advertising claims, each drink has some physiologically unique quality that will replace the body water and salts lost in sweat, thereby improving athletic performance. This has led to some confusion among athletes who are ill-prepared to distinguish between valid claims and promotional statements having little or no factual support. It is for this reason that the following discussion has been prepared: to present a brief summary of the evidence currently available to describe the need for drinks during prolonged exercise, and to detail the prerequisites of a drink for effective body fluid replacement.

(1) WHAT ROLES DO BODY FLUID PLAY DURING MAXIMAL PERFORMANCE?

Water is, by far, the most abundant component of the human body, constituting roughly 60% of the body weight. This large fluid reservoir furnishes an environment that is essential to the survival of the body tissues. In addition to providing a medium to support the internal activities of the cell, body fluids offer a vehicle for transporting nutrients to the active muscles, eliminating waste products and dissipating excessive body heat. For the most part, these tasks are the responsibility of the circulatory system.

It is well established that the athlete's ability to perform endurance exercise is, in part, limited by the circulatory system's capacity to transport oxygen to the working muscles. Of course, even a highly trained circulatory system has its limits. Since the body does not have enough blood to fill all of the blood vessels at the same time, blood is distributed from inactive to active areas of the body by the constriction and dilation of small arteries. Thus, during prolonged severe exercise, the amount of oxygen which can be delivered to the working muscles depends upon the volume of blood available for distribution and the ability to divert a large part of that volume to the contracting muscles. Any factor which reduces either blood volume or compromises the blood flow to the muscles will impair endurance perfomance.

At the same time, the circulatory system must contend with the demand to eliminate the large amount of heat being produced by the working muscle. As a result, a portion of the blood flow must be diverted to the skin where the heat can be transferred to the environment, principally by vaporizing sweat. The amount of blood that must flow to the body surface is dictated by the amount of heat being produced by the muscles and the environmental capacity to accept the body heat. The hotter the environment and the higher the humidity of the air surrounding the athlete, the more difficult it becomes to eliminate body heat. Thus, more blood must be pumped to the skin and less reaches the muscle.

Under such circumstances more sweat will be formed, thereby using more of the body's fluid. As a result, water is removed from the blood, limiting still further the circulatory system's capacity to transport nutrients to the active muscles and to transfer heat to the body shell. There is little doubt, therefore, that warm weather exercise impairs endurance performance and can even threaten the health of the athlete by causing him/her to store excessive body heat. Marathon runners, for example, have been found to have internal body temperatures in excess of 106º F at the end of a race conducted in a warm (75º F), sunny environment (Costill, et al. 1970).

Studies have shown that dehydrated men and women are quite intolerant to heat and exercise (Claremont, et al. 1976). As can be seen in Figure 1, both heart rates and body temperature are elevated during exercise in the heat when the subjects were dehydrated 3% of body weight. Regardless of the cause of dehydration, its impact on the cardiovascular system is always the same: plasma water is lost, the ability to provide adequate blood flow to the skin and muscles is reduced. Under such circumstances it is common for subjects to collapse, showing the usual symptoms of heat exhaustion (cold, wet, pale skin, low blood pressure).

The answer to our first question therefore, is that body fluids play a crucial role in maintaining the homeostasis of the body (a constant internal environment). With a decline in body water neither the circulatory nor temperature regulatory systems can meet the demands placed on them by the stress of exercise and/or warm environments.

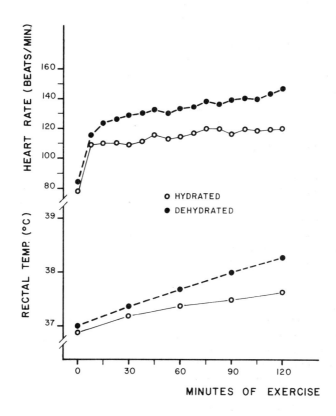

Fig. 1 - Effects of dehydration on heart rate and rectal temperature during 2 hours of cycling. Subjects lost 3 to 4% of their body weight before the "dehydrated" trial. (Claremont, et al., unpublished).

(2) WHAT DO YOU LOSE IN YOUR SWEAT?

Human sweat has been described as a filtrate of plasma since it contains many of the items generally present in the water phase of blood. It should be realized, however, that the sweat glands are capable of selective secretion of specific items, namely, sodium. Of all the minerals lost in sweat, sodium (Na^+) and cloride (Cl^-) are by far the most abundant. Since these particles are electrically charged they are referred to as ions or electrolytes. In the physiology of body fluids, Na^+ and Cl^- are the ions primarily responsible for maintaining the water content outside the cells (extracellular). This is principally accomplished by maintaining a steady relationship between the number of ions and water molecules. If, for example, a great quantity of Na^+ and Cl^- were lost in sweat, the body would lose part of its control over the extracellular water. If you lose the ions, water must be redistributed to maintain the water-ion relationship.

Sweat is, in fact, quite dilute when compared with other body fluids. Thus, during heavy sweating you lose disproportionately more water than body ions. This can be illustrated in Table 1 by comparing the concentration of some ions in both sweat and plasma water. This table demonstrates that the concentrations of sodium and chloride in sweat are roughly one-third that seen in plasma. Sweat, therefore, has substantially fewer ions and is considered to be hypotonic.

If, during a long bout of exercise-sweating, you lose 9 pounds of sweat, the body electrolyte losses would constitute roughly 140 to 150 mEq of Na^+ and Cl^-. Since the body generally contains 2600 mEq, these losses would decrease the body's total NaCl content by 6 to 8%. At the same time, the potassium and magnesium losses would lower the body content by less than one percent. The physiological importance of these losses is not well established; however, it has been theorized that large electrolyte losses may be responsible for muscle cramping and intolerance to the heat.

One point, however, should not be overlooked. Since more water than ions is removed from body fluids, the remaining electrolytes become more concentrated. Thus, as far as the cells are concerned, there is an excess of electrolytes, even though the total quantity of electrolytes has decreased. This may seem a bit confusing, but the point to be made is that during prolonged heavy sweating the need to replace body water is greater than any immediate demands for eletrolytes.

Does everyone sweat the same? The obvious answer is no. Both the quantity and electrolyte content of sweat varies between individuals. The sweat of highly trained, heat acclimated subjects, for example, is even more dilute than previously mentioned. Women pro-

TABLE 1

ELECTROLYTES (mEq/liter)

	SODIUM	CHLORIDE	POTASSIUM	MAGNESIUM	TOTAL
PLASMA	140	100	4	1.5	245.5
SWEAT	40-60	30-50	4-5	1.5-5	75.5-120

duce significantly less sweat than men exposed to similar heat-exercise conditions. As a result, we observe wide variations in the rate of sweating and the quantity of minerals excreted by the sweat glands.

Although the purpose of sweating is to aid in body heat dissipation, some individuals sweat at rates which exceed the environmental capacity to evaporate the water from the skin. Thus, their sweating response is inefficient and their water and electrolyte losses may be extremely large. These athletes, therefore, may have unusually large water and electrolyte requirements. Still, the primary need of these people is water.

(3) WHAT ARE THE ADVANTAGES OF TAKING DRINKS DURING PROLONGED EXERCISE?

There are three very obvious benefits to be experienced by ingesting fluids during prolonged exercise, especially during hot weather conditions. First, drinking fluids will minimize the degree of dehydration which normally results from prolonged heavy sweating, thereby reducing the stress placed on the circulatory system. In Figure 2 you can see that drinking water during two hours of cycling in the heat has a dramatic effect on the volume of plasma when compared to the same conditions performed without fluids. Thus, there is more blood available to permit the transport of essential nutrients to the working muscles and to transfer heat to the body's shell.

This brings us to the second major advantage in taking fluids during warm weather exercise. The threat of overheating is substantially reduced with fluid ingestion (see Figure 3). Although studies have shown that the rate of sweating is unaffected by fluid intake (Costill, et al. 1970), there is apparently a greater flow of heat from the deep body tissues to the body shell as a result of preventing a large loss of body water. Even fluids ingested at near body temperature can help to prevent overheating. Drinking cold fluids

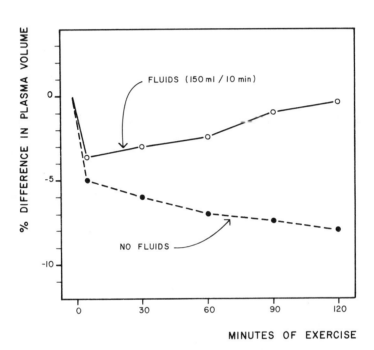

Fig. 2 - Changes in plasma volume during 2 hours of cycling in the heat (100° F, 35% RH) with and without fluid intake. (Balon, et al., unpublished).

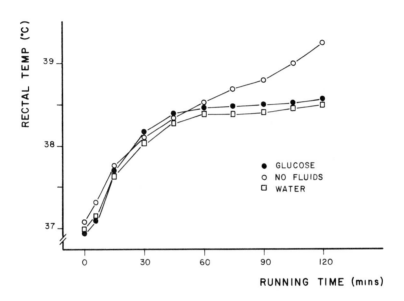

Fig. 3 - Effects of fluid administration on body temperature during 2 hours of treadmill running in six marathon runners. (Costill, D.L. 1970).

seems to enhance this body cooling effect, since it takes away some of the deep body heat in the process of warming the drink to the temperature of the gut.

The third major advantage of taking drinks during endurance performance is that it provides an opportunity to supplement the body's need for carbohydrates. In the course of several hours of exercise, the liver is responsible for releasing sugar (glucose) into the blood for the benefit of use the active tissues of the body (muscles, nerves, etc.). Since this glucose is primarily obtained from larger sugar molecules (glycogen) that are stored in the liver, eventually the liver will become depleted of sugar and unable to maintain the blood glucose at the level desired by the working cells of the body. The absorption of glucose from the intestinal tract offers another route whereby blood glucose levels can be maintained and sugar can be restored to the liver. Thus, a drink containing sugar (glucose, fructose, or sucrose) can supplement the carbohydrates that are of limited supply in the liver.

In light of our previous discussion on electrolyte needs during acute bouts of heavy exercise-sweating, it seems that replacing such items as sodium, potassium chloride, magnesium, and calcium during the exercise is of minimal importance. There is no empirical evidence to demonstrate that electrolyte intake during exercises will enhance performance or eliminate the occasional problem of muscle cramps. In the hours following a heavy bout of sweating, it may be necessary to make some effort to replace electrolytes as body water losses are being restored. This is generally satisfied by the foods in a balanced diet.

This then leads us to the question:

(4) WHAT ARE THE FLUID REQUIREMENTS OF PEOPLE WHO ARE SWEATING HEAVILY ON REPEATED DAYS?

In the late 1930's Adolph reported that men who were exposed to prolonged marching in the desert heat were able to fully rehydrate in 24 hours if food and water were provided ad libitum (Adolph 1938). Recent studies (Costill, et al. 1975) have also shown that during repeated days of heavy sweating the kidney adequately reduces its excretion of water and electrolytes to prevent the subjects from becoming chronically dehydrated or losing excessive body electrolytes. This water and electrolyte (principally sodium) conservation by the kidney is so effective that during repeated days of dehydration and ad libitum replacement the body actually stores water and sodium in excess of the sweat loss. This is illustrated in Figure 4

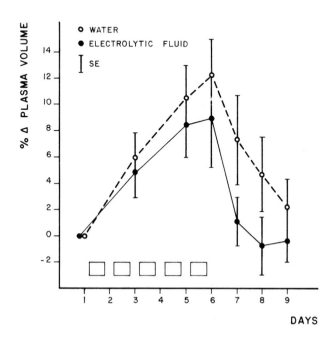

Fig. 4 - Changes in plasma volume during five days (noted by black bars) of heavy exercise. During one nine-day sequence the subjects drank only water (open circles). In the second treatment the subjects drank a glucose-electrolyte solution (filled circles) (Costill, et al. 1974)

by the marked increase in plasma volume. It should also be noted that when the daily exercise bouts were terminated the subjects (men and women) quickly lost this excess water and sodium.

Of course, these results were obtained when the men and women were provided sodium, chloride, and potassium in quantities that met their demands. What would happen to these subjects if they did not ingest sufficient quantities of sodium or potassium? Some indirect evidence has been accumulated to demonstrate that men training in hot climates may become potassium deficient, thereby reducing the muscles' capacity for work (Knochel, et al. 1972; Rose 1975). In efforts to directly measure the effects of repeated days of heavy sweating on muscle potassium content, more recent studies were conducted using both low and normal potassium diets during 4 days of repeated bouts of dehydration (Costill 1976). As a result of measuring the potassium ingested and lost (sweat, urine), and by sampling muscle from the subjects, it was found that there was very little change in either whole body or muscle potassium content.

These studies clearly demonstrate the body's capacity to minimize electrolyte losses during repeated bouts of exercise and dehydration. One of the points to be remebered is that during repeated days of dehydration and heavy exercise, plasma volume generally increases in proportion to an increase in body sodium storage, thereby demanding an increase in extracellular water. At this point it seems appropriate to mention the effects of such an increase in plasma water on the concentration of various items in blood. Since red blood cells and hemoglobin are confined to the vascular space, the concentration of both may decrease significantly as a function of the dilution of more plasma. This may, in part, explain the apparent false anemia reported among athletes undergoing intensive training. It is also possible that such dilution may produce low concentrations of plasma potassium, which might be falsely interpreted as indicating a deficit of body potassium. In any event, some caution should be used in the clinical interpretation of plasma concentrations of various blood items among endurance trained athletes.

With regard to the question of possible body potassium depletion during repeated days of heavy exercise, calculations of potassium balance (intake minus excretion) and measurements of plasma and muscle potassium content fail to confirm the large deficits previously estimated by Knochel, et al. 1972. Despite low dietary potassium and increased potassium losses in sweat, little change in total body potassium (less than 2%) has been reported. Thus, we can conclude that depletion of body potassium stores is not a common problem among athletes, but probably exists only in persons with abnormal kidney function, since very little potassium is lost in sweat (less than 7 mEq/liter). Generally speaking, diet and the electrolyte conservation by the kidney adequately compensate for electrolytes lost in days of heavy sweating.

WATER AND ELECTROLYTE BALANCE IN EXERCISE 677

(5) WHAT TYPE OF FLUIDS SHOULD BE USED TO REPLACE SWEAT LOSSES DURING EXERCISE?

As has been pointed out several times in the preceding discussion, the most important item to be considered in replacing sweat is water. Remember, it is water alone that compensates for the loss of plasma volume and effectively reduces the threat of overheating. Therefore, the real question to be answered is "what is the fastest way of replacing your water losses?"

Since little exchange of water occurs directly from the stomach, the fluids must pass into the intestine before entering the blood. Once in the intestine, absorption is rapid and unaffected by exercise (Fordtran and Saltin, 1967). The major limitaion to fluid replacement, therefore, seems to be how quickly the drink leaves the stomach. Many factors can affect the rate of gastric (stomach) emptying, including the volume, temperature, and sugar content of the drink (Costill and Saltin, 1974).

Although larger volumes (up to 600 ml) empty more rapidly from the stomach than small ones, athletes generally find it uncomfortable to exercise with a nearly full stomach, which might interfere with the movement of their respiratory muscles. Thus, it is probably wise to drink 150-250 ml (5 - 8.5 ounces) at 10-15 minute intervals to minimize gastric filling.

Cold drinks have been found to empty more rapidly from the stomach than do warm fluids (Costill and Saltin, 1974). It has been shown that the ingestion of cold water can reduce the gastric temperature by 7 - 18º C (12.6 - 32.4º F), which may require 5 - 60 minutes to return to normal. Such lowering of the temperature apparently increases the activity (mobility) of smooth muscle in the stomach wall, thereby causing a more rapid flow into the intestine. A common question often asked is "won't cold fluids cause stomach cramps?" None of the studies conducted with cold fluids observed any cases of stomach cramps. It seems likely that the gastric distress reported by some individuals may be more related to the volume of the drink than to either its temperature or composition. What about the idea that cold fluids can upset the normal electrical activity in the heart, thus threatening the health of the athlete? Changes in the electrocardiogram have been observed in some individuals following the ingestion of ice cold drinks (2-4º C), but the clinical significance of these changes has not been established. It appears that drinking cold fluids poses no threat to a normal heart.

The single ingredient having the greatest effect on the rate of gastric emptying is sugar. Regardless of whether the drink contains glucose, fructose, or sucrose, all sugars have a retarding effect on the rate at which solutions leave the stomach. If, for example, a person drinks 400 ml (13.5 ounces) of water, fifteen minutes later only 30-40% of the volume is still in the stomach. When, however, an equal volume of a soft drink containing 40 g of sucrose is ingested, only about 5% of the volume will have left the stomach in fifteen minutes (Costill, D.L. and B. Saltin, 1974). Thus, consuming a strong sugar drink is a poor method of replacing the water lost in sweat. As can be seen in Figure 5, only with very weak sugar drinks can we hope to deliver the maximal quantity of water to the system. These data suggest that the sugar content of the replacement drink should not exceed 2 - 2.5 g per 100 ml of water.

Will this small amount of sugar provide any assistance to the liver and performance of the athlete? Yes. Since the solution empties quickly from the stomach it can enter the blood at a rapid rate, thereby delivering about as much sugar to the liver as a strong sugar drink that remains in the stomach for longer periods (Costill, et al. 1973).

Thus far we have concerned ourselves only with the problem of exercising in the heat where large amounts of sweat are formed. What about competition in a cold environment where overheating and dehydration are not our major concerns? Should these athletes take fluids and, if so, what about the sugar content? Admittedly, running or cycling in a cold climate places lesser demands on water replacement, but does little to reduce the need for a carbohydrate supplement. Under these conditions the athlete should drink small quantities of a strong sugar (15 - 40 g per 100 ml of water) solution, since the slow release of the solution from the stomach will provide a steady source of carbohydrate for the liver and active tissues (Costill, D.L., et al. 1973).

Several of the commercially available athletic drinks make a point of mentioning that their replacement fluid contains the essential minerals lost in sweat. First of all, any solution with the same composition of electrolytes as sweat could not be palatable to the athlete, thus impairing his/her capacity to remain hydrated. This point of palatability is an important one since human thirst is a bit tricky. The mechanisms of thirst are delayed in humans so that the athletes must drink in excess of desire if they hope to minimize dehydration. For that reason, it is important that the rehydrative solution be appealing to the athlete's taste <u>during exercise</u>. The words <u>during exercise</u> are emphasized because human taste preference seems a bit modified during exercise. Most exercising participants dislike a strongly flavored drink, preferring a solution that might appear somewhat tasteless to the non-exercising drinker. For that reason, the replacement drink should not have a strong flavor. In that way the athlete is more likely to drink larger volumes, thereby reducing the degree of dehydration.

Another claim often made by commercial producers of athletic drinks is that their solution is "isotonic" as compared to body fluids. What they are really saying is that the number of solid

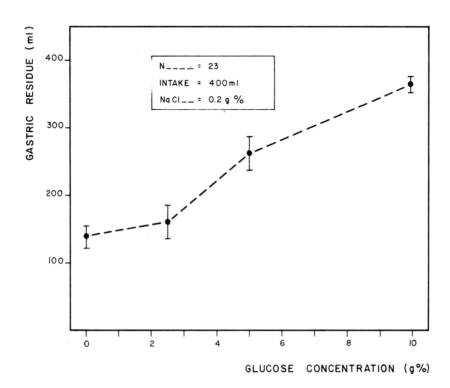

Fig. 5 - Effects of the glucose concentration in a solution on the rate of gastric emptying. Subjects drank 400 ml of solution containing varied sugar contents. After 15 minutes the residue remaining in the stomach was removed through a gastric tube.

particles (electrolytes, sugar, etc.) in their drink is the same as the solids to water ratio in body fluids. For many years it was felt that a drink should be isotonic in order to be absorbed at an optimal rate. Recent studies have shown that this is not the case (Costill, D.L. and B. Saltin, 1973). Solutions that are hypotonic (fewer particles per unit of water than body fluids) empty more rapidly from the stomach than either isotonic or hypertonic solutions. Thus, the rehydrative drink should be quite dilute, another reason for the drink to contain little sugar and electrolytes.

(6) WHAT ABOUT DRINKING BEFORE AND AFTER AN ENDURANCE EVENT IN THE HEAT?

Some efforts have been made to determine the benefits, if any, of taking fluids before competition. There is adequate evidence to support this practice, since it provides a bit of a reservoir of water to compensate for the sweat that will be lost before the first opportunity to drink. It seems logical, therefore, to drink 400 - 600 ml (13.5 to 20 ounces) of water or an athletic drink 30 minutes before the competition. Fluids taken earlier than this may be spilled off as urine before the start, or worse yet, in the early minutes of competition. Soon after the athlete begins exercising, the kidney sharply reduces its operation, thereby producing little urine.

Following the exercise-dehydration session, the task of rehydration is adequately handled by thirst and dietary intake, provided 24 hours of ad libitum feeding is permitted before the next exercise bout. It is during this period that both water and electrolytes are restored to the body. Excessive electrolyte losses are adequately met by adding a bit more salt (NaCl) to the foods. If there is any fear of becoming potassium deficient, then a 250 ml (8.5 ounce) glass of orange juice should take care of most of the potassium lost in 2-3 liters of sweat. At the same time, this will replace most, if not all, the calcium and magnesium excreted in sweat. Other drinks, such as tomato juice, are equally effective in supplementing the dietary intake of most essential minerals.

GUIDELINES FOR THE DRINKING ATHLETE

During hot weather competition and training the athlete should use the following guidelines:

(1) The drink should be:

    a. hypotonic (few solid particles per unit of water)

    b. low in sugar concentration (less than 2.5 g per 100 ml of water

c. cold (roughly 45-55º F or 8-13º C).

   d. consumed in volumes ranging from 100 to 400 ml (3 to 10 ounces).

   e. palatable.

(2) Drink 400-600 ml (13.5 to 20 ounces) of water or the above drink 30 minutes before the start of competition.

(3) During the competition 100 to 200 ml (3 to 6.5 ounces) of the above drink should be taken at 10 to 15 minute intervals throughout the activity.

(4) Following the competition, modest salting of foods and the ingestion of drinks with essential minerals can adequately replace the electrolytes lost in sweat.

(5) The athlete should keep a record of his/her early morning body weight (taken immediately after rising, after urination, and before breakfast), to detect symptoms of a condition of chronic dehydration.

(6) Drinks are of significant value in races lasting more than 50 or 60 minutes.

REFERENCES

[1] Adolph, E.F. "Physiology of Man in the Desert". Wiley, New York, 1947.
[2] Adolph, E.F. and D.B. Dill. Am. J. Physiol. 123;369-378, 1938.
[3] Balon, T., E. Coyle and D. Hoopes. Ball State University Graduate Research Project, 1976.
[4] Claremont, A., et al. Med. Sci. Sports (In Press), 1976.
[5] Costill, D.L., et al. Arch. Environ. Health 21:520-525, 1970.
[6] Costill, D.L., et al. J. Appl. Physiol. 40:6-11, 1976.
[7] Costill, D.L. and W. J. Fink. J. Appl. Physiol. 37:521-525, 1974.
[8] Costill, D.L. and K.E. Sparks. J. Appl. Physiol. 34:299-303, 1973.
[9] Costill, D.L., et al. Aviat. Space Environ. Med. 46:795-800, 1975.
[10] Costill, D.L. J. Appl. Physiol. 37:912-916, 1974
[11] Costill, D.L. and B. Saltin. J. Appl. Physiol. 37:679-683, 1974.
[12] Costill, D.L. and B. Saltin. "Metabolic Adaptation to Prolonged Exercise". Birkhäuser Verlag Basel, 1975. pp. 352-360, 1975
[13] Costill. D.L. Muscle water and electrolytes during acute and repeated bouts of dehydration. Presented at the International Symposium on Sportsmen's Nutrition, Warsaw, Poland. 1975. (Proceedings in Press).
[14] Costill, D.L., et al. J. Appl. Physiol. 34:764-769, 1973.
[15] Fink, W.J., et al. Europ. J. Appl. Physiol. 34:183-190, 1975.

[16] Fordtran, J.S. and B. Saltin. J. Appl. Physiol. 23:331-335, 1967.
[17] Knochel, J.P., et al. J. Clin. Invest. 51:242-255, 1972.
[18] Nielsen, B. Acta Physiol. Scand. 91:123-129, 1974
[19] Pitts, R.F., et al. Am. J. Physiol. 142:253-259, 1944.
[20] Rose, D.D. Physician Sports Med. 3:26-29, 1975.

PREDICTING NUTRITIVE VALUE OF

FOOD PROTEIN BY IN VITRO METHOD

D. Petitclerc, G. Goulet and G.J. Brisson

Centre de Recherches en Nutrition and Departement de
Zootechnie, Université Laval
Quebec, Canada

INTRODUCTION

There have been many attempts to develop a rapid and simple method for predicting in vitro the nutritive value of food proteins. Several of these methods were based on the amino acid content of the protein sources; let us mention, for example, the chemical score using egg protein or the FAO/OMS reference protein as a basis for comparison (Block and Mitchell, 1946; FAO/OMS, 1963) and the Oser essential amino acid index as modified by Mitchell (1954). Other investigators have used growth rate of proteolytic microorganisms as an index of the nutritive value of food proteins (Ford, 1960). Attention is now being given to enzyme hydrolysis in vitro as a technique for predicting nutritive value of food proteins. In this case, many different criteria have been used to measure the degree of hydrolysis: liberated amino acids, growth of microorganisms, soluble nitrogen, formol titration (Akeson and Stahmann, 1964; Buchanan, 1969; Camus et al., 1973; Evancho et al., 1977; Ford, 1962; Melnick et al., 1946; Sheffner, 1964). In this work, we used formol titration following a tryptic proteolysises as a basis for predicting the nutritive value.

The nutritive value in vivo can be measured by several methods. Let us mention, for example, the digestibility coefficient, the biological value, NPU, nitrogen balance, PER, slope ratio, etc. Each of these methods has its limitations but the PER is widely used and appears to be a good measure of the overall nutritive value of food protein. Therefore, PER was used in this work as a measure of the nutritive value.

The objective of this work was to develop a simple method for predicting PER of animal and vegetable protein concentrates and their mixtures.

MATERIALS AND METHODS

Protein Sources

The animal protein concentrates were beef round, pork loin and purified casein utilized as a reference protein. Beef and pork were freeze-dried, defatted with ethyl ether in a Soxlet type apparatus and ground in a Wiley Cutting Mill fitted with a 20 mesh sieve. The vegetable protein concentrates were a rapeseed protein isolate, promine F and wheat gluten. The protein content of these products was determined according to the Kjeldhal method (AOAC, 1970) and 6.25 was used as a conversion factor.

The beef protein concentrate and the three vegetable sources were used to prepare the mixtures. These mixtures were prepared on a protein basis in the proportion beef: vegetable protein, 2:1 and 1:2.

Tryptic proteolysis

Preliminary studies indicated that a tryptic proteolysis compared to a peptic or a peptic-tryptic proteolysis would permit a better differentiation between the nutritive value of the different concentrates and mixtures. The following procedure was then adopted for predicting the nutritive value of the various protein sources:

1) Weigh exactly 175 mg of protein equivalent (N x 6.25) and place in 35 ml Virtis bottle; add 20 ml of 0.1M phosphate buffer, pH 8.0 and one drop of thirmerosal solution 1% W/V to prevent microorganisms growth.

2) Incubate for 1 h at $37^{\circ}C$ in a water bath without shaking.

3) Add 7.5 mg of trypsin suspended in 5 ml phosphate buffer 0.1M, pH 8.0 and incubate for 18 h with moderate shaking.

4) At the end of the incubation period, add 5 ml of sulfosalicylic acid 30% W/V to stop proteolysis and precipitate protein. The final concentration of sulfosalicylic acid is 5% W/V.

5) Place the stoppered bottle in a room at $5^{\circ}C$ and let stand overnight.

6) After this period, transfer into appropriate tubes and centrifuge at 2 600 g for 10 min.

7) Transfer the supernatant into small beaker for formol titration of the soluble amino groups.

Formol Titration

The following procedure was chosen for formol titration:

1) Adjust supernatant to pH 9.0 with NaOH.

2) Add 15 ml of formaldehyde 40% W/V. The final concentration of formaldehyde is then higher than 5% W/V.

3) Titrate back to pH 9.0 with 0.02N NaOH.

The amino groups concentration was calculated by the following equation:

$$\text{Soluble amino groups (\%)} = \frac{\text{NaOH 0.02N} \times 0.28 \text{ mg NH}_2 \times 6.25}{175 \text{ mg}} \times 100$$

Amino acid composition

The amino acid composition of the animal and vegetable concentrates was determined with a Technicon Sequential Multi-sample Amino Acid Analyser (TSM) following 24 hours of acid hydrolysis. The concentration was calculated in g/16gN with an Autolab integrator, system AA. With acid hydrolysis, the data for thryptophan were not available.

RESULTS AND DISCUSSION

The PER values varied between 0.25 for wheat gluten to 3.66 for beef protein. In this work, the PER value for casein was 2.83. The observed PER were not corrected for casein at 2.5 because in our laboratory the PER for casein varied from one lot to another and from one experiment to another.

The observed PER were plotted against the percentage of amino groups as obtained by formol titration and a quadratic equation was calculated. The results were compared after 640 min. and 18 h of proteolysis. After 640 min., the correlation coefficient was 0.907 with a standard error of the estimate of 0.425. After 18 h, the equation gave a correlation coefficient of 0.923 with a standard error of the estimate of 0.388. In both cases, the correlation coefficient was greater than 0.9 but an 18 h incubation period appear slightly better. Furthermore, 18 h or overnight is more convenient for routine tests. Therefore, 18 h proteolysis would be preferable.

The formol titration method was compared with the Oser essential amino acid index as modified by Mitchell (1954). This index is based on the essential amino acid content and is the geometric mean of the egg ratios for all the essential amino acids. To calculate the index of each mixture, the amino acid content was computed on the basis of the amino acid content of the protein components. The Oser amino acid index was plotted against PER and the quadratic equation gives a correlation coefficient of 0.941 with a standard error of the estimate of 0.342.

At this point, it would appear of interest to calculate a multiple regression equation using the results obtained with both techniques and see to what extend this would improve the prediction of PER. The regression equation, then, gave a correlation coefficient of 0.974 with a standard error of the estimate of 0.243.

Thus, the essential amino acid index gives a slightly higher correlation coefficient than formol titration. This index method, however, has the disadvantage of requiring expensive and sophisticated equipment and highly trained technical assistance to determine the amino acids content of the proteins. Furthermore, they do not take into account the amino acid availability in vivo. This is often a great disadvantage of mere chemical analysis. Since, the formol titration technique requires inexpensive equipment and simple manipulation and may take into acount the amino acid availability, this technique would appear better suited for routine quality control.

Under conditions where the results of both techniques were available, the multiple regression equation PER = $-4.244 + 0.053x_1 + 0.10x_2$ would permit a better prediction of the nutritive value of protein concentrates and their mixtures. In the equation, x_1 and x_2 are respectively the results obtained with Oser amino acid index and 18 h of tryptic proteolysis.

CONCLUSIONS

1. The PER of protein concentrates and their mixtures can be predicted by in vitro methods.

2. Formol titration after 18 h of tryptic proteolysis gave a correlation coefficient (r) of 0.923.

3. Oser essential amino acid index gave a correlation coefficient of 0.941 and the multiple regression with formol

titration and Oser amino acid index gave the highest correlation coefficient that is 0.974.

BIBLIOGRAPHY

Akeson, W.R. and Stahmann, M.A., 1964, A pepsin pancreatin digest index of protein quality evaluation. J. Nutr., 83: 257-261.

A.O.A.C., 1970, "Official method of analysis of the Association of Official Agricultural Chemists", Association of Official Agricultural chemists, Washington, D.C.

Block, R.J. and Mitchell, H.H., 1946, The correlation of the amino acid composition of protein with their nutritive value. Nutr. Abstr. Rev., 16(2): 249-278.

Buchanan, R.A., 1969, In vivo and in vitro methods of measuring nutritive value of leaf protein preparations. Brit. J. Nutr., 23: 533-545.

Camus, M.C., Laporte, J.C. and Sautier, C., 1973, Protéolyse trypsique in vitro de divers aliments. Ann. Biol. Anim. Biochim. Biophys., 13(2): 192-202.

Evancho, G.M., Hurt, H.D., Devlin, P.A., Landers, R.E. and Ashton, P.H., 1977, Comparison of Tetrahymena pyriformis W and rat bioassays for the determination of protein quality. J. Food Sci., 42(2): 444-448.

FAO/OMS, 1963, Protein requirements. Report of a joint FAO/OMS expert group. World Health Org. Tech. Rep. Ser., 301.

Ford, J.E., 1960, A microbiological method for assessing the nutritive value of protein. Brit. J. Nutr., 14: 485-497.

Ford, J.E., 1964, A microbiological method for assessing the nutritive value of protein. 3. Further studies on the measurement of available amino acids. Brit. J. Nutr., 18: 449-461

Melnick, D., Oser, B.L. and Weiss, S., 1946, Rate of enzymatic digestion of proteins as a factor in nutrition. Science, 103: 326-329.

Mitchell, H.H., 1954, in: "Die Beurrtung der Futterstoffe und andere Probleme der Tierernahrung", Vol. V., Nehring, K. (ed.), Berlin. Deutsche Akademie der Landwirtschoft wiss Abhandl.

Sheffner, A.L., Eckfeldt, G.A. and Spector, H., 1956, The pepsin-digest-residue (PDR) amino acid index of net protein utilization. J. Nutr., 60: 105-120.

PROTEIN QUALITY OF FOOD MEASURED WITH

TETRAHYMENA THERMOPHILA AND RAT

Herman Baker

Dept. of Medicine - New Jersey Medical School

88 Ross Street, East Orange, NJ 07018 - USA

Many complex methods have been proposed to determine the biological value of protein containing foods[1,2]. Test systems of this sort cannot be applied to man directly; therefore various biological tests with animals, summation chemical scores of amino acid, and microbiological procedures (mainly with bacteria) have been tried[1,2]. A practical, quick, reasonably accurate screening method for assaying protein quality is needed to replace the costly, slow rat assay. Most methods used today are poorly suitable for routine large-scale samplings because of complexities as well as time and cost[3].

Microbial methods have been advocated[2]; none has proven practical since many microbial amino-acid requirements can be spared or met by non-specific metabolites. Others[4] and we[5] have found that strains of the ciliate protozoan Tetrahymena have a mammal-like requirement for amino acids; hence we used Tetrahymena to evaluate protein quality[5]. Because particulates from protein-containing foodstuffs interfered with growth turbidity measurements, as found by other workers using Tetrahymena[6-8], as did color formation[9], the assay proved impractical even though the organism obviously could ingest and digest particulate foods. However Tetrahymena could be used as an assay organism if growth could be measured turbidimetrically as customary in microbiology. Enzymatic solubilization of proteins to peptides and amino acids was enough[5]: the organism adequately met its amino acid needs from solutions rather than particulates.

Growth medium and methods for maintenance of T. thermophila have been described[10]; procedures for preparing the foodstuffs for protein assay are also given[10]. The method used for determining the protein efficiency ratio (PER) is detailed in Official Methods

TABLE 1

COMPARISON OF FOOD SAMPLES – PER VALUES BY THE RAT METHOD WITH TETRAHYMENA THERMOPHILA

No.	CONTENT OF EXPERIMENTAL CRACKERS AND COOKIES	PER-Rat (%)	TETRA-HYMENA (%)
1	Gluten + Lysine (.05%)	37	37
2	Gluten Cookie + Lysine (0.15%)	58	54
3	Gluten Cracker + Lactalbumin (3.2%) + + Wheat Germ (1.3%)	24	27
4	Gluten Cracker + Lactalbumin (2.8%)	6	< 25
5	Gluten Cookie + Casein (2.8%)	46	39
6	Gluten Cookie + Casein (1.8%) + + Soy Protein (2.4%)	51	49
7	Gluten Cracker + Lactalbumin (3.2%) + + Wheat Germ (1.3%)	25	27
8	Gluten Cracker + Lactalbumin (6.5%)	45	43
9	High Gluten Cookie + Lactalbumin (5.9%) + + Soy Protein (5.9%) + Fish Protein Concentrate (4.3%)	64	62
10	Bran Cereal Cookie + Whole Milk	93	92
11	Soy Flour Cookie, Flaked	60	64
12	Durum Wheat Cookie	35	36
13	Yeast Protein Fortified Experimental Cookie	72	68
14	Egg White Fortified Experimental Cookie	106	113
15	Soy Protein Fortified Experimental Cookie	101	101
16	Whole Egg	152	142
17	Lactalbumin	106	96
18	Egg White	152	145
19	Whole Fish Protein Concentrate	120	119
20	Skim Milk Powder	130	135
21	Gelatin (calf skin)	0	< 25
22	Casein (ANRC-standard)	100	100

of Analysis of the American Association of Official Agricultural Chemists[1]. Our results using T. thermophila and the rat are given in Table 1.

Tetrahymena values closely parallel PER units, within 10 PER units for experimental cookies and crackers containing casein, cereals, wheat, powdered milk, eggs, gluten, i.e. both high- and low-protein potencies. Some familiar protein-containing foods, e.g. whole egg, lactalbumin, gelatin, etc. showed the same close agreement to rat PER.

A quick, accurate assay is needed by plant breeders, food processors, and nutritionists to replace the costly, sometimes misleading official 28-day rat assay — the PER (protein efficiency ratio) method[3, 11, 12]. The present PER assay, judged as if it were a microbiological assay, would be deemed outdated since the basal diet specified is inadequate in respect to trace elements. Moreover, with natural foodstuffs, the rats might respond appreciably to the inorganic and organic stimulants brought in with the assay sample since these stimulants would be suboptimal or lacking in the basal ration, the casein standard, or both. Under these circumstances, especially with low-protein potency foodstuffs, the rat assay might overrate low-potency foodstuffs since they would be responding to more than the protein in the samples. Tetrahymena on the other hand responds significantly only to amino-acids or protein.

We use predigestion with bromelain + an SH-activator (mercaptosuccinic acid) to solubilize proteins for Tetrahymena; it does not yield an objectionable blank attributable to enzyme protein itself[10]. We tried papain, pepsin, trypsin, ficin, pronase, diastase, bacterial and fungal proteinases — singly and in combination. Bromelain emerged as far the best, thus providing a simple one-step way to solubilize test samples, thus enabling growth estimations by rapid turbidimetry, obviating wearisome cell counting or adjustments of electronic particle-counters[7, 8].

Comparison with the PER assay (Table 1) revealed few discrepancies between Tetrahymena and rat for common and good-quality foodstuffs and proteins, e.g. egg white. Also, when the protein quality of a common protein such as gelatin — virtually valueless as a complete protein — was assayed, the Tetrahymena value as well as the PER indicated that gelatin was grossly incomplete (Table 1, #19).

Use of Tetrahymena for assaying protein quality usefully classfies proteins as poor, acceptable, or good. If any essential amino acid is present in extremely low concentration, Tetrahymena will divide fewer times indicating low protein potency foodstuffs — a result which eluded other workers[13]. The results given here presumably summate the responses of this ciliate to limiting amino acids in foodstuffs much as higher animals do.

REFERENCES

[1] Biological Evaluation of Protein Quality. Official Methods of analysis of the Association of Official Analytical Chemists, 12th edition, 1975, p. 857.

[2] Sheffner, A.L., in: "Newer Methods of Nutritional Biochemistry", Albanese (ed.), Vol. III, Academic Press, New York, p. 125.

[3] The Midland Conference: New concepts for the rapid determination of protein quality. Nutr. Reports International, 16, 211 (1977).

[4] Kidder, G.W., and Dewey, V.C., The biochemistry of ciliates in pure culture, in: "Biochemistry and Physiology of Protozoa", A. Lwoff, ed., Vol. 1, Academic Press, New York, p. 323 (1951).

[5] Frank, O., Baker, H., Hutner, S.H., Rusoff, I.I., and Morck, R.A., Evaluation of protein quality with phagotrophic protozoan Tetrahymena, in: "Protein Nutritional Quality of Foods and Feeds", M. Friedman, ed., Part I, Marcel Dekker, Inc., New York, p. 203 (1975).

[6] Dunn, M.S., and Rockland, L.B., Biological value of proteins determined with Tetrahymena geleii H., Proc. Soc. Exp. Biol. Med., 64, 377 (1947).

[7] Teunisson, D.J., Elutriation and Coulter counts of Tetrahymena pyriformis grown in peanut and cottonseed meal media. Appl. Microbiol., 21, 878 (1971).

[8] Evancho, G.M., Hurb, H.D., Devlin, P.A., Landers, R.E., and Ashton, D.H., Comparison of Tetrahymena pyriformis W and rat bioassays for the determination of protein quality. J. Food Sci., 42, 444 (1977).

[9] Fernell, W.R., and Rosen, G.D., Microbiological evaluation of protein quality with Tetrahymena pyriformis W. Characteristics of growth of the organisms and determination of relative nutritive values of intact proteins. Brit. J. Nutr., 10, 143 (1956).

[10] Baker, H., Frank, O., Rusoff, I.I., Morck, R.A., and Hutner, S.H., Protein quality of foodstuffs determined with Tetrahymena thermophila and rat. Nutr. Reports International (in press), (1978).

[11] Vaghefi, S.B., Makdani, D.D., and Mickelson, O., Lysine supplementation of wheat proteins. A Review, Amer. J. Clin. Nutr., 27, 1231 (1974).

[12] Waterlow, J.C. and Payne, P.R., The protein gap. Nature (London) 258, 113 (1975).

[13] Kharatyan, S.G., Kerimora, M.G., Ignatev, A.D., and Rozhanskii, M.E., Comparative characterization of microbiological and biological assays of the protein quality of some foods, Nahrung, 17, 243 (1974).

ENZYMATIC HYDROLYSIS OF FOOD PROTEINS

FOR AMINO ACID ANALYSIS

R. Öste and B.M. Nair

Dept. of Applied Nutrition - University of Lund

Box 740 - S-220 07 LUND 7, Sweden

Concept of protein quality has been a subject of extensive research for many years producing numerous publications. Protein quality, though influenced by many factors, depends primarily on the amino acid composition and amino acid availability. This means that the analysis of the amino acid content of a thought protein will always be of importance whenever assessing protein quality. Some amino acids can be determined by specific actions while they are still attached to the protein molecule. However, complete assessment of the amino acid pattern requires hydrolysis of the protein before separation by ion exchange chromatography[1] or gas-liquid chromatography[2].

Current methods of protein hydrolysis for amino acid determination indicate treatment with strong acid at high temperatures for a long time. A separate basic hydrolysis for acid labile amino acids, especially tryptofan, and an additional acid hydrolysis after performic acid oxidation for sulfur amino acids are also necessary.

The yield from these hydrolytic procedures will vary with the duration of the hydrolysis and will also be influenced by the presence of other food constituents. One has to accept a standard deviation of 10% for the yield of some of the amino acids, namely alanine and valine. The aromatic amino acid tyrosine will suffer heavy losses upon hydrolysis with high amounts of carbohydrates. In addition, glutamine and asparagine are converted to their respective acids and will not at all be analyzed.

For the utilization of a dietary protein, the component amino acids have to be made available through hydrolysis by enzymes. Only those amino acids which could be released are of any value. An in vitro method of enzymatic hydrolysis of proteins and determination of the amount of amino acids recovered will avoid the earlier mentioned problems met with the acid or base catalyzed hydrolysis of proteins. The yield would also most likely better reflect the actual amount of biological available amino acids. In recent years there have been a number of reports on successful application of enzymes in hydrolysing peptides and smaller soluble proteins. These results are mainly due to the relatively new technique of immobilizing enzymes, which can be done with many enzymes with retained activity and enhanced stability[3,4,5]. This technique makes it possible to use high enzyme-substrate ratio without any appreciable contamination of the sample. The use of excess amounts of enzymes drives the hydrolytic reaction to the completion. However, it is necessary that the proteins are in solution.

Figure 1 shows the developed procedure for a method of enzymatic hydrolysis. In the first step the food sample is subjected to papain and urea by which the proteins are made completely soluble. In the next step the solubilized proteins are treated with immobilized peptidases for complete hydrolysis.

Urea, by virtue of its ability to form hydrogen bonds and affect hydrophobic bonds, can unfold the molecule causing considerable denaturation. Figure 2 shows the solubility of food proteins in urea of various concentrations. In 8 M urea 76% of the protein was soluble after 24 hours at 37°C.

Papain is a comparatively cheap proteolytic enzyme with broad specificity and it, as shown in Figure 3, retained 45% of its activity after one hour, and prolonged exposure to 8 M urea at 37° for 24 hours caused only slow decrease in activity to 35%. The combination of papain and 8 M urea (Fig. 4) results in complete solubilization of the food protein[6].

Figure 5 shows the solubilization procedure with papain in detail. After incubation papain is inactivated by addition of acid and the insoluble residue is separated from the urea solution by centrifugation.

Figure 6 shows the molecular weight distribution of the protein hydrolysate obtained from gel filtration on Sephadex - G25. As you can see the majority of the proteins are split to small peptides. To get a complete hydrolysis an aliquot of the incubation mixture is subjected to further hydrolysis with the aid of Sepharose-bound aminopeptidase M and Sepharose-bound prolidase. If these enzyme preparations were active in urea a separation step for removing urea could be conveniently avoided. Figure 7 shows the activity of the immobilized peptidases in urea. The activity of the peptidases are markedly reduced with increasing urea concentration. At a urea concentration of 0.3 M aminopeptidase M and prolidase retained about

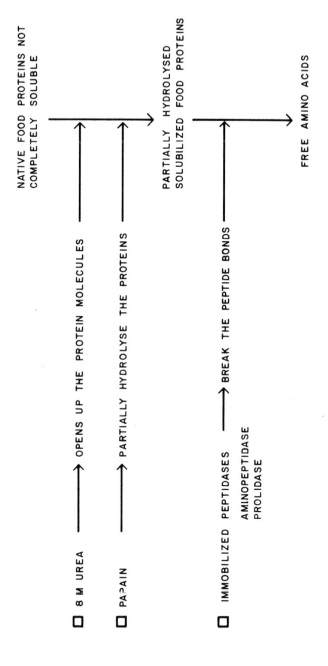

Fig. 1 – Enzymatic hydrolysis of food proteins.

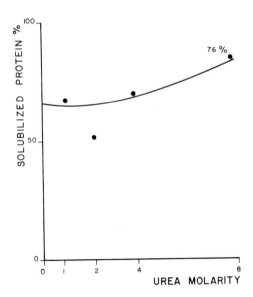

Fig. 2 - Effect of urea concentration on the solubility of food proteins

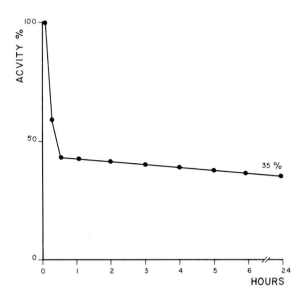

Fig. 3 - Effect of incubation time on the activity of papain in 8M urea

ENZYMATIC HYDROLYSIS OF FOOD PROTEINS

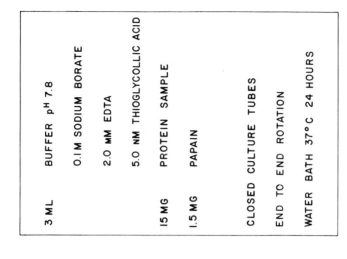

3 ML	BUFFER p^H 7.8
	0.1 M SODIUM BORATE
	2.0 MM EDTA
	5.0 NM THIOGLYCOLLIC ACID
15 MG	PROTEIN SAMPLE
1.5 MG	PAPAIN
	CLOSED CULTURE TUBES
	END TO END ROTATION
	WATER BATH 37°C 24 HOURS

Fig. 5 – Incubation – I (solubilization)

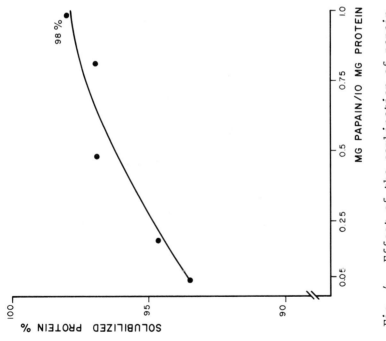

Fig. 4 – Effect of the combination of papain and 8M urea on protein solubility

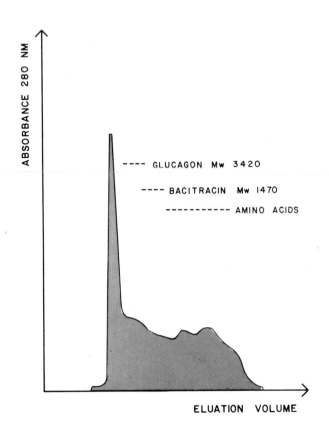

Fig. 6 - Molecular weight distribution of the protein hydrolysate obtained from gel filtration on sephadex - G25

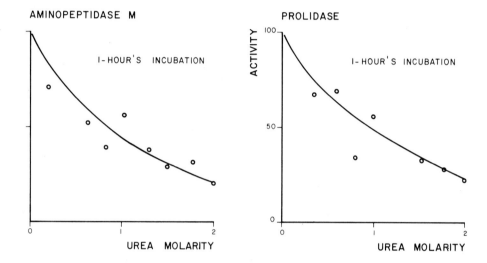

Fig. 7 - Activity of immobilized peptidases in urea.

TABLE 1

ABILITY OF IMMOBILIZED PEPTIDASES TO EFFECT COMPLETE HYDROLYSIS

AMINO ACID	RECOVERY RESIDUE/MOL PROTEIN					
	LYSOZYME			RIBONUCLEASE		
	Enz. Hydro.	Acid Hydro.	Known	Enz. Hydro.	Acid Hydro.	Known
ASP	5.6	20.7	8	5.3	16.1	5
ASN	13.4		13	8.0		10
THR	7.1	6.6	7	9.5	10.0	10
SER	9.2	9.4	10	14.1	14.4	15
GLN	2.8	4.9	3	4.6	13.0	7
GLU	3.0		2	4.0		5
PRO	5.1	2.0	2	3.8	4.3	4
GLY	11.9	11.3	12	2.7	2.8	3
ALA	11.0	10.7	12	11.9	11.4	12
VAL	6.0	4.5	6	9.0	10.2	9
METH	1.9		2	3.7		4
ILE	6.4	5.0	6	3.2	1.8	3
LEU	8.7	7.0	8	2.6	1.9	2
TYR	2.9	3.1	3	5.2	3.4	6
PHE	2.6	3.0	3	2.4	3.1	3
LYS	4.4	4.7	6	9.1		10
HYS	1.0		1	3.0		4
TRP	4.9		6			
ARG	8.8		11	3.1		4
CYSH	–	–	–	–	–	–
CYS	–	–	4	–	–	4

TABLE 2

AMINO ACID RECOVERY FROM PROTEIN OF A 1-DAY MIXED
FAT EXTRACTED FOOD SAMPLE

AMINO ACID	RECOVERY, µmol/mg protein	
	Enzymatic hydrolysis	Acid hydrolysis
ASP	0.19	0.53
ASN	0.27	
THR	0.23	0.31
SER	0.39	0.31
GLN	0.74	1.38
GLU	0.35	
PRO	1.08	0.77
GLY	0.34	0.43
ALA	0.38	0.35
VAL	0.44	0.40
METH	0.11	
ILE	0.36	0.30
LEU	0.68	0.58
TYR	0.34	0.18
PHE	0.32	0.20
LYS	0.22	0.26
HIS	0.09	0.11*
TRP	0.04	0.05
ARG	0.17	0.17

*Determination by spectroflurometric method.

```
2 ML  BUFFER pH 7.8
                0.1 M  SODIUM BORATE
                0.02 M  MANGANEASE CHLORIDE
                0.03 %  TOLUENE
                0.28 M  UREA
                0.18 mM  THIOGLYCOLLIC ACID
                0.07 mM  EDTA
0.350 MG  PROTEIN (PARTIALLY HYDROLYSED)
   2    MG SEPHAROSE - AMINOPEPTIDASE
   2    MG SEPHAROSE - PROLIDASE


CLOSED  CULTURE  TUBES
                              WATER BATH 37°C
END  TO END  ROTATION
                              24 HOURS
```

Fig. 8 - Incubation - II

50% of their activity after one hour of exposure to the urea solution. Further exposure to the urea solution did not diminish this activity. This means that by a simple dilution of the incubation mixture from the first step one obtained a solution in which sufficient amounts of peptidase activity for hydrolysis are retained.

Figure 8 shows the procedure for incubation of solubilized proteins with immobilized aminopeptidase M and prolidase. When the incubation is over the content of the tubes are filtered to remove the enzymes. The filtrate is then subjected to amino acid analysis by ion exchange chromatography

Two pure proteins with known amino acid composition were used to study the ability of immobilized peptidases to effect complete hydrolysis. Ribonuclease and lysozyme were incubated according to the described procedure. Results of this analysis are presented in Table 1.

Most of the amino acids show near to theoretical recovery. Among the essential amino acids lysine from lysozyme and arginine were not recovered completely. Besides, no peak for cysteine of cystine was obtained on analysis of lysozyme and ribonuclease. Glutamine and asparagine are released in native form.

Table 2 shows the amino acid recovery from protein of a 1-day mixed fat extracted food sample. Amino acid values obtained after enzymatic hydrolysis were corrected by substraction of the amino acid blank.

In comparison, the results of the amino acid analysis of 1-day mixed food samples recovery of tyrosine after enzymatic hydrolysis were much higher than the value after acid hydrolysis. Recovery of the other amino acids show good agreement. Even here cysteine recovery was very low, and no peak for cystine was observed. However, the results show that it is possible to hydrolyze protein in food samples with immobilized peptidases once they are in solution.

REFERENCES

[1] Moore, S., and Stein, W.E., J. Biol. Chem., 192: 663, (1951).
[2] Kaiser, F.E., Zumwalt, R.W., and Kuo, K.C., J. Chromatogr., 94:113, (1974)
[3] Bennet, H.D.J., Elliot, D.F., Evans, B.E., Lowry, P.J. and McMartin, C. Biochem. J., 129:695, (1972).
[4] Chin, C.C.Q., and Wold, F. Anal. Biochem. 61: 379, (1974).
[5] Royer, G.P., and Andrews, J.P., J. Biol. Chem., 249: 1807, (1973).
[6] Nair, B.M., Öste, R., Asp, N-G., and Dahlqvist, A.J., J. Agric. Food Chem. 24: 386, (1978).

PROTEIN QUALITY MEASUREMENTS ON A BACTERIAL SINGLE -

CELL PROTEIN (SCP) GROWN ON METHANOL

H.F. Erbersdobler and R. Müller-Landau

Inst. of Animal Physiology, Biochem. and Nutr., Fac. of
Vet. Med., Univ. of Munich, Veterinärstr. 13, D 8000
München 22, Germany/W.

Having done many experiments with several SCP-products in the last years, mainly on yeasts, bacteria and algae (see e.g. 1, 4), we started three years ago a new project with a bacterial biomass, grown on methanol (strain methylomonas clara). The SCP is produced by the companies Hoechst/Uhde in Germany in a continuous fermentation system using a loop fermenter. This paper deals only with our protein quality experiments with rats. Results of other groups in our Institute working with the lipids and the nucleic acids of the SCP or on safety problems are reported elsewhere.

As can be seen from Table 1 the bacterial biomass is — on dry basis — rich in crude protein containing 80% Nx 6,25. The content in nucleic acids is considerably high summarizing up to 12% in the dry matter (Günzel and Gaebler, 1978). The part of amino acid nitrogen plus NH_3-nitrogen (NH_3 derives in the hydrolysates mainly from the amide groups of asparagine and glutamine) amounts to 60-65% in the dry matter. Of the minerals phosphorus is predominant (more than 3%); the crude fat fraction contains mainly the fatty acids C_{16} and $C_{16:1}$ (Rambeck & Beck, 1978). The product itself is a very fine, dusty flour of greyish-green colour and a slightly bitter taste.

The amino acid composition shows methionine plus cystine to be the first limiting amino acids and arginine the second one (for birds only). Generally the amino acid composition of the SCP is very similar to the one of the bacterial SCP produced by the company ICI (Pseudomonas on methanol basis)*. The availability of lysine

---

* Own analyses in several samples.

TABLE 1

CONTENTS IN CRUDE PROTEIN, LIPIDS (ETHER EXTRACT), SOME MINERALS AND THE AMINO ACIDS IN THE BACTERIAL SCP (METHYLOMONAS)

Dry matter	97,3%
In % of the dry matter	
Crude protein (N x 6,25)	81.1
Amino acid nitrogen x 6,25*	64,3
Lipids (ether extract)	8,3
Ash	10,4
Phosphorus	3,0
Calcium	1,0
Potassium	1,7
Iron	0,13

AMINO ACIDS	IN g/100 g DRY MATTER	IN g/16 g NITROGEN	IN g/16 g AMINO-ACID-NITROGEN*
Methionine	1,7	2,1	2,7
Cystine	0,4	0,5	0,6
Lysine	4,6	5,7	7,2
Available Lysine**	3,7	4,6	5,8
Tryptophan	1,6	2,0	2,4
Threonine	3,5	4,3	5,5
Valine	4,7	5,8	7,3
Isoleucine	3,5	4,3	5,4
Leucine	5,6	6,9	8,7
Phenylalanine	3,3	4,0	5,1
Tyrosine	2,5	3,1	3,8
Arginine	3,6	4,4	5,5
Histidine	1,4	1,8	2,2
Aspartic acid	7,0	8,6	10,8
Glutamic acid	6,7	8,3	10,4
Serine	2,7	3,4	4,3
Glycine	4,4	5,4	6,8
Alanine	5,4	6,6	8,3
Proline	3,1	3,8	4,8
Diaminopimelic acid	0,4	0,5	0,6

* Sum of amino acid-Nitrogen plus NH_3-Nitrogen
** Fluor-dinitrobenzene-method

as measured with the fluoro-dinitro-benzene method is high and comparable to other food proteins. The available methionine was measured to be 2,3 g/16 gN using the streptococcus zymogenes test of J.E. Ford (1964)* Since the total methionine was microbiologically and chemically determined to be 2,1 g/16 g N the methionine content of the SCP seems to be fully available for the test microorganism.

In balance studies with rats the so-called amino acid digestibility was calculated. This technique was recently often criticized (see e.g. 3,7) since misleading results can be obtained because of the changes of the nitrogenous compounds in the hind gut by microorganisms. Nevertheless we decided to carry out these experiments in order to get values for the digestibility of the amino acid nitrogen separate from the nucleic acid nitrogen. The Table 2 shows the experimental data from the digestibility trials using casein as a reference protein. As can be seen from the table the digestibility values of the SCP-amino acids are about 10 percent lower than in the casein. The reason for this difference may lie in the protein structure of the SPC or in some cell wall properties.

Nitrogen balance studies were made with growing rats using a protein (N x 6,25) concentration in the diets of 10%. The results are given in Table 3. The true digestibility of the nitrogen was about 10%, the Biological Value about 20% lower than the values found in methionine supplemented casein. After supplementation of the SCP with methionine (1% of the protein) the difference to the casein in the Biological Value decreased to 3%. Besides the N-balance data, Table 4 shows that the N-(food)-intake of the SCP-fed rats was significantly lower than in the casein fed rats. This phenomenon, which increases with increasing SCP-contents in the diet (Beck et al., 1978) remains unexplained until now. Some of the problems seem to be connected with the fine and dusty structure of the SCP material and can be partially eliminated by pelleting the diets (see e.g. Erbersdobler et al., 1975 and J. Gropp, unpublished results).

The food intake problems with the SCP containing diets could be found also in the experiments with growing rats. Table 5 gives results of an experiment with growing rats given ad libitum diets containing equally 10% (Amino acid + NH_3 - Nitrogen) x 6,25.

The results of the balance trials and the experiment with growing rats leads to the assumption that the lower protein quality of the SCP mainly results from a methionine deficiency, the lower protein digestibility and an unknown factor reducing the food intake.

---

* We thank Dr. J.E. Ford, National Institute for Research in Dairying, Shinfield, Reading G.B. for carrying out the measurements.

TABLE 2

APPARENT AND TRUE DIGESTIBILITY OF THE AMINO ACIDS
FOR ADULT MALE RATS IN CASEIN AND THE SCP (n = 6 in
CASEIN, 9 in SCP AND FOR METHIONINE ONLY 4)

	APPARENT DIGESTIBILITY (%)		TRUE DIGESTIBILITY (%)	
	Casein	SCP	Casein	SCP
Dry matter	92±1	89±1	–	–
Crude protein (N x 6,25)	88±2	83±2	93±2	85±2
Methionine	92±2	86±2	94±2	86±2
Cystine	–	79±2	–	82±5
Lysine	96±1	85±1	99±1	89±1
Threonine	90±2	79±2	94±2	84±2
Valine	91±1	82±2	94±1	86±2
Isoleucine	86±2	83±2	90±2	86±2
Leucine	94±1	82±2	97±1	86±2
Phenylalanine	94±0	78±2	97±1	81±2
Tyrosine	92±1	78±2	95±1	82±2
Arginine	92±1	86±1	95±1	90±1
Histidine	96±0	88±1	99±0	92±1
Aspartic acid	87±1	82±2	93±1	86±2
Glutamic acid	91±0	84±2	94±0	88±2
Serine	82±2	74±2	86±2	78±2
Glycine	81±5	80±2	86±5	84±2
Alanine	81±4	85±1	84±4	88±1
Average value	90	82	93	86

The rations contained the protein sources giving 10% (amino acid- + NH_3 - N) x 6,25 to the diet; 10% sugar; 3% Cellulose; 4 - 5% soybean oil; adequate minerals and vitamins and corn starch ad 100%.

TABLE 3

TRUE DIGESTIBILITY (TD), BIOLOGICAL VALUE (BV) AND NET PROTEIN UTILIZATION (NPU) FOR RATS OF METHIONINE SUPPLEMENTED CASEIN, SCP AND THE METHIONINE SUPPLEMENTED SCP (EACH VALUE = 10 ANIMALS)

	N-INTAKE IN mg/RAT AND DAY	TD (%)	BV (%)	NPU (%)
Casein + Methionine*	280	99±0	88±2	87±2
SCP	204	92±2	69±5	64±5
SCP + Methionine*	206	93±3	85±2	80±4

The rations contained the protein sources giving 10% crude protein to the diet. The rest of the diet was comparable to the composition given in Table 2.

*Supplemented with 0,1% DL-methionine in the diet (1% in the protein).

TABLE 4

THE GROWTH OF WEANLING RATS GIVEN DIETS* WITH CASEIN, SCP AND SCP + METHIONINE RESPECTIVELY AD LIBITUM (EACH VALUE AN AVERAGE FROM 10 RATS)

PROTEIN SOURCE	CASEIN + METHIONINE**	SPC	SCP + METHIONINE**
Weight gains in g (3 weeks)	116±9 (100)a**	65±9 (56)b	91±8 (78)c
Food intake in g (3 weeks)	302±21 (100)	232±25 (74)	255±15 (84)
PER*** (3 weeks)	3.9±0.2 (100)a	2.8±0.2 (72)b	3.5±0.2 (90)a

* The same diet composition as in Table 2 given.
** 1% DL-methionine in the protein (amino acid-+NH_3-N)x6,25.
*** The values with different letters are statistically different (Wilcoxon-test 2α = 0.01).
**** Protein Efficiency Ratio (g weight gain/g protein intake) compared at the basis of (aminoacid-N + NH_3-N) x 6,25 (10%).

TABLE 5

THE GROWTH OF WEANLING RATS GIVEN DIETS* WITH CASEIN + METHIONINE
AND THE SCP, SUPPLEMENTED WITH DIFFERENT LEVELS OF METHIONINE
(EACH VALUE IS AN AVERAGE FROM 10 RATS)

PROTEIN SOURCE* METHIONINE-SUPP.**	CASEIN 1,0	SCP -	SCP 0,3	SCP 0,6	SCP 0,9	SCP 1,2
Weight gains in g (3 weeks)	79±5 (100)a*	65±3 (82)b	66±4 (84)b	69±3 (87)bc	71±4 (90)c	72±4 (91)c
g Food per g weight gain (3 weeks)	2,9±0,1 (100)a	3,5±0,2 (121)b	3,5±0,2 (121)b	3,3±0,1 (114)bc	3,3±0,2 (114)bc	3,2±0,2 (110)c
PER*	3,4±0,2 (100)a	2,8±0,1 (82)b	2,9±0,2 (85)b	3,0±0,1 (88)bc	3,1±0,2 (91)bc	3,1±0,2 (91)c

* see foot-notes in Table 4.
** g DL-methionine per 100 g of protein (amino acid-N+NH$_3$-N x 6,25) in the diets.

In experiments with restricted feeding the quality differences caused by the "palatability factor" should be eliminated. For this reason such an experiment testing at the same time graded levels of methionine supplementation was conducted. The results of the experiment show as the digestibility and balance trials before that after the correction of the food intake problem with SCP and after methionine supplementations the difference to the methionine supplemented casein decreased to about 10%.

SUMMARY

The single cell biomass of a new bacterial strain (methylomonas) produced by the companies Hoechst/Uhde was tested in protein quality experiments with rats. Amino acid analyses showed methionine to be the first limiting amino acid. The true digestibility/biological value/net protein utilization were 99/88/87 for casein + methionine; 92/69/64 for the SCP and 93/85/80 for the methionine supplemented SCP, respectively. The digestibility of the amino acids in the SCP for adult rats was 10% lower than in casein. In feeding trials with young rats the SCP yielded relatively low weight gains when fed ad libitum. Methionine supplementation improved the growth rates of the SCP rats, which reached 78% of the casein + methionine group. In experiments with restricted feeding, the differences between the methionine supplemented SCP and casein + methionine were only 10% and can thus be compared with the results achieved by the balance trials. It is concluded from these experiments that the lower quality of the SCP mainly results from the methionine deficiency, the lower protein digestibility and an unknown factor reducing the food intake.

BIBLIOGRAPHY

Beck, H., J. Gropp and H. Erbersdobler, Bakterienprotein in der Geflügelmast; landwirtsch. Forschung 31/II, Sonderheft 66 (1975).
Beck, H., K. König and J. Gropp, Schnellprüfung des Futterwerts von Single Cell Protein mit Hübner- and Wachtelküten; Fette - Seifen - Anstrichmittel, 80, Sonderheft, 601 (1978).
Erbersdobler, H., Amino acid availability, in: "Protein Metabolism and Nutrition", D.J.A. Cole et al., (ed.), Buttenworth, London-Boston 1976, p. 139-158.
Erbersdobler, H., J. Gropp and H. Beck, Utilization of proteins for growth and egg production. Proc. Nutr. Soc., 34, 21 (1975).
Ford, J.E., A microbiological method for assessing the nutritional value of proteins, 3. Further studies on the measurement of available amino acids; Brit. J. Nutr. 18, 449 (1964).
Günzel, R. and S. Gaebler, Nucleinsäuren im Single Cell Protein;

Fette - Seifen - Anstrichmittel, 80. Sonderheft, 606 (1978).
Proc. of the 2nd Int. Symposium on Protein Metabolism and Nutrition. European Association for Animal Production - Publication Nr. 22. Ed. Centre for Agric. Publishing and Documentation, Wageningen 1977, p. 90-91. Discussion on the determination of the digestibility of amino acids in monogastric animals.
Rambeck, W. and H. Beck, Charakterisierung und Verträglichkeit der Lipidfraktion im SCP; Fette - Seifen - Anstrichmittel, 80. Sonderheft, 604 (1978).

METHIONINE-SULPHOXIDE CONTENT OF FISH PROTEIN CONCENTRATE (FPC) AND AVAILABLE METHIONINE DETERMINED MICROBIOLOGICALLY WITH <u>STREPTOCOCCUS ZYMOGENES</u>

L.R. Njaa and E. Lied

Government Vitamin Institute

Bergen, Norway

Pieniazek et al. (1975) suggested that the content of unoxidized methionine in a protein source may be taken as a measure of available methionine. There is general agreement that methionine sulphone is unavailable as a source of methionine (Njaa, 1962; Slump & Schreuder, 1973; Sjöberg & Boström, 1977), but the opinions are divergent as to the biological availability of protein bound methionine sulphoxide. Ellinger & Palmer (1969) and Ellinger (1978) found poor utilization in oxidized casein, whereas Slump & Schreuder (1973), Gjöen & Njaa, 1977, Sjöberg & Boström (1977) and Cuq et al. (1978) found that protein bound methionine-sulphoxide was utilized to practically the same extent as methionine.

Availability of amino acids for animals may be assayed with rats and chicks. The methods are laborious and time-consuming and simpler methods are obviously needed. Ford (1962) showed that results obtained with a microbiological test with Streptococcus zymogenes correlated well with animal tests. The method has been useful for assay of methionine availability.

In order to be able to decide whether the proportion of unoxidized methionine relative to total methionine (total methionine = unoxidized methionine + methionine sulphoxide) is important for methionine utilization, it is necessary to make differential determinations. A method was recently described by Njaa (1977): the protein source is hydrolyzed with barium hydroxide with a small amount of cadmiumacetate present. During this treatment cysteine/cystine is destroyed and will not interfere with the colorimetric determination of methionine with a iodoplatinate reagent (Sease et al., 1948; Awwad & Adelstein, 1966). Total methionine is determined after reduction of methionine sulphoxide with titanium trichloride

TABLE 1

TOTAL METHIONINE DETERMINED CHEMICALLY (T) AND THE PORTION OF THE TOTAL PRESENT AS UNOXIDIZED METHIONINE (U/T), RATIOS BETWEEN AVAILABLE (a) AND TOTAL (t) METHIONINE DETERMINED MICROBIOLOGICALLY (a/t) AND AVAILABLE METHIONINE DETERMINED IN CHICK ASSAYS[1] (M) IN SOME SAMPLES OF FPC

	T g/16 g N	U/T	a/T	M g/16 g N
16/74*	3.4	0.50	0.37	
10/75	3.0	0.80	0.63	2.81
14/75	3.0	0.90	0.86	2.86
16/75	3.0	0.87	0.74	2.84
38/76*	3.5	0.83	0.53	2.85
40/76*	3.3	0.79	0.41	2.90
41/76*	3.4	0.76	0.37	2.35
66/76	3.1	0.90	0.67	2.73
72/76	3.1	0.90	0.64	2.85
44/77	3.1	0.83	0.50	2.96
47/77	3.0	0.87	0.65	2.81
51/77	3.1	0.77	0.68	2.92
52/77	3.0	0.83	0.67	2.60
53/77	3.2	0.91	1.00	2.98
57/77	3.1	0.90	0.59	2.79
4/78	3.0	0.96	0.83	2.89
5/78	3.2	0.91	1.00	3.15
1*	3.4	0.91	0.56	

[1]These assays were done by Opstvedt (1975) and Opstvedt and Mundal (1977).

Samples marked with (*) were used in a growth experiment with rats.

(Gawargious, 1971), unoxidized methionine is determined directly without the reduction step. The colorimetric procedure has been automated using the Technicon system.

We have determined unoxidized (U) and total (T) methionine in 18 samples of FPC, and available (a) and total (t) methionine microbiologically with Ford's (1962) method. In Table 1 the results are presented as ratios, U/T and a/t. Sixteen of the samples were assayed for available methionine with chicks at the Norwegian Herring Oil and Meal Industries Research Institute by Opstvedt (1975) and Opstvedt & Mundheim (1977). Their results are also given in Table 1.

The 16 results from the chick assays plotted against U/T and a/t are given in Fig. 1. The correlation coefficient between available methionine for chicks and a/t was significant ($p < 0.05$) whereas the correlation coefficient with U/T was not significant. On the other hand the correlation between a/t and U/T was significant ($p < 0.01$). There was no significant correlation between methionine available for chicks (M) and total methionine (T).

Five samples of FPC, chosen because they showed clearly different values for a/t, were tested in a simple growth experiment with young rats: the FPC's were given as the only protein sources in diets containing 100 gram protein/kg. The mean weight gains of the 6 rats per group are plotted in Fig. 2 against the a/t and the U/T values. There was a significant correlation between the weight gains and the a/t-values whereas no clear relationship was obvious between weight gain and U/T. FPC No 41/76 obviously gave lower rat growth and available methionine determined with chickens than could be predicted from the a/t, U/T and T values found for that meal. Our results confirm the results of Ford and other groups that the microbiological assay of available methionine may be useful in evaluating protein quality when methionine is the limiting amino acid.

As to the usefulness of the chemical determination of the proportion of unoxidized methionine to total methionine, the results are difficult to interpret. On the one hand the U/T ratio showed no significant correlation with the chick or rat assays. On the other hand U/T values correlated significantly with the a/t values obtained microbiologically.

Taken together the results indicate that biological unavailability of methionine is not primarily caused by oxidation to methionine sulphoxide, but the problem requires further studies.

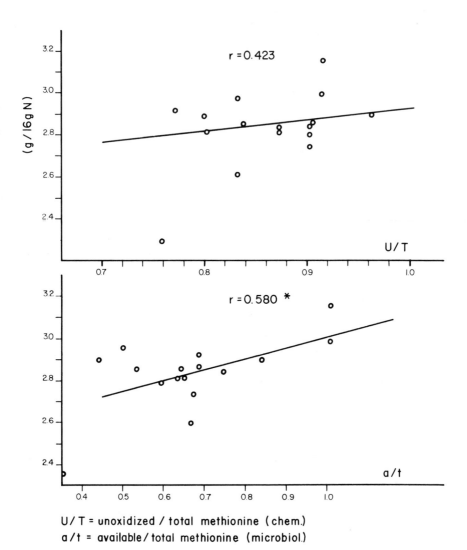

U/T = unoxidized / total methionine (chem.)
a/t = available/total methionine (microbiol.)

Fig. 1 - Methionine available to chicks (g/16gN).

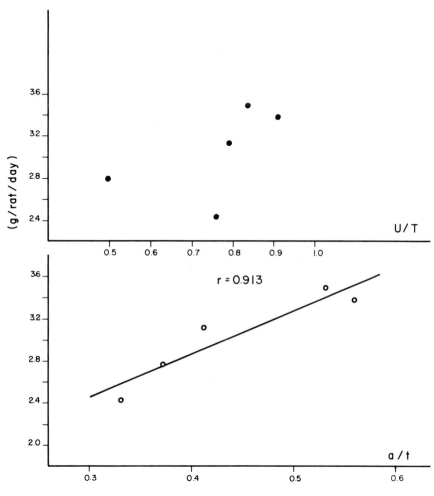

U/T = unoxidized / total methionine (chem.)
a/t = available / total methionine (microbiol.)

Fig. 2 - Weight gain (g/rat/day)

BIBLIOGRAPHY

Awwad, H.K. and Adelstein, S.J. (1966), Anal. Biochem. 16, 433.
Cuq, J.-L, Besancon, P., Chartier, L. and Cheftel, C. (1978), Fd. Chem. 3, 85.
Ellinger, G.M. (1978). Ann. Nutr. Alim. 32, 281.
Ellinger, G.M. and Palmer, R. (1969). Proc. Nutr. Soc. 28, 24A.
Ford, J.E. (1962), Br. J. Nutr. 16, 409.
Gawargious, Y.A. (1971). Microchem. J. 16, 673.
Gjøen, A.U. and Njaa, L.R. (1977). Br. J. Nutr. 37, 93.
Njaa, L.R. (1962). Br. J. Nutr. 16, 571.
Njaa, L.R. (1977). Poster 11th F.E.B.S. Meeting, Copenhagen.
Opstvedt, J. (1975). Acta Agric. Scand. 25, 53.
Opstvedt, J. and Mundheim, H. (1977). 5th Intern. Symp. on Amino Acids. Budapest.
Pieniazek, D., Rokowska, M. and Kunachowicz, H. (1975). Br. J. Nutr. 34, 163.
Sease, J.W., Lee, T., Holzman, G., Swift, E.H. and Niemann, C. (1948). Anal. Chem. 20, 431.
Sjöberg, L.B. and Boström, S. (1977). Br. J. Nutr. 38, 189.
Slump, P. and Schreuder, H.A.W. (1973). J. Sci. Fd. Agric. 24, 657.

DEFINING DIETARY FIBER IN FOOD TECHNOLOGY

D.R. Schaller

Kellogg Company

Battle Creek, MI - USA

I'd like to share with you some thoughts on measuring dietary fiber. I'll try to cover two main topics:

1. What should we measure about the dietary fiber;
2. What methods are available to make these measurements.

There are a lot of different reasons for wanting to analyze dietary fiber and many ways of getting that information.

I've tried to outline some of the facts that may be interesting to know about the dietary fiber in foods:

First, and easiest, knowing the total amount of dietary fiber in different foods is, of course, essential. Why? Well, a couple of the more obvious reasons are:

- To be able to know the amount in different foods or formulated food products to determine how much is present in the diet and to accurately describe foods for labeling. We have been providing dietary fiber information on our cereal products first in Canada and more recently in the United States.

- To be able to measure changes in the amount of dietary fiber during processing, cooking or preparation of foods, e.g. water soluble constituents can be leeched out of foods during preparation or processing or lignin-like non-enzymatic browning adducts can be formed during heating, which would increase the level of dietary fiber.

- To evaluate potential sources of dietary fiber for use in foods. The crude fiber analysis, since it analyzes primarily for cellulose and does not measure other components, can not be used. Table 1 shows some examples — evaluating these samples on anything but a dietary fiber basis provides a distorted view of their usefulness in the diet.

- To know the amount in typical diets of different populations, which would allow better interpretation of epidemiological data.

- And, especially important, to define the foods used in controlled experiments. Too often, researchers in nutrition have not fully described their experiments, or inappropriate measurements such as crude fiber have been used which provide no useful information.

Once the total amount is known, the next question that comes up is how much of each of the constituents of dietary fiber is present.

Table 2 shows some of the constituents of dietary fiber. There are large differences in the composition of dietary fiber when comparing different sources of dietary fiber. There are also differences among varieties of the same type of plant and even differences within the same variety, dependent on maturity of the plant at harvest. So, it becomes especially important to appreciate the complexity involved in trying to understand what constituents of dietary fiber are responsible for various physiological effects in the body. But, even knowing the amount of each of these constituents may not be enough, since each of these differ greatly, chemically and physically, in different plant sources. The chemistry of these classes of constituents is only now being unraveled in detail. But again, especially in defining foods used in controlled experiments, it's important to know what was present in the diet and to measure chemical and physical changes in these constituents that occur during food processing. The exact chemical composition of the constituents of dietary fiber can be related to important chemical or physical properties.

Some of the physiologically important properties that can be measured are shown in Table 3. These are some of the properties which logically relate to clinically observed effects. They are: Water Holding Capacity — which can be related to how much "bulk" is added to the diet; Fermentability with intestinal microorganisms — which can be related to how much of the dietary fiber actually survives the entire digestive process; and Ion exchange properties such as Mineral or Bile acid binding ability — which can be used to estimate the effect of the dietary fiber in hindering intestinal absorption. Also, nutritionally important "impurities" should be considered since dietary fiber often contains such things as important trace minerals which may or may not be bioavailable and which add to the confusion in interpreting results.

TABLE 1

COMPARISON OF CRUDE FIBER TO INSOLUBLE DIETARY FIBER
OF VARIOUS FOOD INGREDIENTS

	Crude Fiber (CF)	Dietary Fiber (DF)	Ratio DF/CF
Deffatted Wheat Germ	10.3%	37.3	3.62
Rice Bran	8.1	21.8	2.69
Corn Bran - a	19.0	88.6	4.66
Corn Bran - b	13.1	62.1	4.74
Beet Pulp	19.8	37.4	1.89
Pea Hulls	36.3	51.8	1.43
Cellulose	72.5	94.0	1.30

TABLE 2

CONSTITUENTS OF DIETARY FIBER	
Cellulose	Gums
Lignin	Waxes
Hemicelluloses	Sterols
Pectins	

TABLE 3

PHYSIOLOGICALLY IMPORTANT PROPERTIES
OF DIETARY FIBER

Water holding (Bulking) Capacity

Fermentability

Mineral Binding Capacity

Bile Acid Binding Capacity

Nutritionally Important "Impurities"

Measurements of these properties should be made on dietary fiber for a number of reasons. Among them are:

1. To compare different sources of dietary fiber. Alternate sources of dietary fiber used in food products can be evaluated and inferior ones rejected. Results from water holding capacity data can even be used to predict some processing characteristics.

2. To develop the rationale for observed differences in physiological activity of different sources of dietary fiber. Coherent theories must be developed to explain the action of dietary fiber in nutrition.

3. To screen sources of dietary fiber for a particular property that has been shown to have nutritional significance and then using that source in foods.

4. And again — to measure the critical properties of dietary fiber used in controlled experiments.

But, it's important to remember that these properties should be measured on the isolated dietary fiber, since digestible components can have an effect. I'd like to show you two examples:

The water holding capacity of dietary fiber may be a way to estimate the ability of different sources of dietary fiber to provide bulk in the diet. The method of measuring water holding capacity has been pioneered by Martin Eastwood and provides good estimates as long as the digestible portion of the food is either negligible or has been removed prior to the measurement. Table 4 shows some of the problems that can occur with foods high in digestible components.

There, the water holding capacity of the acetone extracted food is compared to the water holding capacity of the dietary fiber isolated from the food and what the water holding capacity of the food would be based only on the dietary fiber. The point I want to make is that for foods that contain starch, protein or large quantities of digestible components, the dietary fiber has to be isolated before accurate estimates of the property of water holding capacity can be made. Digestible components also can hold water but are not important contributors physiologically. That point is re-emphasized in Table 5 where one of the ion exchange properties — here the ability to bind a bile acid — was measured. Digestible soy protein and wheat starch both could bind the bile acid, yet would not contribute to bile acid binding physiologically. White wheat bran, which clinically appears to bind little bile acid, also is shown to bind little bile acid if only the dietary fiber is considered.

TABLE 4

COMPARISON OF METHODS OF OBTAINING WATER HOLDING CAPACITY DATA

	Ground, Whole, Acetone Extracted gH$_2$O/g sample	NDF + Enzyme gH$_2$O/g fiber	Calculated water holding capacity for the fiber in 1 g of whole sample
Corn Flakes	3.7	5.7	< 0.1
White Wheat Bran	3.2	8.5	3.0
Processed Wheat Bran Product	2.2	8.5	2.3
Corn Bran	3.1	5.0	3.5
Bagasse	6.0	9.8	7.8
Beet Pulp	5.8	28.1	10.5
αCellulose	3.4	3.4	3.4

TABLE 5

BILE ACID BINDING OF FOOD INGREDIENTS

	% glycocholic acid absorbed (pH 8)
Soy protein isolate	30%
Corn starch	26%
Microcrystalline cellulose	0
Wheat Bran (whole, ground)	32%
Wheat Bran diluted with wheat starch 1:1 (whole, ground)	31%
Wheat Bran (calculated from bile acid binding of the NDF + enzyme residue only)	7%

20 gm glycocholic acid/1 gm sample in 20 ml at pH 8

I've tried to emphasize the things about dietary fiber that should be measured and started to talk about methods. How to measure the amount and constituents of dietary fiber is becoming more and more easy to discuss, since analytical methods are being developed. To help keep them clear in my own mind, I've been dividing them into 3 rough groupings, although there is considerable overlap.

The three groupings I've made are: Biochemical methods, Enzyme methods, and Chemical methods.

Biochemical methods, such as Dr. Southgate's, carefully separate dietary fiber into component parts and analyze each part. Thus, for example, detailed information on the uronic acids in hemicellulose can be available, or knowledge on different hexose or pentose sugars can be gained. This ability can be extremely important to explain why one type of dietary fiber has different physiological effects than others.

The other two types of methods isolate dietary fiber by removing digestible components and measuring what remains. Enzyme methods such as the one used by Hellendoorn (outlined in J. Sci. Food Agric. 26,1461, 1975) using pepsin and pancreatin try to simulate actual digestion. Chemical methods such as the detergent methods of Van Soest use chemical solutions to remove digestible components. Van Soest's methods were designed for forages, but his neutral detergent method has been modified by the American Association of Cereal Chemists for use in foods and now appears as an official method of estimating insoluble dietary fiber.

Each of these 3 types of methods begin by separating dietary fiber into soluble and insoluble constituents. The biochemical methods then begin a detailed breakdown of all of the constituents. The enzyme methods provide the total amount of dietary fiber. The detergent methods can be used to give total dietary fiber, and extended to provide information on the amounts of hemicelluloses, cellulose and lignin, but provide no detailed information on chemical composition of each class. In all of the methods, the soluble constituents have to be isolated either by dialysis or precipitaion with organic solvents.

The enzyme and chemical methods are valuable in another way. We talked before about physiologically important properties that can be measured. These techniques provide the isolated dietary fiber devoid of digestible constituents for further analysis. Thus, the physiological properties can be measured without interference from starch, protein, or other digestible components.

Which method for measuring dietary fiber you choose depends on what you decide you need. For detailed information, the biochemical methods are those of choice. For totals, or isolating dietary fiber for further study, the enzyme or chemical methods.

If you have experience and feel comfortable working with enzymes, you can choose that route. If you now do crude fiber analyses, which we all know are entirely unacceptable and misleading, then the chemical methods such as that of the Cereal Chemists will appeal to you, since it uses the same equipment and requires about the same time and skills. This method is the one we have been using to provide information on dietary fiber in our cereal products. In the food industry routine measurements require accurate, simple methods. Personally, I think the ease of the crude fiber analysis is one reason that analysis is still in use.

In conclusion, the thought I want to leave you with is that there are methods of estimating dietary fiber. Which one is chosen depends upon need. The important thing is to begin using the methods so that inofrmation on food products and processing techniques is built up and the data from nutrition experiments can correctly be interpreted so that we begin to learn fully what exactly dietary fiber contributes to human nutrition.

FRACTIONATION AND CHEMICAL CHARACTERIZATION OF DIETARY FIBRE COMPONENTS

O. Theander

Swedish Univ. of Agric. Sc., Dept of Chemistry

S-750 07 Uppsala, Sweden

It is now generally accepted that the term "dietary fibre" includes the water-insoluble plant cell wall components as well as the water-soluble, unavailable polysaccharides (those not degraded by amylases). The main component in the former group is usually cellulose, the most abundant molecule in nature, built up of so-called β-1,4-linked glucose units (as a comparison can be mentioned that the main linkages in starch are α-1,4-) forming largely crystalline microfibrils which build up fibres in a cement of largerly amorphous matrix polysaccharides, and more or less of lignin, some protein, silica and other associated components. Matrix polysaccharides are usually separated into pectic substances and hemicelluloses (the former extractable with EDTA or oxalate and the latter with dilute alkali).

Pectic substances or pectins are a group of polysaccharides found universally in the primary cell walls and intercellular layers in land plants, and in some very abundant, with galacturonic acid as the main constituent, but which also may contain rhamnose, galactose and arabinose. Xylans, strongly related to cellulose, are often the main hemicellulose components, having a basic chain of xylose units, to which often are linked arabinose and/or glucuronic acid (or its 4-O-methyl ether) units. Another group of hemicelluloses are glucomannans and galactoglucomannans but there can also be other types of polysaccharides present in the insoluble part of the dietary fibre.

In the water-soluble part one finds often a complex mixture of so-called gums and other polysaccharides, for instance arabinogalaktans, xyloglucans, but also pectins and other polysaccharides related to those present in the insoluble part.

Lignin is a very complex, branched polymer, built up of various kinds of phenylpropanoid units and interconnected in various ways. It is in some plant cell-walls an essential component, probably partly linked to polysaccharide components and of importance in connection with dietary fibre properties.

We have since some time work in progress on fractionation and chemical and physical characterization of insoluble and water-soluble dietary fibre components of various foods. Here we will briefly present some results on wheat bran, potato, carrot and white cabbage.

The main fractionation (Figure 1) starts with removal of hydrophilic (water-soluble sugars and other components) and lipophilic (chloroform-soluble) material by extraction with 80% ethanol and chloroform, which results in a fraction A, with all polysaccharides including starch present. By treatment with an aqueous amylase solution and subsequent dialysis, starch and water-soluble dietary fibre components are removed and a fraction B is obtained. This residue, containing the insoluble structural material, is then further divided into EDTA- and alkali-soluble material, where pectins and hemicellulose respectively are enriched and a final residue, with cellulose as a major component. As is indicated by the coming tables there is extensive overlapping between fractions of polysaccharides of related composition and protein and/or other nitrogen-containing components are found in all polymeric fractions.

A graphical presentation of the distribution of the dry matter, according to this fractionation scheme of the four food sources, wheat bran, potato (including skin), carrot and white cabbage is given in Figure 2. The large difference between the four samples is notable already from this schematic comparison and it is still more obvious when we in the coming tables penetrate the chemical composition of different kind of fractions a bit closer. Wheat bran is thus very rich in cellulose- and hemicellulose-containing fractions, potato is rich in starch and hydrophilic extractives and carrot and white cabbage contain mainly hydrophilic extractives but also significant amounts of cellulose and hemicellulose.

The analytical methods used (Figure 3) involve GLC of TMS-ether derivatives of low-molecular carbohydrates and of sugar alcohol acetate derivatives of the neutral sugar constituents (released after acid hydrolysis) in the various polymeric fractions. The quantitative determination of the sum of uronic acids is based uppon decarboxylation and the semiquantitative determination of individual uronic acids uses paper electrophoresis of the hydrolysates. Crude protein (N x 6,25) is based on Kjeldahl analysis. The relative composition of neutral sugar and uronic acid constituents in the total hydrolysates of the water-soluble fractions and the three insoluble fractions respectively, as presented in the next four tables, gives a good idea how different fibre sources may differ in the polysac-

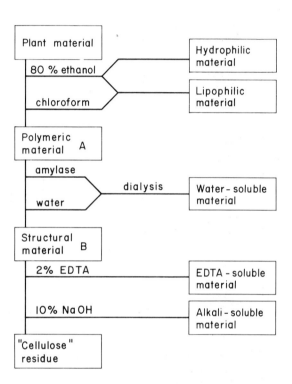

Fig. 1 - Fractionation scheme.

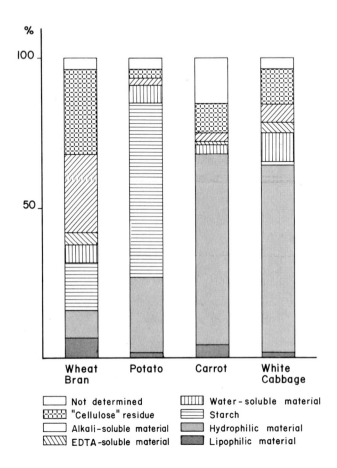

Fig. 2 – Distribution of dry matter contained in different foods.

a. Low-molecular weight carbohydrates

GLC after conversion to volatile TMS-ethers.
PC.

b. Starch

Determination of glucose with glucoseoxidase after hydrolysis with amyloglucosidase.

c. Polysaccharides (neutral sugars)

GLC after hydrolysis with H_2SO_4, neutralization ($BaCO_3$), reduction (KBH_4) and acetylation (Pyr/Ac_2O).

PC after hydrolysis with H_2SO_4 and neutralization with $BaCO_3$.

d. Uronic acids

Conductivity measured after decarboxylation with HI.

Electrophoresis after hydrolysis (H_2SO_4), neutralization ($BaCO_3$) and treatment with cationic resins.

d. Protein

Kjeldahl analysis.

Fig. 3. Analytical Methods.

charide pattern. In order to get a true picture of the differences between different food sources a more complete chromatographic fractionation (for instance in individual neutral and acidic polysaccharides from the water-soluble fractions) is necessary. If we consider further structural analysis of individual polysaccharides of interest, we use methylation analysis and other modern techniques.

Table 1 presents a chemical characterization of the water-soluble material after starch removal, thus involving the unavailable soluble dietary fibre part. Of the four foods discussed this water-soluble fraction is largest for cabbage and smallest for carrot, but both these sources have in common that the protein content is low and that the amount of uronic acid (galacturonic acid) in particular but also of rhamnose, arabinose and galactose is high indicating mainly pectins and associated polysaccharides as components. The other two foods have important amounts of water-soluble proteins and the wheat bran obviously also arabinoxylans and potato glucans and pectins as main components.

We are now going to the first water-insoluble fraction, the material soluble in EDTA-solution (Table 2). As expected at least for potato, carrot and cabbage, pectins are important components, but the protein contents are also significant and in bran, arabinoxylans are the main components and the potato polysaccharides are characterized by a very high galactose content.

In the alkali-soluble material, where the main part of the hemicellulose polysaccharides is to be found (Table 3), we have also protein present in significant amounts. The uronic acid values, obviously representing both acid hemicelluloses and some pectins, are low in these sources. The composition of neutral sugar constituents in the hydrolysates indicate a clear preponderance of arabinoxylans for wheat bran, galactans and glucans for potato and a very complex pattern of hemicellulose polysaccharides for carrot and white cabbage.

In the "cellulose" residue (Table 4), obtained after all extraction steps, the protein- and uronic acid values are low and cellulose is obviously the predominant component in the residues from carrot, cabbage and potato, but in that of the latter and particular in the residue from bran the residual hemicellulose content is important. This illustrates that from some sources the complete hemicellulose extraction (also with the ADF-reagent of van Soest's) is very difficult. In these residues we also find lignin, here estimated as Klason lignin. By this method the amount of native lignin is often somewhat over-estimated, as part of protein, cutin, tannin and other components may be included.

As a comparison the amount of the main low-molecular carbohydrates and starch are given in Table 5. As expected starch is the

TABLE 1

CARBOHYDRATES AND PROTEIN IN THE WATER-SOLUBLE MATERIAL (% OF DRY MATTER OF THE FRACTION) AND RELATIVE COMPOSITION OF INDIVIDUAL SUGARS AND URONIC ACIDS RESPECTIVELY

COMPONENT	WHEAT BRAN	POTATO	CARROT	WHITE CABBAGE
Polysaccharides				
neutral sugars	28	42	26	15
uronic acids	3	16	60	65
Protein	16	29	2	7
Rhamnose	2	6	19	26
Fucose	traces	traces	-	traces
Arabinose	28	5	39	30
Xylose	41	2	2	9
Mannose	traces	1	traces	1
Galactose	10	9	41	34
Glucose	19	77	traces	traces
Glucuronic acid				traces
Galacturonic acid		+	+++	+++

TABLE 2

CARBOHYDRATES AND PROTEIN IN EDTA-SOLUBLE MATERIAL
(% OF THE DRY MATTER OF THE FRACTION) AND RELATIVE
COMPOSITION OF INDIVIDUAL SUGARS AND URONIC ACIDS
RESPECTIVELY

COMPONENT	WHEAT BRAN	POTATO	CARROT	WHITE CABBAGE
Polysaccharides				
neutral sugars	41	59	39	28
uronic acids	3	9	24	ND
Protein	22	15	9	17
Rhamnose	traces	3	9	13
Focose	traces	traces	traces	traces
Arabinose	26	8	36	37
Xylose	55	traces	1	10
Mannose	traces	traces	4	2
Galactose	2	83	45	26
Glucose	17	6	5	12
Glucuronic acid	+	+	+	+
Galacturonic acid	+	++	+++	+++

ND = Not determined.

TABLE 3

CARBOHYDRATES AND PROTEIN IN THE ALKALI-SOLUBLE
MATERIAL (% OF THE DRY MATTER OF THE FRACTION)
AND RELATIVE COMPOSITION OF INDIVIDUAL SUGARS
AND URONIC ACIDS RESPECTIVELY

COMPONENT	WHEAT BRAN	POTATO	CARROT	WHITE CABBAGE
Polysaccharides				
neutral sugars	53	48	33	39
uronic acids	3	2	4	4
Protein	10	19	21	12
Rhamnose	traces	traces	1	1
Fucose	traces	1	1	3
Arabinose	39	9	13	7
Xylose	37	9	20	26
Mannose	traces	3	20	12
Galactose	3	48	11	12
Glucose	2	30	34	40
Glucuronic acid	+	+	+	+
Galacturonic acid	+	+	+	+

TABLE 4

CARBOHYDRATES, PROTEIN AND KLASON LIGNIN IN THE "CELLULOSE" RESIDUE (% OF THE DRY MATTER OF THE FRACTION) AND RELATIVE COMPOSITION OF INDIVIDUAL SUGARS AND URONIC ACIDS RESPECTIVELY

COMPONENT	WHEAT BRAN	POTATO	CARROT	WHITE CABBAGE
Polysaccharides				
neutral sugars	66	65	68	76
uronic acids	3	2	1	3
Protein	2	2	3	2
Klason lignin	11	9	6	5
Rhamnose	traces	1	1	1
Fucose	traces	traces	traces	traces
Arabinose	20	2	3	3
Xylose	45	traces	1	3
Mannose	1	traces	1	1
Galactose	3	20	2	4
Glucose	31	78	93	88
Glucuronic acid	+	+	+	+
Galacturonic acid	+	+	+	+

TABLE 5

NONSTRUCTURAL CARBOHYDRATES (% OF DRY MATTER OF ORIGINAL MATERIAL)

CARBOHYDRATE	WHEAT BRAN	POTATO	CARROT	WHITE CABBAGE
Fructose	0.8	0.7	4.0	6.1
Glucose	0.5	0.8	4.9	11.7
myo-Inositol	traces	traces	0.1	0.1
Sucrose	1.2	0.3	8.1	3.0
Raffinose	0.1	0.1	0.7	–
Stachyose	traces	traces	0.6	–
Starch	16.1	55.7	0.2	0.9

TABLE 6

DRY MATTER AND WATER-HOLDING CAPACITIES

	WHEAT BRAN		POTATO		CARROT		WHITE CABBAGE	
Dry matter % of raw material	91		26		10		6	
	A	B	A	B	A	B	A	B
Amount of dry fraction as % of dry raw material	81	55	68	3	32	13	36	24
Amount of dry fraction as % of fresh raw material	74	50	18	0.8	3	1.4	2	1.4
Water-holding capacity (g water/g dry weight)	8	12	2	62	53	18	66	20

A and B are the fractions obtained, according to Fig. 2 after removal of 80% ethanol and chloroform soluble extractives (A) and further removal of starch and other water-soluble material (B) respectively.

main component in potato (in this sample also the skin is included), but the starch content in a conventional bran is as seen also quite high. Notable is the large content of low-molecular carbohydrates in both carrot and white cabbage and particular the high fructose- and glucose-values of the latter source.

From our studies of physical properties, involving studies on water-holding capacities and ion-exchange capacities for various cations are here given (Table 6) the water-holding capacities of the fractions, obtained after removal of 80% ethanol and chloroform soluble extractives (A) and further removal of starch and other water-soluble material (B) respectively. It is notable that for potato the water-holding capacity is much inriched in the water-insoluble dietary fibre part but for carrot and cabbage it is mainly found in the water-soluble part (obviously pectins and associated polysaccharides). On the other hand the dry matter content of these foods are much lower than that for bran.

EVALUATION OF METHODS SUGGESTED

FOR ASSAY OF DIETARY FIBRE

Nils-Georg Asp, M.D.

Dept. of Nutrition, Chemical Centre, Univ. of Lund

P.O. Box 740, S-220 07 LUND 7, Sweden

More or less detailed quantitative and qualitative analysis of dietary fibre components can be carried out by means of fractionation methods ending up with gas chromatografic analysis of the monosaccharide composition of various fractions. Separate assays of uronic acid in pectin and of lignin has to be added to estimate these components. Such methods are necessary for a proper characterization of dietary fibre in various food stuffs, but remain tasks for the qualified carbohydrate chemistry laboratory. Obviously there is a need also for simpler routine methods for dietary fibre assay.

Since dietary fibre includes a vast number of molecules with widely divergent properties, such as water-solubility, molecular size, monomeric composition, and charge, gravimetric approaches — i.e. weighing the dietary fibre after removal in some way of the non-fibre components — are preferable. Simple colorimetric carbohydrate reactions such as anthron, orcinol, and carbazol, or a combination of these methods is an alternative, but when the monomeric composition of dietary fibre is unknown, which is most often the case, the lack of specificity of these methods makes proper standardization for quantitative analysis impossible. The colorimetric methods seem useful, however, for a rough estimate of the composition of isolated dietary fibre or fractions of this.

Gravimetric assay procedures can be classified according to the method by which non-fibre components are removed.

The crude fibre method is listed for completeness — it is generally agreed that this should be abandoned.

The detergent methods of Van Soest employ sodium — lauryl sulphate for solubilization of protein. This reagent, however, is not efficient in starch solubilization, which is a problem in materials rich in starch. Modifications with amylase treatment before or during extraction have been elaborated to avoid problems due to incomplete starch solubilization.

Enzymatic removal of starch and protein was employed already in the 1920's and 1930's by several different workers. Weinstock and Benham, 1951, used a crude fungal enzyme called Rhozyme S, which is in fact an amylase preparation intended for starch technology.

Physiological enzymes were used by Salo and Kotilainen 1967, and by Hellendoorn and coworkers 1975. Hellendoorn most consistently simulated the physiological digestive events by using first pepsin at acid pH and then pancreatin at neutral pH.

All the gravimetric methods so far published employ filtration through a porous glass filter as the final step to separate dietary fibre from other food components. A fundamental question in method evaluation is therefore whether the nature and amount of the water-soluble dietary fibre components that are lost at the filtration is such that they can be neglected in this way.

To compare and evaluate different methods we analysed 35 samples from a duplicate portion study. The samples were representative of the food consumed on seven different days by 16 men and 19 women, 65 years of age and non-institutionalised (Borgström et al., 1978).

To assay also the water-soluble dietary fibre components an extensive digestion with papain and amylase followed by dialysis was used (Asp et al., 1978).

Surprisingly, roughly half of the dietary fibre thus assayed was water-soluble.

The Weinstock method gave much higher values than the water-insoluble fibre components by the papain-amylase procedure. The Hellendoorn method gave the lowest values, and the neutral detergent fibre method gave intermediate values.

To find out why the different methods gave such divergent results we determined the composition of the material analysed as dietary fibre with the different methods (Asp et al., 1978). Protein was assayed with the Kjeldahl method and/or the Lowry method. Remaining starch was analysed with glucoamylase. The monosaccharide composition of the remaining "true" dietary fibre was estimated by hexose analysis with anthrone corrected for the starch, and by pentose and uronic acid analysis with orcinol. Considerable quantities of protein and starch remained in both the water-soluble and water-insoluble fraction in spite of the extensive digestion with papain and amylase. After correcting for this material, however, almost half of the true fibre material was still water-soluble. The water-soluble fraction contained — as expected — pectin and possibly some pentose based hemicellulose, whereas the water-insoluble fraction contained mainly hexose based material.

The high values obtained with the Weinstock method could be partly explained by remaining protein. There was also much unidentified material, however. Even after correction for remaining protein the Weinstock method obviously does not give a satisfactory estimate of water-insoluble dietary fibre.

The main problem with the detergent method is starch solubilization. Especially in some samples NDF values were obviously much too high due to remaining starch. Treatment with amylase either before the detergent treatment of by inclusion of a thermo-amylase in the detergent reagent will help to remove the starch. In our investigation, however, there was also a considerable contamination with unidentified material — possibly remaining detergent.

An obvious conclusion of our investigation is that assay methods for dietary fibre should assay water-soluble as well as water-insoluble components. The main methodological problem is then to separate the water soluble components from solubilised starch and protein. When the neutral detergent fibre method including amylase is used the proteins are solubilised as highly polymeric detergent complexes. These cannot easily be separated from the water-soluble fibre components. Enzymatic methods, on the other hand, not only solubilise but also hydrolyse both the protein and the starch to low molecular weight fragments, that should be possible to separate by means of dialysis, ultrafiltration or alcohol precipitation.

Our group is trying now to elaborate an enzymatic method for assay of both water-soluble and water-insoluble dietary fibre components. We have chosen to start with the Hellendoorn procedure, since this gave the lowest residues of protein. The amylase step has been improved so that we have now also very low starch residues in most materials. Another advantage with the Hellendoorn method that only physiological enzymes are used. It was recognised already 1935 by William and Olmsted that microbial amylase preparations — as for instance Takadiastase — contained considerable hemicellulase activity. This might be a problem also in more purified microbial enzyme preparations.

Finally I would like to mention two examples of results obtained by the Hellendoorn procedure with recovery of the water-soluble fraction by alcohol precipitation of the filtrate.

In rye bran about one third of the total dietary fibre was water-soluble. The neutral detergent fibre method overestimated the water-insoluble fibre in this material and in fact gave a value almost as high as total fibre. This was due mainly to some unidentified material contaminating the fibre residue. In a carrot preparation also about one third of the dietary fibre was water soluble. In this case neutral detergent fibre underestimated the water-insoluble dietary fibre by 25%.

In conclusion methods using physiological enzymes seem superior to detergent methods due to higher specificity and possibilities to assay also the water-soluble components of dietary fibre.

BIBLIOGRAPHY

Asp, N-G., Carlstedt, I., Dahlqvist, A., Johansson, C-G. and Paulsson, M. (in prep.), Assay of dietary fibre in mixed food samples.

Borgström, B., Norden, A., Akesson, B., Abdulla, J., Jägerstad, M. (eds.), Chemical analysis of food consumed by old people. Scand. J. Gastroent. (in press).

Hellendoorn, E.W., Noordhoff, M-G., and Slagman, J. (1975). Enzymatic determination of indigestible residue (dietary fibre) content of human food. J. Sci. Fd. Agric. 26: 1461.

Salo, M-L. and Kotilainen, K. (1967). Approximate determination of the cell-wall complex of vegetables by enzymatic digestion.

Van Soest, P.J. (1963). Use of detergents in the analysis of fibrous feeds: I. Preparation of fibre and lignin. J. Ass. Off. Agric. Chem. 46: 825-829.

Weinstock, A. & Benham, G.H. (1951). The use of enzyme preparations in crude fibre determination. J. Cereal Chem. 28: 490.

IV. Nutrition Education

FOOD AND NUTRITION EDUCATION

WITHIN NATIONAL EDUCATIONAL SYSTEMS

Francis Aylward, Ph.D., D.Sc.

Queen Elizabeth House, 21 St. Giles

Oxford OX1 3LA, ENGLAND

1. INTRODUCTION

The national system in most countries can be considered in four sectors:

Sector	Age Group
i) The Primary Sector	5/6 to 11/12
ii) The Secondary Sector including vocational schools	11/12 to 17/18
iii) The Tertiary Sector including: - universities and colleges of university rank - colleges of education (teacher (training centers) - agricultural, other vocational and general establishments of sub-university status	17/18 +
iv) 'Continuing' or 'Adult Education' or 'Extension'	17/18 +

The organization varies from country to country; there are, for example, differences in the age of transfer from one sector to another. In many countries a Ministry of Education (or equivalent) has ultimate responsibility for the whole spectrum of public education, but in some countries other Ministries are involved directly or indirectly — for example, the Ministry of Health for Medical Education, the Ministry of Agriculture for Agricultural Education.

In most countries there are special arrangements to promote a degree of autonomy for universities, and provision also for the work of private or non-governmental bodies in the different sectors. In countries with a Federal System there is usually some devolution to provincial or state organizations.

This paper will be concerned mainly with the primary and secondary sectors, together with colleges of education for the training of teachers. Other aspects of national patterns will, no doubt, be dealt with in greater detail in other parts of this Symposium.

2. PRIMARY AND SECONDARY SCHOOLS

Objectives

The reasons for including food and nutrition studies in the curricula of schools may be summarized under five headings:

i) to make a contribution to general education;
ii) to assist in the promotion of better food habits and of better nutrition standards by individuals and within families and communities;
iii) to encourage some proportion of the school population to take up professional or para-professional careers related directly or indirectly to nutrition;
iv) to ensure that those who take up careers in other areas (e.g., in public administration or in industry) are aware of the food and nutrition dimension in public health, agriculture and economic development; and
v) to assist on a long-term basis in promoting a sound public opinion on food and nutrition questions and the establishment in each country of food and nutrition policies.

These objectives are important in all countries, but are of special relevance to those developing countries which have serious nutrition problems arising from shortages of food, poor distribution systems, or other causes.

Stage at Which Teaching Can Begin

UNESCO and other reports on developing countries reveal the following facts:

i) a significant (although declining) proportion of children do not follow a complete primary school course;
ii) the proportion of children entering a secondary school is low;

iii) a significant proportion who enter secondary schools stay for only 1 or 2 years and do not complete the course;
iv) in general, in respect to (i), (ii) and (iii) girls may be less favourably placed than boys;
v) a high proportion of the population lives in rural areas and children in such areas, because of transport and other difficulties, may fare badly compared with children in urban schools.

The above points provide strong arguments in favour of the introduction of some food and nutrition topics in primary, as well as in secondary schools, especially in developing countries, and in rural as well as in urban areas.

Types of Curricula

Few people will advocate the introduction of nutrition as an 'independent' subject in primary or secondary schools. In practice the successful introduction of food and nutrition topics is likely to come from links with other subjects already established in the school curriculum. We can consider these under three main headings:

i) general science (in primary schools) or science subjects such as biology (botany/zoology), chemistry and physics in secondary schools;
ii) 'arts' subjects such as history, geography, the social sciences and the visual arts;
iii) applied subjects such as agriculture and home economics.

These areas will be briefly discussed.

Science Curricula

In many industrialized countries, typical science curricula, although stressing basic concepts of science, often illustrate these concepts from sectors of industry. This has been particularly true in chemistry; elementary textbooks have emphasized topics such as the extraction of minerals and the manufacture of sulphuric acid. Older botany and zoology texts have usually stressed classification and detailed morphological differences.

Over recent years, a change of outlook has taken place in several industrialized countries; many concerned with science education have advocated that curricula for science teaching should be more relevant and related to problems of everyday life, with a greater stress on human, ecological and environmental issues. A wide variety of new curricula have been produced with textbooks that endeavour to combine the basic approach to scientific theory and principles with new applications to illustrate these principles.

Some of the main changes have taken place in biology teaching, with less stress on systematic classifications and a greater emphasis on what is now described as human biology. Under this heading principles of general and human physiology can be introduced and hence an introduction to the physiological basis of nutrition. In chemistry too, new curricula have included some aspects of elementary biochemistry and food chemistry; the role of atmospheric oxigen and humidity in promoting chemical changes can be illustrated by the oxidation of edible oils, as well as by the rusting of iron nails.

Science curricula, which in many country have retained fixed patterns for half a century, are now being subject to scrutiny and change. Examination boards and governmental bodies are now much more ready to accept such changes. This situation provides an opportunity for professionally qualified nutritionists to make suggestions and to put forward ideas to be incorporated in new curricula.

In Table 1, I have indicated some topics which can be included without undue difficulty in syllabuses in different branches of science. It is easier to make proposals for secondary schools which have established science curricula, but efforts are being made in many countries to introduce some science teaching in primary schools and new types of curricula are being worked out.

Arts, Social Studies and General Subjects

I have listed in Table 2 some topics related to food and nutrition that can be linked with geography, history and other subjects.

Geography provides many opportunities especially when it is taught in terms of human geography — the adaptation of man to his local and regional environment. This approach provides an opourtunity of stressing the types of crop production and animal production both locally and in different regions of the world. This historical approach can indicate the effects of explorations and new means of transport in promoting the transfer of crops and animals and their adaptation in regions far from the countries of origin. It would be interesting to know how many school textbooks in Europe note the implications of the voyages of the sixteenth century with the introduction into Europe of crops, hitherto known only to the indigenous peoples of Central and South America. The list is formidable; maize (Indian corn), potato, tomato, several types of beans, manioc (cassava) and the cacao bean. And many of these crops later found their way by direct or indirect routes into Africa and Asia.

Curricula for Social Studies can include various topics of importance in nutrition, for example, food habits, food preferences

TABLE 1

FOOD AND NUTRITION STUDIES
IN RELATION TO SCIENCE SUBJECTS

BRANCH	SOME TOPICS
Botany	- Origin and structure of vegetable foods - Types of plants: importance of leguminous crops - Plant nutrition - Effects of deficiencies of essential growth factors - Micro-organisms: types and activities
Zoology	- Origin of animal foods including fish - Introduction to physiological principles - Insects and pests
Chemistry	- Food components - Composition of foods - Changes in foodstuffs by oxidation and other reactions - Hydrolyses and syntheses - Effects of acidity and alkalinity on chemical reactions - Catalysts, enzyme reactions - Respiration
Physics	- Effects of heating and cooling on food and other systems - Effects of light
Mathematics	- Use of examples from different foods; food purchases and distribution

TABLE 2

FOOD AND NUTRITION IN RELATION
TO SOCIAL AND ECONOMIC STUDIES

BRANCH	SOME TOPICS
History (including history of science, technology and agriculture)	- Food production and utilization at different stages in history - Interaction of food supplies and other events; explorations, opening up of new regions; famines
Geography	- Food production and food patterns in different localities, countries and regions - Effects of local environmental conditions (including climate and soils)
Social Anthropology and Psychology	- Food habits, including taboos and their relation to local cultures - Food preferences and food acceptability - Factors leading to changes in habits
Economics	- Relation of food supplies to general economic conditions. Relations between food, health and income levels for individuals, families and communities - Problems of rural and urban development and migration from country to town - Food marketing - Patterns of internal and international trade
Statistics and Epidemiology	- Application of statistical methods to agricultural production and to food distribution and consumption - Comparisons of health and disease in different localities; correlation with food supplies and food patterns and other factors

and acceptability and food patterns within a family.

In a number of countries, the visual arts have been used — both in school and adult education programs — to illustrate nutritional topics and ideas.

Applied Subjects

Many schools in rural areas provide a basic course in agriculture or in agricultural science and such courses can (but frequently do not) provide an introduction to human nutrition. Home economics is taught as a subject in many types of schools and has often been an important vehicle of nutrition teaching; it suffers from the disadvantage that such teaching is usually confined to girls' schools; there is, moreover, a difficulty that in some of the newer curricula (aimed at ecology) the science basis of a course may be slight.

In some schools, health education is taught as a specific subject, in which case efforts should be made to ensure that the curricula includes an introduction to human nutrition.

Many countries include within their secondary education systems a network of vocational schools of different types (e.g. for agriculture, and for different trades and industries). These vocational centers should be made to ensure that students in agricultural centers (and in centers for the food industries), are aware of the role of agricultural produce as foods for man.

3. THE COLLEGE OF EDUCATION (Teacher Training Colleges)

The Potential Contribution of the College

Proposals for the introduction of food and nutrition teaching in schools cannot be implemented unless there is an adequate supply of teachers. Teachers are normally recruited at two levels, graduates who come from universities or institutions of university rank, and non-graduates who are trained in non-university colleges or education. The pattern varies from country to country, but in most developing countries there will inevitably be, for a long time to come, a high proportion of teachers without a university qualification.

Debates have taken place for many years regarding the place of nutrition within the university system; thare has been far less discussion regarding the non-university colleges of education. Yet these colleges are often key centers for the introduction of changes in curricula. They can make a contribution to food and nutrition teaching in four ways; firstly through courses for students preparing for teaching posts in schools; secondly through the provision of vacation or other special courses and seminars ('in-service training') for serving teachers; thirdly by acting as centers of information and advice; and fourthly by offering facilities for adult education or extension seminars.

Subjects of Study

In Table 3 I have noted topics which may be included in courses in colleges which intend to promote nutrition education. These topics may be summarized under three headings:

    i) food and nutrition topics;
    ii) subjects related to actual teaching practice; and
    iii) related activities.

Under (i) there will clearly be special opportunities when a college is offering instruction in depth on subjects such as agriculture, home economics, health education and physical education. However, full use should be made of curriculum changes in the natural sciences, and in the social sciences and related subjects, on the lines already noted for schools.

Under (ii) are courses of instruction to provide competencies specifically related to the teaching of nutrition, with special reference to educational techniques, and to textbooks and visual aids, class demonstrations and practical work.

Under (iii) three areas should receive special attention, namely school health services, school meals and school gardens. Many publications have advocated the linking of nutrition teaching with these areas of activity; if projects are to be successful, the groundwork must be prepared in the colleges of education. School health services may be limited, and school meal services even more limited, in developing countries, but colleges of education are frequently residential and have opportunities to promote links between different subjects and activities. School garden projects are potentially of importance in demonstrating aspects of food and nutrition teaching. Often such projects have failed because of practical problems such as provision for the care of plots (and small animals) during school vacation periods.

The Need for Local Information and Local Materials

Educationists are usually agreed that, in order to arouse interest and enthusiasm, teaching should be made relevant to the ex-

TABLE 3

FOOD AND NUTRITION STUDIES IN COLLEGES OF EDUCATION

SECTIONS	SOME OBJECTIVES
A. Food and Nutrition Topics	
(i) Natural Sciences	To supply as required a scientific background at some appropriate level in chemical and microbiological aspects of foods and in biochemical and physiological aspects of nutrition.
(ii) Social Sciences including history, geography and social anthropology	To supply as required the social science backgroung to food and nutrition with illustrations from the life of the individual, the family and the local and national communities.
(iii) Agriculture, Family and Health Aspects (including physical education)	To relate (i) and (ii) to local problems and to provide surveys of the effects of under-nutrition and/or malnutrition in local communities.
(iv) Local and National Aspects	To provide a detailed picture of the local, regional and national situation within a given country, including surveys of local foods and food habits.
(v) International Aspects	To consider food and nutrition problems in their regional and world setting and to outline the activities of the United Nations agencies and of international scientific and professional bodies.
B. Educational Competencies	
(i) General	To provide knowledge and understanding of pedagogical principles underlying aspects of educational development.
(ii) Special	To provide competencies specifically related to the teaching of nutrition with particular reference to educational methods and materials.
C. Related Activities	
(i) School meals	To provide practical experiences in support of classroom learning activities.
(ii) School gardens	
(iii) School health services	

periences of the student; this precept applies in both colleges of education and in schools. The teacher in industrialized countries now has available a relatively large number of reference books and texts on different aspects of food and nutrition science, and an increasing number of publications on sociological aspects of nutrition. The teacher in a developing country will have a much more limited choice of 'home-produced' material. There is a natural tendency to make use of imported texts without modification; we may find that the importance of cows' milk is stressed in areas where none is available, and that wheat is emphasized with the neglect of other cereals.

A basic section of local curricula should cover the nutritional values and contributions to the diet of the local food supplies, whether from organized agriculture, small-holdings, the kitchen garden or compound, or wild life. Information is required on the composition of these foods, and on the nutritional changes, for good or ill, in traditional and domestic preparation methods, and in newer methods used outside the farm or home. Many other topics could be listed in terms of changes in food habits in the past, methods of bringing about improvements — for example the better use of complementary foods and mixed diets.

The physiological approach to nutrition can be linked to local and national statistics on sickness or mortality rates in different sectors of the population with a stress on vulnerable groups (pregnant and nursing mothers and pre-school child), the varying nutritional requirements for individuals in such groups, and the necessity of special provision for those in need. Under such headings the importance of breast feeding can be emphasized.

Time does not permit a listing of other topics which, with the minimum of chemical formulae or bio-physical equations, can be incorporated in introductory courses in nutrition. Local examples may well include deficiency diseases arising from shortages of intake of vitamin A and carotenoids, or from the absence of other nutrients. Some of this information may not be readily available in textbooks but full use should be made of national reports and other publications prepared by various UN agencies such as FAO, WHO and UNESCO.

Curricula in industrialized countries may cover rather similar points but there will be different areas of emphasis including the relationship of diet to obesity and other 'diseases of civilization'.

Associated Courses in Microbiology and Hygiene

One final matter should be mentioned in connection with curricula. The nutritionist, whether in industrialized or developing countries, rightly draws attention to the importance of diet in relation to health and emphasizes that this relationship is much more obvious now that many of the once-endemic parasitic and related diseases are under better control. Nevertheless, it would be wise to stress the points (i) that in many areas diseases such as malaria and problems of intestinal infestations are still important in contributing to low vitality, poor health and high death rates and (ii) that many of the improvements in industrialized countries have come about through improved standards of hygiene — in terms of safe water supplies, more efficient effluent disposal and sanitation, and more adequate safety measures in food transport, storage and distribution. In tropical and semi-tropical countries, good standards of hygiene are more difficult to achieve and poor standards undoubtedly contribute to high sickness and mortality rates. It is very desirable, therefore, that topics related to the effects of parasites and micro-organisms, be included in food and nutrition teaching.

4. UNIVERSITIES AND OTHER TERTIARY LEVEL INSTITUTIONS

University courses in nutrition have been considered at several IUNS meetings, but it may be useful to make a few points here to provide a link with earlier sections of this paper. Universities and institutions of university rank often achieve their reputations as centres of research, but one important function of universities, especially through their science and arts facilities is to produce graduates for service as teachers in a wide range of institutions, including schools and colleges of education.

Nutrition education and training in universities has in many countries been considered primarily in terms of medical education. In practice, however, very few medically qualified people teach in schools or colleges of education; this is especially true in developing countries with their shortages of medical and public health personnel. It can be concluded therefore, that parallel to programs for the improvement of nutrition teaching in medical and public health departments, greter efforts should be made to establish (especially in developing countries) food and nutrition courses, which can provide a supply of teachers for other institutions.

Such science graduates, in their courses of training, should be made aware of the social and economic implications of nutrition science. They should recognize also the importance of a qualification in education as well as in nutrition; in most countries there is a great shortage of people with such dual qualifications.

Time does not permit any discussion of the place of nutrition education in tertiary level institutions of sub-university rank. Such institutions include centers for training in subjects such as in dietetics, home economics, institutional management, hotel and cattering, food processing and engineering, and agriculture. Dietetics courses must clearly have a high content of nutrition, but in all the other subjects listed it is desirable that food and nutrition topics be regarded as essential parts of the curricula.

5. CONTINUING OR ADULT EDUCATION AND EXTENSION

This area will be considered in other sections of the symposium. Two points deserve emphasis; firstly that a supply of trained and qualified teachers is essential for the success of adult programs; and secondly that special techniques (such as the effective use of visual and audial aids) may make a significant contribution.

6. SOME GENERAL COMMENTS

In summarizing this paper, I would stress four points: (i) the importance of a supply of teachers trained in nutrition and also in education; (ii) the potential role of the college of education; (iii) the opportunities in schools of different types, and (iv) the special needs of developing countries.

In respect to schools and colleges of education, I have noted the changes in science and other curricula that are now taking place, in part as a result of pressure from groups concerned with ecological and environmental problems. The IUNS through its Commission on Education and Training has in recent years stimulated discussions on many aspects of education. However, it is probably true to say that there is scope for more positive action by national groups affiliated to the IUNS to establish relations within different countries with organizations concerned with curriculum development in schools and colleges.

Over 40 years ago, Novel Laureate Alexis Carrell in his classic Man the Unknown noted how scientists had become bewitched with the beauty of the sciences of inert matter and had neglected the study of man. Attitudes have changed a little in recent years, but much remains to be done. We can legitimately press the claim for the inclusion of food and nutrition studies at different levels in the educational system. Such studies are essential for the study of man in both the industrialized and the developing worlds.

WHAT CAN BE EXPECTED FROM NUTRITIONAL EDUCATION

DIRECTED TOWARD THE POPULATION OF THE THIRD WORLD

        Miriam M. de Chávez

        c/o Adolfo Chávez - Inst. Nacional de la Nutrición
        Av. San Fernando y Viaducto Tlalpan
        Mexico, 22, D.F. - MEXICO

1. WHY EDUCATE IN THE ARE OF NUTRITION?

The majority of the Third World countries are still in a process of territorial integration; in many instances, these countries have come into existence as the result of the combination of what were originally different groups which, in a manner relatively imposed upon them, have assumed a single nationality, a single language and, to a certain extent, have been forced to adopt a new culture. Many of their cities are new; they have been established on the basis of recent tendencies towards industrialization and toward public administration and services. In general, these changes have taken place among groups of humans who had no significant previous scientific knowledges or propensities, especially as regards nutrition, health or welfare. The majority of the people that one sees moving about the cities, towns and roads come from small isolated communities separated from the stream of the scientific and techonolgical revolution that has taken place in the developed countries.

It is only very recently that the majority of the people of the Third World have begun to hear about the bacterial theory of diseases and something about the physiological factors connected with foods. They are barely beginning to learn what calories, proteins, and vitamins are. Until just a few years ago their diet was based on limited local resources and on a series of customs peculiar to the community or to the family.

After World War II, changes in many of the Third World countries have been dramatic. Many have gone through the process of integration, communication has been improved in an explosive manner both in a direct form (roads) as well as in an indirect form (radio and television). Many areas have undergone profound changes in their sociopolitical-economic structures. Among all these changes, the most important that touch upon the matter of nutrition is the fact that, progressively, vast numbers each year are entering into the field of what is known as commercial economy. Many of the local products are now being sold to the foreign market, and, conversely, basic domestic necessities must be satisfied by purchasing them from other countries.

All this has caught the majority of the population of the Third World totally unprepared. They don't know what to sell or buy; they don't know if what they hear on the radio is true or not; they don't know what to eat or what to feed the child for better growth. They really find themselves caught up in a whirlwind of change; they want change; they want to become part of the new world which they sometimes succeed in seeing and about which they sometimes only hear. They want to eat the new products which appear in the stores and which receive so much advertising on radio. They suddenly find that their traditional diet is monotonous and, in the eyes of foreigners, has very little prestige. They are native, cheap, and badly presented foods.

Unfortunately, many of the authorities are leaving this process of change to chance and sometimes in the hands of business people or industrialists whose sole interest is profit and not the health of the populace.

In the Third World countries, food and nutrition are more important for their economy and welfare than they are traditionally considered in the rich countries. It is possible that in a poor country as much as 80% of the socio-economic activity is given over to foods. The men spend the greater part of their time producing, transporting, and storing foods while the women, in great measure, dedicate their efforts to preparing them and distributing them among the members of the family. The activity on the part of everyone is concentrated on food, which is, consequently, the very center of family and community life. Also, the health and the physical development of the children depend strictly upon the availability of food. Yet, as mentioned previously, those in power leave the matter of change to chance and in the hands of those who have vested economic interests at stake.

For these reasons, nutritional education should be looked upon as a very important instrument in the hands of governments, not only to protect those who are weakest and to prevent malnutrition, but also to guide the changes in such form as to benefit collective social health and economic development. From a certain point of view, in many developing countries at the present time the importance of nutritional education far surpasses that of traditional education, such as the written language or mathematics, for example.

In reality, the latter are of no functional importance in many Third World countries because the people in these communities have very little necessity for reading and nobody to whom to write, nor are their circumstances such that they are called upon to use multiplication tables. Furthermore, there are now many new means of communication, radio and television, for example, which render unnecessary the use of the written language.

2. IS IT REALLY DIFFICULT TO CHANGE FOOD HABITS?

Many treatises on nutrition insist that the food problem of the countries in the Third World is economic in origin; therefore, only by increasing income is it possible to better diet. While that may be true from a general point of view, it is by no means applicable when we are dealing specifically with a population in the process of change, especially insofar as concrete families and people are concerned. We are sure that bettering diet is much easier than is usually believed. There is an inherent tendency toward change in all cultures, although this is not always readily apparent because of specific socio-economic circumstances; it is not readily apparent particularly because the first research projects carried on in this respect were always very short and very structuralistic in nature. It is true that most of the studies undertaken by anthropologists in folk-communities showed that there did exist a close interaction between systems of production and handling of food and diet, and that this interaction did suggest that change was almost impossible. But it is also true that the communities selected for these studies were by no means representative of majorities in the Third World; they always worked on a transversal cut of the life of the community, which led to a very static view of the situation. In the first place, in reality, with the changes that have taken place in the world during the last 30 or 40 years, these folk-communities are now very rare. In the second place, if the same anthropologists were to return to the communities that they had originally studied, they would also find them changed; they would find that from among all the value considered, diet would prove to be the most variable.

Many researchers who did their studies during World War II days placed too much emphasis on taboos regarding foods, and this was repeated, like an echo, in many educational-type publications. Furthermore, taboos were used as a pretext to explain the many failures in the social programs of the colonial structures. The truth is that there was a great deal or exaggeration; there are not that many taboos, nor are they so deeply rooted. They are sometimes more of an attitude similar to that of the fox and the grapes. They say that the people do not eat this or that food because they are not supposed to while the real circumstance is that these foods cannot be eaten.

Here we are, talking about whether food habits change or do not change. In the meantime, many large commercial companies have been able to put their products on the market with the greatest of ease. And while we are talking about theory, thirty million bottles of soda pop (carbonated flavored drinks) are being sold in Mexico daily, and millions upon millions of small cellophane-packaged cakes, cookies, potato chips, and similar items enjoy wide popular consumption; at the same time, the popular diet is changing from the traditional Mexican taco to the hamburger, and from the traditional Mexican bean to dough-products (noodles, etc.).

In effect, many experiences of the Department of Nutritional Education of the Mexican National Institute of Nutrition have shown how easy nutritional education is and how much is accomplished in the field of health. The first experimental study of a rural community was undertaken 18 years ago. Observations were not made of how the community was, but rather of how it was changing. In spite of the fact that the project suffered from many deficiencies due to the inexperience of the personnel, it may be said that the results were really impressive. Later, in approximately 5 towns, further evaluational experiences were gathered as to the feasibility of change, and in all cases positive results were obtained. The main findings in this regard are as follows:

1) People accept education very well and learn what is taught to them with relative ease. A high level of credibility is always found, and the people are well disposed to follow indications.

2) The possibilities of change are always very high in the matter of infant feeding, especially in regard to the introduction of varied foods during the first year of life.

3) The economic situation is a real obstacle to change only in the excessively poor communities. Even at the very low income levels, it is relatively easy to get the people to divert their spending toward improvement in the acquisition of foodstuffs.

4) There is a very appreciable dynamic element in the understanding in the communities, and dietetic innovations spread rapidly along the existent family-relationship lines.

5) Mass communication media are very effective when they succeed in awakening the interest of the people. Charts, or posters, and pamphlets distributed by Nutrition and Public Health personnel have proven to be very useful.

6) It is especially easy to bring about improvements among the people through greater possibilities of income or employment.

7) It is possible to counteract ill-directed advertising and

the use of undesirable foods through the educational process.

It has been possible to apply several of these experimentally-learned nutritional education principles, especially in the course of two extensive nutritional education programs being carried on in the rural environment of Mexico. A third program, covering a wide span, is also being initiated at the present time.

The first undertaking consisted of a program based on using the voluntary work force of women in the rural environment. The first stage involved the training, on a national scale, of 2,000 female personnel (promoters) to promote infant feeding; the second stage involved the regional training of groups of 50 volunteer women recruited by each of the original 2,000 promoters. This gave a total of approximately 100,000 women. The objective was to transmit a message by word or mouth, reinforced by the use of a flip-chart folio to serve as a reminder to provide supplementary feeding to children following the third month of life, administering the quantities demanded by the child, always remembering to grind the food and to maintain hygienic conditions.

An evaluation made 12 months after the initiation of this rural program indicated that it had had a great impact in 14 of the 17 areas that were evaluated. A moderate projection of the evaluation allows for the supposition that 60% of the rural population was reached, and that positive changes were achieved in approximately 35% of that population. This program, by itself, can point up various facts: the widespread acceptance of the topic of food, the active participation that it is possible to achieve among the women, and the great possibilities for change that do exist.

Some time later another program was put into effect, this time using instructors in the area of nutrition. This program covered a total of approximately a million people in 20 areas. In this instance, not only infant feeding was covered, but family feeding as well. Once again the results proved to be very impressive. The education carried on did succeed in penetrating. In many of the areas very noteworthy changes were achieved, especially in the tropical zone bordering on the Gulf of Mexico and in some parts of the high plain region.

Through a combined action on the part of the System for the Integral Development of the Family, the National Campaign of Public Subsistence, and the National Institute of Nutrition, the Mexican government is currently carrying a new program under which, in addition to the educational process, another element is being added: the distribution of high nutrition/low cost goods. It is anticipated that through this program even better results will be obtained.

These, as well as otherexperiences, clearly show that if many programs do not succeed in bringing about changes, it is because there is generally insufficient effort in concrete action — that is, on the community level. Too often such programs are based on mere theory, they are devised in some office, they are sometimes extremely sophisticated — but they fail when it comes to actual contact with the populace and in the activities that should really be a part of field work. Many of these projects cancel themselves out because they set up highly complex educational programs which reach very small groups of the population. They want to give a lot to a few, rather than follow what should be a fundamental principle in nutritional education: give a little to many. Sometimes, changing only one single habit, or introducing only one single new food element, may make the difference between life and death for a child.

Educators in the field of hygienic practices are not aiming at having a perfectly equipped bathroom and a totally equipped dining room installed in every dwelling; their hope is that by bringing about small improvements they can better to some degree the living conditions of the people. In similar fashion, we nutritionists should take into consideration the fact that, because of limited resources, these families cannot possibly make radical dietary changes, such as including foods of animal origin every day. But it most certainly is possible to make small changes which will lead to significant results.

As far as nutrition is concerned, a poorly fed people can be compared with a group of persons who are walking along the edge of an abyss: many will fall into the abyss while others will continue walking ahead under these precarious conditions. If these persons can made to walk at least a short distance on the inner side of the path and away from the edge of the abyss, the situation can be totally different. In this regard, the lesson of China has been determinant. China continues to be a country of scarce resources, and its per capita income is two and a half times less than that of Mexico. Nevertheless, they have done away with malnourishment. With relatively few resources, the children are sound and their growth is good. This shows that Western dietary principles are false when they claim that for a diet to be a good one it must contain an abundance of meat, milk, and many other things. These principles of the so-called Normal Diet are frankly exaggerated and are even harmful to health.

It is now possible to accept the idea that relatively small modifications in the basic diet of many of the Third World countries, including Mexico, can result in important changes in health, that a survival diet can, with relative ease, become a prudent diet. Perhaps this can be done through an increase in quantity and quality of no more than 20% over what is right now being consumed in many areas of the world. If this is not being accomplished, it is because the decision makers don't want to do it. We have not been able to convince them of the feasibility and of the social importance of the results that can be expected, not only insofar as health is concerned, but for development itself.

3. WHAT POSSIBILITIES ARE THERE OF FORMULATING EXTENSIVE PROGRAMS

In most of the countries there are very good possibilities of using education as a means of improving nutrition. The only thing that is really needed is a well-defined socio-political decision. Nutritional education, using the existing resources and means, is, in fact, really cheap. In most cases it is not true that resources are lacking, either quantitatively or qualitatively. What happens is that, in many countries, a socio-political decision is not made, perhaps because they are not accustomed to programs of this type. The Third World countries habitually copy their organization and their programatic activities from the developed countries. The latter do not carry out nutritional programs in an organized fashion, undoubtedly because, due to their present level and form of development, they do not need such programs. In the face of these conditions, the underdeveloped imitators do not take into consideration the possibility of carrying out programs of this type. If, in the rich countries, hundreds of millions of dollars are spent to teach their people to read and write, which is very logical in function of the systems of national communication, the poor countries do the same even though their people have nothing to read and nobody to whom to write. If, in the rich countries, millions of dollars are spent on luxury hospitals, the poor countries, too, sometimes build even bigger and more luxurious hospitals. But since the rich countries do not involve themselves in nutritional programs, the poor countries do not consider them necessary either.

However, nothing can be more useful for development than a good nutritional program, not only because it leads to improved health and productive capacity, but also because of the importance that nutrition represents within the economic systems of the poor countries. Within this interaction of factors, the importance of nutrition encompasses consumption as well as production. In many of the poor countries there are large tracts of land that merely lie fallow because there is no assurance as to the commercial possibilities of the products that could be cultivated. There are always terrible differences between the prices that are paid to the producer and the prices that the consumers are charged. Consequently, food production is limited and commercial speculation becomes even worse, resulting in a very reduced consumption.

The poor countries are not accustomed to formulating educational programs, and when they attempt to do so, such programs are very limited or are made subordinate to other programs. Teaching the people better eating habits is not considered important. And so it is with teaching the mothers to feed their children better, in spite of the knowledge that the problem of the interaction between nutrition and infection constitutes the main cause of death and expecially of physical, mental and social underdevelopment in individuals as well as in the masses.

4. WHAT CAN BE EXPECTED OF MASS COMMUNICATION MEDIA?

In theory, mass communication media should be the very backbone in the educational process regarding nutrition and many other things in the poorly developed countries. In practice, though, it has been used on a very limited scale. Furthermore, there are quite widespread opinions to the effect that they are helpful.

It is true that those people at the bottom of the socio-economic ladder, especially in rural areas, either do not listen to the radio or do not watch television, or, if they do, they do not understand the vocabulary or the messages being transmitted. In effect, mass communication media aim their sights at consumers, and these people know, or at least feel, that they are not consumers at all. In any event, the advertising techniques have not moved the people, and the messages go in one ear and out the other ear. Most certainly the reason for this is that the people believe that what the radio or television set is transmitting does not concern them.

But the situation could be different. A study done by our Institute in three small towns in the highlands of the country showed that if mass communication media were correctly used, and if it succeeded in awakening the interest and the attention of such communities, it was possible to achieve results totally similar to those achieved by offering direct education to these groups.

The groups referred to, who did have access to mass communication media, displayed a good deal of credibility; and when the messages transmitted were brought down to their own level, not only in terms of comprehension but also in terms of their socio-economic level, their response was equal to, or better than that achieved at all other socio-economic levels.

The study mentioned above proved that when there was success in capturing the attention of the people, and in combination with radio, posters, and pamphlets, education or advertising is as effective as the use of nutrition promoters and instructors, and at a very much lesser cost. This may very well suggest the more extensive use of programs ot this type.

However, up to the present it must be recognized that mass communication media have been turning out to be more harmful than beneficial to nutrition among the masses in the Third World countries. Mass communication media have been popularizing a series of products derived from sugar, very attractive in presentation and flavor, but having a really low nutritive value.

Since the intensity of advertising a product is proportional to the profit that the product brings in, we see that on television and on radio there is saturation-bombing of the consumer with messages urging him to buy bottled drinks, candy, small cellophane-wrapped cakes, etc., in large quantities. This kind of advertising has been very successful, at least in Mexico, where the sale of such products has increased at least twenty times and is seriously threatening nutrition among the masses. The vendors of these products have found a population very desirous of change, a population that really finds its head spinning what with the elegant presentation of the merchandise and the new flavours. They have found the people who had never consumed these things before and who now have access to them at what seems to be a low price when compared to their spending capacity. Despite the fact that these products are of low nutritional value, they are consumed to excess. They are not consumed merely to satisfy an occasional sweet tooth, but rather as a daily type of food, for breakfast or to be given to children throughout the day.

A recent survey made in 9 poverty-ridden sections in Mexico City showed incredible levels which these people have reached in the consumption of industrialized products of low nutritional value. Advertising is exploiting a series of characteristics very much accepted by the people of this level: the packaging, the flavor, the prestige aspect created by the advertisements, and the facility in the preparation of these items. All of these factors have brought about a situation in which these products have an acceptance which surpasses the traditional consumption of maize and beans and are avidly consumed by adults and children alike.

Nutritional education must bring about a change in these tendencies; it must make the people conscious of the importance that a good diet has for good health; it must work for a better orientation in the process of change. It is a responsibility that the governments must assume to control these tendencies and to better the consciousness and the understanding on the part of the people.

It is for this reason that nutritional education among the poorest sectors must have a double objective: (a) to improve levels of consumption in order to prevent malnutrition, and to improve development in the vulnerable sectors, and (b) among the sectors that are not so poor, that is, low- and middle-income workers, to develop a consciousness toward orientation of consumption that will coincide with the best individual as well as social interests.

Fortunately, some progress is constantly being made to bring the decision-makers to an awareness of the importance of the relationship between nutrition and welfare; there is also more penetration regarding the concepts of an adequate diet among the people themselves in spite of the limitations of their possibilities. Important programs are already being tried out in many countries. Many countries are already moving to free themselves of the chains of hunger and malnutrition which weigh so heavily on mankind.

EVALUATION OF NUTRITION EDUCATION PROGRAMS

Richard Wolf, Ph.D.

Teachers College, Columbia University

New York, U.S.A.

I would like to begin by offering a few observations I have made over the past several years of work with people in the field of nutrition and nutrition education. First, workers in the field of nutrition and nutrition education are enormously capable. Second, people in your field are extremely hard working and persistent in their endeavors. Third, the quality of research and scholarly work is sound, thoughtful and often imaginative. Fourth, there is a very high level of concern, commitment and dedication on the part of workers in the field of nutrition and nutrition education. This is all very good and you have a lot to be proud of. Since you are also keenly aware of what further work needs to be done, there is no need to go into that in the brief time that I have available.

The issues I want to raise with you today involve what happens after you have done the basic and applied research in nutrition, assessed the needs of the groups you seek to serve, and developed programs of nutrition education that are intended to help targeted groups. An educational program is usually instituted at a particular site with the aim of helping a particular group of people in some specified ways. It seems to me that this is the critical point in the process of nutrition education since all the efforts of all the people in the field will not count for much unless the program succeeds in reaching its goals. Program evaluation is the one way we know of determining how well a nutrition education enterprise is succeeding. In the time that remains I want to address myself to this topic.

The role of evaluation in nutrition education can be stated quite simply: it is to improve the quality of nutrition education programs. As such, it is an _integral_ part of nutrition education rather than a peripheral enterprise undertaken to satisfy donor organizations or to produce impressive looking reports. Before proceeding to a consideration of some of the issues surrounding the evaluation of programs in nutrition education, it is necessary to state what evaluation is and what it is not. It is also necessary to indicate, albeit briefly, what is involved in evaluating a nutrition education program.

Evaluation is defined as a <u>deliberate</u> and <u>systematic</u> effort to determine the worth of an intentional effort whether it is an educational enterprise, a social action program or some other helping effort. It involves the collection and analysis of five major classes of information. These classes of information are:

1. <u>Initial status of those who are to be served by the program</u>. Two kinds of information are required here. First, it is necessary to know who are the people to be served by the program. What are their ages, sex, social background, previous educational history, etc.? Such information is necessary to adequately describe the group to be served. Second, it is necessary to know the initial status of such people with regard to what they are supposed to learn. Are they already proficient? If not, what is their initial degree of proficiency? Unless such baseline information is obtained, it will not be possible to adequately estimate the effect of the program.

2. <u>Proficiency of those served by the program after a period of treatment</u>. A program of nutrition education is typically undertaken to bring about some changes in those who are to be served by the program. Such changes could involve what people know, what they feel and/or how they act. Any evaluation program must systematically gather information about the extent to which these desired changes have occurred. Detailed information about such evidence gathering and how it can be accomplished can be found in <u>A Guide to the Evaluation of Nutrition Education Programs</u> published by UNESCO.

3. <u>Execution of the Program</u>. Information about the way in which a nutrition education program has been carried out is crucial. One can normally expect that there will be some discrepancy between a designed or intended program and an actual or executed program. If such discrepancies are small and arise out of the need to make some minor adjustments, then there is no problem. If, however, the program that was carried out differs markedly from what was intended, then it is important to know this and the reasons for it. Failure to gather information about how faithfully the intended program has been carried out can lead to the issuance of evaluative judgments about a program that may be substantially different from what nutrition educators thought was being evaluated. This has happened far too frequently for comfort in the United States, for example. The only antidote is to make sure that the execution of the program is examined so that one knows what is being evaluated.

4. <u>Costs</u>. It is important to know how much an educational program actually costs. A nutrition education program that is tried out on an experimental basis could be so costly that no matter how effective it is in bringing about desired changes in people, large scale implementation would be prohibitive. While educators often find it distasteful to consider the matter of costs in evaluating a program. it is imperative that they do so.

There are a number of ways of reckoning how much a program costs. Procedures range from basic accounting for only those additional costs that an institution actually incurs from having a particular program (ignoring fixed institutional costs) to a cost-benefit analysis. This latter approach to cost determination is exceedingly complex and requires the use of highly skilled personnel. On the other hand, the determination of additional costs needed to mount a particular nutrition education program can be done rather easily by a person with elementary bookeeping skills. The magnitude of the program being evaluated will have considerable influence on the extent of a cost reckoning operation. The important point is that some effort at costing be undertaken.

5. <u>Supplemental information</u>. There are three sub-classes of information involved here. These are: (1) opinions and reactions of various persons concerned with the program being evaluated, (2) supplemental learnings, and (3) side efforts of the program. Opinion and reaction information about a nutrition education program being evaluated can be critical. A program, no matter how successful it is in achieving its goals, is likely to be short lived unless it is accepted by learners, teachers, administrators and the community. Accordingly, information regarding the acceptance of the program must be sought. It is conceivable that a well intentioned program may run up against local customs and taboos. If this is so, it needs to be known. An evaluation enterprise must make provision for securing information regarding local acceptance of the program.

Attempts should also be made to obtain information about what people are learning as a result of a nutrition education program other than what was specifically intended. It may be that people are learning considerably more about nutrition than was intended by the program. If this happy state of affairs exists, then it too should be known. Similarly, if people's attitudes are adversely affected by the program this too should be known, even if attitude change was not a goal of the program. Whatever efforts can be made to secure information about things that people are learning in a program other than what was specifically intended will add to the comprehensiveness of the evaluation.

The matter of side effects of programs is an elusive issue. We know from history of pharmacology that drugs often have effects other than those that were intended. Identifying such effects is a difficult business since one doesn't know what to look for. Educational programs can have side effects too. Follow-up studies are needed to identify what might be happening to persons who have gone through a particular nutrition education program. Sensitivity, tenacity and patience are essential ingredients for such work.

Once the five major classes of information outlined above have been gathered, it is necessary to analyze the resulting data both qualitatively and quantitatively to determine what is being evaluated, the effects it is having and the costs associated with the enterprise. The procedures for analyzing data are generally well known and described in a number of standard textbooks in statistics. Once analyzed, however, it is necessary to synthesize the results into a series of judgments about the worth of the program and to make recommendation with regard to future action. The process by which this is done is not described in textbooks since it requires a set of carefully reasoned judgments, based on the available information, on the part of those responsible for evaluation of the program. There is, at present, some disagreement as to how such judgments can and should be made. One must recognize, however, that the process is judgmental in nature and cannot be handled by standard empirical methods and procedures.

There are a number of unresolved issues in educational evaluation on which debate continues. Some of them are as follows:

First, there is debate on how extensive the evaluation of a nutrition education program should be. Technically trained evaluation specialists are strongly inclined to restrict their evaluation to those cognitive learnings that can be readily assessed at the conclusion of a program. The reason for this is very simple: it makes the job of the evaluation specialist easier. This may not be acceptable to the nutrition educator who is seeking to produce effects that go beyond the program itself. In fact, the nutrition educator may be principally interested in how people behave in their daily life after the program has concluded. If this is the case, the evaluation worker will need to collect information from the field regarding out-of-program behavior. In addition, it will be necessary to find out what are the competing influences that the program has to contend with. For example, there may be strong local customs operating on people in directions that are exactly opposite from those of the program. Or, as in the case of the United States, there may be considerable persuasion in advertisements on TV and in other media that attempt to get people to behave in ways that are in direct opposition to those being fostered by the nutrition education program. In the area of nutrition education, it seems that the situation is extremely complex and will require efforts that go considerably beyond the boundaries of the customary teaching-learning situation. It should also be noted that programs that aspire to have effects that transcend the teaching-learning situation will need to develop learning experiences that also extend beyond the actual learning situation.

A second issue on which there is considerable debate involves the use of comparison or contrast groups in evaluation studies. Stated succinctly, one group of evaluation specialists argues that the only effective way of determining program effects is through the judicious use of comparison groups. The modal for such work is the classic experimental laboratory study in which one group receives a treatment while another does not. There are other evaluation specialists who maintain that not only are such elegancies unrealistic but also unnecessary. Withholding a presumed benefit cannot be easily done. Setting aside the practical difficulties, there is still considerable question about how necessary it is to have comparison groups are in many kinds of evaluation studies.

Finally, there is a fundamental debate surrounding the distinction between evaluation and research. This is reflected throughout the literature on research and evaluation. Some see evaluation as a special form of applied social research while others regard it as a separate and distinct enterprise. While this debate may seem foreign and academic to many, it has enormous implications for the design, conduct, analysis and interpretation of results of evaluation studies. One such instance was noted above, the need for comparison groups in evaluation studies. There are other more far reaching questions such as, how much confidence can one place in a program that was tried and found to work in one location with one group of people? Will it work elsewhere if conditions differ? Must nutritional educational programs always be particularistic or can one build a worldwide knowledge base about the field? In other words, is there something to be gained by having meetings such as this or must we stay within our own borders and try to work out local solutions to the problems of nutrition education as best we can? Perhaps the deliberations of this weel will have furnished at least partial answers to these questions.

I would like to close by again reminding you that nutrition and nutrition education has a vital role to play in the improvement of human welfare. Much good work has already been done but much also remains. As you move out of laboratories, agencies, educational institutions and governmental organizations to directly help those who you serve, it is imperative that a careful monitoring of your efforts be undertaken to determine how well they are succeeding. Evaluation can contribute to this important task.

REACTOR'S COMMENTS ON NUTRITION EDUCATION

William J. Darby, M.D., Ph.D.

The Nutrition Foundation

489 Fifth Avenue, New York, N.Y. 10017 - USA

Our speakers have described in general terms the complex needs and opportunities in nutrition education. I wish to underscore by reiteration and to particularize some of the points they have made:

1. Nutrition education is part of all education;

2. Integration of nutrition into any subject area requires the interest of the non-nutrition specialist and recognition by him of the relevance of nutrition to his subject;

3. In order to be receptive to nutrition education, students and other target groups must have an appreciation of the significance to their career or personal interests of the subject matter taught.

Twenty years ago few medical educators in the U.S. were concerned with nutrition and intense efforts to stimulate such an interest by the American Medical Association and the Nutrition Foundation were not widely successful. Today, two decades later, the majority of medical schools offer formal nutrition instruction within the curriculum and at least one-third of the schools have required courses!

Why this change? This change reflects a variety of altered interests and attitudes on the part of both faculty members and students, including:

- An increasing interest by medical students and young physicians in societal problems and needs, coupled with the general post-World War II awareness of the gravity of malnutrition and hunger in the developing world. A new generation has rediscovered malnutrition as a health problem.

- Realization by physicians and surgeons that many new therapeutic advances in medicine have a potential that obviously is limited by a failure to apply nutritional principles:

- The increased availability of sensitive informative laboratory diagnostic tests for assessing levels of nutriture (state of nutrition).

- Increased recognition of the beneficial influence of a few outstanding academic centers with recognized programs in clinical nutrition.

- The renewed appreciation within all specialities of medicine of the need to treat the patient rather than the disease.

Nutrition is supported especially by those with interest in total parenteral nutrition, renal dialysis, growth and development, cardiovascular disease, diabetes, rehabilitation, gastroenterology, pediatric feeding, community health, geriatrics and international nutrition. It is taught in the context of the case or patient under consideration with the active involvement and support of senior clinicians and a competently staffed nutritional biochemical laboratory that makes available a variety of laboratory measures useful for assessing nutriture. Nutrition consultation and laboratory assessments are as integral a part of patient care as are similar services in endocrinology, hematology, roentgenology and others.

There is a lesson to be learned here for teaching at other levels:

1. The general college student is concerned with "issues" — an interest comparable to the medical student's interest in a patient or case. The "issue" excites his interest and gives a purpose to the learning of basic information. It also provokes exchange between students and instructors.

2. Nutritionists cannot dictate where or what subject is to be taught in a curriculum, the design of which is the responsibility of the other professionals offering the courses. The nutritionist can help the professional understand and interpret nutrition information and, of course, control or design nutrition and food courses per se.

Nutritionists often tend to isolate their subject by overzealous assumption of expertness in and responsibility for other disciplines related to food. The knowledgeable nutrition scientist properly holds that he should control his subject area and does not accept that, for example, a physicist, even one who may be a Nobel Laureate, should determine content or policy concerning nutrition teaching. He finds it difficult sometimes, however, to recognize that:

- dietary habits and food use are behavioral subjects;
- food distribution and marketing are economic areas;
- food presentavion belongs to food science and technology;
- nutrition education belongs at many levels to the educator; and
- national policy is a political matter.

Nutrition concepts and content are importantly relevant to all of these fields. Successful presentation of nutritional aspects of them requires cooperative efforts and joint planning with the receptive experts in these disciplines. But the nutritionist <u>cannot</u> be expert in all of these areas nor expect to set the priorities in them.

Learning about food and nutrition should be a continuing process because both of (1) new knowledge generated and (2) changes in individual and community positions and resources that occur and which affect life style, food values, and value judgments.

Time laters both individuals and countries — hence, the importance of continuous, targeted public education with the nutrition messages sharply focused on the learner's interests, be this art, cooking, marketing, ancient cultures, beverages, gardening, health or wants and needs of changing life style.

The educated development of chemical senses and appreciation of the aesthetics of food can be valuable adjuncts to teaching.

Finally, no single approach to education can be universally effective. <u>The Teacher</u> is the key to success of any curriculum or program.

CHANGING NUTRITION AND HEALTH BEHAVIOR

THROUGH MASS MEDIA: THE PHILIPPINE EXPERIENCE

Delfina B. Aguillon, B.S.H.E., MPH

National Nutrition Council, P.O. Box 1646

Makati Com. Center, Makati, Metro Manila, PHILIPPINES

INTRODUCTION

The Malnutrition Problem

In the Philippines as in the other developing countries, malnutrition is a major problem affecting a large portion of the population, the solution of which, has been the concern of both the government and the private sector. The most important nutritional problems in the country as identified through nutrition surveys conducted by the Food and Nutrition Research Institute of the National Science Development Board, are protein-energy malnutrition (PEM), Vitamin A deficiency, iron and iodine deficiency. The group most seriously affected are the pre-school children, pregnant women and the nursing mothers.

The 1976 Results of Operation Timbang (OPT), a nationwide weight survey, showed 30.6% of pre-school children suffering from third and second degree malnutrition or an estimated 2.9 million out of 9.5 children pre-school population. Among school children 7 t0 14 years of age, nationwide surveys conducted in 1975 showed a 14% prevalence of third and second degree malnutrition or an estimated 1.3 million children. One out of every 4 children are third or second degree malnourished.

Calories and protein deficiency have likewise been found evident in pregnant and nursing mothers and also among heavy manual laborers, particularly the agricultural workers. Iron deficiency anemia affects 62.4% of 0.3 yer olds as against 22.4% of the 4.6 year olds, 48% of pregnant mothers, and 36% of non-pregnant mothers as against 7% of males with endemic goiter. FNRI surveys showed a prevalence of 12% among pregnant women and 8% among lactating mothers. In known goiter endemic regions, particularly, in landlocked areas, prevalence runs as high as 60%.

Causes of malnutrition are traceable to poverty, maldistribution of food within the family and among the various regions; inadequate food availability at both household and farm level; large families and lack of information on correct food habits; poor dietary practices and infectious diseases.

Nutrition Education Strategies

Nutrition education effort in the Philippines dates back in the early years prior to the formulation of a coordinated Philippine Nutrition Program and the establishment of the National Nutrition Council. Varied approaches were used all of which were aimed to reach the families in general and the mothers in particular. One such approach is the interpersonal, face-to-face and/or group approach which makes use of home visits, homemakers' classes, demonstrations, and inclusion of nutrition in classroom subjects. The other approach which is less personal, is primarily done through the mass media - the utilization of the print, broadcast, and audio-visual medium.

The need for coordinating all these efforts later on became necessary in order for all these activities to produce greater impact. An Inter-Departmental Committee on Nutrition Communication (ICNC) composed of 20 members agencies was therefore created. An inventory of nutrition education activities and media facilities done by the Committee showed that the print medium was the most used with about 130 printed materials dealing with various subject matter. Comic and magazine inserts on nutrition messages and the radio were also used but only to a limited extent. School-on-the-air programs, spot announcements and interview session were some of the formats used in radio.

The need for developing a set of priority messages around which all nutrition information campaigns would focus became a major concern. A set of priority messages was therefore developed covering 10 major areas of information in nutrition namely: breastfeeding; supplementary feeding; pre-school feeding; care of the sick child; family diet; pregnant and nursing mothers; family planning; food production; food sanitation; detection of nutritional status.

Since the formulation of an integrated nutrition program, the use of mass media as vehicles for changing nutrition and health behavior, all the more, became apparent.

Mass Media Potentials

Further analysis of the media resources shows that the Philippines enjoye an excellent system of mass media suitable for disseminating nutrition messages.

A survey conducted by an advertising firm in 1975 revealed that there are nine major dailies (5 English, 2 Philipino, 2 Chinese) in the country with a circulation of 870,000. There are eight (8) major magazines and 40 comic magazines with a total circulation of 622,000 and 800,000 respectively.

The radio system reaches throughout the country and remote areas are within broadcast range of at least one station. Research studies have identified radio as a popular medium in the rural areas and it has been utilized to air the messages. As of 1978, there are 215 AM stations, 31 FM stations operated by 105 company/network. Of the 246 stations, 129 are situated in Metro Manila and Luzon (Northern Philippines), 58 in the Visayas (Central Philippines) and 59 in Mindanao (Southern Philippines). As of 1976, out of a total of 7.7 million households sampled in 1978, 54% owned radio sets. Of this, 50% of rural households and 63% of urban households had radio sets.

Statistics show that some 804,495 or 11.2% of the total Philippine households are serviced by 23 TV stations all over the country, 5 of which are based in Manila, 6 in Luzon, 8 in the Visayas, and 4 in Mindanao.

Film is also a potential medium for developmental messages. Filipinos are generally film lovers and this is evidenced by the number of movie houses which have multiplied since 1952. In 1952, for instance, there were 649 cinema houses all over the country, in 1970, the number has risen to 300. Metro Manila alone has more than 108 cinema houses, most of which are first-run with an average rating capacity of about 1,364.

A number of institutions have, likewise, realized the great potentials oq video tape as an instructional medium. To a limited extent, the VTR, is currently, being utilized as a tool in nutrition communications.

THE MASS MEDIA STUDY IN THE PHILIPPINES

Rationale

The key to the success of advertising, one form of commercial open broadcasting, has been the short, pre-recorded message, aired at times a particular audience has been previously determined to be listening to radio, watching television, or reading a magazine.

Its effectiveness is based on knowing an audience in order to "reach" it, and on reaching the audience repeatedly. Hence, the name of the technique, "reach and frequency".

The advertising approach or the "reach and frequency" technique in health and nutrition education programs has been experimented in the Philippines and Nicaragua by Manoff International Inc., a full-service advertising agency, and this was funded by USAID. Pilot tested in Ecuador, the results of this approach showed the promise of its use for nutrition and health education.

Project Design and Development

- Objectives

    The experiment served two main purposes:

    1. To test the effectiveness of carefully designed and frequently broadcast, short, pre-recorded radio messages to bring about changes in food habits, attitudes and knowledge in large groups of rural mother.
    2. To suggest an approach that may be practical and reasonable for other developing countries to adopt and weave into their existing nutrition and health education programs.

The National Nutrition Council has identified specific priority messages. From the interviews conducted with various nutrition authorities, a campaign to encourage increased consumption of calorie and protein dense foods by newly weaned infants, was strongly recommended. Hence, the supplementary leeding messages.

The following specific behavioral objectives were formulated:

- Increase the number of mothers who begin supplemental feeding by at least the infant's sixth month.
- Increase the number of mothers who add chopped fish, green vegetables, and cooking oil to the supplemental food (lugaw or porridge).
- Increase the number of mothers who introduced the enriched suplemental food by at least the sixth month.
- Decrease the number of mothers who believe that small amount of oil, fish and green leafy vegetables are indigestible by six-month old babies.
- Increase the number of mothers who recognize that breast-milk must be supplemented by "Lugaw" enriched with cooking oil, etc. by at least the sixth month.

Selection of Test Area

After consultation with nutrition, marketing, and survey specialists, rural Iloilo Province and the southern portions of Cebu Province were selected as test and control areas, respectively.

The families in both areas were generally similar but two differences exist. First 72% of the test group mothers had some education compared with 52% of the control mothers. Second, the families in the control areas tended to feed their infants more of the ingredients - fish, vegetables, and fried foods.

Drafting and Production of Messages

Two writers conceptualized the creative treatment of the material for the message. The preliminary scripts were modified based on a series of meetings with local education, extension workers, media people, and university nutrition departments of the test area (Iloilo). The following things were thus taken into consideration:

- Introduction of the ingredients, especially cooking oil, might cause diarrhea
- Cooking oil could be scarce in some areas
- Mothers may not have utensils to measure the ingredients of the enriched weaning food
- There are strong beliefs against the introduction of these foods at an early age since most mothers believe that they will cause of stomach upset.

To lend more credibility to the messages, field investigation which included observation and interviews of rural mothers, store owners, and rural workers, was conducted. Important data on the following were obtained: feeding habits and special purchase of foods for infants; availability of measuring utensils in the home; mother's reaction to supplementary feeding practices; radio station preferences and listening schedules; and prices and availability of ingredients.

Based on findings from these interviews and the reactions of nutritionists in Iloilo, the messages were modified in two ways. First, information in any single message was reduced, six different aspects of the recommended way of infant feeding were given in six separate messages. Second, the recommendation about adding oil was changed so that mothers without teaspoons would not add too much. Instead of saying "add a teaspoon", the messages were changed to a "little oil", in drops, and "up to a teaspoon".

A final version of the six messages was prepared, translated, and produced. Local radio personalities were used in the production and the translation was checked with several people who were equally fluent em Ilonggo (the dialect of the rest area) and English, and who also understood the rural culture to which the messages were directed.

- Broadcast Message Testing with the Target Group

The message is the most vital element in any mass media capaign and testing with the target group is essential to assure its effectiveness.

Approximately 65 mothers were interviewed throught the rural areas of Iloilo, using the same field research firm that later conducted the project evaluation interviews.

The message as recorded, were played for the mothers as part of an interview which included questions on the acceptability of characters; under standability and appropriateness of the words, used; effectiveness of motivational elements; consonance to traditional beliefs; and plausibility of concepts.

No drastic changes in the messages were indicated by the tests, but the following refinements were made to focus the messages on the campaign objectives:

- Mothers were urged to breastmilk and to feed with the enriched lugaw. Some of the mothers thought that the lugaw would replace breastmilk.
- "Green" vegetables were specified as opposed to just "vegetables" which included squash or vegetable soup.
- A "drop of oil" was recommended instead of "a little oil".
- Mothers were reassured that their infants would not get diarrhea from the oil if only a few drops up to a teaspoon is given.

- Development of Media Plan and Monitoring System

Media planning refers to the selection of participating stations, the hour and frequency at which the spots will be broadcast, and the duration of the campaign.

Information about the radio listening habits of the target families was obtained through interviews with station owners and

and marketing firms servicing the area, as well as from household surveys undertaken during message testing and evaluation. The audience and frequency with which they were exposed to the messages were increased by requestiong the spots to be played at more popular hours.

The National Media Production Center the government's production arm, monitored the broadcasts of all the cooperating stations during the year. On the average, about one-third of the requested spots which is 3-10 times a day, were broadcast.

- Research Design

From a list of all municipalities in the test area (Iloilo), a random sample of 14 were chosen from which all those close to the provincial capital were excluded. From the remaining municipalities, 70 barrios or barangays were selected randomly. In each barrio, 10 respondents were chosen. Table 1 shows the experimental areas.

Households in the control area were selected in the same manner, except that municipalities with special nutrition and health programs were excluded as were those from Northern Cebu because of differences in socio-economic characteristics.

Qualified respondents were women who were 30 years or younger, or any age and pregnant, or any age and mother of a child 12 months old or younger. The messages were presented through a dialogue between a young mother and her mother with the latter telling the former how to enrich a 6-month old child's rice porridge with oil, fish and vegetables for calories, protein and vitamins. The messages were broadcast for approximately one year.

The impact of the messages was determined through interviews with the target groups before the broadcasts began, at the end of 6 months and at the close of the campaign. Health and other community workers were also interviewed to determine their observations of the projects impact.

The interview instrument was designed by Manoff International Inc. with the assistance of Consumer Pulse, Inc., the field interview firm in the Philippines and approved by the National Nutrition Council. The questionnaire was identical for the first two wave, at the close of broadcasting, additional questions were added so that the respondents' participation in and knowledge of health and nutrition assistance programs could be more fully determined.

Of the 82 questions, 35 were open on unended allowing the mother to respond freely with no prompting from the interviewer.

TABLE 1

SAMPLING PLAN OF THE STUDY

	WAVE I BASELINE STUDY	WAVE II AFTER 6 MONTHS OF BROADCAST	WAVE III AFTER 12 MONTHS OF BROADCAST
Sample Size	1,000 completed interviews	960 completed interviews	951 completed interviews
Test Area	700	674 (175 from the base study)	660 (166 from the base study or Wave II)
Control Area	300	296 (99 from the base study)	291 (76 from the base study or Wave II)
Community Workers (Test Area Only)	99	130	99

Self-reported behavior changes rather than those verified by observation, anthropometric measures, or clinical records were used because of their lower cost. Household interviews reflect the extensive geographical coverage of radio, rather than the more limited catchment areas of health clinics. While there is some danger in using the testimony of the mothers themselves, the independent observation of the community workers and the structure of the questionnaire are checks against exaggerated claims of acceptance.

Self-administered questionnaire were distrubuted to rural community workers at the same time household interviews were conducted.

Consumer Pulse, Inc. a Manila-based marketing research firm, provided field interview, coding and keypunching services for the message testing, and household and community worker evaluation interviews. Manoff Iternational Inc and NMPC staff supervised each of the evaluations.

Findings

- Message Recall

    Table 2 shows mentions of message elements comparing Wave II and and Wave III.

TABLE 2

MENTIONS OF MESSAGE ELEMENTS,
EXPRESSED AS A PERCENTAGE OF TOTAL RESPONDENTS,
COMPARING WAVE II AND WAVE III OF THE TEST GROUP

	WAVE III (N = 660)	WAVE II (N = 674)
Complete Recipe	18%	13%
Oil Mentions	57%	39%
Vegetable Mentions	50%	37%
Fish Mentions	40%	15%
Child Should be Fed Lugaw at 6 Months to 1 year	17%	16%
Correct Quantity of Oil	8%	18%

By the end of the campaign 57% of the mothers could recall without prompting that the messages recommended oil for the baby's food, an increase of 18% over the recall of this element in Wave II (after 6 months). Other parts of the messages show similar increases in recall during the last 6 months of the project. Recall of oil is particularly encouraging since education about the benefits of oil and persuading the mothers to add it were the highest priority objectives of the project. Complex instructions such as how to prepare flaked fish for the infant received less emphasis in the messages and were recalled by few respondents.

- Attitude Change

Positive attitudes toward all of the enrichment ingredients increased during the course of the project. Table 3 shows the shift in positive attitude during the 12 month study.

The change in attitude about adding oil, the highest priority, was particularly dramatic. This is especially important as it tends to demonstrate the effectiveness of short radio messages in effecting change in practice behavior of those being reached. The radio messages were the only source of information about the benefits of oil during the test period, although their recommendations may have been repeated and reinforced by doctors, teachers, neighbors, and others. Vegetables and fish, on the other hand, a far more acceptable part of the food culture of rural families.

TABLE 3

ATTITUDE CHANGE:

EXPRESSED AS A PERCENTAGE OF TOTAL RESPONDENTS

ATTITUDE CHANGE ITEMS	TEST			CONTROL		
	WAVE III	WAVE II	WAVE I	WAVE III	WAVE II	WAVE I
N =	(660)	(674)	(700)	(291)	(296)	(300)
Oil is good	74%	74%	15%	28%	26%	29%
Fish is good	81%	80%	48%	48%	54%	56%
Vegetables are good	79%	82%	49%	72%	81%	76%

* $p = 0.05$

Between Wave II and Wave III, a period of 6 months attitude change did not take place. This leveling off may by attributed to the decline in the frequency with which the messages were broadcast during the last months of the project, or to messages fatigue. It is more likely, however, that in the first 6 months of broadcasting the most easily convinced were converted by the messages, and the remainder are more intractable.

In either case, continual broadcasting and a redesign of the messages is recommended to deal with some of the continuing points of resistance to change.

- Knowledge Change

Increasing knowledge about the right age to begin feeding an infant different foods was not an objective of the campaign, but later analysis showed that this was most significantly associated with behavior change as shown in Table 4.

It was noted that there was a general shift toward knowing the more correct age (5 to 7 months) over the course of the campaign. Fewer mothers in Waves II and III than in the benchmark said that they would never feed these foods to a child or said that they would

TABLE 4

RELATIONSHIP BETWEEN KNOWLEDGE OF AGE TO INTRODUCE NEW FOODS AND INGREDIENTS AND BEHAVIOR CHANGE: TEST AREA ONLY WAVES II AND III COMBINED (N = 278) *

BEHAVIOR CHANGE	KNOWLEDGE = r =	INGREDIENTS + r =	AGE r =
Add fish	0.39	0.15	0.38
Add vegetables	0.27	0.17	0.26
Add oil	0.38	0.17	0.37
Add fish & oil	0.36	0.15	0.29
Add vegetables & oil	0.29	0.15	0.29
Add fish & vegetables	NA	NA	NA
Add all ingredients	0.31	0.11	0.31

* All figure significant at $p = 0.05$

wait until the first birthday. This lack of significant change is not surprising since this information was not stressed in the messages. The analysis of factors showed that knowledge about the correct age to introduce new foods is strongly associated with behavior change. This suggest that the belief that infants cannot eat even specially prepared adult foods must be dealt with more directly and forcefully in new campaign.

Changes in knowledge about vegetables and fish were not as sharp as for learning about oil. This is to be expected since the messages did not concentrate as much on fish and vegetables as they did on oil.

- Behavior Change

During the 12 months that the messages were broadcast in Iloilo, substantial numbers of mothers began feeding their infants logaw enriched with oil, fish, and vegetables. In the control area, during the same period, no change was recorded. Table 5 shows the percentage of mothers enriching lugaw with oil.

TABLE 5

PERCENTAGE OF MOTHERS ENRICHING LUGAW WITH OIL

	TEST			CONTROL		
	WAVE III	WAVE II	WAVE I	WAVE III	WAVE II	WAVE I
N =	(136)	(142)	(157)	(58)	(61)	(58)
Add Oil	24%	23%	0.00	12%	13%	12%

* $p = 0.05$

The number of mothers who reported adding oil to the lugaw sharply increased after 6 months of broadcast. The amount and frequency with which oil was added vary widely, from a few drops once or twice per week to a teaspoon daily.

Verification of this particular behavior change was also done. Interviewers asked all mothers to show them the containers in which they stored cooking oil. It was found that 41% of all sample families had cooking oil at the time of the interview, compared with 46% of those claiming to add oil.

Before the broadcasts began 46% of all respondents reported that they purchased oil weekly. This increased t0 67% by the end of the project. Those who added oil to the lugaw follow a similar pattern, with 71% buying oil weekly by the end of the project.

Enrichment of lugaw with both fish and vegetables likewise, increased significantly during the first 6 months, but leveled off during the remainder of the experiment. This is probably due to the fact that some of the mothers were accustomed to giving these foods even before the project began.

By the end of 12 months all three ingredients were being added, on the average, significantly greater quantities and served on the average with significantly greater frequency than before the test began. Some of the mothers may have been adding enrichment foods in such small amounts and in such low frequency so as to be nutritionally insignificant. However, their intermittent and trial use may be precursors of more generous enrichment in the future. These findings support the recommendations to continue broadcasts.

Also, a secondary objective of the experiment has been to insure that the mothers who began enriching their infants' lugaw would continue breadfeeding instead of perceiving this new food as a substitute.

Analysis of diet recall data showed that there was no significant difference in the habits of breastfeeding between the mothers who reported that they enriched the rice porridge and those who did not. On the average, 85% of all mothers in the target groups nurse their 6-12 month old infants.

- Radio a Source of Information

Only 3% - 4% of the respondents mentioned radio as the source of information about child care. However, when mothers were asked where they had heard advice on enriching lugaw, they overwhelmingly cited radio as the source on information. Responses to this question were correlated to the behavior change variables as shown in Table 6.

Enrichment with oil, the most novel recommendation of the messages, undersdably is most strongly associated with the radio messages. Promotion of fish and vegetable consumption for infants had preceded the radio campaign by several years.*

Summary

A mass media study was conducted in Iloilo province of the Philippines involving the airing of short, pre-recorded messages on enrichment of rice porridge (lugaw) with oil, fish and vegetables.

The messages were directed to mothers of children under 12 months of age and broadcast for approximately one year. Evaluation data were gathered through questionnaires administered to mothers in their homes in baseline studies, 6 months after broadcast began, and 12 months thereafter. There were a total of 700 respondents in the test group and 300 in the control group.

---

* A positive relationship was found by Dr. Zeitlin and Ms. Formacion when they correlated enrichment with oil and mothers who cited sources in addition to radio. This suggest the importance of reinforcing, where feasible, the radio message with other sources of information. However, since many of the sources cited by mothers were neighbors and friends, it does not necessarily follow that radio programs must be supplemented by other education programs.

TABLE 6

RELATIONSHIP BETWEEN RADIO AS
THE SOURCE OF ENRICHMENT RECOMMENDATIONS
AND BEHAVIOR CHANGE: TEST AREA ONLY
WAVES II AND III COMBINED (N = 278)

BEHAVIOR CHANGE	P =	COEFFICIENT
Add fish	0.02	0.2265
Add vegetable	0.02	0.1278
Add oil	0.02	0.2902

Recommendations

Based on the data obtained, several recommendations were made to increase the effectiveness of future information campaigns in the Philippines.

The messages could be changed. By recommending more strongly than was done in the original messages that the enrichment ingredients be added to all the foods given to infants, the likelihood that they would receive them would be increased.

The frequency with which the oil should be added can be more strongly emphasized. The experiment's messages mentioned "every day" four times in total, but apparently this was not enough to overcome the resistance to giving it more often. This recommendation does not ignore the severe constraint imposed by the high cost and relative unavailability of cooking oil.

Mothers should be taught that the oil for a child is the same oil they normally buy for their families, rathers than a special oil, as reported by some of the interviewers. The messages must also persuade the mothers to set a side some portion of the family's oil supply especially for the infant since it appears that many of those who want to give it regularly cannot do so since the supply is exhausted in the first few days after purchase.

From the experiment, it is clear that radio messages and other instruction about preparation of foods ought to be as exact as possible. Little doubt should be left in the mother's mind about the best way to prepare the new food or medicine.

USE OF THE STUDY

The National Nutrition Council, the government agency coordinating the implementation of the PNP, realizes the importance of the research findings and its implications in the nutrition information campaigns. It also recognizes the promising potentials of using radio as a medium for bringing about knowledge, attitude and behavior changes, especially if combined with more conventional outreach and education programs.

The first step taken by National Nutrition Council to disseminate information on the study is the conduct of two conferences.

In April 1977, an International Conference on Nutrition Education was held. Co-sponsored by the two agencies involved in the study - NMPC and USAID, the conference was attended by participants from Bangladesh, Indonesia, Korea, Malaysia, Pakistan, Thailand, and the Philippines. Various nutrition education strategies adopted by the named countries were discussed. The mass media study in the Philippines was discussed by the project proponent. The participants came up with a recommendation based on the mass media study. A formal and practical course in Mass Communication and Advertising Technique for Nutrition Education (including media strategy and message design) was recommended. Course participants will include media representatives and nutrition educators.

Following the conduct of the International Conference was the National Conference on the Use of Radio in Disseminating Nutrition Messages. This was held in Iloilo, the study test area. The conference was organized primarily to disseminate the results of the mass media research project on the use of radio, to the agencies involved in the Nutrition Information and Education campaign. Participants to the conference recommended that the advertising technique be utilized in the nutrition information campaign. Likewise, they suggested that the broadcast messages used in the study be aired nationwide.

Taking heed of such recommendations, the National Nutrition Council and the Kapisanan ng mga Brodkaster sa Pilipinas (KBP), the largest association of broadcaster in the Philippines, signed a Memorandum of Agreement to jointly undertake a nationwide broadcast campaign. The messages used in the study were recorded and distributed by KBP to all its member-stations.

The Philippines has learned from such an experience. The present nutrition information campaign takes into consideration the reach-and-frequency technique and the use of the mass media, particularly radio as a medium for knowledge, attitude and behavior change. However, mass media alone cannot do all the wonders of changing beliefs and practices. While various media mix can lead to significant gains in knowledge, increase in positive attitudes and changes in behavior, the need to synchronize this approach with the more conventional education methods, is still one fact to be reckoned with.

BIBLIOGRAPHY

Cundal, Allan W., "Why Radio", an advertising commentary printed in the August 20, 1973 issue of broadcasting.
National Nutrition Council, Information and Education Subsector. Paper prepared for the IEC Committee.
KBP - Radio Ownership.
KBP - List of Radio/TV Stations (1978).
Manoff International Inc., "Changing Nutrition and Health Behavior Through The Mass Media". Volume one; Philippines Final Report. October 3, 1977.
Manoff International Inc., "Radio Nutrition Education A Test of the Advertising Technique: Philippines and Nicaragua". Final Report December 1977.
Roque, Francisco. "The Mass Media in the Family Planning Program: The Philippine Experience" (Speech delivered at the Research Seminar of Asian Science Writers, November 24, 1977, Manila).
NMPC List of Standard Broadcast Stations.
The Philippine Nutrition Program 1978-1982 prepared by the National Nutrition Council in coordination with NEDA and the various cooperating agencies, September 1977.

NUTRITION EDUCATION IN THE EARLY CHILDHOOD

GRADE LEVELS IN ISRAEL

R. Lipsky, Ed. D.,

Coll. of Nut. & Home Econ., Ministry of Education

Jerusalem, ISRAEL

Nutrition education has usually been conceived as part of home economics, science education or health education. In Israel, however, nutrition education now has an identity of its own - it stands as a separate subject area in the elementary school curriculum and is taught to boys and girls. In the space of one generation, Israel's cultivated area was increased threefold, with a sixfold increase in yield[1]. Yet at the same time, incidence of obesity, diabetes, coronary heart disease, hypertension and arteriosclerosis all have increased, together with the dramatic changes in Israel's food supply[2]. Nutrition education therefore has as a long-term goal the prevention of the current health problems in Israel, which are the same of those of other affluent nations. But nutrition education at its start in this country faced quite different conditions.

Nutrition education in Israel actually began sixty years ago, thirty years before the establishment of the State of the United Nations. In 1919, Palestine has just been mandated to the British by the League of Nations, after the defeat of the Ottoman Empire in World War I, and the usual pattern of post-war conditions prevailed: food and medical services scarce and water supply contaminated. The Hadassah Medical Organization, a voluntary American relief organization installed a new medical unit in Palestine and not long afterwards, the director of the unit sent a cable from Jerusalem to Henrietta Szold, Hadassah's founder, in New York. The cable read: "Please bring with you a nutritionist"[3]. Julia Aaronson, a nutritionist with the American Red Cross in New York, answered the call. With her arrival in Palestine, nutrition education was begun, directed in the main, at the time, at doctors and nurses, and with special attention to the sanitary aspect of food.

In 1921 Sarah Bavly arrived from the Netherlands and it was her effort that established the school lunch program[3]. It was intended to provide a nutritious meal to all school-aged children and was probably the main source of nutrients for these children in those years. Nutrition education was begun for teachers supervising the school lunch in the summer of 1930 and its sponsorship passed, in 1950, two years after the establishment of the State, to the Ministry of Education and Culture[4]. The nutrition education aspect of that school lunch program, as it developed over the years was reported in 1956 at the Second International Congress of Dietetics by Bavly, and Ritchie, in her book, included a description of the program and a photograph of the children involved in it[5].

By 1961, Israel's cultivated land was more than double that in 1949, producing a larger and more varied food supply[6] and by 1968, McLaren and Pellett reported a higher per capita consumption than in the United States of cereals, pulses, vegetables, fruits, eggs and fish[7]. Almost ten years previous, as early as 1959, Evang had noted the beginning of over-feeding[8]. With an increasing food supply and an increasing incidence of over-feeding, it would seem obvious, therefore, that a change in focus for nutrition education was needed, from the earlier concern for the children's adequate intake to the current concern for optimal intakes without overfeeding.

In 1973-1974 an experimental curriculum in nutrition education was developed, jointly sponsored by the Ministry of Health and the Ministry of Education. Working in cooperation with the home economics teacher, the classroom teachers, the principal and the school nurse in a public school in a deprived area of Jerusalem, this researcher presented nutrition in this pilot project as a distinct subject area, related to the content of the other subjects in the curriculum but divorced from the preparation of school lunch[6]. It started at the First Grade Level where other earlier programs had started at Seventh Grade or Fourth Grade Levels and two consecutive hours per week set aside for its instruction with half the class for half a year. The second half of the class was accomodated in the second half of the year. The lessons took place in a "learning kitchen", a room for a special kind of learning and activity, just as the gymnasium or the library of an elementary school is a specialized kind of room for a different kind of activity. The "learning kitchen" provides the opportunity to concentrize the theoretical learning so that the children see the immediate application of the theory, and its relevance to the food preparation they do. Particularly because the children themselves are involved in food preparation, working in groups of four or five, can they experience this learning with all their senses. According to Tyler[9], the more senses are involved in learning, the more lasting the learning experience. Nutrition education, in our opinion, is one of the few disciplines that has the capacity to involve all the senses: not only sight, sound and touch, but also smell and taste!

As a consequence of the budget cuts following the Yom Kippur War in October 1973, wider application and testing of this experimental curriculum was impossible: the specialist, that is the home economics teacher, was not budgeted for any hours below the Fifth Grade Level, nor was the classroom teacher prepared with the specialist's background. To address this problem, the Ministry of Education and Culture this year invited a number of people, this researcher among them, to develop a nutrition curriculum for general classroom teachers. Starting this school year 1978-1979, teachers-in-training will be able to choose nutrition as a possible area of specialization, from several available. When these students graduate from the teacher's seminary in two to three years' time, they will have sufficient background information to teach nutrition in the early childhood grades. At that point, wider testing and evaluation of the nutrition curriculum for the early childhood grade levels will be possible.

We consider the long-term objectives of nutrition education to be the improvement of the nation's health by preventive rather than therapeutic measures. We try to do this 1) by giving people the tools to help them define an individual's nutrient need based on age, sex, physiological state, physical activity and environment, and then 2) by providing them with the knowledge of the variables in food choices - individual preference, nutrient content, cost, convenience, cultural pattern, among others. Equipped with this information it is hoped that people will be able to make food selections appropriate for themselves and others dependent on them for such decisions. But progress toward this long-term objective is made in small steps and in Israel this now starts in Grade 1. We undertook this revolutionary step based on Bruner's premise that any subject can be taught to any child at any age provided "that more complex ideas can in fact be rendered in... form that comes within reach of any learner"[10]. In putting nutrition in a form within reach of the early childhood grades, we were guided by Piaget's theory that a child at the First, Second, and Third Grade Level - the concrete operational stage - is perceptually oriented. In other words, at that age a child needs concepts to be concretized, so he can perceive them, rather than have them transmitted linguistically. For this reason, this author felt that it was essential that children have physical contact with food as the experimental material and that they themselves create the changes rather than have someone else do it and tell them about it. As Piaget said, "a child learns very little indeed when experiments are performed for him"[11].

And so children observe physical characteristics of food. They note color, size, texture, smell and taste of unprepared food and then after effecting some change in the food they note if there are any changes in the physical characteristics previously noted. Changes can be brought about by exposure to heat, cold or the use of some utensil like a grater or a knife. As an example, to see an apple coming out of the oven brown and wrinkled after seeing it go into the oven red or yellow and round, is quite a different experience in learning about the effect of heat on certain foods than having the teacher tell you about it. Children eat when they prepare at the end of the lesson and get a chance to judge and evaluate the results of their own, or others' efforts. If we want children to grow up being able to look at things objectively, including looking at themselves and their own efforts objectively, we must create the opportunities for them to practice this skill when they are young. They are encouraged to express preferences, to give reasons for their preferences, and to rank their preferences, if they have more than one. Again this is a way of refining these skills for use as adults when, perhaps, they will have to consider several options and rank them according to some criteria. Classifying food by many different criteria is done in the Second Grade as are many other types of classifying activity. This will be dealt with at somewhat greater lenght later in the paper. At the same time, food preparations continue and again there is observing and recording of food and its physical characteristics before and after some preparation. In the third grade the focus shifts to the tooth, a very relevant subject for third-graders. Discussion centers on changes brought about in food as the result of the action of teeth, or the changes required in food in the case of no teeth, for example infancy or old age. The concept of the relationship of structure and function is introduced as the children examine their own teeth and their function and the teeth of grown-ups in the classroom - parents or teacher. They also examine the structure of teeth of classroom pets and animals in the zoo or on a farm and relate the structure of those teeth to the type of diet the animal eats. The subject of teeth was chosen for focus because of its relevance to the pupil and also to build an awareness of preventive dental health, currently lacking in Israel. Since dental health, nutrition and general health are closely intertwined, the school team agreed with the suggestion of the researcher that this topic be included in the curriculum at this grade level.

The curriculum for the early childhood grades is designed with several skills and attitudes in mind intended to prepare the children for successful functioning in contemporary society. Among the skills and attitudes which it is hoped this curriculum will impart are 1) language skills, 2) logical thinking, 3) concept formation, 4) motor skills, 5) social interaction and communication, 6) creativity and 7) sensitivity to and respect for differences. I would like to review briefly the special contribution of each of the items mentioned:

1) Language skills: the contact with equipment, utensils, food and methods of preparation brings about the enlargement of vocabulary.

As a result of this increase in vocabulary, a child is enabled to ask for something he wants by name, for example, "broccoli" instead of "green stuff", or "strainer" instead of "that thing". Building vocabulary is part of early childhood education in any case but particularly in Israel, a country of immigrants, with languages brought along from abroad, many of the terms used, particularly at home, are not translated into the new languages for lack of contact with its Hebrew equivalent. This curriculum gives a very special kind of vocabulary to the children and the children bring it home to adults.

Of particular concern among professionals working with disadvantaged children is the low level of verbalization in the home[12]. The language aspect of this program gives the pupil and the family the vehicle for increased verbalization through repetition of the classroom activity, whether verbally or practically, at the same time that it brings nutrition information into the home. Language is the key to communication and this program seeks to develop communication. The home, furthermore, is seen as a reciprocal partner in this venture: although parents are not in the classroom on an on-going basis, we invite parents in from time to time and ask for their recipes and special holiday preparations so that children are learning the language of other cultural groups, in addition to their own.

2) Logical thinking: different kinds of mental operations are built into the curriculum. Mental conservation, reversibility, one-to-one correspondence, classification, multiple classification and hierarchies are among the mental operations involved. Piaget feels that basic classifying abilities are established around seven-eight years of age - the Second and Third Grade Level - and that later problem-solving is enhanced by the child's ability to classify the problem. The use of multiple classification particularly is intended to move the child away from the steriotypic that characterizes much adult thinking, in general and particularly on the question of food. For example, bread is thought of popularly only as a rich source of starch but it is also 10% protein; carrots are popularly associated with eyes as a cure for night-blindness but the same vitamin A is required to maintain the integrity of the mucous membranes - and maybe that would do more for the common cold than vitamin C. In other words, the same food can have more than one importance for us and the same nutrient can serve more than one purpose. In reverse, several foods can combine to provide increments of a total daily intake. At this grade level, multiple classification does not deal with nutrients but with food. In Second Grade, for example, we classify foods by where they grow relative to the earth's surface. In the group of fruit foods, which is one kind of classification, we have fruits with white insides, such as the apple and pear, we have foods with orange insides, such as oranges and apricots, we have foods with yellow insides like lemons and grapefruit. The children are learning that an object can have more than one quality, color

and type of food, a concept that is sometimes difficult for young
children to attend to. Even adults often do not attend to two qualities in one object and tend to stereotypic thinking. For example,
few of us think of a policeman as the father of a family or an ele-
vator operator as a grandmother. We take them for granted in the
role in which we see them and do not think of other roles they may
fill. Such a break with past stereotypic thinking has important
implications for nutrition learning at a higher grade level. In
Israel it also has implications for a multi-cultural society, where
it is important to get away from the characterization of members of
the "other" cultural groups stereotypically and see them objectively.
We hope the development of logical thinking will help.

3) Concept formation: many concepts that are presented in other
subject areas are re-enforced in this curriculum or presented from
another perspective, as for example heat-cold, size-volume, natural-
-manufactured. Concepts related to heat, for example, are differentiated, concretized and named: cooking with water and heat or "boiling" is different than cooking with dry heat or "baking or roasting"
and both are different from cooking with oil and heat or "frying".
Concepts related to size or weight are practically applied in following the measurements from a recipe or through asking the child
to make a choice of a utensil to be used. "Shall we use the shal-
lower baking pan or the deeper one?" "Is this ladle long enough to
reach to the bottom of this deep pot"? are examples of concretizing
volume and size.

4) Motor skills and eye-hand coordination: The safe use of appropriate tools, utensils of all kinds is an integral part of this
program, since the children are involved in food preparation. It
may or may not be a repetition of activities done at home. If it
is a repetition, then the "success" of the outcome is assured, an
important ingredient in classes of disadvantaged children. If the
skills have not been gained at home - not all mothers are tolerant
of children's beginning efforts - then this program presents the
child with the opportunity to use tools he will be expected to
master as he grows to adulthood. According to Bruner[10], man has no
special skills except as an effective tool user and so a social system evolved in which the skillful use of tools was rewarded. This
is no less true today - think of athletic competitions or the preparation of a gourmet meal and the value society accords the achiev-
ers. In Israel, cultural identity is reflected in food preparation
traditional to that group. However, the skill to prepare these
foods, despite their positive emotional value and their positive
nutritional value (in contrast to the Western-type diet) is fast
disappearing[13]. This program seeks to encourage the continued use
of cultural foods and to develop the motor skills needed to prepare
them.

Eye-hand coordination is an inseparable part of opening an egg, pouring oil into a measuring cup, peeling and slicing potatoes. It is conjectured by some researchers that accident-prone adults never had a chance to develop fully their motor skills in their growing stage[16]. Many of us are familiar with the expression, "He's all thumbs" meaning he is clumsy, or in other words he has trouble with fine motor skills. For children whose bones and muscles are still in the developing stage, the opportunities for practice are very important and this program provides such an opportunity.

5) Social interaction and communication: the organization of the class into small working groups requires pupils to interact with each other, to choose a group leader, to divide responsibilities, to follow instructions. By placing the responsibility for group organization on the children with guindance and continued practice, we cultivate democratic group-decision-making, a skill and an attitude necessary for political participation at a later age. Such interaction requires communication which is predicated on the language skills mentioned earlier. Such interaction according to Almy[12] provides the children with opportunities to exchange information which may clear up ambiguities for learning are made available.

6) Creativity: food preparation, table decorations and table setting certainly lend themselves to the creative genius of children. Attention in meal planning to color, shape, and texture are concrete enough considerations for the early childhood grades even if calories, protein, mineral and vitamins cannot be handled at this level. Nutrient content in meal planning is left for a more advanced grade. But festive celebrations with contributions flowing from cultural differences broadens everyone's experience, including the teachers'. In addition, exercise like table setting can be seen not only as an expression of creativity and aesthetics but as an opportunity for the practice of one-to-one-correspondence. Lavatelli[14] sees it also as an exercise in spatial transformation and should be used by teachers as such. She noted

> children who set the table and in doing so must transform the place-setting 180º, keeping spoon, napkin, and cup in the same position relative to the diner, are building into their motor a concept of relativity. Perhaps the reason it is difficult for adults to make certain transformation in thought is because a sensormotor underpinning for the operation is missing. (p 41)

Additionally, this program addresses the problem of "powerlessness" in disadvantaged groups. By allowing children to create, to control, to plan and execute activities in school, children develop the feeling that to some extent they can predict and determine outcomes. These same activities can be duplicated at home and add to the selfesteem and sense of control disadvantaged children need. Our cultivation of the children's ability to predict and determine, or at least influence outcomes has important implications for their health care as adults. They will learn, hopefully, to predict that certain food choices will determine or at least influence the outcome - their health - and act in the interests of their own and their children's health.

7) Sensitivity and respect for differences: food preparation is particularly reflective of differences, whether the difference is economics, cultural or personal. In a multi-cultural society such as ours it is important to cultivate a sensitivity to such differences and an appreciation or respect for them, rather than merely a toleration of them. In the context of holidays and special festive occasions, we bring these differences into the classroom. The need to move away from stereotypes was mentioned earlier and the same need is served by presenting the food of various cultures. This presentation brings each child into contact with an aspect of other cultures he may not have previously experienced. it is intended, thus, to give the child a broader perspective of other cultural groups. We hope in this way to help the children recognise and esteem values contributed by others just as each child would like others to recongnize and esteem hers or his.

Through the use of this curriculum, and units that will be developed sequentially, we hope to help the students achieve better health and an openess to things unfamiliar to them - whether foods, equipment or cultural patterns - that will help them to function in an integrated society without the loss of their own individuality.

REFERENCES

[1] Israel Information Center, Israel Briefing, 197/28.5.78/1.01, Jerusalem, 1978.
[2] Kark, S.L., "Epidemiology & Community Medicine", Appleton-Century-Crofts, New York, 1974.
[3] Duskin, Julia, Transcribed interview dated August 2, 1961, Oral History Section, Department of Current Jewish History, Hebrew University, Jerusalem.
[4] Bavly, S., "Evaluation of Nutrition Education Programmes in Israel", College of Nutrition & Home Economics, Ministry of Education & Culture, Jerusalem, n.d.
[5] Ritchie, J., "Learning Better Nutrition", Food & Agriculture Organization of the United Nations, Rome, 1967.
[6] Lipsky, Rachel, Nutrition Education in Israel: a Sequenced Curriculum for Grades One, Two and Three, and unpublished doctoral diassertation, Teachers College, Columbia University, New York, 1975.
[7] McLaren, D.S., & Pellett, P.L., Nutrition in the Middle East, World Review of Nutrition and Dietetics, 12:43-127, 1970.
[8] Evang, K., Report on a general evaluation of the health services in Israel, November 1959 - January 1960, World Health Orga-

nization, Regional Office for the Eastern Mediterranean, 1960.
[9] Tyler, R.W., "Basic Principles of Curriculum & Instruction," The University of Chicago Press, Chicago, 1949.
[10] Bruner, J.S., "The Relevance of Education", W.W. Norton & Company, Inc., New York. 1971.
[11] quoted (page v) in Almy M. with Chittenden, E., & Miller, P., "Young Children's Thinking", Teachers College Press, New York, 1966.
[12] Passow, A. Harry & Elliott, D.L., The nature and needs of the educationally disadvantaged, in "Developing Programs for the Educationally Disadvantaged, A.H. Passow (ed.), Teachers College Press, New York, 1968.
[13] Bavly, S., Changes in food habits in Israel, J. Amer. Diet. Ass., 48:488-495, 1966.
[14] Lavatelli, C.S., Moore, W.J., & Kaltsounis, T., "Elementary School Curriculum", Holt, Rinehart & Winston, New York, 1965.

CASE STUDY ON NUTRITION EDUCATION

IN THE UNITED KINGDOM

Dorothy F. Hollingsworth

The British Nutrition Foundation

2, The Close, Petts Wood, Orpington, Kent BR5 IJA - ENGLAND

Thirty years ago the international nutrition world was expressing its surprise at the success of the British food administration of the Second World War. This administration included a policy for nutrition education. It is generally believed, though it has not been proved because no evaluation was attempted, that this policy was successful. Wartime nutrition education, in common with most other contemporary nutrition teaching, was designed to counteract the effects of food deprivation[8]. People in Britain were urged to eat all their rationed foods, which were: carcase meat, bacon, cheese, butter, margarine, cooking fats, sugar, preserves, chocolates and sweets, and tea together with special allowances of milk, eggs, orange juice and cod liver oil for the nutritionally vulnerable groups. Beyond that, people were advised to eat plenty of green and other vegetables (to compensate for the lack of imported citrus fruit) and to fill up on bread and potatoes (which were not rationed until after the war when there were serious food shortages).

This system was one of three food groups, easy to understand: rations, green and other vegetables, bread and potatoes. The foods available provided sufficient energy, but not too much. Foods which provide energy and nothing else, particularly sugar and cooking fats, were in short supply. The extraction rate of flour was raised to conserve wheat supplies. Concentrated foods, such as boned meat, dried eggs, dried milk and cheese were imported. In addition, there was full employment and food prices were controlled. Thus, people in all sections of the community could afford their fair share of the restricted food supply, which was of good nutritional value. And health improved.

In many ways the problem of informing the public about nutrition was easy. The message was relatively simple. The Government owned nearly all the foods available during at least part of the distribution chain and could dispose of the food according to the sound nutritional advice at its call. The ratios supplied good nutrition. Bread of good nutritional value, potatoes and vegetables were readily available. The Government "looked after the calories" and the protein, minerals, vitamins and dietary fibre locked after themselves. The Government spent a great deal of money on the dissemination on information about food and nutrition.

However, there was not much choice and people got bored with the diet. When the war was over the feeling grew that food restrictions should be eased. To quote from the second edition of The Englishman's Food[6], which I prepared, "The population was vocal in its discontent, and the importance of palatability and acceptability of food supplies became abundantly clear". There was a pent-up demand for sweet foods and fats, which the public satisfied as soon as the sought-after foods became available.

An important point about British wartime nutrition education policy was that it was relevant to the needs of the British people at that time and to the composition of the foods available. Also, it was coherent and a consistent story was told throughout the country, and at all levels of education and instruction.

Since that time there has been in Britain, as in other countries, much controversy about nutrition, particularly about diet and certain diseases. This has led to a general distrust of nutritionists, a malaise in nutrition education, and a lack of agreement on what, in 1978, should be taught about nutrition and how it should be done. And much conflicting, and some misleading, information has been put out both in formal education and through the mass media.

Since 1970, the British Nutrition Foundation (an independent, industrially-financed body) has posed questions about what to teach, expressing the need for nutrition education in terms, first, of the obvious contemporary health problems caused by malnutrition — the extremely high incidence of dental caries, the excessive number of overweight people, the incidence of nutritional deficiency among some old people, the existence of health cranks who endanger their health by extremes in faddism, and the occurence of rickets and osteomalacia, particularly among Asians. Secondly, the Foundation thinks that people should have a sound and sensible attitude to food so as to be able to cope with change, either in diet or life style, for financial or other reasons. Thirdly, the Foundation has advocated more nutrition education in the hope that people will increasingly hold informed and responsible opinions on food topics of national importance[9].

That there has been malaise in nutrition education is widely acknowledged, as has the need for rectification. As a step in the process of rectification, in 1973 the British Nutrition Foundation, the Department of Health and Social Security (a government Ministry) and the Health Education Council (a government-funded body) set up a working party to examine the state of nutrition education in Britain and to consider ways and means by which it might be furthered and improved. The working party identified five main groups of people who between them probably have the greatest influence on the nutrition of the rest of the community, either by the example they set or the information they provide, and thus are themselves most in need of nutrition education. They are: housewives or homemakers; caterers or food administrators in hospitals, schools, canteens, restaurants, hotels, etc.; food manufacturers; professional health-care workers, including those in the medical, dental and nursing professions; and teachers in schools and other educational establishments. The working party emphasis particularly the importance of teaching children good food habits at an early age. They identified the organizations which at present provide nutritional information in Britain.

The working party's report[3], which was published in 1977, is about the machinery of nutrition education and not about its content. Indeed, the working party stressed in its first recommendation the urgent need "for a point of reference that would provide simple and accurate information on nutrition. The British Nutrition Foundation and the Health Education Council go some way towards meeting this need, but their capacity is limited by present resources. We (the working party) think that the work of both these bodies would be strengthned by the appointement of a committee constituted so as to benefit from the experience of practising dietitians, health education officers, teachers and the relevant government departments". In a letter to the British Medical Journal of 13th May 1978 the Chief Medical Officer of the Health Education Council announced that the Council is "in process of joining with the British Nutrition Foundation in a nutrition education committee"[2]. This is an important announcement. Developments are awaited.

There have been at least two other positive developments. Cuncurrently with the considerations of the working party on Nutrition Education, the Schools Council, an independent body financed equally by local education authorities and the Department of Education and Science (a Government Ministry), whose purpose is to develop curricula, theaching methods and examinations in schools and to help teachers to decide what to teach an how to teach it, has undertaken to try to improve nutrition education[1]. To this end, it has funded from September 1975 a three-year project to examine the place of home economics in the curriculum of schoolchildren between the ages of 8 and 13 years. A two year dissemination of the project proposals is to start in September 1978. The project suggests planning through a consideration of <u>concepts.</u> Assuming the focus to be the home and family, the project team have identified within home economics, five key concepts, one of which is nutrition. Within each concept, the team have identified main ideas and suggested a pattern of sequencing them for learning for children between 8 and 13 years. The four stages in the nutrition sequence are: (i) Food and the factors affecting food — to develop a positive attitude to food; (ii) Meals including food patterns and habits, and such points as variety of colour, flavour and texture; (iii) Food and its function in the body, including the concept of differing needs of individuals of different ages; (iv) nutrients and diet — to lead to understanding that foods are mixtures of nutrients, and the relation of this to function and the concept of diet, including the idea of personal body responses to types and amounts of food. It is hoped thus to build on concrete experience in the first three stages towards the abstract idea of nutrients in the fourth to a real understanding of nutrition.

The other contemporary development is that the government health and education departments and the Health Education Council have launched preventive medicine campaigns. The government departments have published Prevention and Health[5], which makes brief mention of nutrition and the Health Education Council has started a better health campaign which is about exercise and diet[7]. Of greater importance for the content of nutrition education, the Department of Health and Social Security has announced the preparation of a simple but comprehensive booklet on Eating for Health[4]. This will give guidance for individual eating behavior on the general line that prevention is better than cure.

One of the difficulties of this approach will be that the advice one must give in a time of abundance of food will appear to those not learned in nutrition to contradict the advice given when food is in short supply. An important task for nutrition educators is to unravel an apparent paradox for those who do not have a complete understanding of the principles of nutrition.

REFERENCES

[1] Anon. (1978). Changing the concept of nutrition education. Brit. Nut. Found, Nut. Bull. 23 (4) 293-296.
[2] Cust, G. (1978). Nutrition education. Br. med. J. 1,1280.
[3] Department of Health and Social Security (undated). Nutrition Education. Report of a Working Party, Department of Health and Social Security, London
[4] Department of Health and Social Security. "Eating for Health", in the press.
[5] Department of Health and Social Security, Department of Education and Science, Scottish Office, Welsh Office (1977). "Prevention and Health", Cmnd. 7047, Her Majesty's Stationery Office, London

[6] Drummond, J.C. and Wilbraham, A. (1957). "The Englishman's Food: A History of Five Centuries of English Diet", revised and with a new chapter by Dorothy Hollingsworth, Jonathan Cape., London

[7] Health Education Council (undated). "Look after yourself - a simple guide to exercise and diet", Ron Bloomfield and Freddie Lawrence, ed., Health Education Council, London

[8] Hollingsworth, D.F. (1977). Translating Nutrition into Diet. Food Technology February, 38-44.

[9] Hollingsworth, D.F. and Morse, E. (1975). Nutrition Education in an Affluent Society. J. Inst. Hlth. Educ. 13, 71-78

NUTRITION EDUCATION IN THE NETHERLANDS

K. Clay

Netherlands Bureau for Food & Nutrition Education

P.O. Box 85700, 2508 CK The Hague - NETHERLANDS

The Netherlands Bureau for Food and Nutrition Education and Information is a non-profit corporation, subsidized by the State. Being a corporation makes the Bureau independent from political or economic demands made by the subsidy giver.

The Bureau's task is to supply information in order to promote good nutrition among the Dutch population. In view of the number of its staff (50) and the size of the target group, our contacts with the population have always run through groups of experts. For experts those persons, bodies or groups qualify, who have contacts with the consumers and are able to transmit the information directly.

Direct contact with the Dutch population takes place through lectures, the mass media and other channels, but is is only incidental. In 1976 the Ministers of Agriculture and Public Health together put a subsidy of 112.000 US dollars at our disposal for a "campaign to change the housewife's purchasing behaviour".

The method the Ministers had in mind was nationwide advertising through the dailies. This wish of the Ministers was based on the success of a campaign to save energy, which had been conducted during the energy crisis and had drawn the attention of many people - because of the population's great need for further information. The Bureau thought that advertising in the dailies would have little sense. Is is impossible to draw any parallels with the situation at the time of the energy crisis. When information about nutrition is concerned, people are not tapping their feet with impatience. Moreover, the nutrition message is complicated and can hardly be formulated concisely, which is a must for advertising.

In addition, the Bureau thought the budget too low to construct a really good programme. This design would not have guaranteed continuity either, a demand that must be made on any form of mass communication in order to ensure success in the social marketing sector. The Bureau dismissed as unrealizable the idea that it is possible to effect changes in human behaviour in the available period of three months.

Nevertheless, after ample considerations we accepted the assignment as a trial project subject to a number of conditions:

1. with regard to the objectives of the campaign: "changing behaviour" should be replaced by "awakening people's consciousness"
2. Because of the limited budget it is better to give the campaign a regional character. In this case the region should be representative for the whole of the country and show a sufficient coverage rate for the regional dailies.
3. The campaign must be evaluated. Its regional character facilitates this, because a control group can be found elsewhere.
4. Advertising in the dailies must be underpinned by calling in the help of professional people in the region, i.e. doctors, school--doctors, school-dentists and other public health workers, and teachers.

These groups were later extended to include shopkeepers.

Both Ministries could approve of our proposals. After a preliminary inquiry North-Holland to the north of the North Sea Canal was chosen as the testing region. The campaign was to consist of six advertisements, each of which was to appear once a fortnight. Each advertisement was to contain different information, which was dealt with more extensively in a brochure that could be sent for by filling in a coupon from the advertisement and be obtained at the doctor's, the policlinic, the grocer's, etc. Moreover, every applicant for such a brochure was offered a pamphlet containing general information on nutrition.

The schools were approached with separate material and there was a colour contest for children, which was announced by means of a poster.

The advertisements are full page. Each advertisement bears the same slogan "eating sensibly can be tasty as well" and invariably our Bureau's symbol, the "Basic Five", is to be found in the same corner. The following subjects are treated:

thirst   attention for the necessity of taking enough moisture, information about alcoholic drinks, soft drinks and alternatives.

proteins	such as replacement of meat by other products that are rich in protein, also vegetable products, on budgetary considerations and to meet the growing interest in eating without meat.
vitamins	as an antidote against the advertising campaign for vitamin preparations that is launched in the cold season.
vegetables	variation, advice for their preparation, pesticides.
eating sweets	one moment of sweets a day, alternative "sweets" and playing in-betweens; not all in-betweens are bad, energy value.

We were conscious of the fact that a relatively great number of subjects were dealt with in such a short period, but we hoped that repetition of the slogan "eating sensibly can be tasty as well" and our symbol would be sufficient guarantee for recognition. Our problem was that we could not formulate our aims more clearly than by the phrase "furtherance of the consumer's nutritional consciousness". This problem became manifest at the briefing of the advertising agency. We could not supply our agency with much concentrate fotting. As yet hardly anything is known about the eating habits of the Dutch. Our assumptions and starting-points were pragmatic and based on our own experiences. It is common knowledge that the Dutch consume too much fat and sugar; the gross consumption figures show it. We assumed that the housewife is the central figure in the family as regards shopping and preparation of the meals. Further we supposed that the housewife mostly makes her purchases impulsively. Therefore we chose the shopkeeper as in intermediary.

Our advertisements answered both the objective and the subjetive need for information and we wavered between nutritional information and information about products.

In view of the short time that was available for preparations, we neglected certain public relation aspects, which may also have had an adverse effect on the campaign. For example, we were unable to coach the shopkeepers and had to confine ourselves to a few meetings with professional people and the press to inform our hinterland adequately.

As I said before, this information campaign has been evaluated. Before the campaign started in the province of North-Holland, 500 housewites were interviwed there about their knowledge, attitudes and behaviour with respect to good nutrition it was checked to what extent the main parts of the campaign had been noticed.

The aim of the campaign was: to make the housewife more conscious of the importance of good nutrition for our health. After the final evaluation, however, we had to accept the fact that this goal had not been reached. It appeared that only ten per cent of the interviewed housewives had been in touch with the campaign at all. The standard we applied for this was that they had seen at least one of the six advertisements and taken home one brochure. Only in this group might any effect of the campaign's message be expected. However, it has not been checked if they have really become more conscious about nutrition.

From the following data it appear that contact with the campaign remained superficial. Over one tenth of the housewives could recall spontaneously one of the six ads about good nutrition, and only half of that number could mention one or more topics that had been raised in them. Over one and a half per cent had clipped the coupon from one of the ads and sent it to our Bureau to ask for further information. It appeared that the topic winter-food in which the superflous use of vitamin pills was discussed, had yielded most reactions, relatively speaking. One per cent of the interviewed persons managed to recall the title of this campaign, viz. "eating sensibly can be tasty as well".

So much for the results of the evaluation-research. Instead of the rather sad conclusion that the campaign has not had any effect at all, a more positive interpretation of the whole enterprise is possible as well. In the course of this campaign quite a number of data were gathered that might be taken as a starting-point for some future campaign. Next time a new set-up can have a considerably more solid basis, while this first experiment has been based on a lot of presuppositions and common sense.

I would like to mention briefly a few of the research data. It appeared that there is no reason to complain about people's knowledge of good nutrition. The housewife can answer questions about the energy value and nutritional value of well-known foodstuffs reasonably well. She knows roughly how to compose a healthy diet. At least, she manages to pick out the best reply from a number of possibilities on paper.

In reply to the question what things are given special attention at shopping, freshness and the date on the package stand out very clearly.

A tasteful appearance, too, weights very heavily with most housewives, whereas in practice the role of energy value and nutritional value appears to be very moderate. On the other hand it is found that the preference of husband and children and the cost of the article are much more important criteria.

In the opinion poll the women were also asked what they thought they should give attention to shopping, if they did everything in accordance with nutritional theory. Comparison of the replies reveals an enormous gap between norm and reality. Here something becomes visible of the stings of conscience many housewives must feel. They realize their responsibility for serving the family good food on the other hand they know perfectly well that they do not nearly always buy what would be best. The points of neglect are nutritional value and energy value. As I have said before, it is rather the preference of husband and children that turns the scale. The order of the things considered to be important changes; only freshness, date and appearance remain at the top of the list.

Although the poll showed a fair knowledge of good nutrition and nutritional value, it apperas that the relation between nutrition and health is not nearly perceived by everyone. We tried to find out how nutrition ranks among ten factors that can each be detrimental to health. It is not among the first three mentioned. At the top of the list we find: much alcohol, stress in the family and smoking a lot. Eating rich food follows in the fourth place, whereas apparently eating a lot and eating sweets are not associated directly with health concern. They take the ninth and tenth place respectively.

After this, the next outcome reflects something of the housewives' conflict once more: over four-fifth of them indicate that their children should not distribute sweets on their birthday. No more than half the number of housewives are willing to push their sense of responsibility so far as to prepare once in a while food that is not popular in the family, but that contributes to health.

By now it has become generally accepted that rich food is harmful for the heart and blood vessels and that brown bread is better than white bread. Yet, a number of stereotyped opinions are still widely held, such as the idea that one needs vitamin C tablets in winter, that potatoes make you fat and that tinned vegetables contain fewer vitamins than fresh ones.

Finally, it has been asked whom housewives would ask for advice on good nutrition. About half of them chose the physician for this and one-third preferred the dietician. It seems that relatives and friends have no role in this matter. Shopkeepers, too, are regarded by hardly anyone — 2.5 per cent — as authorities on this. It is not quite clear how much weight should be attached to this — very low — percentage. As has already been said in the beginning about the set-up of the campaign, due to lack of time little attention was paid to coaching of grocers and to the introduction of the campaign material. The fact that we don't know how many of the 960 retailers actually placed the brochure stand in their shops, and if they did, where they put it, is another weak spot in the evaluation-poll. As a result, it can be checked if the campaign has influenced the grocer's existing image as an expert on nutrition.

What role does nutrition play in the housewife's price policy and how important is it for her own image?

The results of the poll indicate that women have a very good knowledge of food prices. On the whole, prices are watched closely, but the tendency to economize on food is very small.

The fact that two-thirds of the interviewed persons answer the question "what makes a good hosewive recognizable" with "the fact that she pays attention to nutrition" will surprise nobody.

Finally, for a Bureau that is engaged in giving health education it must be a confort to know by now that only a quarter of the housewives thinks far too much attention is being paid to the promotion of good nutrition.

On the basis of this campaign we would like to make the following recommendations:

- Hosewives are too limited as a targetgroup: Husband and children should be aimed at.
- Information should be limited; repetition of one subject might be more helpfull than a variety of subjects.
- Practical information might be more succesfull than theoretical, as basic knowledge in nutrition seem to be sufficient.
- In the Netherlands a nation-wide campaign offers more possibilities than regional approach because of free publicity.

COMMUNITY NUTRITION EDUCATION

Margot Moya de Medina, MD

Ruta 7, Res. Julia Apto. 2, Col.,

Sta. Monica, Caracas 104 - VENEZUELA

For several years, a series of programs have been developed in Venezuela aiming to improve the level of population nourishment, especially that of less affluent groups. Some of the methods have been to provide supplement to pregnant women, pre-school and school children and adolescents. Likewise, important measures affecting agriculture, housing, education employment, health, commercialization are being put into effect. Whithin a certain period of time, these measures will produce lasting and wide effects on living standards and, therefore on the nourishment situation of the population.

Nutritional education has received wide support which has allowed us to carry out very interesting programs and projects. We are expecting medium and long term results from them. I believe the following points should be mentioned:

BREAST FEEDING CAMPAIGN

It has been carried out in different phases, such as mass media communication; publishing bulletins for physicians, dieticians, nurses; publishing literature for mothers; giving lectures at health service centers, schools and universities; printing posters, etc.

TELEVISION PROGRAMS

We have already produced up to this date, 213 programs with a duration between 5 and 10'. These programs are intended mainly to housewives. We have presented simple recipes to stress that a high-family budget is not essential in order to feed the family; these recipes include the preparation of products in season, for special ocassions, picnics. Also meals for different ages, physiological conditions, and for some frequent illnesses (diarrhea, rash). We have also included invitations to specialists in different related fields to participate in this program. We offered cook books to those requesting them in writing, therefore delivering approximately 20.000 copies in just one year.

SALES PROMOTION OF NUTRITIONAL FOOD AT SCHOOL CAFETERIAS

These cafeterias have been operating for a long time in almost all schools, under different conditions starting from the simple selling of soft drinks and candies, up to a cafeteria where complete lunches are sold. Through a survey, it was determined that the products being sold were not sufficiently adeauate, neither in nutritional value, nor in price, and therefore, by joint resolution from various ministeries, this was declared as a "first need service". A permanent commission was created to study in depth the situation. Meanwhile, in order to start up tht program activities, lists of foods permitted and their prices were prepared as well as organization and hygenic measures. Courses have been given to food handlers. It has been decided to carry out an inspection program with the help of the Instituto Nacional de Nutrición (National Institute of Nutrition), the M.S.A.S. (Health Ministry), and the Superintendencia de Protección al Consumidor (Consumer's Protection Superintendency) to supervise, make sanitary inspection and to control prices, respectively. Publicity has been provided through communication media, lectures to teachers as well as to students. The community is expected to be the best supervisor of the measures, while shcools would gradually be incorporated into the program with the requirement that they achieve the desired objectives within 1 year.

BASIC NUTRITION EDUCATION COURSES

Courses consisting of 12 classes with audio-visual aids have been organized. Simple and economic recipes to support the classes have been prepared for mothers and adolescents. We have achieved improvement in the knowledge of the participants for short and medium term, evaluated through questionnaires. We also expect to carry out a survey to judge knowledge and ability at the long-term participation (1 1/2 years). This is currently under analysis. Through these courses we expect to attain an adequate selection, purchases and preparation of food, to promote early pregnancy medical check-up, acceptable breast feeding, early medical chek-up of the sick child, immunization, childbirth — assistances, adequate behaviour at different ages, and more frequent physiological and pathological conditions.

COMMUNITY NUTRITION EDUCATION

RESEARCH RELATING TO THE EFFECT OF CHILDREN'S TELEVISION COMMERCIALS

This research was conducted to nutritional habits. We have come to the conclusion that there is a high percentage of harmful messages involving health, particularly nourishment: likewise children influence their mothers to buy the advertised products. Mothers purchase products influenced by: TV 70%, market 15%, friends and relatives 12%, other ways 3%.

The National Nutritional Institute is supporting a law to control certain advertisement, such as foods, alcoholic beverages and cigarretes.

THE ORCHARD: SOURCE OF NUTRITIOUS FOODS

I am going to present in detail a simple and worthwhile project: "The Orchard: Source of Nutritious Foods", that is being carried out this year.

Justification

Growing fruits and vegetables at schools, homes and in the community permits convenient utilization of space. While certain space limitation exists, it may very well be used for the production of some crops.

The orchard itself, represents various advantages. It favors work groups and the enjoyment of an acitivity which integrates school, family and community. It is a mean against food scarcity, high prices, monopoly, etc. It improves the nutritional situation since it enables the use of fresh, diverse and good quality products.

Those items that may succesfully be cultivated on the so called "conucos de balcón" (home balconies) are vegetables, which are characterized by its high vitamin, mineral and fiber (cellulose) content. These are regulating elements extremely necessary for the utilization of other essencial nutritional substances. Vegetables and fruits make our meals more tasty and colorful.

The low or no consumption of fresh vegetables and fruits, found in suerveys could bring a series of disorders limiting health and activity. Constipation skin, visual, bones, teeth and growth problems, anemia and others, may be prevented by a balanced diet which obviously must include fresh fruit and vegetables.

With this project the following objectives were proposed:

General: That students and the educational community actively participate in the promotion and development of school, family and community orchards.

Specific:

- To create in public and private schools an orchard on ground, in boxes, cans and pots, etc., according to the physical space they have.

- To practice growing and care of vegetables and fruits.

- To utilize products of the orchards to improve the student and family nourishment.

- To develop consciousness about health problems resulting from lack of vegetables and fruit consumption.

Description

At the beginning of January of this year, we prepared a project to be submitted for the consideration of the National Institute of Nutrition, Education Ministry, Environmental and Natural Resources Ministry, Agriculture and Livestock Ministry, with the purpose of obtaining their support, advise and cooperation.

The project was engaged mainly in the promotion to create school orchards for which the following steps were considered:

- To prepare a simple leaflet for distribution in schools to encourage and demosntrate the activities to be carried out and the basis for the constest to reward the best effort. (Januray 1978).

- Due to the fact that in the cities, a great majority of schools do not have appropiate ground for this task, suggestions were given to plant either in boxes, pots, or any other container available at home.

- A conference was given by a specialist in organic agriculture to the directors of the School Nutritional Clubs (SNS) so that they could carry the message to the other students in each school and to their families.

    During the conference, envelopes containing seeds as a sample of what could be planted, were provided. Additionally, leaflets about hygiene, preparation and advantages of vegetable and fruits, and posters, were distributed.

- A period for the reception of the working plans was started. A special event was carried out on the International Health Day. Children delivered their orchard plans and the types of crop they will plant, and in turn they received a certificate of participation at the contest.

- Simultaneously, as a demonstration we made an orchard using a variety of containers at the terrace of the INN headquarters (National Institute of Nutrition). Contest participants having questions or doubts about the program were invited to visit the demonstration orchard.

- During the month of May of this year, we carried out an intensive visiting program to the participating schools, to observe the work being carried out.

- At the beginning of June, a special ceremony took place to reward the best orchards, based on the children's work and resources used.

- The products of the orchard were used by the students in preparing foods at school and at home.

I think it is important to underline that one of the implied objectives is to encourage children to take the idea to their homes. Likewise, in urban zones this small experience, which from now on will be a permanent program of the SNC will fulfill an educational need through the introduction of vegetables consumption and the active participation of the children. At the rural zones, where schools as well as houses have more physical space for planting, the tendency is to help school (lunch program) and family food supply.

From the 120 schools visited, 44 (36.6%) deserved special recognition, which consisted of a trip for the best working team (5 children and one teacher) while the rest received tools for home planting. The trip had an educational objective too, as children will go to a farm of great agricultural importance and to Uverito, an experimental forest of caribes pines. This important project has acclimatized a foreign specie in Venezuela, modifying the ecology in warm and dry lands, changing its raining periods. This is a promising project for the lumber and paper industry and will be a flora and fauna reserve.

This will be an unforgetable experience for the children and a fair reward to their efforts which will encourage them to continue working in the orchard, while benefiting form its products.

THE MARKETING APPROACH TO NUTRITION EDUCATION

IN DEVELOPING COUNTRIES

Richard K. Manoff[a], Thomas M. Cooke[b], Delfina Aguillon[c]

[a]President; [b]Vice-President - Manoff Intern. Inc.
1511 "K" St NW Suite 438 - Washington, D.C. 20008
[c]Deputy Exec. Director - Nat. Nutr. Council
P.O. Box 1646 - Makati Commercial Center
Makati, Metro Manila - PHILIPPINES

Six years ago at this same Congress in Mexico City, one speaker opened his remarks with the playing of a Coca-Cola jingle which was mockingly referred to as the "International Anthem of Nutrition". The jest was intended to make us face the absurdity of the world of nutrition education in which our well intentioned efforts were overwhelmed by the commercial exploitation of the mass media. Our thesis then, as now, was that nutrition education should use the mass media. The opportunity to test this thesis was afforded Manoff International by the Governments of the Philippines and Nicaragua.

The experiences in the Philippines and Nicaragua used the mass media technique commonly associated with commercial advertising — short messages, aimed at specific behavioral objectives, broadcast over many stations over extended period of time. The experiments tested the thesis that the radio — excluding other forms of planned communication — could have an impact on nutrition knowledge, attitudes and practices of large numbers of rural families. The findings of the studies support this thesis, and suggest, as expected, that the mass media in combination with existing health and nutrition education and service programs can have an even more profound impact.

In Nicaragua, the Manoff International project taught mothers how to prepare a simple rehydration formula for the treatment of diarrhea in infants.

THE PHILIPPINES

In the Philippines, the key objectives was to teach mothers to enrich the lugaw — the traditional weaning food — a starchy porridge of rice and water.

Our goal, set by the Philippine National Nutrition Council, was to teach mothers to add small amounts of cooking oil for calories, chopped fish for protein and green vegetables for vitamins and minerals.

Project development steps included the following :

- Briefings by the National Nutrition Council in Manila and regional nutritional authorities in Iloilo, the project site on the island of Panay.

- Identification and study of the target group, visiting communities throughout the province. The visits helped us identify resistance points to the idea of enriching the rice porridge in the new way and they confirmed information gathered in briefings and literature reviews. Local stores were checked to ascertain the availability of the ingredients — fish, rice, and cooking oil.

- Refining specific attitude, knowledge, and behavior objectives.

- Preparation and production of radio scripts, using local radio writers, anthropologists, and actors to insure authentic scripts.

- Preparation of a broadcast media plan.

The following script is the English translation of one of the six messages that were prepared, tested, and broadcast on a rotating basis for the 12 months experiment period. Each message focused on one aspect of this objective.

> Increase the number of rural mothers who begin supplemental feeding by at least the sixth month with a food consisting of rice porridge, cooking oil, chopped fish and vegetables.

LITA : Mama what are you putting in my baby's lugaw?

MOTHER : A drop of oil, some chopped green vegetables and fish.

LITA   : Where did you get this strange idea?

MOTHER : From the doctor on the radio. Listen!

DOCTOR : (FILTER) After 6 months a baby needs lugaw as well as breast milk but lugaw must be mixed with fish that gives protein for muscles and brain. Green vegetables for vitamins. Oil for more weight on his body.

LITA   : But mama, a 6-month-old baby can't digest such foods.

MOTHER : Sh-h. Listen to the doctor on the radio.

DOCTOR : A 6-month-old baby can digest these foods. Just wash the salt from the dried fish, chop the vegetables and cook them well, add a little oil and mash with the lugaw.

LITA   : But, mama, you dind't feed me like that.

MOTHER : How could I know? I didn't even own a radio. Times change. You live and learn.

LITA   : Mama, you must be sad that the old ways are changing.

MOTHER : Not all the old ways are changing.

MOTHER : But only a fool remains with an old way when there is a new, better way.

Evaluation was done through a random sample of 1,000 household interviews throughout the rural Iloilo province, the test site, and in southern Cebu province, a separate island that served as a control. Interviews were conducted before broadcast, at the end of 6 months, at the end of 12 months.

In addition, non-random self-administered questionnaires were completed on the same schedule by doctors, nurses, rural schollteachers and others in the experimental areas. All research was done by Consumer Pulse, an independent research organization, well known in the Philippines.

The findings supported the hypothesis that radio messages alone have a significant effect on attitude, knowledge and behavior. This leaves little doubt that the impact would have been even greater if the activities of community workers had been integrated.

Attitude change was widespread throughout the province. The newest idea, adding oil, met with the greatest initial resistance (Wave I), and by the end of 12 months (Wave III) had achieved a level of acceptance almost equal to the more traditional ideas of giving fish and vegetables (Table 1).

Attitude change was largely the same regardless of the location of the mother in the province, radio ownership, her education, income, occupation of the husband. This suggests that the radio messages appealed to all segments of the mother population and that they spread beyond the reach of the radio through word-of-mouth.

Knowledge change was pronounced in disabusing mothers of false ideas, a very important objective, as can be seen from the decline in "incorrect answers" from Wave I to III (Table 2).

We were less dramatically effective in giving them positive motivation for adoption of the new ideas suggesting, perhaps, that the latter lagas behind the former. Note that the most frequent incorrect answer, "causes loose bowels", a common objection to giving oil, dropped from 48% at the beginning of the project (Wave I) to 6% by the end of the project (Wave III).

To measure behavior change we had mothers report through open-ended questions the foods usually given infants (Table 3).

These reports, supported by the survey of rural health workers and others, were the basis on which an estimation of behavior change was based. Of the approximately 50,000 target group families (i.e. having young children) we project that 32,000 to 38,000 heard and remembered the message, and 11,000 learned why small amounts of cooking oil are good for the infants. Of the 12,000 families with infants between 6-12 months, approximately 2,750 or 23% began to add oil to the rice porridge, while no change was recorded in the control area.

NICARAGUA

Nicaragua was the site of the second experiment. In Nicaragua, the government estimates that diarrhea and the dehydration and malnutrition that accompany it account for 40% of all pre-school infant deaths.

It is common for low income and poorly educated mothers to withhold fluids and essential foods from sick babies as well as to administer purges, thus exacerbating the diarrhea and dehydration.

Our work procedures were similar to those described for the Philippines. We worked closely with government health authorities in consultations, field visits, interviews with mothers, doctors and others.

TABLE 1

ATTITUDE CHANGE:
EXPRESSED AS A PERCENTAGE OF TOTAL RESPONDENTS

	TEST			CONTROL		
Attitude Change Items	WAVE III	WAVE II	WAVE I	WAVE III	WAVE II	WAVE I
N =	(660)	(674)	(700)	(291)	(296)	(300)
Oil is good	74%	74%*	15%	28%	26%	29%
Fish is good	81%	80%*	48%	48%	54%	56%
Vegetables are good	79%	82%*	49%	72%	81%	76%

* $p = 0.05$

TABLE 2

KNOWLEDGE ABOUT ADDING OIL TO LUGAW EXPRESSED
AS A PERCENTAGE OF TOTAL RESPONDENTS

	TEST			CONTROL		
	WAVE III	WAVE II	WAVE I	WAVE III	WAVE II	WAVE I
N =	(660)	(674)	(700)	(291)	(296)	(300)
Correct Answers						
Makes baby fatter	15%	18%	3%	6%	3%	4%
Makes baby livelier	22%	26%	4%	3%	—	4%
Gives heat & energy	3%	1%	1%	1%	—	—
Incorrect Answers						
Causes loose bowels	6%	12%	48%	48%	48%	29%
Causes stomach upset	6%	6%	36%	10%	15%	19%

TABLE 3

BEHAVIOR CHANGE:

MOTHERS WITH INFANTS 6-12 MONTHS OF AGE

	WAVE III After 12 Months	WAVE II After 6 Months	WAVE I Before Broadcast
	(N = 136)	(N = 142)	(N = 157)
Add oil	24.0%	23.0%	0.0%
Add fish	27.1%	26.8%	16.7%
Add vegetables	16.9%	13.3%	5.0%
Add fish and oil	13.0%	13.0%	0.0%
Add fish and vegetables	14.0%	11.0%	2.0%
Add vegetables and oil	10.0%	8.0%	0.0%
Add fish, vegetables and oil	8.0%	8.0%	0.0%

The communication and education objectives were finally selected under the direction of the Comité Técnico de Nutrición of the Government of Nicaragua.

Knowledge

- Recipe and dosage of rehydration fluid
- Recommended foods
- Dangers of giving purges

Attitude

- Confidence in recommended treatment

Behavior

- Give fluid in correct amount
- Continue feeding
- Stop giving purges

Since low-cost rehydration fluid or salts are not available in Nicaragua, our objective was to teach mothers to prepare it. The formula consists of:

- 1 liter water
- 2 tablespoons sugar
- 1/2 teaspoon salt
- lemon juice

For purposes of increasing awareness and ease of remembrance, we made an important marketing and communications decision involving the name. What we had in our formula was actually a popular beverage — limonada or lemonade — plus salt. So, we decided to call our formulation Super Limonada — a very positive name for a very serious task. Bicarbonate of soda was not included in the recipe because it was not readily available in the countryside.

The recipe and its name reflect important communication principles. They are:

- Single-minded
- Actionable — the mothers had the ingredients
- Culturally relevant — an adaptation of a traditional recipe
- Memorable

Six different messages were prepared, each treating a different aspect of the objectives. The format of a minidrama is based on a common reality of rural villages — an older woman, Doña Carmen, giving advice on child care to a younger mother.*

Using 10 to 13 local radio stations and 3 national stations, the messages were broadcast for 12 months during hours when the rural families were most likely to listen.

Data for evaluation of the project were gathered through a random sample of rural mothers throughout Spanish-speaking Nicaragua with children under 5 years of age. Interviews were conducted before broadcast, after 6 months and after 12 months.

The campaign was successful in transmitting new information, changing attitudes and behavior. By the end of 6 months, 60% of the mothers could recall some portion of the messages. But knowledge gains were more widespread. After 6 months, 79% of the respondents knew that Super Limonada was for treatment of diarrhea, increasing to 86% at the end of 12 months. 65% of the homes owned working radios, so there was some word-of-mouth communication prompted by the radio messages.

The radio messages were successful in teaching the recipe for Super Limonada, with approximately 50% of the respondents knowing the key ingredients of one liter of water and 1/2 teaspoon of salt.

By the end of six months 43% of the mothers, compared to 21% at the baseline, said that a baby could consume a liter of fluid daily. By 12 months this had increased to 52%. By the end of the experiment, 50% of all mothers correctly identified the proper way to administer the rehydration fluid — by spoon, cup, or "little by little", as Doña Carmen said in the message.

Behavior change was widespread throughout the country — reported by the mothers in the interviews and supported by information from health workers.

After 6 months of broadcast, 25% of the mothers whose children had been sick with diarrhea during the preceding 4 months, reported giving Super Limonada (Table 4).

Had a low-cost commercial product been available, eliminating most of the task of preparation, adopting might have been more wide-

---

* The text of the messages has been ommited for purposes of brevity.

TABLE 4

PERCENTAGE OF MOTHERS ADOPTING SUPER LIMONADA:
EXPRESSED AS A PERCENTAGE OF MOTHERS WHOSE
CHILDREN HAVE HAD DIARRHEA RECENTLY

III After 10 Months	II After 6 Months	I Baseline
N = (984)	(942)	(926)
22%**	=5%*	2%

* p = .95 between Waves I and II
** p = .95 between Waves I and III

spread. This suggests that the World Health Organization and Pan American Health Organization program of home oral rehydration salts could have a widespread impact if combined with a professsional marketing and education program.

Reported adoption was also impressive because only the radio was used. Had health workers and others been involved rather than excluded for experimental purposes, adoption would presumably have increased.

Other communication objectives included more appropriate feeding practices for sick infants and children and cessation of purging during the diarrhea. In each category, major improvements were noted, but the findings suggest that the limitations of the available and affordable food supply greatly restricted the food given to the infants and children. Although there was a reported decline in the incidence of giving purges, their ready availability and constant promotion of commercial preparation such as Milk of Magnesia contributed to their continuing use.

We were able to project that Super Limonada was adopted by approximately 18,600 households in rural Nicaragua with children under 5 years of age, out of a total of 60,000 households, affecting approximately 31,000 young children.

Monoff International Inc. has worked in nearly a dozen coutries in Latin America, Africa, and Asia. On the strenght of this experience we have found that highly effective nutrition and health education programs can be developed if these communication principles are followed:

- Identification of the target group;

- Identification of specific health and nutrition problems;

- Understanding the target group and their perceptions, beliefs, habits, economic and cultural characteristics as they relate to the problem;

- Determining what aspect of the problem can be influenced by education and persuasion — not everything can: economic constraints are considerable;

- Selection of specific objectives, actionable by the target group and shared by a consensus of health and nutrition authorities throughout the target area;

- Preparation of education and communication materials that reflect an understanding of the target group and their perceptions of the problem;

- Testing the materials with the target group before dissemination;

- Maintaining consistency of information;

- A media plan that responds to the characteristcs of the target group;

- Long-term exposure of the messages;

- Integration of the mass media communications with interpersonal education, acknowledging that most mothers will be seen infrequently by health or nutrition workers;

- Constant verification with the target group that the messages are being received and understood as planned and that unforseen circumstances have not changed the circumstances of the target group so as to warrant changing the education strategy.

NUTRITION TRAINING OF HEALTH WORKERS

IN DEVELOPING COUNTRIES

Dr. K. Bagchi

Nutrition Unit, WHO

1211 Geneva 27 - SWITZERLAND

Health Care Systems of developing countries have important responsibilities and roles in combatting malnutrition which is, in all these countries, a grave problem. It is taken for granted that health workers, i.e., all professional workers in the health care systems, perform certain tasks as a part of their job to promote nutrition and to prevent malnutrition.

Who are these health workers who provide nutrition services? Though the model would vary from one country to another in the developing world, one can safely make the following categorization:

1. Doctors: including public health workers in various disciplines, e.g., nutrition, maternal/child health, family planning, etc.

2. Nurses: including health visitors, public health nurses, mid-wives, etc.

3. Other health workers: not included in the above two categories working mostly in peripheral areas.

The health workers in categories (1) and (2) are usually based in health infrastructures which is their main site of operation. The third category of workers are usually, or for the major part of their work, mobile, offering services "at the doorstep".

While the workers in the first two categories are well defined profession-wise, those in the third category vary considerably from one contry to another, not only in their designation, but also in their role and in their training. However, even in the midst of all these diversities, it is increasingly realised that these are the people who are in the first line of contact between the national health care system and the community, and, more importantly, these are the people who, in almost all developing countries, will provide essential health care to millions of people in peripheral and rural areas. Most of these people at the present moment have no access to any form of health care. The importance of primary health care stems from this realization. The primary health care workers or the community health care workers already exist in many countries, and with the increasing adoption of this approach, more and more countries will have such workers to provide essential health care to all. In the developing countries, malnutrition is probably the most important health problem, and in this context, effective and realistic nutritional activities will be a cornerstone of primary health care. Obviously, the community health worker, or whatever name they are designated, will have to undertake these activities. Let it be made clear at this stage that the health workers in the other two categories also carry out important tasks for nutrition promotion and prevention of malnutrition. However, the community health workers' area of operation, and the number of people they serve, will far outstrip those of the doctors and nurses.

Nutrition training in the schools of medicine and nursing have already attracted considerable attention. Efforts have been, and are being, made to make nutrition training in these professional courses realistic and practical. Other speakers in the panel today will discuss these in detail. I propose to present the global trend in the nutrition training of the health workers in the third category mentioned previously. For obvious reasons WHO is presently giving high consideration to the training of those workers who operate at the first line of contact between the community and the health care system, irrespective of the designation of such workers. For years to come these workers will shoulder the major responsibility of providing nutrition services to millions of people in the rural outlying areas, however little these services might be. The strategy is to cover as many people as possible and not to improve the quality of service which, as it is, is only being provided to a very few.

Why should the nutrition training of such workers be different from others? In order to get a reply to this question, one should remember the following characteristics of such a worker:

a) they are usually selected form the community where they are expected to work;

b) they have very limited formal education, usually a few years of schooling;

c) they are multi-purpose workers;

d) they have to provide both curative and preventive services and as such have very limited time for preventive work.

Obviously, nutrition training for these workers will have to be considerably different from the conventional nutrition courses, e.g.

i) nutrition content will have to be short. Nutrition promotion has to be viewed as a component of total health care;

ii) the training has to be task-oriented;

iii) each skill has to be backed by knowledge, to be imparted through appropriate techniques;

iv) the training to be given should, as far as possible, be in the vicinity of the community to be served;

v) the length of training is best determined in the light of the educational aims and the results of preliminary testing of individuals, since training has to be adapted to their degree of literacy.

NUTRITION THROUGH INTEGRATED APPROACH

Nutrition training has also to take advantage of integrated or packaged approaches. Nutrition interventions, as commonly adopted in the past, were most often isolated approaches, e.g., feeding programmes. Health sectors in most developing countries were developing and implementing feeding programmes, making distributions of nutrient tablets, nutrition education, and in some countries conducting nutrition surveys. The tendency in most cases, was to develop "vertical" programmes. Fortunately the trend is now changed and there is a growing realization that the improvement of nutrition is an objective, and in view of its multi-factoral nature the objective cannot be reached through activities mentioned before.

Instances are not rare where feeding programmes for young children did not produce any impact since, it was realized later, gastroenteritis was extremely common among the beneficiaries and no attention was directed to the prevention and control of diarrhea. Again, providing nutritional care to suffering malnourished children will produce, in many cases, no impact until and unless preventive measures are taken against too many closely-spaced pregnancies in the mothers. This clearly indicates the need for an integration of family planning and nutrition activities. Control of infectious diseases and regulation of fertility should be closely linked with nutritional activities. The family health approach for nutrition promotion, vigorously sponsored by WHO and other agencies, is an outcome of this approach. In fact, the three components: nutrition, family planning and the control of communicable diseases "packaged" together for delivery will have much more than additive effects.

TRAINING OF PRIMARY HEALTH CARE WORKERS

Coming specifically to nutrition promotion, two basic questions have to be answered:

1. What should the nutrition content in such training be?

2. How should the training be imparted?

Nutrition Content

Since the training should be job-oriented, the content obviously would depend on the job the worker is expected to perform. To identify the nutrition problems in the community and, on the basis of these, to select the appropriate tasks are the essential steps in developing training content, and serve as the learning objectives of the course. For the performance of these tasks, the worker will have to have some knowledge and skill. Utmost care is to be taken to decide how much knowledge is necessary to acquire the required skill and perform the task in an intelligent manner. The educational background of the worker should be kept in view when finalizing these contents.

Educational Techniques

How do we impart the knowledge and skill to the health worker? Possibly this is where the conventional didactic method will need drastic modification. In the conventional training, the trainees are "passive" agents who are "taught" most probably in a classroom. An example will clarify the limitations of such classroom teaching. Growth of infants and young children is a topic which is "taught" to all health workers. Use of growth charts to monitor growth is also considered as a task of these health workers. However, what is lacking is the skill to use appropriately the growth chart. What the trainees do not learn is that this simple task consists of three separate components:

a) weighing the child;

b) recording his weight on the card;

c) interpreting it for action.

Each of these components implies distinct but related tasks for which one needs skill, which cannot be acquired through conventional classroom teaching. It is no wonder that in actual situation this apparently simple task of weighing children for nutritional monitoring is done, in most situations, in a haphazard manner. For workers at this level, it is essential that during the training the tasks should be well-defined for a particular problem, and the skills for performing these tasks should be acquired by actual task-performance under the supervision of trainers.

In other words, there is a definite need to replace the commonly accepted method, where teaching alone preponderates, by a system where teaching is to be combined with a learning process. There should be more emphasis on learning than on teaching — the teaching-learning system. There is general realization that learning is a dynamic process in which the behaviour and experience of the student are vital components. His perception of what is happening is just as important as the perception of his teachers.

Coming specifically to the training of the primary health care workers in nutrition, the need for the adoption of the teaching-learning approach is obvious. The format will be as follows:

Problem identification at the community level
↓
Selecting appropriate learning objectives
↓
Identifying skills necessary to perform the required tasks
↓
Developing contents for knowledge and attitudes
↓
Identifying learning activities
↓
Selecting simple indicators for evaluation

It is obvious that the development of an appropriate training module will depend basically on the tasks that the primary health care worker is expected to perform, which are basically the learning objectives. Utmost care is taken to identify these tasks taking into consideration the limitations of such workers. This is the base on which the contents of training, comprising of knowledge, skill and attitudes have to be properly built up. Similarly, instead of the conventional classroom teaching interspersed by a few demonstrations, the learning activities will have to be identified, through which the trainees will develop the skill to perform the prescribed tasks.

Learning activities enable the trainees to learn and not be taught under the supervision of the trainers.

WHO'S ROLE IN NUTRITION TRAINING OF COMMUNITY HEALTH WORKERS

It would be useful to consider first the reasons as to why nutrition interventions through the health sector of most developing countries have not made much impact. An analysis of the situations indicate two main reasons:

a) inadequate health coverage of those populations who are desperately in need of such care. Nutrition intervention, however well-planned, will not be able to reach the needy population:

b) the tendency to implement a few commonly adopted activities, e.g., supplementary feeding programmes as single interventions without any link or integration with other related health activities. Many times the effect of supplementary feeding programmes has been very limited of even nullified because simultaneous activities for the control of widespread gastro-enteritis were not undertaken.

The primary health care approach being sponsored by WHO and other agencies offer the best opportunity for maximum coverage in health care, and through which nutrition activities integrated with other related health activities could be delivered to those who are in greatest need and who at the present moment have no access to any form of health care. Through a number of consultations and workshops involving experts form various parts of the world, the common nutritional problems encountered at the community level have been identified[1]. Though this would vary from one country to another, there is a common trend in most developing countries. Appropriate nutrition activities for these problems have been identified and for these suitable guidelines have been developed. Some very commonly adopted activities like nutrition education have been taken up for in-depth study to determine the factors which usually lead to failures or successes in these activities[2]. For some non-nutritional activities with considerable nutrition implications, e.g., weanling diarrhea and oral rehydration, guidelines for conducting these activities at the primary level have been and are being developde, and subsequently field tested[3].

The stage is now set to take a careful look at the training pattern of community health workers in nutrition. There are already good experiences from which one can learn a lesson. The West Azerbaijan Project in Iran, in which WHO was closely involved, is one such experience. Efforts are being made even now to rationalize and upgrade nutrition training in this project. There are other good examples also in other parts of the world, e.g., Nigeria, Ghana, Philippines, Jamaica, etc. WHO organized a Consultation of experienced trainers from this area in August 1978 with the general objective of developing methodologies and guidelines for the training of primary health care workers in nutrition, after reviewing the experiences already available. The report shows very clearly the urgent need to reorient our approach to nutrition training on lines mentioned earlier. The broad areas under which the nutrition and nutrition-related activities of the community health workers can be grouped are the following:

a) getting to know the community;
b) motivating and organizing community groups;
c) nutrition surveillance;
d) nutrition education;
e) identification and management of cases of malnutrition;
f) control and management of infection;
g) collaboration with other workers, e.g., for birth-spacing.

Training modules for each of these areas have been worked out for testing at national levels.

Training of trainers

Needless to say, this approach to training will only succeed if the trainers are fully oriented, which unfortunately, is not the case. A list of trainers of community health workers in nutrition was prepared and it was observed that the background of the trainers varies considerably. At one end of the spectrum is the medical officer and at the other end are health professionals such as medical assistants. In between are the nurses, sanitarians, nutritionists and home economists.

The common factors are the conventional process of learning through which they acquired knowledge and nutrition, their lack of knowledge in educational techniques and, more serious, their ignorance about the role of community health workers in nutrition, especially exactly what they do and how.

Who has already started the process of training the trainers through a series of inter-regional training courses. The first course is to be held in October this year and will be attended by about 30 trainers from about 18 countries. This will be followed by other inter-regional courses, both in English and in French, in different parts of the world, through which key national trainers will be exposed to this approach. It is expected that this will lead to national courses with a "multiplier effect". Several countries are in the process of producing simple manuals outlining training modules for each task encountered by the community health worker.

The guideline and methodologies developed by WHO is now being tested through these Courses for Trainers, and will be appropriately modified. It is expected that before very long a pattern will be developed which can be easily adapted or adopted for national training manuals.

REFERENCES

[1] WHO. Inter-regional Consultation on Strategies for Nutrition through Local Health Services, NUT/77.4, World Health Organization, Geneva, 1977.
[2] WHO. Report of the Consultation on Nutrition Education through Health Care Systems, NUT/78.3, WHO, Geneva, 1978.
[3] WHO. Treatment and prevention of dehydration in diarrhoeal diseases: a guide for use at the primary level. WHO, Geneva, 1976.

NUTRITIONAL EDUCATION FOR MEDICAL AND

OTHER HEALTH SCIENCE PROFESSIONALS IN EAST EUROPE

E. Morava

Institute of Nutrition

1097 Cialy-ut 3/a - Budapest, HUNGARY

In our paper we summarize the present situation of nutrition education in Hungary and discuss some special features of the education systems in other East European countries.

At the medical universities in Hungary there are no chairs of nutriton. Nutrition education is given by chairs of biochemistry, physiology, pathophysiology, internal medicine and hygiene. This system is advantageous in integrating nutrition and basic physiological sciences as well as nutrition and clinical medicine, but unfortunately it results in an incomplete and fragmentary picture of nutrition as a whole. A similar type of nutrition education exists at the medical universities of Bulgaria, Czechoslovakia, German Democratic Republic, Rumania and USSR. Some medical universities in USSR have chairs of nutrition hygiene delivering integrated courses on this topic. For example, nutrition hygiene is a compulsory one year' subject with examinations at the Faculty of Mecical Hygiene in Leningrad.

At the Postgraduate Medical School in Budapest functions the Chair of Human Nutrition. It organizes courses in nutrition and dietetics, food chemistry, food toxicology, food poisoning and food born infections for physicians, pharmacists, food chemists and biologists. The Chair of Human Nutrition is an integrated part of the Hungarian Institute of Nutrition, which provides the laboratory and research background for postgraduate teaching.

In Hungary 7 months long, full time courses are organized for medical doctors who wish to get a specialist degree in hygiene and epidemiology. It includes 4 weeks lectures and 4 weeks practice in nutrition and related subjects. For hospital doctors and pharmaceutists interested in nutrition 1-2 weeks long special courses are organized on selected topics of nutrition. At present, however there is no specialist degree of nutrition and dietetics for medical doctors in Hungary.

In Czechoslovakia graduates of the Medical Faculty of Hygiene can get a grade I specialization in hygiene and epidemiology, and later a grade II specialization in food hygiene and nutrition. Specialists in internal medicine can attend postgraduate courses in dietetics and diabetology and get a grade II specialization in this field.

In Poland however, unregular courses are organized on selected problems of clinical dietetics and bromatology in the framework of postgraduate medical teaching.

In GDR the postgraduate teaching program of physicians includes special lectures on nutrition. The topics are selected according to the line of specialization.

For hospital nurses there is no compulsory special training in nutrition in Czechoslovakia and Hungary. In GDR 24 hours lectures on general nutrition and 18 hours lectures on basic dietetics are included into the training program. In Poland there are 36 hours lectures and 18 hours practice in nutrition and dietetics for hospital nurses.

In Hungary the establishment of the Faculty of Advanced Paramedical Training is a recent achievement in the education of health science professionals. This faculty includes inter alia Dietitian Training School, Public Health Inspector Training School and Public Health Nurses Training School. Students are admitted to these schools at the age of 18, and can get a diploma after 3 years' full time training and examinations. The teaching program bestows sufficient time upon nutrition education. The training program for dietitians in GDR is similar to the Hungarian one. It contains 375 hours lectures in nutrition and dietetics. In Poland 480 hours lectures and 800 hours of practice are included into the teaching program of dietitians. In Czechoslovakia however dietitians are trained in special secondary schools.

A field of growing importance is nutrition education for food chemists and biologists, as many of them work at the Hungarian Public Health Stations, research laboratories and also in health education.

In Czechoslovakia, GDR and Hungary there is no university diploma in human nutrition at present. At Warsaw Agricultural University, Faculty of Human Nutrition and Home Economics has been established recently. At this institution a special degree in human nutrition is granted to the students after 4 1/2 year study.

There are several textbooks on nutrition and dietetics in every country of the area. In Hungary a pediodical "Health Education" is published bimonthly with several papers on nutrition education.

Summarizing the situation of nutrition education in the East European countries we conclude that during the last 30 years when food supply became more than sufficient and the living standards of the people improved significantly in every country of the area, the development of nutrition education was slower than that of general education. Further improvement of nutritional education could significantly promote the prevention and cure of diseases.

NUTRITION EDUCATION FOR MEDICAL AND OTHER

HEALTH SCIENCE PROFESSIONS IN ASIA

Aree Valyasevi, M.D., D.Sc.

Mahidol University

Bangkok 4, Thailand

I. NUTRITION EDUCATION FOR MEDICAL AND
   OTHER HEALTH SCIENCE PROFESSION IN ASIA

Survey of nutrition education in curriculum of Schools of Medicine, Pharmacy and Dentistry has been conducted in Asia in 1977. Eight countries responded to our survey which included India, Japan, Korea, Malaysia, Pakistan, Philippines, Sri Lanka and Thailand. The questionaires were sent to one contact person for each country (from the recommendation of Drs. Fidanza and Gopalan) who will collect information from schools of medicine and other health sciences in their own country. The number of schools that responded to the questionnaires are shown in Table 1. The data from Japan collected from 37 medical schools and their nutrition problems are different from the rest; therefore, data from Japan will be presented separately.

Prevalence of Nutritional Problems and
Responsibilities of Health Professions

The four most common problems are protein-energy malnutrition, vitamin A deficiency, riboflavin and other vitamin B deficiency and nutritional anemia. Other problems, including iodine deficiency, rickets, scurvy, beri-beri, calcium deficiency and urinary calculi, are also listed by these countries (Table II). In Japan, obesity, diabetes mellitus and cardiovascular disease including atherosclerosis are considered to be major problems.

In regards to profession and/or organization responsible for prevention and treatment of nutritional diseases, all countries indicate that physicians and other members of health team are res-

TABLE 1

COUNTRIES WHERE DATA WAS OBTAINED

COUNTRY	NUMBER OF SCHOOLS	
	Medicine	Pharmacy
India	2	–
Japan	37	–
Korea	5	2
Malaysia	1	–
Pakistan	1	–
Philippines	1	1
Sri Lanka	1	–
Thailand	3	1

TABLE 2

PREVALENCE OF NUTRITIONAL PROBLEMS
IN SURVEYED COUNTRIES

PROBLEMS	PERCENTAGE
PCM	28%
Vitamin A deficiency	16%
Nutritional anemia	13%
B_2 deficiency	11%
Other vitamin deficiencies	10%
Calcium deficiency	5%
Iodine deficiency	5%
Others	12%

ponsible for these tasks. Referring to the organization, the Ministry of Health has the primary responsibility for prevention of these nutritional problems (Table 3). Therefore, it is essential for the school of medicine and other health professions to prepare our graduates for their future responsibilities in prevention and treatment of various nutritional diseases.

How Nutrition is Taught in the Schools of Medicine and Pharmacy

As shown in Table IV, nutrition is included in the curriculum as a topic of its own in 68 percent of the schools. The topics include human nutrient requirement, mineral and vitamin metabolism and nutrient deficiency in infants and children. In another 26 percent, nutrition is integrated into biochemistry, physiology, preventive and social medicine, and public health and hygiene.

Table V shows the methods used for teaching nutrition. Forty-seven percent are using lecture and another 41 percent are using case presentation and small group discussion. Other methods are used for the rest.

According to the available information, there is no data on the number of hours allocated for teaching this subject as well as the contents and objectives of the course. Therefore, it is difficult to

TABLE 3

PROFESSION AND/OR ORGANIZATION RESPONSIBLE
FOR PREVENTION AND TREATMENT
OF NUTRITIONAL DISEASES

PROFESSION AND/OR ORGANIZATION	PERCENTAGE
Physicians	22%
Other health personnel	25%
Nutritionists	16%
Ministry of Health	25%
Ministry of Education	6%
Ministry of Agriculture	3%
Food and Nutrition Research Institute	3%

TABLE 4

TEACHING NUTRITION IN SCHOOLS OF MEDICINE AND PHARMACY

	PERCENTAGE
A. Nutrition included in the curriculum as a topic of its own (human nutrient requirement, mineral and vitamin metabolism, nutrient deficiency in infants and children)	68%
B. Nutrition is integrated in other topics - Biochemistry and Physiology, Preventive and Social Medicine, and Public Health and Hygiene	26%
C. Unknown	6%

TABLE 5

METHODS USED FOR TEACHING NUTRITION

METHODS	PERCENTAGE
Lecture	47%
Case presentation	23%
Small group discussion	18%
Others	12%

judge regarding the efficiency and effectiveness of the existing nutrition course in Asian countries.

Opinions Regarding Changes or Modifications of Nutrition Teaching

Eleven of fourteen schools (excluding Japan) recommended that nutrition teaching should be changed because they are not satisfied with the existing curriculum. The suggested changes are as follows:

    a. Nutrition course should be in the curriculum starting from the first to senior year.

b. Nutrition should be integrated into practical experience (ward round, case study, field work, etc.).

c. More qualified staffs to teach nutrition are required.

It is their opinions that the International Agencies, such as UNESCO, WHO, FAO and IUNS should give full support to improve nutrition education.

Nutrition Teaching in 37 Medical Schools in Japan

These informations were collected by Dr. Fujiwara in 1972. Table 6 shows the nutrition topics and lecture hours which demonstrate that over 60 percent of lectures concentrated on basic nutrition, about 20 percent on catering and food production, and another 20 percent on clinical nutrition.

Table 7 demonstrates the name and numbers of various departments responsible for teaching nutrition in the Japanese medical schools. As expected, basic nutrition, and catering and food production are taught by Basic Science Department and Department of Hygiene and Public Health; and clinical nutrition by Clinical Departments. Number of years lectures are given in various nutrition topics are shown in Table 8. Most of the teachings were carried out during the third and fourth year and clinical nutrition during the fifth year.

As far as we can tell from the available informations, lectures are major means for teaching nutrition in various academic departments. Major emphasis is on basic nutrition. Professor Yoshimura of Hyogo College of Medicine has informed us that the Committee of Nutrition Sciences in the Science Council of Japan would like to strengthen nutrition education in the Japanese Medical Schools.

Conclusions

Survey of nutrition education was carried out in medical schools from eight countries, namely India, Japan, Korea, Malaysia, Pakistan, Philippines, Sri Lanka and Thailand. All countreis except Japan are faced with common nutrition problems including protein-energy malnutrition, vitamin A and B_2 deficiency, and nutritional anemia. The ill effects of malnutition to poor health are generally recognized and the health profession, especially doctors, have the primary responsibility in solving these nutritional problems.

Nutrition education in the curriculum of most medical schools in the present survey require major changes or modifications. Short-

TABLE 6

NUMBER OF LECTURE HOURS IN VARIOUS FIELDS
OF NUTRITION IN 37 JAPANESE MEDICAL SCHOOLS

SUBJECTS	AVERAGE HOURS	Percent
Introduction	1.2	
Classification of Foods	8.6	
Digestion and Absorption	7.2	34.4
Energy	4.7	
Nutritional Requirement	3.6	
Food Sanitation	5.9	28.7
Nutritional Status of Japanese	1.8	
Communal Catering	1.0	
Food Production	0.8	7.3
Obesity	2.2	
Undernutrition	1.3	
Nutrition in the Aged	1.0	9.1
Nutrition for Mother and Infant	2.8	
Nutrition Related to Various Diseases	1.7	9.1
Cooking Instructions	0.0	
Observation Study	1.9	
Practical Experiments	3.7	11.3
TOTAL	49.4	

TABLE 7

NUMBER OF DEPARTMENTS RESPONSIBLE FOR TEACHING VARIOUS FIELDS OF NUTRITION

SUBJECT	Hygiene & P.H.	Bio-Chemistry	Nutrition	Physiology	Clinics
Introduction	7	8	4	–	1
Classification of Foods	6	26	3	3	1
Digestion and Absorption	6	23	2	17	1
Energy	15	18	3	8	6
Nutritional Requirement	24	15	4	–	–
Food Sanitation	36	–	–	–	2
Nutritional Status of Japanese	30	2	3	–	1
Communal Catering	11	–	1	–	1
Food Production	11	2	2	–	–
Obesity	17	–	–	–	12
Undernutrition	12	3	–	–	15
Nutrition in the Aged	5	–	–	–	8
Nutrition for Mother & Infancy	5	–	–	–	20
Nutrition Related to Various Diseases	3	–	–	–	20
Cooking Instructions	–	–	–	–	1
Observation Study	15	–	–	–	–
Practical Experiments	15	3	3	–	–

TABLE 8

NUMBER OF YEARS LECTURES GIVEN IN VARIOUS NUTRITION TOPICS

SUBJECTS	YEAR OF CLASSES					
	1st	2nd	3th	4th	5th	6th
Introduction	0	2	10	7	1	0
Classification of Foods	1	4	26	8	3	0
Digestion and Absorption	0	3	30	8	4	1
Energy	0	4	24	14	8	1
Nutritional Requirement	0	2	18	24	10	1
Food Sanitation	0	0	7	24	12	3
Nutritional Status of Japanese	0	0	6	20	9	0
Communal Catering	0	1	3	14	7	0
Food Production	1	1	3	9	5	0
Obesity	0	2	4	15	16	5
Undernutrition	0	1	6	8	13	4
Nutrition in the Aged	0	0	1	6	8	0
Nutrition for Mother & Infancy	0	0	2	12	14	2
Nutrition Related to Various Diseases	0	2	4	15	16	5
Observation Study	0	0	2	11	3	2
Practical Experiments	0	0	5	9	3	1

age of qualified staffs for teaching nutriton is generally recognized. All countries suggest that the international agencies, such as UNESCO, WHO, FAO and IUNS should play an active role in supporting nutrition education.

II. NUTRITION TEACHING IN RAMATHIBODI MEDICAL FACULTY, BANGKOK, THAILAND

The Mahidol University has two medical faculties, the oldest one is at Siriraj Hospital and the newest at Ramathibodi Hospital. The undergraduate medical curriculum in Thailand requires six years after twelve years of primary and secondary education.

The Ramathibodi Faculty of Medicine in Bangkok, Thailand, founded in 1964, developed a teaching and research program aimed at bringing the institution into close association with the health needs of the country. Our medical graduates in Thailand are required to serve in the government services for two years after completion of one year internship. Therefore, about 80 percent of our graduates will work in either provincial hospitals or district health centers or hospitals. Since nutrition is one of our major health problems especially in the rural areas, teaching of nutriton has been emphasized in our medical curriculum.

Schematic description of undergraduate medical curriculum is as follows:

```
Year
I
    }  Basic Sciences or Pre-medical
II

III
    }  Basic Medical Science or Pre-clinical
IV

       Community Health Program

       - Health and Demographic Survey        }
       - Analysis of Community Health Problems }  112 hrs.
       - Planning for Community Health Care      26 hrs.

       Clinical Clerkships

V      - Medicine and Pediatrics              9 months
       - Surgery, OB-GYN, EENT and
           Anesthesiology                    10-1/2 months
       - Community Health                     1-1/2 months
VI     - Electives                            1-1/2 months
```

Nutrition teaching can be divided into three stages as follows:

Basic Biochemical Nutrition (Year III)

This course is taught in the Department of Biochemistry, Faculty of Sciences, Mahidol University. It includes organic compounds established as dietary essentials and its normal physiological functions as well as the results of an impairment in the functions. Knowledge of the biochemical functions of these essential factors, the amount required to maintain normal health and their sources in food are also stressed. Awareness of the presence of toxic substances in local foods and their interaction with other nutrients are also included.

At the end of the course, it is expected that the students should be able to:

a. explain what substances are required for growth, maintenance of good health and reproduction;

b. tell how much of these are required and describe the results when the requirement is not met of the consumption is excess;

c. explain the biochemical function of each nutrient and its relation to the deficiency signs;

d. tell the food sources and how much of these foods are required.

Community Health Program (Year IV)

This program is divided into 3 courses as follows:

Course   I - Health and demographic survey
Course II - Analysis of community health problems
Course III - Planning for community health care.

Our Community Health Course is designed to introduce our students to health problems in a given community with a population of 50,000. In Courses I and II, the students will have opportunities to learn about field survey techniques, community approach and community diagnosis. They will also learn by themselves under close supervision of our Faculty Staffs about assessment of nutritional status, factors influencing nutritional status — such as increasing magnitude of malnutrition in a family with many children, and etc.

In Course III, the students will choose five or six topics for exercise in health planning. There have always been one of two topics involving nutriton problems which assigned students to plan for solutions at a community level. Common topics are protein-energy malnutrition, family planning and diarrheal diseases.

During these courses, there are three sessions (three hours for each session) allocated for nutrition seminar. The topics for these sessions include Nutrition Problems in Thailand with emphasis on ecology, nutrition requirements, Methodology for Assessment of Nutritional Status, Food Group and Food Hazards, Food and Nutrition Planning and its implementation and many others as requested by the students.

At the end of the Community Health Course, as far as nutrition is concerned, it is expected that the students will:

a. be aware of an able to recognize nutrition problems as one of the major health problems;

b. be able to apply the appropriate methods for nutrition assessment in individuals as well as in groups;

c. develop positive attitude in solving nutritional problems in all three levels: individual, family and community.

Clinical Clerkships (Year V and VI)

In Ramathibodi Hospital:

During these two clinical years, especially in medical and pediatric wards, the students will learn from their patients who come to the hospital with varieties of medical and health problems. About 30 to 40 percent of cases admitted to our wards, protein-energy malnutrition is also an associated problem. In addition, it is not rare to find cases who come to the hospital with nutrition deficiency diseases, such as beri-beri, vitamin A deficiency, severe iron deficiency anemai, niacin deficiency and riboflavin deficiency.

The nutrition deficiencies, particularly P.E.M., are found in about 60 percent of hospitalized patients after either chronic illnesses or post-operation which further complicate the patient's problem. Such cases will be illustrated to the students. In addition, cases which require nutritional management such as diabetes mellitus, obesity, hyper-lipidemia, gout and hypertension will also be brought to the attention of the students.

Learning experiences provided to our students include bedside teaching, attending nutrition clinics, well-baby clinics, seminars and conferences. The topics included in the Pediatric Nutrition Teaching are PEM, beri-beri, xerophthalmia, breast-feeding, infant formula, weaning food and nutrition management. Additional topics include enteral and parenteral nutrition, dietotherapy, nutrition management in obesity, hyperlipidemia and coronary heart disease also discussed during the medical clerkships.

In small District Hospital and Health Center:

During this six-week period, in addition to the clerkship in the out-patient-clinics including well-baby, and in-patient wards, the students will have an opportunity in (i) giving health and nutrition education to school children, (ii) giving health and nutrition intervention program in the community; and (iii) carrying a field research program in nutrition (nutrition is one topic to be chosen).

At the end of the Clinical Clerkships, it is expected that the students will be able to:

a. take history, do physical examination and select suitable lab test as well as its interpretarion;

b. diagnose common nutritional problems and able to solve, especially for individual patients;

c. select a suitable and practical type of food for the solution of nutrition problems, and

d. give simple nutrition education to a group or general public.

Organization of Nutrition Teaching

There is no Department of Nutrition in the Ramathibodi Medical Faculty. Nutrition activities including teaching and research are inter-departmental consisting of staff members from the Departments of Pediatrics, Internal Medicine and Clinical Research Center.

Nutrition teaching for the whole course (Years IV, V and.VI) is planned together to lay out the course objectives and contents, learning experiences and evaluation. The topics to be discussed in the nutrition course based on the prevalence of the problems occurring in Thailand, such as, in Pediatrics, protein-energy malnutrition, vitamin A deficiency, beri-beri, iron deficiency anemia, riboflavin deficiency, simple goitre and urinary bladder stone disease are seven most common nutrition problems in pre-school children.

Our staffs at the present time consist of three pediatricians, two internists, one food toxicologist and visiting staffs from the ministries of health and agriculture. All medical staffs are well qualified in their specialities and have also earned doctorate degrees in nutrition. Therefore, our staffs are taking care of patients like the other staffs in the departments and are able to integrate nutrition into daily patient care. The students, interns and residents are able to see and practice their nutrition knowledge, skill and experience with their patients.

Summary

In summary, teaching of nutrition in Ramathibodi Medical Faculty can be divided into three stages. In the pre-clinical course (Year III), the students will learn about basic biochemical nutrition which is integrated into biochemistry course. Then, in Year IV, the students will be exposed to the community nutrition which is integrated into the community health course. It is expected that the students will be aware of and be able to recognize nutritional problems as one of the major health problems in Thailand.

During Years V and VI, clinical nutrition is integrated into patient's care in the hospital. Nutrition diagnosis and management in various diseases are stressed during their clinical years in the Ramathibodi Hospital. The sutdents will also learn about community nutrition during their clinical clerkships in the Health Center and small District Hospital.

REFERENCES

[1] Buri, P. et al. Ramathibodi Community Health Program. J. Med. Educ. 49: 264, 1974.
[2] Valyasevi, A. and P. Buri. Making Pediatric Education Relevant to the Health Needs of the Community, in: "Perspective in Pediatrics". O.P. Ghai, ed. Interprint, India, 1977, p. 115.

ACKNOWLEDGEMENTS

I would like to express my special thanks to the following persons who kindly collected the data presented in this report from their respective medical faculties:

1. Dr. S.G. Srikantia, Director, National Institute of Nutrition, Hyderabad, India.
2. Dr. Hisato Yoshimura, Professor of Physiology, Hyogo College of Medicine, Hyogo, Japan.
3. Dr. Jin Soon Ju, Professor of Nutrition and Biochemistry, Korea University Medical College, Seoul, Korea.
4. Dr. Hyung S. Lee, School of Medicine, Busan National University, Busa, Korea.
5. Dr. J.H. Kim, Associate Dean for Academic Affairs, Chonnam University School of Medicine, Chonnam, Korea.
6. Dr. Bum Suk Tchai, Associate Professor, College of Medicine, Seoul National University, Seoul, Korea.
7. Dr. Nak-Eung-Sung, School of Medicine, Ewha Womans University, Seoul, Korea.
8. Dr. In Hoi Huh, School of Pharmacy, Chung-ang University, Seoul, Korea.

9. Dr. Jong Yull Yu, Professor of Biochemistry, School of Pharmacy, Duksang Women's University, Seoul, Korea.
10. Dr. Y.H. Chong, Institute of Medical Research, Kuala Lumpur, Malaysia.
11. Dr. M. Ataur Rahman, Professor and Head, Department of Biochemistry, Jinnah Postgraduate Medical Center, Karachi, Pakistan.
12. Dr. Carmen Ll. Intengan, Food and Nutrition Research Institute, Manila, Philippines.
13. Dr. C.C. Mahendra, Head, Department of Nutrition, Medical Research Institute, Colombo, Sri Lanka.
14. Dr. Pithaya Viriyanondha, School of Pharmacy, Mahidol University, Bangkok, Thailand.
15. Dr. Prakong Posakrishna, School of Medicine, Chulalongkorn University, Bangkok, Thailand.
16. Dr. Poolsak Sambhavaphol, School of Medicine, Chiengmai University, Chiengmai, Thailand.
17. Dr. Sastri Saowakontha, School of Medicine, Khon Kaen University, Khon Kaen, Thailand.

THE TANZANIA EXPERIENCE

T.N. Maletnlema

Tanzania Food and Nutrition Center

P.O. Box 977 - Dar Es Sallam, TANZANIA

The title of this workshop makes it a bit difficult for me to describe what is happening in Tanzania today, because the meanings attached to words like Nutrition Education and Scientific Professionals are likely to be different in different countries. All the same, I shall try to describe the situation as it was in the past, the changes that have taken place and the present situation culminating in what may now be called Nutrition Education in Tanzania.

In the 1940's on to the 1960's, Tanzania, like many other countries, thought of Nutrition as a science dealing with interaction of living cells with nutrients taken. In the case of man, nutrients were described by the professionals as proteins, carbohydrates, fats, vitamins, minerals and water, the latter serving mainly as the media for all metabolic reactions. From this definition of nutrients, malnutrition was understood to be the lack of any one or more of these nutrients. Thus, lack of protein caused kwashiorkor, lack of proteins, carbohydrates and fats caused marasmus, lack of vitamins caused various avitaminosis diseases like xerophthalmia, beri-beri, scurvy, etc., and lack of minerals was responsible for diseases like anemia, goitre, etc.

According to the above concept, the causes of malnutrition were obvious and could be very easily summarized as: "lack or insufficient intake of a nutrient or nutrients". Hence medical and para-medical schools at that time concentrated on this aspect of nutrition. Consequently nutrition science was taught as part of biochemistry, chemistry, paediatrics and occasionally in clinical medicine. Severe cases of malnutrition were studied and treated in hospital like any other cases and discharged as soon as they improved. Some of those dying in hospital had post-mortem carried out by pathologists so as to find out the cause of death and to elaborate further on the interaction between cells and the nutrients in malnutrition. These two sciences (i.e., clinical medicine and pathology supplied more information for the medical schools in addition to that already supplied by chemistry (food composition) and biochemistry (body composition and changes in malnutrition). At this time the major cause of malnutrition was defined as lack of proteins, while diseases like obesity and atherosclerosis did not come under the definition of malnutrition. The relationship between malnutrition and the socio-economic status of the community could hardly be defined in the then recognized scientific terms and very few people even attempted to look into this aspect of malnutrition.

STUDIES OF THE COMMUNITY

Although surveys were carried out in the 1940's and 1950's the main purpose was to find out how much of the nutrients were being taken by Tanzanians and what malnutrition diseases existed.

After independence, however, and due to the ideology of the new government, few nutrition workers began to look a bit further into the factors which led to malnutrition and despite strong opposition and despite from the professionals, a small, but important support, came from the politicians in power. A few general causes of malnutrition began to emerge as important aspects which needed much more research. Poverty, chronic and recurrent diseases, and lack of mental vitality among the people appeared to be strongly linked to cases of malnutrition.

Poverty

Since Tanzania is basically an agricultural country, poverty is more often than not linked to the ownership of land. But owning land alone did not necessarily result in being wealthier. There was a marked mal-distribution of income between the rural and the urban population with the urban dwellers earning on the average five to ten times as much as their rural colleagues. Job opportunities as well as education opportunities were more for the urban dwellers than for the rural dwellers and the urban environment as a whole was more attractive than in the rural areas. Accordingly there was and still is more malnutrition in the rural than the urban areas.

Disease

Medical services were relatively well established in the towns although the sanitary facilities and living conditions were also much better. Communicable diseases like TB, measles, etc., were very common in the rural areas and since medical services were so scarce, the mortality was inevitably high. At that time little was known about the effects of malnutrition on disease and mortality.

Lack of Mental Vitality

This is now described by the party as a state which results from prolonged oppression and control. People get into a state where they believe the poverty, diseases, high mortality, lack of food and the presence of many suffering people is a natural phenomenon, either brought about by God or other supernatural being and for which man has no control. For a community, this kind of attitude, if it affects all people, is fatal. In Tanzania many poor people were thus affected mentally and it is very difficult to revive and make them believe they can solve their own problems. The politicians have done it admirably to a large extent.

POLITICAL AWAKENING

Between the early 1960's and the 1970's the politicians in Tanzania conducted extensive campaigns of words and action. The main question that the nation addressed itself to was why were people poor, ill and unproductive. Confronted with this question, scientists began hot debates on what research they should conduct in order to answer the questions. Training institutions were also debating on what to teach; they immediately started special training and reorientation courses for new graduates and school leavers in order to stimulate new ways of thinking, and "plant" the new ideology among the educated. This included all medical workers trained within and outside Tanzania. The politicians, on the other hand, requested the leaders of the people, starting from the President to the village leader, to abandon their high positions and get down to the level of the people; live with the people, see their life, their problems and work out solutions with them.

Hence the Arusha Declaration which prevented leaders from acquiring a lot of wealth and rendering themselves less and less useful to the people. This led to many other actions by the government: to put the control of the economy in the hands of the Tanzanians and to give them more power to decide on matters that affect them; for example: decentralization of the administrative set up and decision making, nationalization of the main economy activities, villagization of all scattered communities, were among the many actions taken.

The university and other teaching institutions were specifically requested by the Government to reorientate their teaching and research work to the problems facing the country. Now if you put the same questions: What is nutrition science?; What is malnutrition?;

What causes malnutrition? and What is nutrition education?, the answers will be quite different from those given in the 50's and I would like to examine these a bit further.

THE CONCEPT OF HUMAN NUTRITION IN TANZANIA TODAY

After realizing the importance of human nutrition the Tanzania Government created an interdisciplinary Nutrition Institute to initiate and coordinate all nutrition activities in the country. The Institute is semi-autonomous, with a Governing Board and an Executive Director who controls the activities of the institute including the use of funds in bank account. Like all other government bodies the Institute requests funds from the general government funds annually.

Nutrition status is not only looked at as a function of nutrients upon living cells, but as a result of all activities and environmental conditions surrounding man. For example, malnutrition in a cotton farmer has as much, if not more, to do with the international price of cotton, fertilizer, insecticides and other farm implements required for growing cotton, as it has to do with the farmers lack of knowledge in planning and his weakness in bargaining prices, etc. It is thus futile to go on teaching the farmer that he is malnourished because he does not eat a well balanced diet; that his children should get eggs, milk, meat and so on; (this he knows a lot about already; even his greatgrandparents knew it). In those days children were always fed on milk or some other well balanced diet and every family kept animals for milk and meat. Certain foods, especially the rare ones like eggs, were forbidden, but there were alternatives.

Malnutrition is no longer regarded as an entity in itself but rather as a symptom of malnutrition of the society, except when otherwise proven to be due to an individuals' disorder. Likewise fever is a symptom of malaria and other diseases, and treating the fever alone will not cure the person. One has to find out the actual cause of fever and treat that as well. Only then will the person recover completely. Similarly Tanzania has conducted several studies to establish the true causes of malnutrition and all medical and paramedical students are now engaged in these studies in villages and towns.

NUTRITION IN SCHOOLS OF MEDICINE

The Dar es Salaam Medical School is slowly changing from the classical nutrition in Biochemistry, Physiology, Medicine and Pathology to the more socio-economic aspect of medicine now called "Com-

TABLE 1

FOOD AND NUTRITION IN UNIVERSITIES IN AFRICA

UNIVERSITY OF	NUTRITION TAUGHT AS BIOCHEMISTRY, ETC.	NUTRITION TAUGHT IN MEDICINE AND PREVENTIVE MED.	NUTRITION EXAMINED	ARE STUDENTS INTERESTED IN NUTRITION?	IS MALNUTRITION A PROBLEM IN THE COUNTRY?*	WHAT SUBJECTS WOULD TEACHERS LIKE TO BE TAUGHT?
Stellenbosch, S. Africa	No	As Dentistry	No	No	Yes	Basic Nutrition, Prevention treatment of dental & oral diseases
Oral & Dental Hospt. - Pretoria	No	As Preventive Dentistry	With Physiology	No	Yes	Physiology and Dentistry
Khartoum, Sudan	- Physiology - All five nutrients	No	With Physiology	Yes	Yes	Values of local foods
Dakar, Senegal	Medicine	No	as Medicine	Yes	Yes	Nutrition in relation to M.C.H. - Agriculture - Food production - Food distribution - Rural Education
Pretoria	Mixed in all	Mixed in all including Psychiatry	Yes	Yes	Yes	- Nil particular
Witwatersrand S. Africa	Yes	Yes	No	Yes	Yes	- Best taught mixed in Biochemistry, Physiology, etc.

* From my own knowledge.

TABLE 1 (cont.)

UNIVERSITY OF	NUTRITION TAUGHT AS BIOCHEMISTRY, ETC.	NUTRITION TAUGHT IN MEDICINE AND PREVENTIVE MED.	NUTRITION EXAMINED	ARE STUDENTS INTERESTED IN NUTRITION?	IS MALNUTRITION A PROBLEM IN THE COUNTRY?	WHAT SUBJECTS WOULD TEACHERS LIKE TO BE TAUGHT?
Orange Free State, Bloemfontain, S. Africa	No	Yes	No	Moderately	Yes	- Malnutrition - Obesity - Selective malnutrition
Matal, S. Africa	Mixed in all	Mixed in all	No	Yes	Yes	- Physiology, Biochemistry - Clinical disease
Addis Ababa, Ethiopia	Mixed in all	Mixed in all	No	Yes	Yes	- Parental & Infant Nutrition - Community acess to nutrition - Nutrition & Infection
Khartoum, Sudan	No	Yes & in Physiology	No	Yes	Yes	- Composition of local foods - PCM causes & prevention - Vitamins
Assint, Egypt	Mixed in all	Mixed in all including Surgery	No	Yes	Yes	- Metabolism - Nutrition disease - Preventive Medicine
Alexandria, Egypt	Biochemistry	Medicine	Yes	Yes	Yes	- Nutrition principles - Composition of diets - Vitamins - Nutrition disorder & disease - Dietetics - Nutrition & Commonhealth - Field Studies of Nutrition problems

TABLE 1 (cont.)

UNIVERSITY OF	NUTRITION TAUGHT AS BIOCHEMISTRY, ETC.	NUTRITION TAUGHT IN MEDICINE AND PREVENTIVE MED.	NUTRITION EXAMINED	ARE STUDENTS INTERESTED IN NUTRITION?	IS MALNUTRITION A PROBLEM IN THE COUNTRY?	WHAT SUBJECTS WOULD TEACHERS LIKE TO BE TAUGHT?
Tripoli, Libya	Mixed in all	Mixed in all	Yes	Yes	Yes	- Public health - Human development & child health
Cape Town	No	Yes & in Physiology	No Post-graduate Diploma in Therapeutic Dietetics	Yes	Yes	Nil specific
Salisbury, Rhodesia	No	Yes & in Physiology	No	Yes	Yes	Nutrition should be a separate dept. - NIL
Dar es Salaam, Tanzania	No	Preventive Medicine	Yes	Yes	Yes	Growth & Development Infant feeding Nutrition source Nutrition requirements Nutrition disease Nutrition status Causes of malnutrition Preventive measures
Butare, Rwanda	No	Specified in Dietetics	Yes	Yes	Yes	Malnutrition
Liberia	Mixed in all	Mixed in all	No	Yes	Yes	Nil specific

TABLE 1 (cont.)

UNIVERSITY OF	NUTRITION TAUGHT AS BIOCHEMISTRY, ETC.	NUTRITION TAUGHT IN MEDICINE AND PREVENTIVE	NUTRITION EXAMINED	ARE STUDENTS INTERESTED IN NUTRITION?	IS MALNUTRITION A PROBLEM IN THE COUNTRY?	WHAT SUBJECTS WOULD TEACHERS LIKE TO BE TAUGHT?
Ibadan, Nigeria	Mixed in all	Mixed in all	No	Yes	Yes	Basic Nutrition Human Nutrition Nutrition & Disease Health Education
Mondlane, Maputo	No	Ecology of Human Nutrition: - Ecological - Social & Economic aspects of food and nutrition	Yes	Yes	Yes	as in column 3
Nairobi, Kenya	No	Preventive Med. (details not sent to us)	Yes	Yes	Yes	

munity Medicine". The nutrients and their actions are still taught under Biochemistry and Physiology. Malnutrition is still taught under medicine especially in Paediatrics and in Pathology classes, but these alone no longer constitute an important part of human nutrition. The other part which is regarded as the human nutrition proper is taught under Community Medicine. Students study the effects of good and poor food intake on growth and body functions (work, etc.), dietetics based entirely on Tanzanian foods and particular attention is given to infant feeding. What was taught in Biochemistry, Medicine and Pathology is then linked to practical experiences when students spend 2-3 weeks in villages examining the people to find out, among other things, the nutritional status of the community. Back in classes the results of the survey are discussed at length to show the students the closely interwoven relationship between the socio-economic status, diseases, medical facilities, production, the environment, etc., and the nutritional status of the community. During leave some medical students are attached to the Food and Nutrition Center and here they work on specific problems and report back to their fellow students.

From Table 1 (a survey kindly conducted by Prof. Omololu, Nigeria) it is obvious that apart from the Universities of Mozambique, Senegal (to some extent), Ethiopia (to a small extent), Dar es Salaam, Alexandria (to some extent), Nigeria and probably Kenya the rest are still teaching nutrition under the old concept of nutrients and how they affect living cells.

The Tanzania case and I am sure that of Mozambique demonstrate very clearly the relationship between nutrition and the political ideology in a country. This in turn affects what is studied (in research, surveys and philosophy) and taught to students. Universities are or should be the origin of new ideas and ideology and if they are to be useful to a country they must focus on the problems of the majority and they will not miss the real causes of malnutrition.

NUTRITION AND PRIMARY HEALTH CARE SERVICES

II: TRAINING

E.F. Patrice Jelliffe, MPH

Div. Pop. Family & Intern. Health

UCLA, Los Angeles, U.S.A.

The discovery of nutrition as an important discipline by some health professionals is a very recent event.

If nutrition as a subject is neglected in the curriculum of medical and nursing cadres who are the most highly trained health professionals, there is little hope that primary care health workers will receive accurate or professional information from the top health hierarchy. In both technically developed and especially in Third World countries large numbers of individuals are denied the right to health and appropriate nutritions and, in an effort to remedy such social injustice the concept and application of primary health care has become a glimmer of light or a reality in some countries at the end of decades of desnetude and neglect in these areas[1].

TRAINING OF NUTRITION IN HEALTH CADRES

The educational background and length of training of different health cadres may determine in some measure the amount of nutrition included in the various programs, but this is not always the case[2].

TRAINING IN NUTRITION OF PHYSICIANS AND NURSES

Radical changes are often viewed with great suspicion and unease by the painstaking authors of curriculo for health personnel. Minor cosmetic changes may be made over the years with great reluctance, and these usually do not deal with the urgent needs of populations in terms of nutritional improvement but more often there appears to be a pressure to include in the Training Spectacular recent advances in Treatment affordable and benefitting a limited few.

If primary health care is to be successfully implemented, timid and desultory changes in curricula should no longer be tolerated. Scrutiny of curricula at national level is required as parochialism within medical and nursery schools breeds parochialism, and a much broader view of the important health and nutrition issues in the country must be included in curricula. Bennis[3] commenting in curriculum changes and organizational development generally has stated "that trying to reorganize a University if trying to reorganize a gaveyard". Possibly over the next years, if enthusiasm and research on primary health care is maintained such cyniscism will not be merited.

Training for all cadres of personnel should not be subject oriented by community geared[4].

Some behavioral scientists believe that strategies for curriculum change depend on three methods, power, rationality and re-education, and barriers to the latter include territoriality, priority conflicts, inertia and aggressive actions. Certainly nutrition as a discipline has been the victim of these many barriers[4].

The subject of nutrition should be given in the curriculum equal status for example with subjects such as pathology or anatomy, this may require immunizing the importance of more esotenic subjects presently taught which could be better learnt at post-graduate level for those specializing in these fields[5].

If this important and radical approach is taken a careful choice of areas to be covered must be actively guided by the needs of the community in each country and be supportive of any government food and nutrition policy if such exists.

Health personnel trained in medical and nursery schools often have a narrow chemical view of the patient and because of time pressure and their type of training are not concerned with the fate of the patient on discharge from hospital.

Emphasis must be strongly placed on the fallacy of this approach and knowledge of the living conditions of different socio-economic strata is required. This can only be taught by practical field experience gained early in the training process.

Training in nutrition for medical and nursing cadres[6] should include among other areas:

1. The assessment of nutritional status of both hospitalized individuals and the community. This requires detailed training in recording accurate measurements using accurate but simple im-

plements and should include particularly serial weight and height measurements of young children.

2. The nutrition needs of individuals in health and disease, and knowledge of the nutrient value of foods included in the diets and their nutrient cost.

3. The knowledge of prevalent nutritional diseases, their causation and prevention as well as programs existing in the country to combat these disorders.

4. The special needs of vulnerable groups such as pregnant, lactating women and young children.

5. The prevention of proteins calorie malnutrition in Third World countries, by vigorously promoting the use of breast milk and indigenous weaning multimites.

6. The implications of the need for culturally acceptable family spacing when appropriate to promote a better nutritional status among women and their children.

Training in nutrition for these cadres should be undertaken by professionals trained in nutrition (nutritionists, dietitians, etc.) and as much exposure as possible should be allowed to teaching from individuals in nutrition-related fields, e.g., agricultural extension officers, dentists, economists, etc. For example two recent international surveys undertaken by committee V/II of the International Union of Nutrition Sciences, among a variety of nursing cadres showed that nurses ill prepared in nutrition were the primary nutrition teachers in all nursing schools surveyed[2].

Teaching Materials

Should (1) be available; (2) be geared to the level of education of the students and appropriately reflect the nutritional deficiencies in the country.

Assistance should be given to training schools, usually lacking sufficient accurate documentation by local nutrition institutes, international, bilateral and other agencies to reinforce and keep up to date not only the students but the trainers.

NUTRITION TEACHING AMONG COMMUNITY HEALTH WORKERS

In recent years it has been realized that economic growth and ensuing benefits were not shared by the poor of the land and that an urgent need existed for new categories of workers to be trained.

These were usually chosen by the community, received a relatively short training which enabled them to help provide some essential health care to all people at community level[7,8]. This important new step has required both community participation[9] and a change in national health strategies.

However, the nutrition components of these programs must consider local cultural food preferences which are often sound and have evolved over a period of many decades.

Training in nutrition for these cadres will be basic, essentially practical and geared to the needs and nutritional deficiencies prevalent in the community with the assistance of senior health personnel community health workers may encourage villagers to diagnose their primary needs, such as better sanitation, clean water supply, etc., which are all included in the vast nutrition panorama.

Community health workers must:

1. Be taught the use of simple anthropology as a diagnostic aid in malnutrition (i.e. use of weighing machine, the arm circumference measurement for young children;

2. Recognize the value of indigenous crops available to the community and advise the judicious use of mixtures of foods providing essential nutrients, employing local methods of cooking, or the use of appropriate technology if this has been introduced at village level;

3. In rural areas, aided by agricultural extension officers, workers can advise families on the use of simple low cost gardens, fish ponds or small animal husbandry;

4. Community health workers should be knowledgeable regarding the importance and many benefits of breast feeding as well as the value of child spacing;

5. The value of immunization and other methods in the prevention of childhood diseases and their impact on nutritional status should also be taught to these workers.

A simple nutrition manual made available to the literate worker during his or her training would be most useful, and a pictorial booklet would be helpful to the illiterate.

CONCLUSION

Nutrition must be learnt as an integral part of general health care by the health cadres, and each of these has an important role to play. The present artificial division existing in many countries between levels of health workers has a negative impact on the health of nations. A multidisciplinary approach is required, but even more pressing is the recognition by all countries that a practical and inteligent approach to nutrition training is needed to speed up the process of "Justice in Health".

REFERENCES

[1] WHO. Primary Health Care Magazine of the Worls Health Organization, May 1978.
[2] IUNS. Nutrition Training in Schools of Nurse Auxiliaries. Committee V/II unpublished document, 1978.
[3] Bennis, W.G. "The leaning ivory tower". Jossey Blass, San Francisco, California. 1969.
[4] McGoyhie, W.C., G.E. Miller, A.W. Sajid and T.V. Telder. Competency-based curriculum - developmental in medical education - an introduction. Public Health Papers 68. WHO, Geneva, 1978.
[5] Jelliffe, E.F.P. Nutrition in nursing curricula historical perspective and present day trends. Moregrath, J. Trop. Pediat. & Env. Child Health. June, p. 149. 1974.
[6] Report of the Second Meeting of the IUNS Committee V/II. Nutrition education in nursing. J. Trop. Pediat. & Env. Child Health 21 (6): 345, 1975.

REPORT ON THE WORKSHOP ON NUTRITION EDUCATION

FOR MEDICAL AND OTHER HEALTH SCIENCES

The following observations regarding the overall presentation is provided for consideration in a final report:

1. It is agreed that nutrition education and training in all health education aspects should have an established and prominent place in the curriculum. Primary focus should be to improve the situation for nutrition education and training among all cadres working in the field of health, i.e., medical students, nurses, dietitians, home economists and dentists, as well as individuals working in closely allied fields, eg. agricultural extension workers. Nutrition plays a key role in the medical and surgical management of most hospitalized patients and proper therapeutic diets may well serve to hasten recovery and reduce hospitalization time — both critical personal and economic factors.

2. The concept of the establishment of a nutrition support service team should be the goal within hospitals and other institutions caring for patients as well as in community health programs. The critical elements of this team should consist of the following:

   a. A Physical(Clinical Nutritionist) leader who has the ability to coordinate and organize the team and most importantly can can get along with people — both professional and lay administrators and patients. <u>He or she must be above all nutritionally motivated</u>.

   b. Representatives from the discipline of Nursing who will have the basic responsibility for training all categories of nurses (in baccalaureat, diploma programs as well as auxiliaries, aides and community nurses) must be fully aware of the importance of a relevant, culturally acceptable diet for hospital patients, and which is also, as well affordable to them in discharge from hospital, and at community level.

   Prevention of malnutrition in all patients should be the goal of the nutrition support team: for example cancer patients who undergo surgery and subsequently receive irradiation treatment should receive nutritional support and our compassionate understanding and appropriate advice to the family from the combined team of the surgeon, physician, radiologist, nurse and dietetian. In industrialized countries the use of parenteral nutrition has become widely accepted as an important measure to feed malnourished patients who are unable to take food by mouth. Nurses play a key role in this area which requires specialized knowledge and skills.

c. The dietetitian is an essential team member with responsibility for translating the biochemical and nutritional facts into food — the ultimate final focus of nutrition. The dietitian plays a leadership in educating patients, students, staff, doctors, nurses and other medical personnel in nutrition management; plays a leadership role in developing food services which incorporate sound nutritional practices; and provides special education and guidance to patients afflicted by diseases controllable by diet and patients who can be helped to recovery and/or made more comfortable by modified food intakes.

d. The Pharmacist is a key team member — particularly in providing nutritional supplements, special nutrient combinations and the material and components involved in total parenteral nutrition.

e. Surgeons who will cooperate with the team and by placing deep vein catheters for total parenteral nutrition. The surgeon should in fact be a co-team leader with the medical clinical nutritionist component.

3. Community health workers and especially those operating at the first line of contact between the community and the National Health Care system, must receive a type of training enabling them to carry out essential nutrition and nutrition-related tasks at the community level. This worker should be especially equipped to understand the cultural and ethnic backgrounds of the people being served. In order to make a rough diagnosis of the community situation and its relation to malnutrition, such worker should be able to collect information about food availability, prevalence and pattern of infectious disease, availability of safe water. The training should be task-oriented supported by adequate skill, knowledge and change in attitude. The Community Health Worker, and especially the Primary Health Worker of the developing countries, is an important link between the community and the health care system, and in most cases will be responsible for the protection against malnutriton and promotion of nutrition.

4. Preventive medical approaches should be emphasized in conjunction with those of groups concerned with food production, economic betterment, child spacing and family planning. Also preventive medicine and nutrition is viewed as the cutting edge for prevention. The major causes of morbidity and mortality in both developing and developed countries may be related to nutrition, i.e., overnutrition and undernutrition. The preventive approach has the best chance of reducing the escalating health care costs we now are seeing and also to improve individual productivity and quality of living for which all are striving.

5. Nutrition education defined in common with all educational processes involves the following:

   a. Development of concepts, relating health to preventive medicine must be evolved having in mind the cultural context of each community or country.

   b. Developing and implementing a relevant body of knowledge showing the interrelationships between the basic sciences of Biochemistry, Physiology, Dietetics and Food Technology as well as the clinical aspects of nutrition relating to a whole list of clinical problems, i.e., whole list of medical problems such as obesity, atherosclerosis, hypertensive cardiovascular diseases, diabetis mellitus, and more especially in the Third World protein energy malnutrition, vitamin and mineral deficiencies.

   c. Development of skills relating to the application of the above body of knowledge, i.e., the <u>how to do it</u> aspects of nutrition at all levels of health services and sciences.

   d. Attitude development relating to the necessary socio-political-economic aspects of relating nutrition of Governments, Educational institutions and other representing the power structure and the decision makers relevant to the allocation of money.

6. Development of an International Nutrition Faculty and Educational and Training Institute.

To better developed and implement the previous goals and objectives stated above it is recommended that a center for the training of faculty be developed. Here leaders in the various aspects of nutrition science would work (probably as visiting teachers) to develop a broad based nutrition curriculum providing training for nutrition workers and educators and most importantly putting together teams of qualified and credible teachers who would go on short assignments on invitation from various medical schools and from institutions training other health professionals to provide a stimulus for nutrition education in the settings visited. A worthy team effort could be a two weeks educational program on clinical nutrition, basic nutrition and food science plus community nutrition which would be directed to the general public and community health workers. This two weeks intensified effort would involve a condensed and coordinated series of formal lectures extended into smaller workshops involving team members and representatives of the local institutions faculty. Appropriate clinical and grand rounds would be proposed in cooperation with specific clinical departments and their faculty; comparable "field trips" in community settings would be proposed in cooperation with community health care personnel. All clinical and grand rounds activities would be proposed in cooperation with specific clinical departments and also should include the faculty (medicine, nursing, dietetics, dentistry, pharmacy). It was also proposed that meetings would be held with representative students bodies and associations, the schools curriculum committee and official representative administrative officials of the University to discuss development and implementarion of nutrition education and training within the institution. Most important would be the identification of local faculty resources from which the nutrition education would have a continuing impetus. It would be important to "plant the seed" from which to grow the financial resources for nutrition education and training both within the University, the community and region, the country and also carefully selected nutritionally relevant food industries. It is most reasonable to give priority to developing countries in the application of an International Nutrition Education Program. Getting back to responsibilities for the Center for International Nutrition Education several important areas need to be identified:

- The development of a computorized and continually updated data base of potential faculty available for assignment and capable of doing the best possible educational job.

- The development of an audiovisual center for providing the best possible educational facilities that would aid the faculty in their educational efforts. It is anticipated that the audiovisual center would carry on a continuing research effort devising the best possible techniques and tools for "getting the message across".

- The development and maintenance of a publications service in the nutrition field that again would have the prime objective of providing relevant useful printed information to support the nutrition education faculty.

- The establishment of a critical evaluation office that would have the difficult task of setting up measuring guidelines that would be applied to the efforts of the nutrition education center and provide guidance for change as may be needed.

- Of course critical to success would be a planning, development and coordinating committee that would have a primary supervisory responsibility for the Center of International Nutrition Education and Training. Along with this there would need to be appointed, probably for a five year non-recurring ter, a Center Director.

- The organizational impetus for such a nutritional education program, international in scope, should probably be vested in and supported by the United Nations System of Health oriented organizations and derive financial support from such organization and from those countries utilizing the services provided.

During the two days informal discussion after the Workshop, the following further recommendations were made:

1) The need to elaborate the definition of the clinical nutritionist.

2) Expand the concept of Hospital Nutrition Support Team out to the Community working with the Community health team and with emphasis in Preventive Medicine and promoting health.

3) It is striking to note that medical people learn less about human nutrition than students in agriculture and veterinary learn about animal nutrition

4) Emphasis should be made on the role of the Community nurse as a health promoter and home oriented + community oriented nutritionists.

5) The need to include the dentist on the Nutrition Support Team.

6) The need to broaden the scope of distribution of the UN System (WHO) published materials related to nutrition and primary health care. It was decided to forward them for consideration to the next IUNS Committee on Nutrition Education and Training in Health Sciences.

THE NUTRITIONIST-DIETITIAN IN LATIN-AMERICA

Micheline Beaudry-Darismé, Ph. D.

Advisor in Nutrition Education, Pan American Health Org.

525 23rd St. NW, Washington, D.C. 20037 - USA

During this short presentation, I will try and give you an overview of the evolution of the profession of nutritionist-dietitian in Latin America, of what are some of the present concerns or preoccupations of the profession and of the directions beeing taken by their training programs.

Before 1966, there were less than a dozen of 2-3 year training programs for dietitians in Latin-America. These were usually oriented to the training of personnel to function in hospitals or other institutional food services, even though some of them did provide the future dietitians with a very broad perspective in nutrition, including some elements of community nutrition. These programs seldom provided a university degree.

In 1966 a Conference was held in Caracas where the directors and other teaching personnel of the existing schools of nutrition and dietetics in Latin America met with technical representatives from the Ministries of Health, from UNESCO; from UNICEF and PAHO, to review the situation of human resources in nutrition in Latin America and the corresponding training programs. This assessment was done considering the nutrition problems of the countries and the strategies being proposed to deal with them.

Following this analysis, the group recommended that existing courses to train dietitians be modified to four-year university courses that would train professionals to function in both public health and community nutrition programs as well as in institutional services. This new person would be called a "nutritionist-dietitian", and was envisioned to be the basis of any nutrition program in health, whether at the hospital or service level, or in prevention and health promotion. It was recognized at that time that:

1) the nutritionist-dietitian could not be responsible for all the activities of the program and would therefore have to rely heavily on the whole health team;

2) there are sometimes factors that prevent the health team from being responsible for all the nutrition activities and in such cases the development of auxiliary personnel could be justified, if supervised by specialized personnel.

The nutritionist-dietitian was therefore viewed mainly as the person responsible for planning for, orienting and supervising the integration of nutrition in health services from the planning level to the actual provision of the services. That is the nutritionist-dietitian would not himself be generally involved in direct community level service since this was incompatible with a rational use of limited resources.

Recommendations were therefore made concerning specific aspects of curriculum planning for this new professional. A subsequent meeting of all Directors of schools of nutrition and dietetics was held in 1973 in Sao Paulo to review progress. By then some 20 programs existed, following to one degree or another the recommendations of the Caracas Conference. Recognizing the increasing number of schools and the need to further review and expand existing recommendations, the group created a Study Commission which has since met three times for that purpose. Their last meeting was held in Brasilia last August. There are now some 40 training programs in 15 countries of Latin America which attempt to follow the recommendations made by these groups.

In 1976, PAHO made a survey of the then existing training programs to find out to what extent they could or wanted to implement the existing recommendations, and what were the areas they considered merited further study. Among the major concerns expressed by their directors were the following:

1) the need to review and evaluate current programs, curriculum, plan of studies, to see to what extent they reached the proposed objectives and actually contributed to training the professional nutritionist-dietitian for the role described above;

2) the need to review the role of the nutritionist-dietitian and its field of action, to be able to have a strong basis from which to realistically review curriculum and plans of studies.

Although it was recognized early on that the nutritionist-dietitian should also play an active role in other sectors, e.g., education and agriculture, the guidelines for training programs were really made for the nutritionist-dietitians in the health sector, and most graduates are in fact working in that sector.

It is estimated that there are now some 5000 nutritionist-dietitians working in the health structures in the countries of the region. Some 1500 to 2000 of these graduated from the previous programs to train "dietitians" and have not yet been able to revalidate their degree. To meet the norms established in the 10 year health plan for the Americas, it is estimated that another 5000 would be needed in the health sector. It is not known how many other nutritionist-dietitians work in other public or private sectors such as those relating to education and agriculture, but there is an increasing demand in these, and interest on the part of graduates to work in them.

Data concerning the actual roles carried out by nutritionist-dietitians on a region wide basis are still scarce. Present experience in some of the countries seems to indicate that while many have in fact been able to develop a role as a coordinator, planner and integrator of nutrition as described above, several have continued to work mainly in direct community services and have not always been able to function as a resource in an interdisciplinary team.

On the other hand, more and more nutritionist-dietitians recognize the complexity of the nutrition problem in the region and the importance of multisectorial planning for nutrition interventions from an interdisciplinary approach. What is the contribution of the nutritionist-dietitian in this new area of activity, and how is he prepared to carry out the corresponding responsibilities? These are questions which are raised with increasing frequency among nutritionist-dietitians in Latin America and which deserve serious consideration.

Aware of these concerns, and reemphasising earlier recommendations of the 2nd Conference of Directors of schools of nutrition and dietetics, the IIIrd CEPANDAL meeting last year recommended that each country try and do an actual study of the role and competencies needed by present-day practitioners in nutrition and dietetics as a basis to review and evaluate present curriculum. In the province of Cordoba in Argentina, a small study of that type has recently been carried out and discussed by the professional association, who in turn is discussing it with the authorities of the corresponding training programs. In Colombia, the newly created Association of Schools of Nutrition and Dietitics, along with the National Association of Nutritionist-dietitians and with the cooperation of the Ministry of Health, have planned and are about to carry out a similar study. Other countries of the Region are presently considering embarking on similar ones, the results of which can be extremely useful to review present guidelines and programs to train professionals in function of expected roles and competencies based on the needs of national nutrition programs.

In the meantime, more and more training programs are utilizing new methods of teaching, where the emphasis is on developing problem-solving skills rather than on the traditional didactic approach. As part of this change, the student has an earlier contact with the service area and learning experiences are planned in a graduated manner with the participation of students and professionals from other disciplines. This also contributes to bringing the future professionals in closer contact with their future role.

While the countries are training professionals who must have a broad preparation that will allow them to evolve and progress in the course of their professional career, they also recognize the need to train them adequately for the reality that awaits them in the few years following graduation so as to assure the most productive use of this limited resource.

It is evident from all this that while general guidelines can be prepared for Latin America, the specific training programs must vary from country to country on the basis of their national programs and resources. It is also evident, given the changing characteristic of our present day society, that continuing education of graduates is an essential component of training throughout their professional life. In several countries, such continuing education programs have in fact been developed.

In summary, the profession of nutritionist-dietitian in Latin America has shown considerable evolution in the past 12 years to adapt itself to the reality of the countries and even to participate in planning for the modification of this reality, and present evidence indicates that the next 12 years will witness a new and significant evolution to continually improve the quality of the professional and its relevance to the nutrition problems of the Region. A key element of this evolution has surely been the continuous exchange between training and service.

COMMUNICATIONS COMPETENCIES

OF THE NUTRITIONIST-DIETITIAN

>   Susana J. Icaza, M. Sc. in Hyg., Ed. D.

>   INCAP

>   Guatemala City, GUATEMALA

INTRODUCTION

The nutritionist-dietitian plays many roles while fulfilling her professional responsibilities. Some of these roles require scientific expertise and technological skills, in order to carry out specific activities such as program planning, selection and use of equipment, food handling and diet calculations. There are other roles, still more important, for which not only scientific and technical expertise but abilities for human interaction as well, are indispensable. Working with people means communication expertise whether these interactions are oral, written or non verbal.

Communications can be of a direct nature, either through face to face contacts or through written messages; they can also be indirect or meant to reach a group or a community.

Whatever the type of communication, there is a message to be sent to a person or a group of persons, delivered through a specific channel or several channels, which will elicit a response that the nutritionist is greatly interested in receiving.

Messages will vary according to their content, the recipient and the media. In general, their purpose is to convey concepts. Messages will, at the same time, convey attitudes, practices and habits that constitute the frame of reference within which those concepts are presented. Therefore, the message need to be coded according to the language used by the recipient.

ROLE OF THE NUTRITIONIST IN COMMUNICATIONS

In their daily work nutritionists communicate with the people they serve: patients, physicians, nurses, mothers, students, teachers, community leaders, home makers, agricultural extensionists. They all need help in terms of nutrition advice. In this case, the role of the nutritionist is to disseminate the nutrition gospel and the message must be shaped so as to fill the needs of every recipient.

Nutritionists also communicate with the people they work for. In either administrative or scientific roles these persons need to be informed of the purposes of the nutrition program, its development and results. In this situation the nutritionist is the promoter of the program and must convince others of its importance.

On the other hand, nutritionists communicate with the people who work under their supervision. These persons need to know exactly what is expected from them and how well they are doing. For them, the nutritionist is the leader who must orient them, interpret to them the regulations, as well as understand their limitations, and provide them with advice and training.

Communication is purposeful. Nutritionists communicate with people in order to give and to collect information, to coordinate, to obtain cooperation, to motivate, to explain, to demonstrate, and to evaluate. All these communications have a common aim: to improve the nutritional status of the population through the promotion of better feeding practices and the increment of nutrition knowledge.

COMMUNICATIONS IN NUTRITION EDUCATION

After a long period of experience, nutritionists working in nutrition education programs now recognize that the success of the program lies in counting on good communication channels.

Information must flow through the channels and reach every person involved.

At every step of the channel the message must be clearly defined and transmitted in terms that are clear for the receiver.

The triangle of Fig. 1 represents the different degrees of nutrition expertise required to participate in a nutrition program.

At each level, different communication skills are required in order to transmit the nutrition message to the next level.

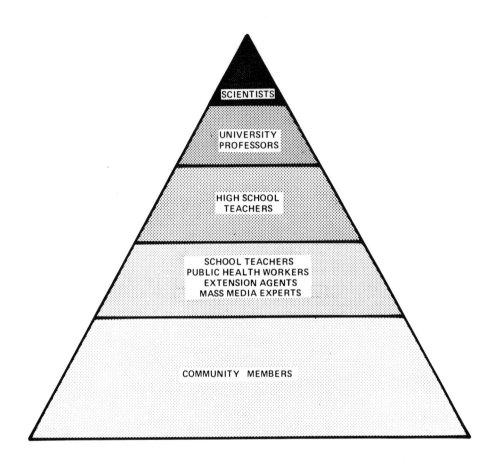

Fig. 1 - Degrees of nutrition expertise needed for a nutrition program. The intensity of the color shows the concentration of knowledge required and the width of the band represents the numbers of transmitters needed in each category.

Thus, a university professor must be able to teach college or high school graduates; a high school teacher must learn to communicate with youngsters; a teacher must know how to talk with children; a physician, a nurse, and a social worker must learn to communicate with patients and parents; an extension agent must know how to communicate with rural families, and people working in the press, radio and T.V. must be able to communicate with the population at large.

Each one at his own level is supposed to possess the necessary communication competencies in order to interact effectively with his receptors. No matter how much knowledge they possess, they must know their audience and share with them the same code; otherwise there will be no communication. It is like trying to use a pay phone without having the right change; no matter how much currency in higher denomination bills people may have, obviously, only the right coin will do the miracle.

If we break this triangle into several smaller triangles, one for each discipline, we find a small triangle for the public health team, another one for the agricultural extension group, and a much bigger one for the educational system. There is also another triangle for the mass media. Although there is some overlapping (in a family some members receive one channel and others listen to another), people in each channel or discipline have very well established habits of communication (see fig. 2).

Faced with this situation, the nutritionist working in a nutrition education program, needs to understand the conditions under which the nutrition message is to be communicated through each one of the channels and serve as an adviser to all of them. In such a program the nutritionist is asked to communicate effectively with people at _all_ levels and through _all_ channels.

In order to ensure good communication some countries have organized information centers or authorized sources located above all channels, which can communicate reliable information on nutrition (see fig. 3).

This measure helps to reduce the number of interfering factors and facilitates communication. This information is released mostly in printed form: bulletins, pamphlets, charts, books, journals, press articles, posters, slides, movies, etc. adapted to the different groups. At each channel, nutritionists help the people in charge of the program to do the necessary adaptations for the receivers at different levels, and act as consultants in the preparation of audiovisual materials. Thus, we see the role of the nutritionist as communicator is a complex and an essential one.

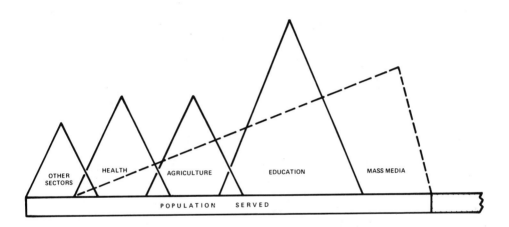

Fig. 2 - Overlap of communication channels.

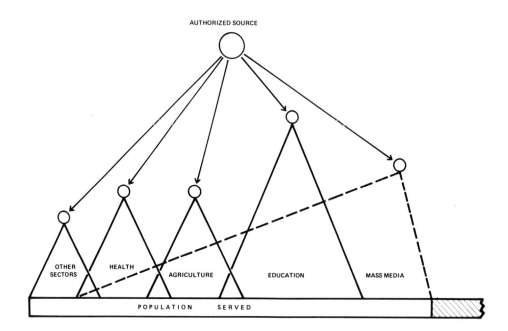

Fig. 3 - Position of an authorized source with respect to different communication channels.

ROLE OF THE SCHOOLS OF NUTRITION

The ability to communicate effectively requires the acquisition of a series of concepts and practical experiences in the communications area that will facilitate the development of skills and proper communication habits and attitudes. Obviously, all of them represent a great challenge to the schools training nutritionist-dietitians.

To enable future nutritionists to communicate effectively, the curriculum of the schools of nutrition must provide a theoretical frame in social and pedagogical sciences, and sufficient practical experience to develop the above mentioned skills, habits and attitudes.

Courses like general psychology, social psychology, educational psychology and social anthropology will help the student understand the human being, his motivations, his way of expressing himself, and its reactions to different types of stimulus. The student will also learn about the individual types of behavior and group dynamics.

Courses in education will teach the student how to organize his subject matter and how to define the different messages. Still more important, these courses will teach the student how to decide on the amount of content in a message, and the frequency of its presentation.

Courses in communications will help the student understand the communication process, the factors affecting it in a positive or negative way, and how to improve communications. Nutritionists should be able to detect communications problems and establish preventive measures in order to communicate effectively.

Courses in audiovisual aids will give the student practical experience in planning and defining the message and interpreting it according to the audience. Teaching experiences and the use of audiovisual materials in mass media programs are excellent opportunities to develop communications skills.

All this knowledge must be integrated within specific experiences in nutrition education, directed to determine nutrition knowledge of target groups in the community, its relation to their eating practices, and the way by means of which nutrition information reaches these groups. Small communications research projects can be carried out by students so as to develop skills in identifying the best channels of communication for nutrition messages in the community, and the factors that affect the communications process.

These experiences must be offered at an early stage of the career and the development of skills observed throughout all of their studies, until graduation.

ROLE OF PROFESSIONAL ASSOCIATIONS

Associations of nutritionists should also plan continuing education programs to improve the communications competencies of their members.

In this way nutritionists will be more effective communicators and, consequently, better educators and happier and more fruitful professionals.

CONCLUSION

Nutritionists work with people. Their role as communicators is essential for their work. They can learn to be good communicators. Schools and associations can provide this training which is a must if they are to play their role with the maximum efficiency.

THE ROLE OF THE NUTRITIONIST IN EXPANDING

COMMUNITY NUTRITION EDUCATION PROGRAMS

Helen D. Ullrich, M.A., R.D.

Director, Society for Nutrition Education

2140 Shattuck Av., Suite 1110, Berkeley, CA 94704 - USA.

I would like to present some ideas on what may be a relatively new role as an advocate to increase programs for nutritionists and dietitians. I would like to describe something about

1. The background of knowledge the nutritionist needs;

2. Some examples of arenas where the nutritionist can be effective in creating programs;

3. The role in assuring effective programs·

4. The need to identify the qualifications of the persons carrying out the program.

The dietitian and nutrition educator today must be a facilitator in many arenas to assure that nutrition information and education is incorporated into community food, nutrition, and health programs at a wide range of levels. We have been hearing a great deal this week about the need for multidimensional programs. This means that there must be nutritionists who develop knowledge and/or expertise in the legislative, economic, agriculture, education, social, consumer, and legal issues in addition to a good basic training in nutritional sciences in order to be effective in increasing the level of community nutrition education programs. Nutritionists who develop these areas of expertise can be instrumental in formulating governmental policy at the national, regional, and local levels. This is a major change in roles which traditionally focused on research, applied nutrition, and dietary roles. However, this does not suggest that all nutritionists should have expertise in all the above mentioned areas, but rather that now there are opportunities for nutritionists to further specialize and direct their efforts in a variety of arenas.

In the time since the Society for Nutriton Education was established ten years ago, there has been a significant growth in community nutrition education programs in the United States and a significant growth of nutritionists who are knowledgeable about a wide variety of issues. The rewarding part of this growth is the involvement of nutritionists in identifying needs, working with key groups, and developing the technical expertise to carry out such planning which has resulted in increased programs.

Probably a significant step in the nutritionists' contribution in this era of expanding nutrition awareness in the United States is attributable to the White House Conference on Food, Nutrition, and Health which took place in December, 1969. This Conference contained a series of panel groups addressing themselves to a wide variety of nutrition and food concerns, including nutrition education. One of the important outcomes of that Conference was the fact that nutritionists gained a greater awareness of the multitude of concerned groups who had a mutual concern about the social need to provide food to people who are malnourished and hungry. By forming coalitions with consumer, civil right, religious, and political groups, the almost immediate result was the expansion of the food stamps program which provided additional income for food, feeding programs in schools and day care centers, supplementary foods for pregnant women, nursing mothers, and children in the beginning years of life, and congregate feeding programs for the elderly which Dr. Mayer has already described. However, there is a growing awareness that to be effective every nutrition intervention program should have an education and training component which addresses the social, cultural, economic, and agriculture policies of the community.

Therefore, in order for the nutritionist to be effective in expanding programs, there is a need for understanding of governmental structure, agricultural production, economics, food distribution systems, and health and education delivery systems. For those nutritionists who are interested in policy making there is also a need to understand the many competing forces in order to demonstrate where nutrition programs fit.

The nutritionist may function at several different kinds of levels to be effective in creating programs.

In the United States this has been done in several ways, such as:

   1. Testifying in public hearings which are conducted by government agencies or interested legislators. These hearings are at local, state, and national levels.

2. Serving on the professional staff or acting as an advisor to concerned legislators.

3. By pointing up the need and establishing the position of a nutrition specialist in an agency such as the Federal Trade Commission which regulates advertising and access of information to the public.

4. By raising the public awareness to the need via mass media, work with city officials, and hospital and school administrators.

If there is to be a nutrition education component in all nutrition intervention programs, it is essential to develop a plan to expand the effectiveness of the program. The education components should reflect the needs of the population involved in the program. The nutritionist can be instrumental in identifying these components at several levels.

First of all, to identify what core of information consumers need in order to function adequately to achieve nutritional well-being. Second, there should be identification of specialized needs for particular intervention programs with focus on needs for periods of stress in the life-cycle and nutrient needs in relation to identified food habits. This information should be integrated with the specialized needs of the particular nutrition education program.

Because nutrition education is a process by which people learn to use information to affect the nutritional well-being, it is necessary to develop mechanisms of coordination with others who have an impact on this learning process. All too often, nutrition education has been considered an end product rather than a process. After the nutrition education needs are identified, it is important to coordinate with the full range of professional and lay persons who would be involved.

Concurrently with the advocacy to expand programs, it is useful to identify the kinds of expertise needed. For example, when funds were appropriated this year for nutrition education in schools, the Society for Nutrition Eudcation were ready with a recommendation of the qualifications of a nutrition education specialist working in schools. SNE recommended that this person have advanced specialization beyond the bachelor's degree. Their first concern is that this person not only have training in advanced human nutrition but also have training in the basic education concepts, including curriculum development, educational psychology, and/or early childhood education. Additionally, there should be some training in the behavioral sciences and knowledge of school and community organizations.

While these qualifications pertain particularly to the nutrition educator in school systems, it is hoped that similar qualifications can be developed for a variety of nutrition education programs.

QUALIFICATION OF THE NUTRITIONIST-DIETITIAN IN ORDER TO

PARTICIPATE IN COMMUNITY BASED HEALTH PROGRAMMES

Sonia Moreira Alves de Souza

Director, Nutr. Inst. of the Fed. Univ. of Rio de Janeiro

UFERJ - Ilha do Fundão - Rio de Janeiro, BRASIL

I. INTRODUCTION

Recognizing the afluence of nutritional problems on the levels of both the population's health and the economic-social development of the countries as well as its multicausuality, Latin-American governments have undertaken great efforts toward the search for their solutions with a multidisciplinar and multisectorial emphasis. Evidence of this fact lies in the number of countries which, have over recent years, shown concerned in formulating their National Food and Nutrition Policy, thus assuring the definition and the execution of a joint of social, technical and legislative actions, with the objective of the permanent improvement of the population's nutritional status.

In the health area, the expansion of primary services with the community participation has been the governments constant concern and in this sector, Nutrition has been affored an importance never given before. However, one of the greatest limitation within this strategy, in relations to the expansion of services in the field of nutrition, lies in both the shortage of specialized personnel and in the lack of nutrition knowledge of non-specialized personnel.

Being aware of the importance of the professional Nutritionits-dietitian as a technical element fot the development of the programs and as a multiplying agent with responsabilities of promoting the better use of Nutrition knowledge and the training of non-specialized personnel in this specific area, the countries of the Continent have, for over a decade, tried to permanently improve this academic training. As an example we have a number of Conference on Nutritionists-dietitian's Training and the CEPANDAL meetings which have been held since nineteen sixty-three (1963) supported by the OPS/OMS.

During the last few years, the Brazilian Government has been given high priority to Nutrition, specially in the Health sector where it has been conducting a large scale program directed to the group of largest biological vulnerability, that is, pregnant and lactating women and preschool children. The government has also been concerned with the proper professional training of the nutritionists-dietitian whose initial awareness has been stamed in 1975, at Seminar promoted by the National Institute of Food and Nutrition (Instituto Nacional de Alimentaçao e Nutriçao - INAN) with the purpose of study and diagnose the situations of the Schools and Graduate Course in Nutrition.

It is worthwhile to point out that, in spite of the regionalization of the health services and of the significant expansion of the nutrition actions integrated to them, which would imply an increase in the demand for the nutritionist-dietitian as a member of the regional multidisciplinar team, it has been observed that this field is the one where the rate of job openings is the lowest and, therefore, has the smallest number of this professional working.

THE NUTRITIONIST-DIETITIAN IN COMMUNITY HEALTH PROGRAMS, THEIR RESPONSIBILITIES AND QUALIFICATION

Any correctly health program considers actions to each one of the health aspects, among which nutrition is one of the most important.

Its execution requires a multidisciplinary team of which the nutritionist-dietitian is a fundamental component in the quality of technical element performing a most important role in the development of all the nutrition actions integrated into health programs.

As a member of the health team, one of the main responsilibities of the nutritionist-dietitian lies in the planning, evaluation and supervision of the nutrition actions included in the general health program, a work that must be jointly performed with the entire team, since nutrition is the fundamental part of a whole: health. Therefore, its activities must be integrated to the health program and not viewed as an independent and isolated supplement. This joint work contributes still further toward promoting, among the others members of the team, the interest and the knowledge about nutritional problems and more adequate activities, in the effort to fight them.

Another important responsilibility of the nutritionist-dietitian lies in the training in nutrition of the health staff. This contributes toward the better development of nutrition actions, and is a way by which the nutritionist-dietitian's knowledge can be multiplied in order to better serve the community.

Research and advising are also included among the professional's technical responsibilities. In regard to research, the professional participates and collaborates, among others, in the planning and execution of research concerning the population's nutritional status and in the operational studies which will afford the availability of useful methods and procedures toward the development of nutrition activities which would also be used as a base in order to establish and/or modify guidelines related to the referred activities. As for advising, within the team, the nutritionist-dietitian is the professional specialized in nutrition, and therefore, one of his duties is to provide advice to the other members of the staff.

In spite that the technical responsibilities have been stablished and accepted, in general lines, by the countries since long years, in practice, it has been observed that these responsibilities have been impaired in favour of administrative ones. We consider to be a most urgent need for each country to define more extensively the nutritionist-dietitian's responsibilities and activities in the Public Health Programs, based on a previous study and according to the characteristics of the existing programs and availability of the professional resources, thus guarantying the fulfillment of the technical responsibilities.

In relation to this training, during the academic education, the nutritionist-deititian must acquire knowledge and develop abilities and skills by which the professional will be qualified to, efficiently, assume the above discribed responsibilities. Such knowledge and skills can be summarized as follows: basic education of food and nutrition sciences; experience in planning, organization and supervision of the different prevention, protection and rehabilitation programs of the community nutritional status, basic knowledge concerning the sciences of Public Health, Applied Nutrition, and Social and Behavior Sciences that will be the foundation for the professional's work on permanent education, both in the Community as well as in the team of which he is a member.

We believe that there is a general consensus in all the countries in regard to the curriculum to be offered, as well as, the educational experiences required for the training of the professional nutritionist-dietitian in public health. The Director's meetings of the Nutrition School and CEPANDAL have greatly contributed in this sense and the schools have not measured efforts toward putting their recommendations in practice.

In Brazil's specific case, as already mentioned in recent years, Government has given great emphasis to the education improvement of this professional. Nonetheless, we are obliged to admit certain limitations in regard to what is specifically referent to the Nutritionist-dietitian in the Public Health area, which we will next consider.

In the first place, the program contents of the subjects in the area of Public Health and Applied Nutrition, need a revision and adaptation to the country's conditions.

Upon revising these contents, emphasis should be given to the following aspects: epidemiological emphasis in the diagnosis of nutrition problems and the selection of preventive measures; concept and integration methodology of the nutritional activities into health progams; health planning specially of mother and child health programs; supervision and elaboration of guidelines for nutrition programs.

Another important point lies in the urgent need to create fields for the practice of teaching in Public Health with the objective of offering the student the opportunity to benefit by supervised professional trainning. This is made even more difficult by the inexistence of the professional nutritionist-dietitian on the Public Health team, depriving the trainee of the contact with the already active professional. In addition to this, the problem is agraved by the shortage of teachers in the Public Health Nutrition area.

Thus, human resources continue to be the critical point in the integral attention of health, and of course nutrition. It is necessary, therefore, to plan the academic training and the utilization of the nutritionist-dietitian so that not only their category, number and functional situation but also their education, experience and availability are understood. This requires precise definitions on the duties of this professional which must be carried out as a process under continuous up-to-dateness.

CONCLUSIONS

Based on the herein submitted in this paper, we can conclude that:

- the Latin-American government have, in recent years, alloted high priority to nutrition;

- in the health sector, nutrition has been given an importance never considered before;

- for the development of nutrition actions, integrated to the health programs, one of the great limitations lies in the shortage of specialized personnel;

- recognizing the importance of the nutritionist-dietitian professional in the development of nutrition programs, the improvement of professional training has been of constant concern;

- the directors's meetings of Nutrition Schools and CEPANDAL, promoted by the OPS/OMS, have contributed toward the improvement of this professional education standard;

- the nutritionist-dietitian responsibilities in Public Health proprams are requiring of revision in order to protect the technical activities which, in practice, are impaired in favour of the administration ones.

- in the specific case of Brazil, the situation is as follows:

  1) the Public Health field has the lowest number of active nutritionist-dietitian, due to the small number of job openings;

  2) the creation of new nutrition schools has over the last few years, been agravated for the shortage of specialized teachers in the Public Health Nutrition area;

  3) with the objective of better qualifying the nutritionist-dietitian professional in the Public Health area, it is required to consider certain aspects specially the creation of fields for supervised professional training;

  4) the non-inclusion of the nutritionist-dietitian in the sanitarian careed prevents this professional from integrating the Public Health team.

RECOMMENDATIONS

1 - Promote the academic training, utilization and distribution of the human resources in nutrition, for health sector, according to the general health planning.

2 - Reformulate the academic training process of the professionals in nutrition, incentivating the performance in multi-disciplinar and multiprofessional teams.

3 - Consider the need to conduct an extensive study of human resources in nutrition as the main objective of subsidizing the redefinition of the existing models in using the personnel and in adjusting their academic training.

ADAPTATION OF THE TRAINING OF

DIETITIANS TO THEIR ROLES

       Y. Serville

       15 Rue Lakanal

       75015 - Paris, FRANCE

It is clear that the training of dietitians must be oriented towards making them competent to carry out the different tasks incumbent to their profession. We must therefore examine a definition of the profession in order to analyze these tasks.

An official definition of the profession, laid down by the International Labour Office in Geneva was coded as follows: Group 0-69 Dietitian dan Public Health Nutritionist. "These workers plan and supervise the preparation of diets for individuals or groups, supervise and evaluate nutrition elements of health programmes and assist in appraising the various factors related to nutrition and food problems in the community. Their functions include: planning and supervising the preparation of therapeutic or other diets for individuals and for groups in hospitals, institutions and other establishments, and for workers in particular sectors; participating in programmes of nutrition education and in nutrition rehabilitation activities; planning and coordinating nutrition programmes and advising on nutrition aspects of community food problems and health problems".

If we analyze this very wide definition, we can see that three types of activity are possible for dietitians:

- organization and supervision of cattering for large groups of consumers (hospitals, schools and workers canteens, etc.);

- organization and supervision of therapeutic diets;

- participation within community and preventive health services (programmes of applied nutrition, education in nutrition etc.).

If, in theory, all dietitians should be able to assume these roles to a greater or lesser extent, it appears in practice that, depending on the country concerned, priority has been given to one or other of these activities. Consequently, the training now received by future dietitians gives more weight to some of these sectors.

But needs are changing and, in certain countries, the training of dietitians must be modified or perfected in order to develop their activity in new fields.

The examples which follow to illustrate this point are drawn from the nine countries in the E.E.C. (Belgium, Denmark, France, Federal Republic of Germany, Ireland, Italy, Netherlands and United Kingdom). The associations of Dietitians of these countries created in 1972 a Work Committee (C.A.D.E.C.) to study in common the problems of the profession in West European countries.

The Committee has met on seven occasions. At the first meeting (1972) we outlined the education and training of dietitians in each country.

We then noticed differences in the length and orientation of trainings. These differences arose precisely from the conception of the role of the dietitian in each country.

Thus, in Denmark, dietitians worked mainly in collectivities to organize and carry out catering; few were employed in therapeutic dietetics. On the other hand, in France and United Kingdom, the vast majority of dietitians supervised the diets of patients and very few worked in collectivities or in community health services.

In 1973-74, the Committee studied the "Role and place of the Dietitian". The report published in each country was aware of the different aspects of our work and encouraged all of us to broaden our approaches to this very wide and fullfilling profession of dietetics. In 1977-78 the Committee discussed the role of the dietitian in prevention and in community health.

These studies led to reports, translated into the different languages of the member countries of the E.E.C. and distributed in each country by the National Association of Dietitians. These reports have already had results by bringing about, in certain countries, an evolution in the tasks entrusted to dietitians which, consequently, has led to changes in their training programmes. For example, in Denmark, a supplementary year's training in therapeutics has been created. It has enabled posts of therapeutic dietitians to be created alongside the already existing posts of dietitians supervising general catering (Table 1).

TABLE 1

TRAINING OF DIETITIANS IN DENMARK

YEARS		TRAINING HOURS	MAIN SUBJECT	OFFICIAL NAME
1/2	Dietetic School	600	Nutrition Food Technology	DIETITIAN
1	Hospital Kitchen		Practical work	
1/2	Dietetic School	500	Dietetics Economics	
1	Working			
1	University Training Hospitals	525 6 weeks	Therapeutic Diet Clinical Work	Therapeutic Dietitian

However, the main problem which still poses itself for the majority of European countries is that of the role of the dietitian in prevention and her insertion in community social services.

Many diseases, in these countries are, more or less, related to food habits. The consumer must take his prevention into his own hands but first, he needs to be aware of this and to be educated and given guidance in order to succeed. In these countries, there can be seen a movement not to confine the dietitian to the hospital but to consign to her duties related to prevention.

This implies changes in the initial training of dietitians or eventually possibilities of complementary training to be acquired during employment. Solutions to these new problems are different depending on the existing systems of training.

Two examples can illustrate this point:

- In Germany, basic training is mainly oriented towards therapeutic dietetics. This training may be preceded by an year at dopitals and in the Community. They have to provide an advisory Serv-

mestic science school or an year's practical work in a hospital. Training lasts two years at a school of dietetics (2,300 hours). The diplomees are "governmental registered assistant in therapeutic diet" and can work in hospitals (Table 2).

After four years of professional experience, those who wish may receive further training at a specialized school (520 hours) and thus become "Diet Kitchen Leader".

After one or two more years experience, those who wish to specialize in Community Health may follow a second period of training (1,020 hours) in public health and become "Nutrition Medical Adviser". At the present time 80 dietitians in Germany are working in these Community Health Services.

In United Kingdom, basic training is longer and more thorough. The present tendency is to prolong these studies and to replace the "Diploma" obtained in 3 years by a "Degree" prepared in 4 years. At present, the two types of training co-exist; 63% of the students still study for the "Diploma" and 27% for the "Degree". But, as from 1984, only the "Degree" will be kept.

The purpose of these training programmes is to prepare future dietitians for different tasks (Table III). The scolarity comprises 3 years at college (2,095 hours) with a large part devoted to basic sciences, nutrition and therapeutic dietetics. This scolarity is followed by an year of practical courses before obtaining a Degree in Dietetics. The holders of this Degree are qualified to become "Registered Dietitians". To date, no further training is envisaged during employment.

Thus, 2 very different types of training can be seen:

- In the first system (Denmark, Germany) the basic studies lead only to posts of limited responsibility, but they can be perfected, for those who wish to and who are capable, by ulterior training leading to new orientations. The advantage of these systems is that they allow a great flexibility in the training programme which can in this way adapt to new needs and new outlets in the profession.

- In the second system (United Kingdom, Belgium, Holland, France) the basic studies are longer and the scientific level higher. All dietitians in the same country receive the same training but, to date, there are no official structures leading to training at a higher level or for the adaptation to new responsibilities.

In the United Kingdom, the National Health Service was reorganized in 1974. This new arrangement has given dietitians the opportunity to work in the Community Health Service. Over 300 district dietitians have been appointed and they coordinate the work in the hos-

TABLE 2

TRAINING OF DIETITIANS IN THE FEDERAL REPUBLIC OF GERMANY

YEARS		TRAINING HOURS	MAIN SUBJECT	OFFICIAL NAME
2	Training Centers in Dietetics	2300	Therapeutic Diet	DIATASSISTENT
4	Working			
1/4	Continuing Training Step I	520	Nutrition Education Administration	DIATKUCHENLEITER
1-2	Working			
3/4	Continuing Training Step II	1020	Community Health	ERNARUNGS MEDIZINISCHER BERATER

TABLE 3

TRAINING OF DIETETIANS IN UNITED-KINGDOM
(Degree in Dietetics)

YEARS		SUBJECTS	TRAINING HOURS
3	College	Sciences	870
		Nutrition and Diet Therapy	480
		Medicine	50
		Food Science	140
		Food Preparation and Catering	250
		Social Studies Education	270
		Statistics	35
			2095
1	Practical Training	Dietetics	24 weeks
		Catering	16 weeks

ice to the local authority and social Service departments and to the local education authorities.

These 300 dietitians have a good experience in therapeutic diet but few in communication and education. The problem is now: how to prepare them for their new duties?

Some other countries of the E.E.C. have similar problems.

In conclusion, it seems that discussion on the training of dietitians should not be limited to the comparison of initial training. We must also take into account the possibilities offered for complementary training after some years of professional experience.

However thorough basic training may be, it is advisable not to prolong it too much; young students with no professional experience have difficulty in seeing problems as a whole and would be unable to assume positions of responsibility. On the other hand, those who have faced the realities of their job would benefit fully from a second period of training.

This is why the latest report of C.A.D.E.C. (1978) includes the following recommendations:

"Dietitians having several years of professional experience can be qualified in order to be effective in the field of Community Health. They must receive complementary training in the following fields:
- Psychology, sociology, economics
- Epidemiology and Community Health
- Legislation
- Organization of large-scale catering and administration
- Communication techniques
- Nutritional education
- Written material.

In the case where certain dietitians have to assume the role of advising, programme organizing, coordinating and higher duties in the Services of Nutrition in Community Health, it is necessary to prepare these higher officers to an advanced level.

As experience and maturity are necessary for occupying these posts the possibility should be envisaged of allowing for preparation for these diplomas to dietitians who have already been practising for a number of years".

The C.A.D.E.C. work-group, has been a help towards a better understanding between dietitians, an improvement in their training and an extension of their work-field.

As the 1976 C.A.D.E.C. report on the ethics of the profession recalled "the essential role of the dietitian is to protect and support health and to contribute to the recovery of health of man by means of good nutrition".

In order to carry out this role dietitians must be able to meet the present-day needs of each of their countries and to adapt their training in consequence.

BIBLIOGRAPHY

CADEC. Report of the seminar on the training of dietitians in the European Common Market, Paris, 1972.
CADEC. Role and place of the dietitian in the E.E.C. Countries, Vittel, 1974. J. of Human Nut., 32: 127-131, 1978.
CADEC. Thoughts on the ethics of the dietetic profession. London, 1976. J. of Human Nut., 32: 132-135, 1978.
CADEC. The role of the dietitian in community health. Copenhagen, 1978.

DIETETICS AROUND THE WORLD —

1977 SURVEY OF THE WORK AND TRAINING OF DIETITIANS

        Joan Woodhill, Ph.D.

        6/18 St. Luke's St. - Randwich

        New South Wales 2031 - AUSTRALIA

The 1977 survey on the work and training of dietitians is an on-going project and in 1977 was conducted by Jean Marr (U.K.) for the International Committee of Dietetic Associations and was presented to the 7th International Congress of Dietetics in Sydney, Australia, May 1977.

1. INTRODUCTION

Data from the Dietetic Associations of 23 countries were collated.

    1.1   The training of dietitians is related to the role of the dietitians required by the health services of their countries.

    1.2   It is expected, accepted and required that there will be diversity in the various roles of the dietitians and therefore the training from the 23 different countries.

2. SUMMARY OF ROLES AND WORK OF
   DIETITIANS IN 23 COUNTRIES

    2.1   Areas of employment.

           2.1.1   Hospitals. Dietitians from all countries work in hospitals.
           2.1.2   Nutrition and Health and Disease. Nineteen (19) of the 23 respondents placed treatment by diet in hos-

pitals, clinics and poly-clinics as the first area of employment: 3 respondents placed hospital food service administration first and 1 respondent gave Ministry and Government the most important area of employment.

2.1.3 Education, in its broadest sense (patients, community, schools, health teaching institutions etc.) was mentioned by 20 countries as an important area of employment for dietitians.

2.1.4 Community Nutrition was included in the replies from 14 countries.

2.1.5 Administration, i.e., Management, Food Service and Catering, was included in the replies from 14 countries.

2.1.6 Industry. Thirteen (13) out of the 23 Dietetic Associations mentioned that dietitians worked in industry.

2.1.7 Research, Investigations and Surveys. Half of the responding associations mentioned applied research as an area of employment for dietitians.

2.1.8 Only 1 country specifically mentioned the dietitians role in rehabilitation.

2.1.9 Two (2) countries specifically mentioned the dietitians' role in the armed services.

2.1.10 Self Employed. Only 2 Dietetic Associations referred to self employed dietitians.

3. TRAINING OF DIETITIANS

3.1 In all countries the dietitians' training commences after 9 to 14 years of schooling.

3.2 The dietitians' professional courses, at the tertiary level varies from 2 to 5 years. Ninety per cent (90%) of the responding associations reported 3 years or more of tertiary training in dietetics.

3.3 Some courses are a three year University award, followed by an appropriate post-graduate training of 1 year or more. Other courses have the professional training integrated into the undergraduate years.

4. ESSENTIAL COURSE WORK

4.1 Many different names are given to the various courses. An analysis of the courses listed showed that applied nutrition is the major subject. Courses are cited with the essential pre-requisites and as expected there is a diversity in the proportion of time given to the various subjects.

4.2 Chemistry and Physics are pre-requisites for the biological sciences. Biochemistry, physiology and microbiology are the pre-requisites to nutrition and food science.

4.3 Behaviourial sciences and social sciences are pre-requisites for essential studies in education and communication.

4.4 Mathematics, statistics and computer sciences are specified in most courses.

5. REGISTRATION

The professional practice of dietetics can be identified and recognized by law, medical practitioners (for referrals), and the community. This type of recognition, based on appropriate qualifications, can serve to identify the qualified person practicing dietetics according to a code of professional practice and so distinguish the registered practitioner from the food quack or person engaged in promotion of commercial products claimed to be cure alls.

5.1 Fifty per cent (50%) of the associations reported legal registration for dietitians.

6. FUTURE DEVELOPMENT OF THE PROFESSION OF DIETETICS

The 1977 survey indicated the current practice in dietetics and there were indications from the respondents that training levels will respond to the rationalization of the profession, i.e., the dietitians in the various countries should be trained not only to do the job required at the present time by their communities but also the profession should take the responsibility of planning effective future developments to provide:

(a) Best health care, preventive and curative, in the broad field of community nutrition recognizing diversity and specific needs.

(b) To identify which of the tasks in the broad field of nutrition is unique to the dietitian-nutritionist and to rationalize what can be done by other members of the health team.

(c) To recognize that the training for the work of the dietitian-nutritionist will continue to rest on the subjects food and nutrition but that the role of the dietitian-nutritionist is that of an educator.

(d) To recognize that constant review and evaluation of work is needed to keep abreast with the needs of communities, awake to the importance of reliable nutrition information. This is particularly important if the ever changing food supply is recognized.

THOUGHTS ON THE ETHICS OF DIETETIC PROFESSION*

Waltraute Aign

Institut für Ernährungswissenschaft

D-63 Giessen, FED. REP. OF GERMANY

An ethical aspect which must avoid descending into a childish or outdated moralism yet which must lead to certain rules of conduct for dietitians.

It seems that, in order to gain general agreement, this aspect of ethics must be based mainly on the actual definition of the profession and on the consequences which arise in practice.

The lines between the ethical aspect and the legal aspect are tenous: the rules of good professional conduct which the Associations can advise to their members do not all lend themselves to being objects for disciplinary action: the borderline is between "allowed" and "forbidden". On the other hand, according to the true spirit of its legislation, each country tends to rely on a detailed code or on custom where no writen law exists.

For all these reasons, it has seemed preferable not to entitle this report "Code of Ethics" but Thoughts on the Ethics of the Dietetic Profession".

GUIDING PRINCIPLES

The essential role of the dietitian is to protect and support health and to contribute to the recovery to health of man by means of good nutrition.

---

* Prepared by the Committed of Dietetic Association of the European Community.

The dietitian must consider himself as being at the service of mankind; his attitudes and deeds must be guided by respect for people and the care to promote health.

Whether concerned with caring for the sick or advising the healthy, the dietitian treats all people in all circumstances with devotion whatever their race, nationality, social condition, political, philosophical or religious opinions. He respects the liberty of conscience in all and pays regard to the food habits which may be linked with each type of society. This is particularly important for dietitians working among immigrants or acting as fellow-workers in a foreign country.

GENERAL CONSEQUENCES

In order to be able to play this role, the dietitian must continually improve his professional knowledge, be receptive to new methods and to progress in Nutrition and Dietetics. He must also know the needs of the individuals or groups with whom he comes in conduct to improve his knowledge of psychology and to be aware of human problems.

The dietitian must assume his proper responsibilities, recognize his limits and know how to collaborate with other professions.

RELATIONSHIP WITH THE PATIENT

The dietitian ensures with the doctor the nutritonal needs of patients requiring a therapeutic diet:
- by assisting for example by means of a diet history;
- by having the responsibility for the supervision and the carrying out of the diet prescribed by the doctor.

The dietitian does not have the right to prescribe a therapeutic diet himself nor to advise on medicines or vitamines.

He tries to make the patient understand the reason for the diet and the methods of carrying it out and to gain his confidence to it. However, he knows how to be firm, precise and effective. He does not give up a patient without ensuring that care can be continued by another professional.

The dietitian has occasion to know numerous facts concerning the health of the patient, and about his personal life. The dietitian is also constrained by professional secrecy as are doctors. It embraces all that the dietitian may see, hear, understand or deduct in the exercise of his profession. It is absolute, and nobody can release him from it.

The dietitian is however, obliged to report to the doctor in charge all that he has been able to learn concerning the health of the patient. In the same way, he can give useful information to the nursing staff where this is necessary for the good of the patient. With regard to the family and those around the patient, the dietitian will reply carefully to the questions he may be asked: collaboration with the doctor and nursing staff is necessary. The basis for this professional secrecy lies in the confidence between the patient and whoever cares for him.

RELATIONSHIP WITH THE DOCTORS

The dietitian does not have the right to prescribe a therapeutic diet, it is for the doctor to indicate precisely the type of diet and its requirements in terms of the relevant nutrients. This prescription must be written and signed by the doctor.

The dietitian should carry out the doctor's request and cooperate faithfully with the doctor. He should maintain contact with the doctor throughout the treatment.

RELATIONSHIP WITH OTHER PROFESSIONS

The dietitian respects the expertise of other paramedical personnel.

He coopeartes with them with loyalty and team spirit for the good of the patient and for maximum combined efficiency.

He cooperates with the administration in order to reconcile the health interest with the necessary financial restrictions. He has responsibility for the people working under his direction and must ensure good organization of work and staff training.

THE DIETITIAN IN PRIVATE PRACTICE

There are very few in Europe — is, just as a doctor, not allowed to advertise for patients. He has, however, the right to inform the doctors and other paramedical workers of the existence of his office.

RELATIONS WITH INDUSTRY AND COMMERCE

The dietitian must always bear in mind the special nature of his profession and safeguard his independence. He must never accept commission from industrial or benefits from commercial firms.

The dietitian cannot advertise food or dietetic products and should not advise on premises where dietary products are on sale. These restrictions are particularly important so that the public differentiates clearly between true professional dietitians and imposters existing in this field.

If the dietitian carries out trials on new food or dietetic products on man, work on the clinical or biological effects must be carried out under the direction of a medical doctor.

The dietitian must be able to safeguard his right of recognition in publications concerning these trials and on the use made of their results.

RELATIONS WITH THE PUBLIC

The dietitian has the right to take action against the abuse of dietetics which is poorly-based scientifically and against wrong information and false ideas. He must give warning to advise against excessive unjustified or magic diets.

The dietitian must inform and educate the public in matters of nutrition and feeding. He can use the various means of communication. He has the right to give advice on the nutrition of healthy people and organize large-scale feeding without medical advice.

RELATIONS WITH OTHER MEMBERS OF THE PROFESSION

The dietitian must feel himself at one with them, collaborate with his colleagues and exchange useful information and experience with them.

He has particular care for training of students and young dietitians. The dietitian must play his part in raising the cultural, moral and social level of his profession, by taking an active part in the affairs of the professional associations.

The associations must arrange for further education and promote mutual aid between them.

They must take the profession known and contribute to its development.

AUTHOR INDEX - Volume II

A

Aguillon, D.B.
Aign, W.
Asp, N.G.
Astofi, E.
Aylward, F.

B

Bagchi, K.
Baker, H.
Bauernfeind, J.C.
Beaton, G.H.
Beaudry-Darismé, M.
Benitez, L.P.
Bhat, R.V.
Birbeck, J.A.
Bourges, H.
Braekkan, O.R.

C

Campbell, V.S.
Castagnolli, N.
Chavez, M.
Chavez, A.
Claesson, C.O.
Clay, K.
Costill, D.
Cowey, C.B.

D

Darby, W.
Delfino, A.H.
Dendy, D.A.V.

E

Erbersdobler, H.F.
Essafara, M.B.

F

Ferrando, R.
Ferreira, F.A.G.

G

Garine, I.
Geissler, C.
Gross, R.
Gurney, J.M
Guzman, M.P.E.

H

Halver, J.E.
Hettiarachchy, N.S.
Hollingsworth, D.F.
Hulse, J.H.

I

Icaza, S.J.
Inone, G.

J

Jambunathan, R.
Jelliffe, E.F.P.
Jerome, N.W.
Jul, M.

K

Kallal, Z.
Kercher, C.J.
Koksal, O.

L

Lam-Sanchez, A.
Lipsky, R.
Luse, R.A.

M

Maletnlema, T.N.
Manoff, R.K.
Marchione, T.J.
Mariani, A.
Medina, M.M.
Milner, N.
Morava, E.
Munro, I.C.

N

Njaa, L.R.
Nose, T.

O

Orraca-Tettch, R.
Oste, R.
Oyenuga, V.A.

P

Pelto, G.H.
Pérez-Gil, R.F.
Petitclerc, D.
Pigden, W.J.
Polychonopoulou-Trichopoulou, A.

R

Rackis, J.J.
Rérat, A.

S

Saio, K.
Salum, R.
Sanahuja, J.C.
Schaefer, A.E.
Schaller, R.
Schmidt-Hebbel, H.
Schulze, H.
Serville, Y.
Souza, S.M.A.

T

Theander, O.
Titus, D.

U

Ullrich, H.D.

V

Valyasevi, A.
Villegas, E.M.

W

Wapinski, J.
Wirths, W.
Wolf, R.
Woodhill, J.

Z

Zerfas, A.

INDEX

Acidophilus milk, 113
Addis Ababa, University of, 14
Adult education, food and
    nutrition courses in, 758
Aflatoxins
  from Aspergillus flavus, 417
  as carcinogens, 417
  food contamination by, 417–419
Africa, see also South Africa;
    Tanzania; West Africa
  anthropological approach in,
      619–635
  cereal grains and legume
      consumption in, 10
  food and nutrition studies in,
      865–868
  food supply problem in, 8
  historical development of, 621
  models for nutrition evaluation
      in, 634–635
  national nutrition surveys in,
      511–513
  nutrition evaluation in,
      619–635, 865–868
  population of, 153
  training of women in, 633–634
  woman as producer and
      reproducer in, 619–635
  women's hours of work in, 624
  women's time and energy
      expenditure in, 623–632
  women's motivation in, 625
Agave juice, 279
Agency for Internation
    Development, 63
Agricultural research
  international centers of, 15–16

Agricultural research (continued)
  investment in, 12
  mixed cropping in, 16
Agriculture
  biological and chemical
      products applied in,
      375–380
  feed technology in, 157–158
  food additives and, 375–380
Alegria, nutritive value of,
    278
Alfalfa, digestibility of, 143
All India Coordinated Project
    on Post-Harvest
    Technology, 452
Amaranthus hypocondriacus, 278
American Medical Association,
    775
Amino acid metabolism, 173
Amino acids
  in cowpeas, 60
  dehydroalanine and, 438, 441
  food protein enzymatic
      hydrolysis for, 693–703
  in legumes, 422
  lysinoalanine and, 442–445
  in maize, 32
  in millet seed, 61
  in pigeon pea, 61
  protein requirements and,
      172–173
  in sesame seeds, 60
  in single-cell protein,
      706–707
  in soybean storage and
      processing, 438–441

Amino acids (continued)
  supplementation of proteins with, 172–173
  in triticale, 33
  in vegetable proteins, 233
Ammonia, in waste processing, 147
AMUL dairy, Anand (India), 125
Andes, Lupinus mutabilis in, 223
Animal carcasses, body composition changes in, 79–81
Animal feeding, see also Cattle feeding; Fish feeding; Livestock feeding; Swine feeding
  allergens in, 398
  animal nutrition and, 153–155
  body composition and, 85–91
  castration effect on, 84
  cellulose in, 87
  cellulose-starch relationship in, 103
  contaminants and additives in, 393–401
  dietary lipids vs. growth performance in, 86
  energy intake and, 85–87
  feed efficiency in, 65, 68, 103–104
  growth and body composition factors related to, 78–83
  human competition and, 102
  modern practices in, 65–72
  pesticides in, 394
  physiological characteristics of, 87–89
  polysaccharide decomposition in, 603
  protein efficiency in, 65
  waste used in, 141–151
  workshop on, 153–155
Animal feedstuffs
  additive residues in, 397
  antibiotics in, 398
  DDT in, 395
Animal food production, waste utilization in, 141–151
Animal growth
  body composition in, 78–79

Animal growth (continued)
  dietary lipids in, 86
  factors in, 78–83
Animal nutrition, animal feeding and, 153–155
Animal production, protein utilization in, 121
Animal protein, see also Protein(s)
  from pork, 73–78
  world totals for, 73
Animal Protein Factor, 396
Animals
  bacterial exchange from, 399–400
  food yields from, 69
Animal wastes, in livestock feeding, 144
Antarctic krill, 209
Anthropological approach or methodology
  action techniques and, 579–580
  in Africa, 619–635
  baseline and evaluation in, 580–581
  in Caribbean, 593–607
  in evaluation research, 563–568
  means and techniques in (Senegal), 573–578
  in nutrition program evaluation, 553–568
  possible contributions of, 584–585
  results in, 578–579
  scope of, 586–587
  shortcomings of, 581–582
  sociocultural environment in, 594
  successes of, 582–583
  teamwork in, 585
  unsolved problems in, 584
Antibiotics
  in animal feedstuffs, 398
  in calf feeding, 113
Antragnose, in lupin cultivation, 223
Arabinogalaktans, 727

Argentina
  family basket for sample
    family in, 410
  food contamination by chlori-
    nated residues in,
    403-416
  Lupinus albus in, 221
Argentine beef, chlorinated
    residues in, 412
Arsenic, fish contamination
    by, 383
Asia
  cereals and legumes in, 11
  health profession problems and
    responsibilities in,
    847-849
  medical school nutrition
    programs in, 850
  nutritional disease prevention
    in, 849
  nutrition education in, 847-859
  nutrition teaching in, 850-855
Aspergillus flavus, 232, 417
Aspergillus parasiticus, 417
Athletic activity, dietary
    intake in, 657-665
Athletic drinks, types of,
    678-680
Attitude change, in nutrition
    education, 787-788, 827
Avitaminosis, in finfish,
    181-182, see also Vitamin
    A deficiency; Vitamins

Bacterial exchange, between
    animals and man, 400
Bagasse, digestibility of, 143
Banana leaves, protein
    concentrate from, 239
Barley straw, digestibility of,
    143
Beans, see also Soybeans
  hard-to-cook, 422
  winged, see Winged beans
Beef
  Argentine, 412
  chemical composition of, 275
  chlorinated residues in, 412
  cost of, 274
  food constituents of, 273

Beef consumption, vs. pork,
    75-76
Beef production
  feed savings in, 71
  milk and, see Tropical milk/
    beef production systems
Behavior change, following
    nutrition education,
    789-790, 829
Bicycling, energy expenditure
    in, 662
Bile acid binding, of food
    ingredients, 724
Biological contamination, see
    Environmental contamina-
    tion; Food contamination
Blood
  bovine, as protein source,
    281-282
  chlorinated residues in,
    403-408
Blue-green algae, Mexican
    studies of, 279
Body composition
  protein intake and, 89-91
  sex factors in, 83
  variations in during animal
    growth, 78-79
Body fluids, exercise and,
    668-674
Body temperature, body fluids
    and, 671-673
Body weight, exercise and,
    667-669
Bovine blood, as protein
    source, 281-282
Bovine trypsin, in trypsin
    activity assays, 305
B.P.C. fish protein
    concentrate, 249-265
  amino acid content of,
    250-251, 254
  as cow's milk substitute,
    253-262
  defined, 249
  growth curves for, 256-261
  mineral content of, 250
  pregnancy growth charts for,
    263-265

B.P.C. fish protein
concentrate (continued)
  pregnant mothers and, 262
  preparation of, 254
  protein content of, 349, 254
  protein efficiency ratio for, 252
  protein utilization for, 252
Brazil
  fish feeding experiments in, 199-202
  Lupinus albus 221
  nonconventional food utilization in, 269-275
  rabbit raising in, 269-276
  soybean production in, 289-291
Bread
  composite flours for, 246
  from triticale, 27
  vitamin A fortification of, 360
Breast milk
  lugaw additive for, 784, 790
  polychlorinated biphenyls in, 388
British Nutrition Foundation, 806-807
Butter, chlorinated defensives in, 411

Cabbage
  dry matter in, 730, 738
  glucosinolates in, 317
Calcutta, see also India
  food prices in (1970), 129
  income distribution vs. milk consumption in, 126-128
Calf feeding
  milk products in, 113
  milk replacements in, 113, 115
Calisthenics, energy expenditure in, 661
Caloric gap, interpretation of, 8
Calorie intake, of weight-lifters, 663
Calorie-protein malnutrition, see Protein-calorie malnutrition

Cardiovascular system, dehydration and, 668
Caribbean Food and Nutrition Institute, 478-490
β-Carotene, in margarine, 351
Carp, test feeding of, 200, see also Fish
Carrots, dry matter in, 730, 738
Cassava
  food quality of, 56
  product improvement programs for, 58
Castration, effect of on animal feed efficiency, 84
Cattle feeding, see also Animal feeding; Livestock feeding
  and chemical composition of milk, 120-122
  and "keeping quality" of milk, 122-123
  for meat production, 109-112
  for milk production, 113-123
  milk quality and, 120-123
  polyunsaturated fats in, 122
CCK, see Cholecystokinin
Cellulose wastes
  digestibility of, 142-143
  in livestock feeding, 141-151
Centro Internacional de Agricultura Tropical, 4, 15, 55-58, 451
Cereal consumption
  in developing countries, 101
  in Morocco, 537
  world totals for, 457
Central Food Technological Research Institute (India), 421, 452
Cereal grains and legumes, see also Legumes
  importance of, 10-11
  increased productivity in, 11
  nutritive value of, 11-12
  vitamin A in, 356-357
Cereal production
  vs. legume production, 453-454

Cereal production (continued)
  losses in, 457-458
Cereal wastes
  in livestock feeding, 141-151
  types and amounts of, 144
CGIAR, see Consultative Group
    for International
    Agricultural Research
Change
  attitudinal, 787-788, 827
  behavioral, 789-790, 829
  in knowledge, 788-789, 829
  systematic nature of, 563
Cheese, chlorinated defensives
    in, 411
Chicken meat
  chemical constitution of, 275
  cost of, 274
  food constituents in, 273, 275
Chick-pea protein concentrate,
    278
Childhood grades, nutrition
    education in, 795-802
Children's television programs,
    nutrition education
    through, 819
Chile, Lupinus species in,
    221-224
Chinook salmon
  liver storage in, 189
  vitamin test diet for, 182
Chitosan, in fish feeding, 202
Chlorinated residues
  in blood, 403-408
  in cow's milk, 415-416
  food contamination by, 403-416
  in human milk, 415
Choco seed study, Ecuador, 238
Cholecystokinin, as tryptin
    inhibitor, 305
CIAT, see Centro Internacional
    de Agricultura Tropical
CIMMYT, see International Maize
    and Wheat Improvement
    Centre
CIP, see International Potato
    Center
Clinical Nutritionist, in
    nutrition education, 877

Coccidiosis, from animal
    feeding, 395
Coconut oil, in swine feeding,
    94
Codex Alimentarius Commission,
    417
Coho salmon, diet vs. growth
    in, 189
College courses, food and
    nutrition education in,
    753-757
Colossoma bidens, 202
Colossoma macropomum, 202
Communications
  in nutrition education,
    888-893
  for nutrionist-dietician,
    887-894
Community Health Aide Program
    (Jamaica), 593-607
Community health programs
  nutritionist-dietician role
    in, 899-903
  social anthropology in,
    593-607
Community health workers,
    nutrition training for,
    835-842, 873-874
Community nutrition education,
    see also Nutrition
    education
  breast feeding programs in,
    817
  children's TV programs and,
    819
  nutritionist's role in,
    895-898
  orchard role in, 819-821
Composite flours, substitution
    of, 245-246
Conlac, 284-285
Consulative Group for International Agricultural
    Research, 4, 13
Contaminants, food, see
    Food contaminants
Contamination, environmental,
    see Environmental
    contamination

Corn meal
   in carp diet, 200
   vitamin A in, 361-363
Cottonseed oil, 228
Cowpea
   amino acid patterns in, 60
   food quality of, 56
   sulfur/nitrogen ratio in, 57
Cowpea improvement program, West Africa, 55-58
Cow's milk, see also Milk
   B.P.C. as substitute for, 262
   chlorinated defensives in, 411, 415-416
   soybean protein as substitute for, 247, 302, 319, 430-431
Crop improvement techniques
   for food legumes, 15, 69
   priorities in, 11-12
Cruciferae family, glucosinolates in, 317
Crude fiber, see Dietary fiber
Cyprinus carpio, 200

Dairy cow feeding, human nutrition and, 113-123
   see also Cattle feeding; Livestock feeding
Dairy projects, nutritional benefits from, 125-140
DDT
   Acceptable Daily Ingestion of, 416
   in animal feedstuffs, 395
   in human milk, 416
   ingestion of in Argentina, 409, 413, 416
Defensives residues (Argentina), 403-416, see also Chlorinated residues
Dehydration, heart rate and, 669
Dehydroalanine, 438, 441
Denmark, dietitian training in, 907-908
Developing countries
   cereal grain and food legume consumption in, 10, 101
   community health workers in, 835-842

Developing countries (continued)
   food and nutrition problems in, 516
   Health Care Systems in, 835
   health workers in, 835-842
   intervention programs in, 498-500
   livestock/grain competition in, 102
   milk production in, 119-120
   nutritional status revision in, 515-519
   nutrition education in, 823-834
   nutrition research in, 483-486, 495-501
   population growth rates in, 7
   protein deficiency in, 101
Dieldrin, ingestion of, 413
Dietary fiber
   analytical methods for, 731
   constituents of, 720-721
   crude fiber method for, 741
   detergent method for, 743
   EDTA-soluble material in, 734
   enzyme method for, 743
   evaluation of assay methods for, 741-744
   in food technology, 719-726
   fractionation and chemical characterization of, 727-739
   isolation of, 725
   measurement of, 719-720
   water holding capacity of, 722-723
   water-soluble material analysis for, 732-733
   Weinstock assay method for, 742-743
Dietetic profession
   ethics of, 917-920
   future development of, 915-916
Dietetics, world survey of, 913-916
Dietitian, see also Nutrionist-dietitian
   ethics of, 917-920
   in nutrition education, 878

Dietitian (continued)
  public and, 920
  role of, 906, 913-916
  training of, 905-911
  work and training of, 906-916
Dietitian-patient relationships, 918
Dietitian-physician relationships, 919
Disease, nutrition and, 534
Drinking fluids, for athletes, 678-680
Dry alkalis, in waste processing, 146
Dry matter
  carbohydrates and protein in, 735-736
  distribution of, 730
  nonstructural carbohydrates in, 737
  water-holding capacities of, 738

Early childhood, nutrition education in, 795-802
Eastern Europe, nutrition education in, 843-845
Economic protein requirement, concept of, 171-172
Ecological analysis, intracultural diversity in, 557
Ecological frame of reference, 557, 562
Ecological perspectives, 556-563
Ecology of Food and Nutrition, 556
Ecology of Malnutrition, 556
Ecuador, nonconventional food source studies in, 237-247
Effort evaluation, 598-599
Eggs, as protein source, 229
Endurance events, fluid intake in, 680
Energy expenditure
  activity type and, 657-665
  body weight and, 645-646, 652-653

Energy expenditure (continued)
  dietary intake and, 657-665
  Max Planck respirometer in, 642-643
  by occupational group, 660
  respiration rate and, 642
  running speed and, 651
  walking speed and, 648-650
  in weightlifting, 657-665
Energy requirements
  in controlled vs. ad libitum work, 641-654
  food consumption and, 641-654
Englishman's Food, The, 806
Environmental contamination, 381-391
  guidelines in, 390
  of milk, 386-387
  organic, 387
  regulations covering, 389-390
  source of, 382
Enzymatic hydrolysis, in protein quality measurement, 693-703
Escherichia coli, transfer of R factors from, 399
Essential amino acids, daily requirements for, 172-173, see also Amino acids
Ethiopia, IDRC activities in, 14
Euryhaline fish, protein requirements of, 171, see also Fish
Evaluation
  defined, 770
  vs. research, 773
  social anthropology in, 593-607
Evaluation research, see also Nutrition evaluation
  anthropological methodology in, 563-568
  data base in, 564
  effort evaluation in, 598-603
  general guidelines in, 554-556
  levels and sublevels in, 596-598

Exercise, see also Energy
        Expenditure
    athletic drinks in, 678-680
    body fluids in, 667-669
    drinking and, 671-674, 680-681
    electrolytic losses in, 674-676
    plasma volume changes in, 672
    sodium and potassium losses
        in, 676
    sugar intake in, 677-678
    sweating in, 668-671
    water and electrolyte balance
        in, 667-681
    weightlifting as, see
        Weightlifters;
        Weightlifting
Expert Consultation on Grain
        Legumes Conservation and
        Processing, 421

FAO/WHO Report on Protein and
        Energy Requirements, 472
    see also Food and Agriculture
        Organization; World
        Health Organization
Far East, population of, 153,
        see also Asia
Farm income, India, 134
Fats, in fish diets, 178-179
Fatty acids, in fish diets,
        178-179
Feed, animal, see Animal
        feedstuffs; see also
        Livestock feeding
Feed efficiency, 65, 68
Feed formulation, 157
Feeding, animal, see Animal
        feeding; Livestock
        feeding
Fertilizers, food contamination
        from, 378
Fetal contamination, by methyl-
        mercury, 384-385, see
        also Food contamination
Fiber, dietary, see Dietary fiber
Fibrous plant materials,
        availability of, 142
Field bean, food quality of, 56,
        see also Winged bean

Finfish, see also Fish
    vitamin deficiency signs in,
        185-186
    vitamin requirements for,
        181-195
    water-soluble vitamins and,
        184-188
Fish
    ascorbic acid deficiency in,
        187
    arsenic contamination of, 383
    avitaminosis symptoms in, 181
    biotin deficiency in, 188
    choline deficiency in, 187
    environmental contaminants of,
        382-385, 390
    fat and fatty acid require-
        ments of, 178-179
    fat-soluble requirements for,
        190-194
    feed formulation and technology
        for, 157-158
    folic acid deficiency in,
        184-187
    inositol deficiency in, 188
    maximum liver storage in, 189
    mineral requirements of, 179-180
    niacin deficiency in, 187
    panththenic acid deficiency in,
        184
    para-aminobenzoic acid in, 188
    as protein source, see
        Fish protein
    pyridoxine deficiency in, 184
    quantitative fat-soluble
        vitamin requirements for,
        194
    quantitative water-soluble
        vitamin requirements for,
        188-190
    riboflavin deficiency in, 184
    scoliosis and lordosis in,
        177-178
    thiamine deficiency in, 184
    vitamin requirements for,
        177-178, 191-194

Fish feed
    nutrients in, 158
    protein in, 158

Fish feeding experiments
    (Brazil), 199–204
Fish feeding technology, 161–168
  ad libitum feeding in, 167–168
  chitosan in, 202
  daily feed allowance in,
    166–167
  diet types in, 161–166
  dry pellets in, 163–164, 202
  energy required for growth
    in, 166–167
  feeding equipment in, 167–168
  metabolizable energy in, 167
  moist pellets in, 162–163
  natural food in, 165
  paste feed in, 164
  pellet feed in, 162–164
  supplemental diet in, 165–166
  wet fresh feed in, 162
Fish food formulation, status
    of, 199–202
Fish nutrition research
    (Brazil), 199–202
Fish protein, 229, see also
    B.P.C.
  in human and animal feeding,
    71
  soybean and, 325–331
Fish protein concentrate, 253
  methionine-sulfoxide content
    of, 713–717
Flours
  composite, 245–246
  vitamin A in, 357–361
Food(s), see also Dietary
    fiber; Nutrition
  phytic acid content of,
    315–316
  post-harvest conservation of,
    451–455
  zinc availability in, 316
Food additives
  in animal feeding, 393–401
  defined, 375
  effect of on quality, 375–380
Food and Agriculture Organiza-
    tion (UN), 4, 9–10, 13,
    417, 421, 451, 503,
    520, 683

Food and Feed Grain Research
    Institute, 451
Food and Nutrition Research
    Institute, 779
Food conservation, stages in,
    455
Food constituents, per 100-g
    edible portion, 273
Food consumption, research on,
    484–485
Food contamination
  aflatoxins in, 417–419
  of animal food, 393–401
  by chlorinated residues,
    403–416
  defined, 376
  effect of on quality, 375–380
  lindano residues in, 414
  metallic, 377–380
  pesticides in, 453
Food cost, for cereal-fed
    animal production, 70
Food demand, Morocco, 537–548
Food education, 747–758, see
    also Nutrition education
Food habit losses, 460–461
Food habits, nutrition education
    and, 761–765
Food ingredients, bile acid
    binding of, 724
Food legumes, in nutritional
    programs, 14–17, 422,
    see also Legumes
Food losses
  habit in, 460–461
  measurement of, 461
  post-harvest, 458–460
  reduction in, 451
  types of, 458–461
Food preparation, international
    research activities
    in, 62
Food prices (Calcutta), 129–130
Food processing, international
    research in, 62
Food production
  from animals and crops, 69
  increase in, 10–11, 15, 110,
    299–320

Food production (continued)
  plant cultures in, 65
  research in, 484
  world totals, 3, 7
Forest production residues, as
    animal feedstuffs, 144,
    148-151
Food proteins, see also
    Protein(s)
  enzymatic hydrolysis of for
    amino acid analysis,
    693-703
  in vitro nutritive value
    prediction for, 683-687
  papain in solubility of, 697
  urea concentration in
    solubility of, 696
Food quality, biological and
    chemical products
    affecting, 375-380
Food samples, PER value
    comparison of, 690
Food sources, nonconventional,
    see Nonconventional
    food sources
Food storage, see also Grain
    storage; Soybean storage
  cost-benefit ratio in, 454
  polythene linings in, 453
Food supply, milk production
    efficiency in, 114-120
Food taboos, in Third World, 761
Food technology, dietary
    fiber in, 719-726
Food utilization and processing
    studies, in nutritional
    evaluation, 55-64
Food yields, animal vs. crops
    in, 69
Ford Foundation, 4, 12
Formol titration, 685
Fortification, iron compounds
    in, 333-336
F.P.C. (fish protein
    concentrates), 253
Frijoles, soybeans for, 279
Fruit-vegetable wastes, in
    livestock feeding,
    141-151

Galactoglucomannans, 727
Geophagus brasiliensis, 202
Glucomannans, 727
Glucosinolates, in plant
    foodstuffs, 317-318
Glucuronic acid, 727
Goitrogens, in soybeans, 317-319
Government policies, in
    nutrition research,
    485-486
Grain legumes, world production
    of, 421, see also
    Legumes
Grain processing, nutritional
    losses in, 457-463
Grain storage
  acidity changes in, 426-427
  generator capacity and, 425
  moisture content and, 424
  nutritional changes in, 428
  post-harvest losses and,
    458-460
Grazing, in livestock feeding,
    144-145
Greece
  age-specific mortality
    rates in, 638, 640
  malnutrition and under-
    nutrition in, 637-640
  nutrional habits in, 637-640
  population of, 637
Green leaves, as protein
    source, 228
Green Revolution, 4, 558
Group for Assistance on System
    Relating to Grain after
    Harvest, 451
Guaje, as protein source, 278
Guamuchil, 278
Gujarat region (India),
    agricultural develop-
    ment in, 139

Hadassah Medical Organization,
    795
Halogens, in environmental
    contamination, 387-388
Hansenula montevideo, 249

INDEX

Heart rate, dehydration and, 668-669
Heat exhausion, 668
High lysine sorghum, 40-43, see also Sorghum
Higureilla flour, protein from, 242
Histomoniasis, in animal feeding, 395
Hog carcasses, see also Swine
  compared with those for other animals, 77
  fat percentage of, 76
  growth and body composition factors in, 78-83
Hog production, 73
Human fat, chlorinated residues in, 403-416
Human milk, see also Breast milk; Cow's milk
  chlorinated residues in, 415
  DDT in, 416
Human nutrition, see also Nutrition
  and cattle feeding for meat production, 109-112
  milk production and, 113-123
Hungary, nutrition education in, 843-845
Hybrid corn, in carp diet, 200, see also Corn meal
Hydrolysis, enzymatic, 693-703

ICARDA, see Internation Center for Agricultural Research in Dry Areas
ICRISAT, see International Crops Research Institute for the Semi-Arid Tropics
IDRC, see International Development Research Centre
IFDC, see International Fertilizer Development Centre
IFPRI, see International Food Planning Research Institute
IITA, see International Institute of Tropical Agriculture

ILCA, see International Livestock Center for Africa
ILRAD, see International Laboratory for Research on Animal Diseases
Immobilized peptidases, in complete hydrolysis, 700
Income distribution (India), nutrition and, 127-128, 133
  agricultural development in (Gujarat region), 139
  city milk consumption in, 126-128
  dairy project in, 125-140
  farm income in, 134
  flood objectives vs. dairy project in, 125-126
  food caloric balance vs. milk production in, 138
  food grain losses in, 454
  food prices in, 129-130
  free milk distribution in, 130-131
  income distribution in, 126-127, 133
  landholding vs. milk production in, 137
  low-income rural populations vs. dairy production in, 137
  malnutrition in, 131
  milk animals in, 135
  milk collection in, 133-138
  milk prices in, 129-130
  milk production in, 137-138
  per capita expenditures in, 136
  small farmer in 134
  World Food Programme in, 126
Infant formulas, soy protein for, 297-298
Infants, PCB contamination in, 388-389
Insects, protein content of, 281-282
Intercropping, defined, 15

International Agricultural
    Research Centres, 12
International Agricultural
    Consultative Group, see
    Consultative Group for
    International Agri-
    cultural Research
International Agricultural
    Research System, 3-6,
    12-13, 55-56
International Center for Agri-
    cultural Research in
    the Dry Areas, 4
International Conference on
    Nutrition Education, 793
International Crops Research
    Institute for the Semi-
    Arid Tropics, 4, 14,
    39-40, 43
International Development
    Research Centre, 12-13,
    451
  contract research projects
    of, 14
  in Ethiopia, 14
  food legumes programs of, 15
International Fertilizer
    Development Centre, 15
International Food Policy
    Research Institute, 8
International Institute of
    Tropical Agriculture,
    4-15, 55-58, 451
International Laboratory for
    Research on Animal
    Diseases, 4
International Livestock Center
    for Africa, 4
International Maize and Wheat
    Improvement Center,
    4, 12, 21
International Potato Centre, 4
Intervention and development
    programs
  biological characteristics of,
    563
  in developing countries, 498-561
  latent and manifest functions
    in, 559-561
  target groups in, 9

Intracultural density, in
    ecological analysis, 557
Iodine deficiency, soybeans
    and, 318
Iran
  agriculture in, 612
  food and nutrition issues
    in, 613-614
  nutrition evaluation in,
    611-618
  population of, 611
  school feeding program in,
    614-618
  sociodemographic information
    on, 611-613
  supplementary food programs
    in, 615
Iron compounds
  bioavailability of, 333
  chemical reactions in,
    334-335
Iron deficiency anemia, in
    Philippines, 779
Iron fortification, technology
    of, 333-336
Israel
  childhood grades curricula
    in, 798-799
  nutrition education in,
    795-802
Italy
  nutrition research in,
    489-491
  rabbit raising in, 270

Jamaica
  anthropological methodology
    in, 593-607
  children's nutritional
    profiles in, 529
  Community Health Aide
    Programme in, 593-607
  food costs in, 526-527
  food supply system in, 526
  nutrition evaluation in,
    593-607
  nutrition indicators in,
    525-531
  population of, 525-526, 531

Jamaica (continued)
  sociocultural factors in, 593–607
Japan
  mixed protein studies in, 325–331
  soy protein isolate in, 325–331
Jogging, energy expenditure in, 661
Joint FAO/WHO Ad Hoc Expert Committee, 8–9

Khombole Project (Senegal), 573–588
Khombole-Thienabe area, Senegal, 575
Kjeldahl analysis, of crute protein, 728, 742
Knowledge change, from nutrition education, 788–789, 828–829

Lactobacillus acidophilus, 113
Lactodif, 284–285
LAL, see Lysinoalanine
Larval fish, water-stable diets for, 157, see also Fish
Latin America
  maize and bean consumption in, 10
  milk/beef production systems in, 102–103
  nutrition-dietitian in, 583–586
Latin American Rabbit Raising Scientific Association, 270
Lead, as environmental contaminant, 386–387
Leaf proteins, 208–209, 281–282
League of Nations Technical Commission on Nutrition, 9
Le Chatelier's principle, 457
Legumes, see also Beans; Food Legumes; Grain storage; Soybeans; Winged beans
  amino acid content of, 422
  defined, 422
  as foof, 422
  importance of, 10–11
  nutritional losses in storage and processing of, 421–428
  nutritional quality of, 422
  principal species of, 422
  protein quality of, 423
  as protein source, 208, 211, 423
  toxic substances in, 422
  trypsin inhibitor content of, 314
  world production of, 421
Leguminous seeds, Mexican studies of, 278
Leucaena esculenta, 278
Leucaena leucocephala, 278
Lignin, defined, 728
Lima bean, food quality of, 56
Lindano residues, 414
Lipid intake, pork fat quality and, 92–95
Lipid levels, in meat of native animals, 78
Lipoic acid deficiency, in fish, 194
Livestock feeding, see also Animal feeding
  animal wastes in, 144
  chemical processing and, 145
  forest product wastes in, 148–151
  grazing and, 144–145
  sawdust in, 150–151
  simplex processing and preservation in, 145
  waste cereal residues in, 141–151
Livestock raising, in tropics, 104–105, see also Tropical milk/beef production systems
Lugaw, as breast milk additive, 784, 790
Lupino tricolor Sodiro, 238
Lupins
  acreage for, 222
  cultivation of, 221–225
  sweet, 221–222

Lupinus albus, 721-723
Lupinus angustifolius, 221
Lupinus luteus, 221
Lupinus mutabilis, 221, 223-225
Lysine
  vs. cereal production, 453-454
  in pearl millet, 51-52
  in sorghum, 40-44
Lysinoalanine
  changes in, 443-445
  in soybean processing, 442-446

Magnesium, in fish diets, 179
Millard reaction, for soybeans, 440
Maize
  amino acid composition of, 32
  broad-based hard endosperm opaque 2, 21
  evaluation tests for, 34-35
  high quality, 21
  historical development of, 20
  nutritional improvement of, 19-22
  protein utilization in, 34-35
  vitamin A in, 357-361
Mallun (Sri Lanka), 211
Malnutrition
  in Greece, 639-640
  in India, 131
  mass media and, 779-790
  milk and, 131
  nutrition education and, 760, 764, 767
Manihot esculenta, 56
Manioc leaves, protein concentrate from, 233, 240
Margarine, vitamin A and β-carotene in, 349-351
Marine protein feeds, 209,
  see also Fish feeding technology
Marketing approach, in nutrition education, 823-834
Massachusetts Institute of Technology, 9
Mass media, in nutrition education, 766-768, 779-794, 817-819, 823

Max Planck respirometer studies, 642
Meat availability, in Tropical America, 102
Meat consumption, in Morocco, 538
Meat extenders, soybeans as, 284
Meat fat, personal preferences for, 77
Meat meal, replacement of by shrimp wastes, 201
Meat production
  animal selection in, 111
  cattle feeding for, 109-112
  feed savings of marketing "good" vs. "choice" beef in, 71
  human nutrition and, 109-112
  increase in, 110-111
  lipid levels and, 78
  rabbit raising as extension of, 269-276
  ruminant animals in, 109-110
  world population growth and, 110
Meat production costs
  cereal diets and, 65-70
  direct and indirect, 70
Medical schools, nutrition education in, 849
Mercury, fat contamination by, 384, 390
Methionine-casein diet, for rats in PER studies, 709-710
Methionine-sulfoxide content, of fish protein concentrate, 713-717
Methylated metals, as fish contaminants, 384
Methylmercury, placental concentration of, 384-385
Mexican wheats, photo-insensitivity of, 25
Mexico
  nutrition education in, 763-764
  soybean cultivation in, 281-283

INDEX

Mexico (continued)
  unconventional protein sources in, 277-285
Middle East, nutrition evaluation in, 611-618
Milk, see also Breast milk; Cow's milk; Dairy cow; Dairy project; Soybean milk
  bacteriological stability of, 122-123
  B.P.C. as substitute for, 253-262
  chemical composition of, 120-122
  environmental contamination of, 386-387
  free distribution of (India), 130-131
  "keeping" quality of, 119-123
  lyophilized, 394
  malnutrition and, 131
  nutritive value of, 120
  as protein source, 229
  smell and flavor of, 123
  from soybeans, 247
  vitamin A in, 350-355
  wholesomeness of, 123
Milkaflatoxin, in animal feeding, 394
Milk animals, distribution of (India), 135, see also Cattle; Dairy cow
Milk/beef production systems, human nutrition and, 101-106
  see also Tropical milk/beef production systems
Milk collection
  food supply and, 135-138
  in India, 133
  nutrition and, 135
  per capita expenditures and, 136
Milk consumption
  vs. income distribution, 126-128
  in India, 126-127
Milk prices, 129-130

Milk production
  by-product use in, 117-118
  caloric balance and, 138
  cattle feeding and, 113-123
  chemical composition changes in, 120-122
  in developing countries, 119-120
  efficiency of in human food supply, 114-120
  energy intake for, 113
  improved efficiency of, 114-119
  input/output ratios in, 114
  net income and, 137
  nonprotein nitrogen supplement in, 118
  yield per cow in, 115
Milk proteins, 122, 229, see also Protein(s)
Milk replacements, in calf feeding, 113, 115
Milk substitutes, 247, 253-262
Millet, pearl, 50-52
Millet seed, amino acid pattern in, 61
Milling technology, food conservation and, 454
Minerals, in fish diets, 179-180
Mineral utilization, soybean protein and, 314-317
Miso (Japan), 325
Mixed cropping, research in, 15-16
Mixed proteins
  nitrogen balance in, 331
  nutritive values of, 325-331
Monosodium glutamate, vitamin A in, 367-369
Morocco
  daily requirement per income class in, 543
  food and nutrition situation in, 537-544
  food consumption in, 538-539
  food demand in, 537-538
  food ration in (1970-71), 540

Morocco (continued)
  per capita energy and nutriment intake in, 540
  present and projected food demand in, 545-548
  protein-calorie deficit in, 542
  recommended daily intake for, 541
  rickets in, 542
MSG, see Monosodium glutamate
Multiple cropping, defined, 15

National educational systems, food and nutrition education in, 747-758
National Institute of Nutrition (India), 452
National nutritional status, periodic revision of, 515-518
National Nutrition Council (Philippines), 782, 785, 793
National nutrition Policy (New Zealand), 508
National nutrition surveys (Africa), 511-513
National Rabbit Raising Association (Brazil), 270
Neonate, environmental contaminants and, 384-386
Netherlands
  nutrition advertisements in, 812-814
  nutrition education in, 811-816
  nutrition opinion polls in, 814-816
New Zealand
  national diet survey in, 505-509
  National Nutrition Policy in, 508
Nicaragua
  marketing approach to nutrition education in, 826-834
  nutrition education in, 823-834

Nicaragua (continued)
  "Super Limonada" campaign in, 830-832
Nitrogen balance, vs. nitrogen uptake, 331
Nonconventional food sources
  in Brazil, 269-275
  development of, 207
  Ecuador studies in, 237-247
  lupins as, 221-225
  as meat substitutes, 238
  as milk substitutes, 247, 253-262
  in Mexico, 277-285
  single-cell protein and, 280
  utilization of, 207-210
  vegetable protein and, 227-234
  winged bean as, 211-219
Non-protein nitrogen supplements, in milk production, 118
North Africa, nutrition research priorities for, 503-504, see also Africa
Norway, milk production improvements in, 116-118
Nurses, nutrition training for, 871-873
Nutrimpi pellets, 284-285
Nutrition, see also Dietary fiber; Food; Human nutrition; Malnutrition
  anthropological approach to, 551-552, 579-585, 619-635
  cancer and, 491
  cereal grains and legumes in, 10-11, 421-428
  defined, 467, 483
  as discipline, 871
  disease and, 491, 534
  fish, see Fish nutrition
  general health and, 515, 534
  League of Nations recommendations on, 9
  national, see National nutrition
  obesity and, 491
  physical performance and, 657-665

Nutrition (continued)
  from plant kingdom, 208, 230, 281–282, 315–316
  proteins in, 173–174, 227
  sex roles and, 623
Nutritional goals, post-harvest conservation in, 451–455
Nutritional health, assessment and monitoring of, 515–524
Nutritional losses, see also Food losses
  causes of, 460
  in grain processing, 457–463
  phytic acid and, 463
  in storage and processing, 421–448
  types of, 461–463
Nutritional status
  centralized monitoring system for, 520–521
  evaluation of, 533–535
  in Jamaica, 525–531
  nutrition indicators in, 525–530
  periodic revision of, 515–524
  simplistic assessment and monitoring of, 521–523
  validity and reliability of, 523
Nutrition education, see also Nutrition teaching; Nutrition training
  action-oriented approach to, 567
  in Africa, 865–868
  in Asia, 847–859
  attitude change in, 787–788, 827
  audiovisual centers for, 880
  behavior change in, 789–790, 829
  breast feeding campaigns in, 817
  case study on, 805–808
  in China, 764
  college courses in, 753–757
  college student and, 776
  communications in, 888–892
  community, 817–821

Nutrition education (continued)
  community health worker role in, 878
  concept formation in, 800
  creativity in, 801
  in developing countries, 823–834
  dietitian in, 878
  in early childhood grades, 795–802
  food habits and, 761–765
  food preparation in, 801–802
  general education and, 775–777
  as government tool, 760
  for health workers in developing countries, 835–842
  hygienic practices and, 764
  in Hungary, 843–845
  International Nutrition Faculty for, 879–880
  knowledge change in, 788–789, 828–829
  logical thinking in, 799
  long-term objectives in, 797
  malnutrition and, 760, 764, 767
  marketing approach to, 823–834
  mass media and, 766–768, 779–794
  for medical and health science professionals in Eastern Europe, 843–845
  national programs in, 765–766
  in national systems, 747–748
  in Netherlands, 811–816
  nutritionist's role in, 895–898
  pharmacist in, 878
  in Philippines, 779–794,
  for physicians and nurses, 871–873
  in primary and secondary schools, 748–753
  radio in, 784–793
  "reactor's comments" on, 775–777
  research and, 486
  school competence and, 818

Nutrition education (continued)
  teacher in, 777
  in teacher training colleges, 753-757
  television programs in, 766-768, 779-794, 817-823
  in Third World, 759-768
  in Turkey, 500
  in United Kingdom, 805-808
  at university level, 757-758
  Workshop on, 877-881
Nutrition education programs, 765-766
  evaluation of, 769-794
  execution of, 770
  side effects of, 771-772
  supplemental information in, 771
Nutrition evaluation, see also Evaluation research
  in Africa, 619-635
  anthropology and, 553-568
  in Caribbean context, 593-607
  food utilization and preserving studies in, 55-64
  of food proteins, 683-687
  in Middle East context, 611-618
  participatory research in, 564-570
  periodic, 553
  priority evaluation and, 599-600, 603-605
  proposed activities for, 59-61
  social anthropology in, 593-607
  structure of research in, 554-556
Nutrition Foundation, 775
Nutrition intervention
  biological characteristics of 563
  in developing countries, 498-500
  target groups in, 9
Nutritionist
  communications role of, 888
  in community nutrition education programs, 895-898

Nutritionist-dietitian
  communications competencies of, 887-894
  in Latin America, 883-886
  qualifications of for community programs, 899-903
Nutrition priorities, integration of into crop improvement programs, 7-17
Nutrition programs
  anthropological perspectives in, 573-588, 619-635
  ecological perspectives in, 556-563
  evaluation of, 553-556, 769-794
  social organization and, 551-562, 579
Nutrition research
  in Africa, 622-623
  defined, 468
  in developing countries, 483-486, 495-501
  epidemiologic observations in, 490
  flexibility in, 493
  government policies and, 485-486
  intervention programs and, 498-500
  in Italy, 489-491
  in North Africa, 503-504
  prerequisites in, 473
  priorities in, 467-469, 471-472, 477-481, 483, 489-494, 496-498, 503-504, 549-600, 603-605
  protein quality and, 492
  resource utilization in, 480-481
  role of, 471
  service of, 478-480
  soft end of, 477-478
  targets in, 489
  in Turkey, 495-501
  women's image in, 622-623

Nutrition schools, communication
in, 893
Nutrition studies, bidisciplinary
focus in, 619-620
Nutrition systems, women's role
in, 619-635
Nutrition teaching
in Asia, 851-855
in Thailand, 855-859
Nutrition training, see also
Nutrition education
for community health workers,
873-874
for health workers in
developing countries,
835-842
integrated approach to,
837-838
for physicians and nurses,
871-873
Nutritive value, in vivo
prediction of, 683-687

Occupational groups, energy
expenditure by, 650
Oilseeds
major crops of, 228
Mexican studies of, 279
proteins in, 208
Oncorhynchus kisutch, 171
Oncorhynchus tschawytscha, 171
Organochlorinated defensives
residues (Argentina),
403-416
Oryctolagus cuniculus, 269

Pan American Health Organization,
480-481, 883-884
Pancreatic secretion
cholecustokinin and, 305
raw soybeans and, 303-305
Panicum miliare, 452
Participatory research, in
nutrition evaluation,
564-567
PCB, see Polychlorinated
byphenyls
Peanut oil, 228
Pearl millet
improvements in, 50-52

Pearl millet (continued)
lysine content of, 50-51
preparations made from, 51
protein content of, 50
Pectins, defined, 727
Pelletizing, in fish feeding
technology, 162-164,
202
PEM, see Protein-energy
malnutrition
PER, see Protein nutrition
qualtiy
Pesticides, food contamination
by, 453
Pharmacist, in nutrition
education, 878
Phaseolus bean lines, product
improvement in, 58
Phaseolus lunatus, 56
Phaseolus vulgaris, 55-56
Philippine Atheletic Amateur
Association, 657
Philippines
attitude change in, 787-788
behavior change in, 789-790
dietary intake vs. physical
activity in, 657-665
malnutrition and deficiency
diseases in, 779-780
marketing approach to
nutrition education in,
824-826
mass media study in, 781-794
nutrition education in,
780-781, 824-826
Phosphorus, in fish diets, 179
Physicians, nutrition training
for, 871-873
Phytic acid
in nutritional loss, 463
in plant foodstuffs, 315-316
Pig, see also Pork; Swine
modern practices in raising
and feeding of, 73-98
population of in South
America, 73
world production of, 74
Pigeon pea, amino acid
pattern in, 61

Pig feeding, 73-98, see also
    Animal feeding; Livestock
    feeding; Swine feeding
Pinol concentrate study,
    Ecuador, 239
Pithecollobium dulce, 278
Planned interventions, latent
    functions of, 560
Plant foodstuffs, phytic acid
    in, 315-316
Plant kingdom, nutritional
    needs served by, 230
Plant proteins, 208
  digestibility of, 232
  genetic improvement in,
    209-210
Plasma volume changes, exercise
    and, 672-675
Plastic containers, toxicity
    of, 379
Polished rice, losses in, 462
Polychlorinated biphenyls, in
    breast milm, 388-389
Polysaccharides, pectins as,
    727
Pork
  chemical constitution of, 275
  consumption of, 73-76
  cost of, 274
  food constituents of, 273
  problems related to, 73-78
  quality of, 96-98
  world supply of, 73-74
Pork fat
  cereal feeding and, 95
  dietary lipids and, 92-95
  quality of, 91-96
Post-harvest conservation
  of cereals and grains, 458-460
  nutritional goals in, 451-455
Potassium balance, exercise
    and, 676
Potatoes
  dry matter in, 730, 738
  as protein source, 228
Potato flour, 245-246
Pregnancy, anemia in, 525
Pregnant mothers, B.P.C. in
    diet of, 262
Prevention and Health, 808

Primary schools, food and
    nutrition education in,
    748-753
Priority evaluation, 599-600,
    603-605
Prochilodus scrofa, 200
Product quality
  CIAT activities in, 58
  IITA activities in, 56-57
  international agriculture
    research centers and,
    55-56
Protein(s)
  amino acid composition of,
    685
  amino acid recovery from, 701
  animal, 73-78
  in B.P.C., 249
  from cereals, 229
  conversion rate in milk
    production, 115
  economic protein requirement
    concept for, 171-172
  from eggs, 229
  enzymatic hydrolysis of for
    amino acid analysis,
    693-703
  fish, see Fish proteins
  in fish feed, 158
  from green leaves, 228
  in high lysine sorghum,
    40-44, 46
  from higureilla flour, 242
  in insects, 281
  interaction of with other
    energy sources, 174-175
  in vitro nutritive value
    prediction for, 683-687
  from leafy plants, 208-209,
    228, 281-282
  from legumes, 208, 423
  "luxus consumption" of, 227
  in maize, 19-20, 23-24
  in manioc, 233, 240
  marine, 209
  in meat, 229
  in milk, 122, 129
  minimal dietary protein
    level for, 171
  mixed, 325-333

Protein(s) (continued)
  nutritional value of, 173-174, 683-687
  oil-seed, 208, 228
  in pearl millet, 50
  physiological needs and, 227
  in potatoes, 228-229
  from pulses, 228
  in rabbit meat, 272
  single-cell, 209, 705-711
  in soybeans, 283, 289, 293-299
  in sugars, 736
  in triticale, 23, 30-31
  tryptic proteolysis and, 684-685
  from unconventional food sources, 207-208, 277-285
  in uronic acids, 736
  vegetable, 227-234
  in winged bean, 212-219, 228
  world production of, 229
Protein-calorie deficiency
  in Morocco, 542
  in Philippines, 779
Protein-calorie malnutrition, sorghum and, 49
Protein concentrate
  from banana leaves, 239
  from sesame oil cake, 241
Protein controversy, 8-10
Protein deficiency, in developing countries, 101, see also Protein-calorie deficiency
Protein-energy malnutrition, 484, 779
Protein fortification, quinoa plant in, 243-244
"Protein gap," 227
Protein hydrolysate, 698
Protein intake, in swine feeding, 89-91
Protein level
  in Brazilian catfish diet, 201
  in rainbow trout diet, 201
Protein nutrition quality, 683-687
  amino acid composition and availability in, 693-703

Protein quality measurements
  on bacterial single-cell protein, 705-711
  Tetrahymena thermophila and, 689-691
Protein requirements, 171-174
  essential amino acids and, 172-173
Protein shortage, socioeconomic, 227
Protein supplies, per capita/per day, 66-67
Protein synthesis, raw soybeans in, 303-304
Protein utilization
  in animal production, 121
  in maize, 35
  in triticale, 36
Protein variation, in soybeans, 46
Proteolytic microorganisms, growth rate of, 683
Psophocarpus tetragonolobus, 211, 228
Pulque, 279
Pulse
  as protein source, 228
  toxic substances in, 422

Quinoa, in protein fortification, 243-244

Rabbit meat
  amino acid content of, 272-273
  Brazilian markets for, 272
  chemical constitution of, 275
  cost of, 274
  food constituents of per 100-g portion, 273
  qualities of, 272
Rabbits
  breeds of, 270
  fecundity of, 270
  feeding of, 270
  as meat source, 269-276
Radio in nutrition education, 792-793, see also Mass media

Radishes, glucosinolates in, 317
Rainbow trout, diet vs. growth in, 189
Rat growth, trypsin inhibitor of, 307–310
Rat pancreas weight, tripsin effect on, 310
Rats
 amino acid digestibility in, 708
 methionine supplemented casein diet for, 709–710
 nitrogen balance studies in, 707
Raw soybeans, see also Soybean(s)
 biochemical effects of, 301–303
 pancreatic enzyme secretion and, 303
 trypsin inhibitor and, 308
Research, see also Nutrition research
 agricultural, 12–16
 defined, 467
 nutrition evaluation, see Evaluation research
 participatory, 564–567
 soft end of, 477–478
Respiratory tests, work expenditure and, 641–654
Retinol, see Vitamin A
R-factors, transfer of from E. coli to man, 399
Rice
 parboiling of, 453, 462–463
 vitamin A and, 361–364
 washing and cooking losses in, 462
Rickets, in Morocco, 542
Rockefeller Foundation, 4, 12
Ruminant animals, see also Animal feeding; Livestock feeding
 feeding of, 154
 in meat production, 109

Salmonides, vitamin requirements of, 190
Salt, vitamin A in, 366–367
SAT, see Semi-arid tropics
Save Grain Campaign (India), 452

Sawdust, in livestock feeding, 150–151
Scandinavia, milk production improvements in, 117
School feeding program, Iran, 614–618
School cafeterias, nutritional food promotion in, 818
Science curriculum, food and nutrition education in, 749–751
Secale, 19
Secondary schools, food and nutrition education in, 748–753
Selenium, in peroxide-destroying enzymes, 396
Semi-arid countries, defined, 14
Semi-arid tropics, sorghum and millet crops in, 39
Senegal, Khombole Project in (1957–78), 573–588
Sesame oil cake, protein concentrate from, 241
Sesame seed
 amino acid patterns in, 60
 production of (Ecuador), 241
Shrimp wastes
 "chitosan" from, 202
 meat meal replacement by, 201
Simonson effect, in work-energy studies, 643
Single-cell protein
 amino acid content of, 706
 protein quality measurements for, 209, 705–711
Skim milk, soybean flour in, 283–284
Small farmers
 India, 134
 Mapuche, 222
Social and economic studies, food/nutrition education in, 752
Social anthropology, in evaluation research, 593–607
Social organization, food behaviors and, 561–562

INDEX

Sociocultural environment, in anthropological approach, 594
Sockeye salmon, vitamin test diets for, 182-183
Sodium, in fish diets, 179
Sodium hydroxide, in waste processing, 145-146
Sodium/potassium levels, in exercise, 674-676
Sorghum
  in carp diet, 200
  chemical composition and seed characteristics of, 41
  germ plasm collection and, 45, 47-48
  high lysine gene of, 40-43
  nitrogen distribution in, 42
  nutritional improvement in, 39-50
  polyphenolic compounds and, 45
  preferred characteristics in, 49
  protein and lysine estimation for 43-44
  protein-calorie malnutrition and, 49
  protein distribution in high lysine variety, 40-43
  protein variation in, 46
  research in, 13-14
  in semi-arid tropics, 39
South America, lupin cultivation in, 221-225
  see also Argentina; Brazil; Chile, etc.
Southeast Asia, cereal grains and legumes consumed in, 10
  see also Asia
Soya bean oil, 228, see also Soybean(s)
Soyacit, 283
Soybean(s), see also Trypsin inhibitors
  aging of in storage, 429-438
  agricultural production of (Brazil), 289-291
  amino acid content of, 294, 438, 441

Soybean(s) (continued)
  as and fiber content of, 294
  in Asia and China, 293
  carbohydrate content of, 294
  composition of, 294
  conversion of to high-quality protein products, 299-320
  defatted or toasted, 306-307
  estrogens in, 319-320
  fish protein and, 325-331
  as food source, 299
  goiter and, 317-319
  heat-labile factors in, 302
  iodine deficiency in, 318
  isopeptide bond in, 440
  as meat extenders, 284
  methionine content of, 290
  Mexican studies of, 278-279
  as milk substitute, see Soybean milk
  mineral bioavailability in, 314-317
  nutritive value of varieties of, 289-291
  oil content of, 289
  protein content of, 214, 289, 293-299, 314-320, 325-331
  as protein extender, 283
  protein extractability for, 296
  raw, 299-320
  solvent extraction for, 299
  structural parts of, 293
  toxic factors in, 290
  trypsin inhibitors and, 290, 300
  U.S. production of, 300
  varieties of, 289-291
Soybean cultivation (Mexico), 281-283
Soybean curd, 325
Soybean extracts, tryptic digestibility of, 311
Soybean flour, 283
  defatted, 309
  types of, 295
Soybean goiter, 317-319

Soybean milk
  for rats, 302
  studies of (Ecuador), 247
Soybean milk storage
  nitrogen changes in, 431
  solids changes in, 430
Soybean oil, refining of, 294
Soybean processing
  changes during, 438-442
  lysinoalanine in, 442-446
Soybean products
  composition of, 294
  goitrogenicity of, 317-319
  in Japan, 325-331
  trypsin inhibitor content and, 310-314
  types of, 293-298
Soybean protein
  allergens in, 319
  in baked goods, 297
  as cow's milk substitute, 319
  estrogenic effect and, 319-320
  in infant formulas, 297-298
  isolated, 296
  in meats, 297
  mineral utilization and, 314-317
Soybean protein isolate, 325-331
Soybean protein products, 297-298, see also Soybean(s); Soybean products
Soybean seed, price of, 283
Soybean storage
  ash and nitrogen losses in, 436
  changes during, 429-438
  experimental design in, 438
  protein crosslinking reactions in, 438-442
  sugar loss in, 435
  water immersion losses following, 433-434
Soybean trypsin inhibitors, see Trypsin inhibitors
Spaghetti, composite flours in, 246
Spices and condiments, composition of (Morocco), 539

Sri Lanka, winged bean as garden vegetable in, 211
Starea, in milk production improvement, 118
Storage, see Grain storage; Soybean storage
Streptococcus zymogenes, in fish protein concentrate study, 713-717
Sugar
  carbohydrate and protein content of, 736
  exercise and, 677-678
  vitamin A in, 363-366
Sugar-cane subproducts, in tropical milk production, 105
Sugar crop residues, in livestock feeding, 144
Sunflower oil, 228
Surimi, soybean protein in, 297
Swan Report, 401
Sweating
  cold or warm drinks in, 677
  exercise and, 668-671
Sweden, milk production efficiency in, 116-117
Swine carcasses, fat percentage in, 76
Swine breeding, performance and outcomes in, 82
Swine feeding
  backfat thickness in, 90
  cellulose/protein variations vs. body composition in, 88
  coconut oil in, 94
  energy intake vs. growth performance in, 85-91
  protein intake in, 89-91
Swine population, changes in, 75
Swordfish, mercury contamination of, 390

Tanzania
  nutritional experience in, 861-869

Tanzania (continued)
  nutrition concept in, 864
  nutrition education in, 864–869
  political awakening in, 863–864
  poverty and disease in, 862–863
Tea, vitamin A in, 355–356
Teacher training colleges, food and nutrition education in, 753–757
Technical Advisory Committee, 4
Television programs, see also Mass media
  nutrition education and, 817–818
  in Philippines, 781–786
Tetrahymena thermophila, protein quality measurement with, 689–691
Thailand
  community health program in, 856
  nutrition teaching in, 855–859
Third World, see also Developing countries
  food taboos in, 761
  milk production in, 119
  nutrition education in, 759–768
TI, see Trypsin inhibitors
Tilapia nilotica, 200
Tilapia rendalli, 201
Timothy, digestibility of, 143
Tocopherols, in fish diets, 193
Tortilla flower, soybean additive for, 279, 283
Tree foliage "muke," as livestock feed, 148–149
Triticale
  amino acid content of, 33
  biological evaluation of, 29
  bread from, 27
  commercial stands of, 25
  defined, 19
  hexaploid, 19
  high-yielding strains of, 26
  industrial quality of, 25, 28

Triticale (continued)
  nutritional improvement in, 22–36
  potential prospects for, 29
  protein utilization in, 30–31, 36
  seed shrivelling in, 25
  yield and adaptation of, 22–25
Tropical America, meat availability in, 102
Tropical countries, livestock raising in, 104–105
Tropical milk/beef production systems, 101–106
  livestock breeds in, 105–106
  market requirements in, 106
  sugar-cane subproducts in, 105
Tropical Products Institute, 451
Trypsin inhibitors
  activity of in winged beans, 217–218
  biological threshold levels of, 306–314
  cholecystokinin as, 305
  defatted soybean flakes and, 306
  of different soybean varieties and strains, 301
  in food legumes, 314
  nutritional assessment of, 300–306
  pancreatic activity and, 305
  rat growth and, 306, 310
  for soybeans and soybean flour, 300, 309–314
Trypsin secretion, regulation of, 304
Trypsin unit levels, in winged beans, 218
Tryptic proteolysis, in protein analysis, 684–685
Turkey
  direct feeding and food distribution programs in, 499

Turkey (continued)
  nutrition research in, 495-501
  supplement and fortification programs in, 499-500
Turkey meat, cost of, 284
Turnips, glucosinolates in, 317

Udy instrument reading techniques, in sorghum evaluation, 44
UNDP, see United Nations Development Program
United Kingdom
  dietitian training in, 908-910
  nutrition education in, 805-808
United Nations, 13
United Nations Development Program, 13
  CIMMYT Global Research Project in, 21
United Nations World Food Council, 3-4
United States, rabbit raising in, 270
University courses, in food and nutrition, 757-758
UN University (Tokyo), 451
Urea, immobilized peptidase activity in, 699-700
Uronic acid, protein-carbohydrate content of, 736

Vegetable foods
  dry matter in, 730-731, 738
  low appeal of, 232
Vegetable protein production
  advantages of, 230-231
  disadvantages of, 231-233
  energy obtained in, 230
  energy required for, 231
  nitrogen fertilizers used in, 231
Vegetable proteins, 227-234, see also Protein(s)
  amino acid content of, 233
  as flour supplements, 237
  for low-protein foods, 237

Vegetable proteins (continued)
  as meat extenders, 232
  toxic substances accompanying, 234
  Veno-occlusive disease, Panicum miliare and, 452
Vigna unguiculata, 56
Vitamin A
  in animal feeding, 396
  application forms of, 341-344
  in baked bread, 360
  in beverages, 350
  in cereal-grain products, 356-357
  chemical and physical forms of, 345
  chemical synthesis of, 337, 339
  dry and wet forms of, 342-343, 352
  in fish diets, 193-194
  in flour, 357-361
  as food additive, 347-369
  intermittent dosing of, 344-346
  lability and stability of, 339-341
  liquid or dry forms of, 342-343, 352
  in margarine, 349-351
  in milk, 350-355
  in monosodium glutamate, 367-369
  in nutrified foods, 347-369
  in oil and fats, 349
  parenteral dosing of, 347
  physico-chemical data on, 341
  premix feeder for, 358
  price pattern for, 338
  retention of in cooked grifts and cornmeal, 363
  retention of in stored gelatin capsules, 346
  in rice, 361-363
  in salt, 366-367
  in seasonings, 365
  in stored water-dispersable parenterals, 349

INDEX

Vitamin A (continued)
  structure of, 337
  in sugar, 363-365
  in tea, 355-356
  technology of, 337-369
  water-dispersable, 348
Vitamin A compounds, structural
    formulas for, 338
Vitamin A deficiency
  in fish, 193
  incidence of, 337
Vitamin A esters, 337
  technical synthesis of, 340
Vitamin A palmitate, 344
Vitamin B_{12}
  in animal feeding, 396
  in fish diets, 179-180, 188
Vitamin C
  in fish diets, 178
  iron fortification and, 334
Vitamin D, in fish diets,
    193-194
Vitamin deficiency diseases,
    in fish, 177-178, 181-195
Vitamin E, 193
Vitamin K, 193
Vitamin requirements, test diets
    for fish in, 182-183

Walking speed, energy
    expenditure and,
    648-650
WARDA, see West African Rice
    Development Association
Waste products, in animal
    food production, 141-151
Waste processing
  alkalis in, 146-147
  ammonia in, 147
  cost per ton for, 147
  sodium hydroxide in, 145-146
Wastes, types and amounts of,
    134-144
Water/electrolyte balance, in
    exercise, 667-681
Weightlifters
  calorie intake and
    distribution in, 663
  daily activities of, 662

Weightlifters (continued)
  daily mean energy intake
    and expenditure for,
    657, 664
Weightlifting, other activities
    related to, 661
Weinstock method, for dietary
    fiber assay, 742-743
West Africa, cowpea improvement
    program in, 55-57,
    see also Africa
West African Rice Development
    Association, 4
West Germany
  aflatoxins in, 417-419
  dietitian training in, 909
WFP, see World Food Programme
Wheat, see Triticale
Wheat bran, dry matter in,
    730, 738
Wheat straw, digestibility of,
    143
Whey, as protein source, 281-282
White bread, vitamin A
    fortification of, 361
Wholemeal bread, losses in,
    461
Winged beans, 211-219
  anti-tryptic activities of,
    216-217
  compared with soybean, 214
  enzyme activity in, 213
  protein content of, 212-219,
    228
  seeds of, 214-218
  toxic effect of, 219
Women, as "producer and
    reproducer" in Africa,
    619-635
Workshop on Nutrition Education,
    Report of, 877-881
World Bank, 13
World Food Council, 3-4
World food problem,
    seriousness of, 7
World food production, growth
    of, 3, 7, 110
World Food Programme, 126

World Health Organization, 9,
386, 417m 528
  community health workers and,
840-842
World population growth, meat
production and, 110

Xastle, Mexican studies of, 279
Xylans, 727
Xyloglucans, 727

Zeby stock, in tropical
livestock raising, 105
Zinc content, of various foods,
316